Spreadsheet Exercises in Ecology and Evolution

Spreadsheet Exercises in

Ecology and Evolution

Therese M. Donovan
U.S.G.S. Vermont Cooperative Fish and Wildlife Research Unit
University of Vermont

and

Charles W. Welden
Department of Biology
Southern Oregon University

SINAUER ASSOCIATES, INC. ▪ PUBLISHERS
Sunderland, Massachusetts U.S.A.

SPREADSHEET EXERCISES ECOLOGY AND EVOLUTION

Cover
The cover image is a royalty-free photograph from Corbis Images.

Notice of Liability
Due precaution has been taken in the preparation of this book. However, the information and instructions herein are distributed on an "As Is" basis, without warranty. Neither the authors nor Sinauer Associates shall have any liability to any person or entity with respect to any loss or damage caused or alleged to be caused, directly or indirectly, by the instructions contained in this book or by the computer software products described.

Notice of Trademarks
Throughout this book trademark names have been used and depicted, including but not necessarily limited to Microsoft Windows, Macintosh, and Microsoft Excel. In lieu of appending the trademark symbol to each occurrence, the authors and publisher state that these trademarked names are used in an editorial fashion, to the benefit of the trademark owners, and with no intent to infringe upon the trademarks.

Library of Congress Cataloging-in-Publication Data
Donovan, Therese M. (Therese Marie)
Spreadsheet exercises in ecology and evolution / Therese M. Donovan and Charles W. Welden
Includes bibliographical references.
ISBN 0-87893-156-2 (pbk.)
1. Ecology--Data processing. 2. Evolution--Data processing. 3. Electronic spreadsheets. I. Welden, Charles Woodson. II. Title.
QH541.15.E45 D66 2001
577'.0285--dc21 2001049730

For Peter and Evan (T.M.D.)

To my students, whose steadfast refusal to take my word for anything has forced me to learn (C.W.W.)

Contents

Preface

This book is about using a spreadsheet program to build biological models. Spreadsheet programs have many uses, such as entering and organizing data, tracking expenses, managing budgets, and graphing. In this book, we use a spreadsheet program to create models to help you learn some basic and advanced concepts in ecology, evolution, conservation biology, landscape ecology, and statistics.

Why build your own models when so many specific, prewritten models are widely available? Because when you program a model from scratch, you learn all aspects of modeling—what parameters are important, how the parameters relate to each other, and how changes in the model affect outcomes. In other words, you not only learn about **models**, you also learn about **modeling**.

Why use a spreadsheet program rather than a dedicated modeling package or general-purpose programming language? In part, because most colleges and universities have a spreadsheet program readily available for their students, and many students are already familiar with basic spreadsheet operations. Using a spreadsheet thus reduces expense and learning time. In addition, using a spreadsheet allows more flexibility than is possible with most prewritten models. Students can easily modify or elaborate a model, once they have mastered the basic versions presented here. Finally, the spreadsheet takes care of much of the tedium of carrying out repeated calculations and creating graphs.

Why do modeling at all? Because modeling is a powerful learning tool. By building and manipulating models, you can achieve a deeper understanding of concepts. Models allow you to explore concepts, examine them from various angles, extend them in various directions, and ask "what if" all in rigorous and objective ways. Many models generate a clear set of predictions that can be tested in a natural or laboratory setting. Models offer a check on your understanding. When you plug values into a model and get unexpected results, you have to ask, "Why?" Answering that "why" leads to deeper understanding.

Acknowledgments

We are grateful to the many undergraduate and graduate students at the University of Vermont, Southern Oregon University, and the State University of New York who worked through early draft spreadsheets, pointed out problems, and offered suggestions. David Bonter (University of Vermont) worked tirelessly through every exercise in preparation for his graduate candidacy exams (he passed). Each exercise also benefited from critical reviews by our colleagues, including Guy Baldassarre, Jeff Buzas, Mark Beekey, John Cigliano, Luke George, James Gibbs, Nick Gotelli, Thomas Kane, Mark Kirkpatrick, Robin Kimmerer, Rollie Lamberson, Kim McCue, Bob McMaster, Madan Oli, Julie Robinson, Erik Rexstad, Robert Rockwell, Nick Rodenhouse, Eric

Scully, Bill Shields, David Skelly, Beatrice Van Horne, Sandra Vehrencamp, and, last but not least, Hal Caswell, who clarified our understanding of reproductive value and sensitivity and elasticity analysis. Steve Tilley provided in-depth reviews and helped sharpen our prose.

We also fully acknowledge the contributions of the co-authors who aided in model or exercise development, including Shelley Ball, David Bonter, Jon Conrad, James Gibbs, Wendy Gram, Larry Lawson, Mary Puterbaugh, Rob Rohr, Kim Schulz, and Allan Strong.

It takes many people to produce a book, and we have been very fortunate to work with Andy Sinauer and his associates. We are indebted to Carol Wigg and David McIntyre for the extraordinary energy and enthusiasm that they brought to the project, and to Susan McGlew, Roberta Lewis, and Joan Gemme. Finally, our families have been a consistent source of support and encouragement.

TERRI DONOVAN
CHARLES WELDEN
DECEMBER, 2001

How to Approach These Exercises

This book is intended to be a supplement to the primary text in an undergraduate or a beginning graduate course in ecology, evolution, or conservation biology. Although there are many excellent texts on the market, two primers were instrumental in helping us develop many of the spreadsheet exercises in this book: Nick Gotelli's *Primer of Ecology* (2001) and Dan Hartl's *Primer of Population Genetics* (2000). Both are extremely well written and helped us fully understand the basic mathematics behind many ecology and evolution models.

Each exercise was written with the notion that an instructor would introduce the basic material, and that the spreadsheet exercises would reinforce the concepts and allow further exploration. **We will assume that you have read the Introduction, "Spreadsheet Hints and Tips," and that you have mastered Exercises 1 and 2, "Mathematical Functions and Graphs" and "Spreadsheet Functions and Macros," before attempting other exercises in the book.**

Each exercise consists of an Introduction, followed by Instructions and Annotations that guide you through the model development, and then by a series of Questions. In the introduction to each exercise, we have tried to include enough background material for you to understand the context and purpose of the exercise, but we have also tried to keep these commentaries relatively brief. The Instructions give rather generic directions for how to set up the spreadsheet, such as "Sum the total number of individuals in the population." The accompanying Annotations provide the actual spreadsheet formulae that we used to accomplish each step, with a complete explanation of the logic behind each formula. Because our formulae are provided for you, you may be tempted to leap to the Annotations section before attempting to work through the problem on your own. Don't. You will learn more about the process of thinking through a model if you struggle through it on your own, and you may come up with a better way of doing things than we did. As much as possible, use the Annotations as a cross-check.

The last portion of each chapter consists of a set of questions that will challenge you to "exercise" your model and explore it more deeply. Some of the questions ask you to change your spreadsheets in various ways. You may want to save your original spreadsheet, and use a copy of the spreadsheet model when answering questions to preserve your original entries. The answers to the questions are posted on the Web site **www.sinauer.com/spreadsheet-ee/**. Although you can double-check your results with those posted on the Web, in reality scientists don't have the luxury of an answer section when developing a new model. If your results look odd to you, an assumption of the model may have been violated, you may have made a mistake in your programming, or the result may be, in fact, correct. Learn to critically interpret your own results—that is what scientists do.

The Web site also contains all of the spreadsheets used in the book. Students have access to "shell" versions, containing only titles, labels, headings, etc. Downloading these before class can save class time. Instructors have access to complete spreadsheet models, which they can use for exploration, modification, or verification. The Web site is also a clearinghouse for errata, instructors' comments, ideas for modifications, and links to related Web sites.

The process of entering formulae, making graphs, and answering questions in each exercise is just the beginning. We have attempted to build models that are very open-ended and encourage you to play with the models and take them beyond the questions posed. Don't be shy about changing parameter values, initial variable values, and modifying formulae. Observe how the model responds, and think about why it does whatever it does. Question, modify, and question again. Think about how you might make the model more realistic, how you might include other processes in it, or how the same model might be applied to a different system. All these ways of thinking will help you understand the models that you encounter in your texts and in the scientific literature.

T.M.D. AND C.W.W.

INTRODUCTION

SPREADSHEET HINTS AND TIPS

This introduction covers procedures that you'll use in the exercises throughout this book. It is intended to be a ready reference, and as such it has a different format than the exercises. The first two exercises, "Mathematical Functions and Graphs" and "Spreadsheet Functions and Macros," apply some of the procedures discussed here to the exercise format and give you an opportunity to practice them.

If you are already familiar with spreadsheets, you may want to skip this chapter, or perhaps just check out any unfamiliar topics. To help you find what you're interested in, here's an outline:

Three warnings: First, this chapter is not a substitute for your spreadsheet user's manual. We base our instructions throughout the book on Microsoft Excel, and most will work as written in other spreadsheets, but there may be differences in the details. If you follow our instructions carefully, and they don't work, consult your spreadsheet user's manual. Second, you should already be familiar with some basic computer skills, such as booting up your computer, starting your spreadsheet program, saving files, and printing. If you're not, consult your operating system user's manual. Third, **save your work frequently to disk!** Few things are as frustrating as spending hours building a model, then losing all your hard work when the computer crashes.

Starting Up

How you start up your spreadsheet program will depend on whether you use a Macintosh, an IBM-compatible computer, or a UNIX computer, whether the computer is on a network or not, and which spreadsheet program you choose. Consult your operating system manual, your spreadsheet program manual, or a local computer expert.

All of the exercises in this book were developed with Microsoft Excel version 98 or higher, which utilizes the "Visual Basic for Applications" code. If you are using an older version of Excel or a different spreadsheet program, make sure the basic functions used in the exercise are available. Some exercises require the use of the Solver function, an optimization function that is within the spreadsheet's Add-In Pak. Your system administrator may need to help you install the Solver

These exercises were written by several authors, using either Macintosh or Windows platforms; most, however, were developed in Windows. Table 1 gives some alternative commands and keystrokes that may help if the instructions are not tailored to your machine.

Menus and Commands

Most spreadsheet programs have graphical user interfaces in which you use a mouse to choose commands from menus across the top of the screen. Many menus have sub-

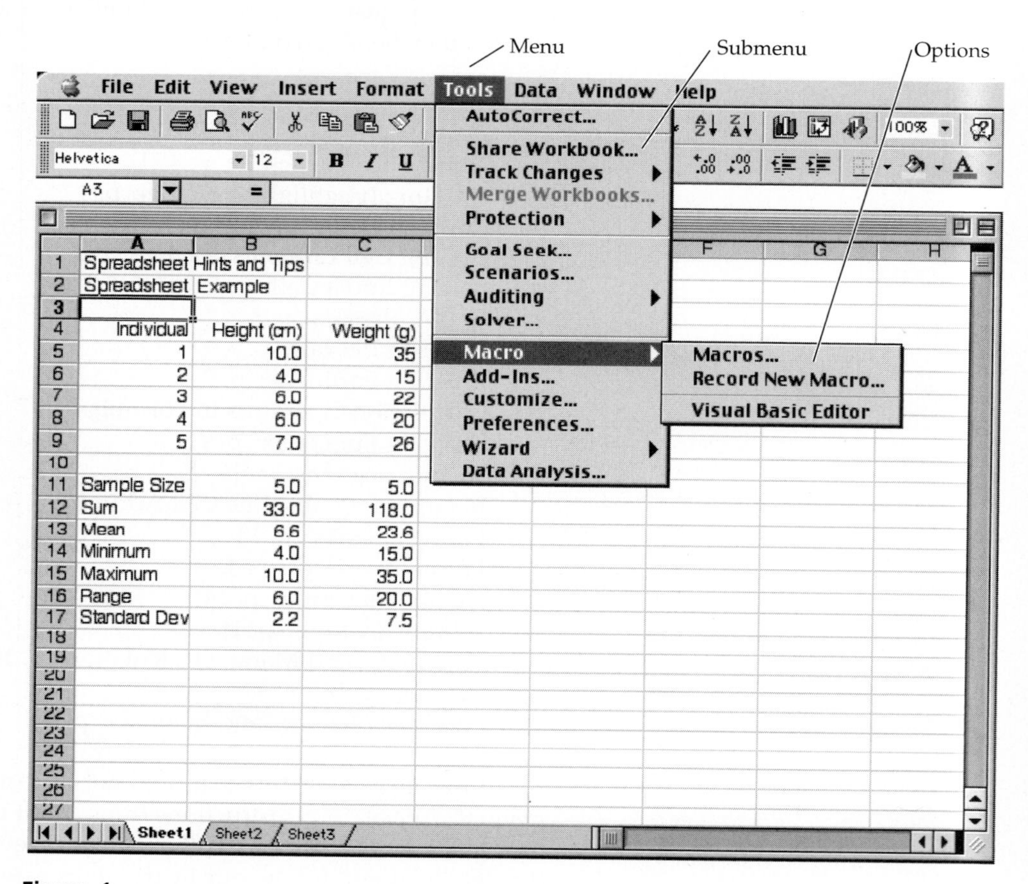

Figure 1

Table 1. Some Commonly Used Keyboard Commands in Microsoft Excel

Windows	Macintosh	Action
Enter	Return	Complete a cell entry and move down in the selection
Tab	Tab	Complete a cell entry and move to the right in the selection
Control+Shift+Enter	+Return	Enter a formula as an array formula
Esc	Esc	Cancel a cell entry
Backspace	Delete	Delete the character to the left of the insertion point, or delete the selection
Delete	Right delete	Delete the character to the right of the insertion point, or delete the selection
Arrow keys	Arrow keys	Move one character up, down, left, or right
Home	Home	Move to the beginning of the line
End	End	Move to the end of the line
Control+Home	+Home	Move to the beginning of a worksheet
Control+end	+End	Move to the last cell on the worksheet.
Control+x	+x	Cut the selection
Control+v	+v	Paste the selection
Control+c	+c	Copy the selection
Control+z	+z	Cancel or undo an entry in the cell or formula bar
Control+y	+y	Repeat the last action
Control+f	+f	Open the Find dialog box
Control+s	+s	Save your work
Control+d	+d	Fill down
Control+r	+r	Fill to the right
Control+F3	+l	Define a name
F1	+/	Opens Help menu
F4	+t	Makes cell reference absolute or relative in the formula bar
F9	+=	Calculate (or re-calculate) all sheets in all open workbooks*
Tools | Options | Calculation	Tools | Preferences | Calculation	Set manual versus automatic calculation

* The Calculate key, F9, is used extensively throughout these exercises. The F9 function key will work on Macintosh machines provided the Hot Function Key option in the Keyboard Control dialog box is turned OFF. If the F9 key does not work on your Mac, use the alternative, +=.

menus, and/or options as shown in Figure 1. Your mouse may have one, two, or three buttons. All operations described in this section are performed with the *left* button. In current Macintosh and Windows operating systems, a single mouse-click will open a menu and keep it open. To execute a command from a menu, move the cursor over the available commands until the one you want is highlighted, and then click the mouse a second time. On Macintoshes running older operating systems, you must click the mouse button and *hold it down* as you move the cursor down the menu options. Release the mouse button when the command you want is highlighted. The command will flash when it is successfully invoked.

For instance, if you wanted to record a macro in your spreadsheet to carry out a set of instructions, you would open the Tools menu, select the Macro submenu, and choose the Record New Macro Option. Throughout this book we will use the vertical bar (|) and sans serif type (Menu) to indicate a menu, submenu, or option. Thus, the instruction above would read, "Open Tools | Macro | Record New Macro." The results of this operation are shown in Figure 1 (and discussed in more detail on p. 16).

Many menu commands also have keyboard shortcuts—key combinations that you can press to execute the command without having to open a menu and sort through its submenus and options. Shortcuts are listed next to the commands in the menus, and always begin with <Control> in Windows and with on a Macintosh, followed usually by a single letter (see Table 1). To use a shortcut, press and hold the <Control> or the key while simultaneously typing the indicated letter. We will represent this simultaneous key-pressing like this: +c on (Macs) or <Control>+c (Windows). This is the shortcut for Edit | Copy. Many people use shortcuts for frequently used commands, and you may find it worthwhile to memorize a few of these, such as the one for copy, and +v (Macs), <Control>+v (Windows) for Edit | Paste.

Don't be afraid to thrash around in the menus. In other words, if you're not sure how to do something, try opening menus and submenus, searching for a command that looks like it might work. Try different commands and see what happens. This is how we learned most of what we know about spreadsheets. However, be sure to save your work *before* you start to thrash—then, just in case you do something that messes up your work, you can close the file *without* saving any of the changes you made and the file will revert to what it was before you started thrashing.

Spreadsheet Structure

A spreadsheet consists of a **matrix**, or **grid**, of **cells**. Any cell can contain information (text, a number, a formula, or a function). The **columns** of a spreadsheet are identified by letters; the **rows** are identified by numbers (although this may vary in different programs). Each cell has an **address** consisting of its column letter and row number. For example, the top-left cell's address is A1; two cells to the right is cell C1; two cells down from cell C1 is cell C3 (Figure 2).

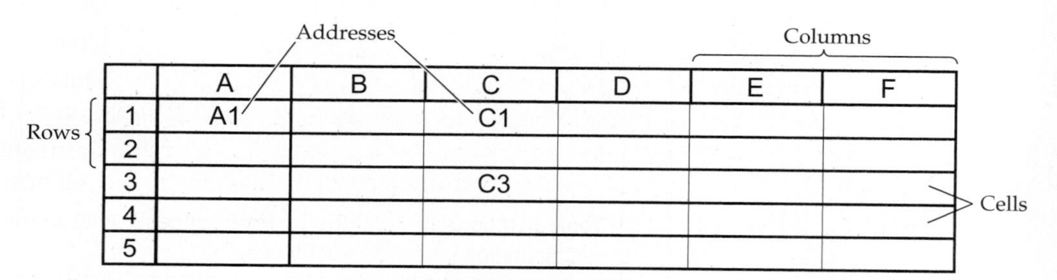

Figure 2

Selecting (Highlighting) Cells

To enter information into a cell, you must first select it by placing the cursor (the on-screen arrow) in it and clicking the mouse button. You can move the cursor either with the mouse or with the arrow keys. You can tell a cell has been selected because it will be highlighted—either the entire cell or its outline will be shown in a different color from other cells. You can simultaneously select more than one cell by any of the following procedures.

If the cells are in a contiguous block:

- Move the cursor to one corner of the block of cells.
- Click and hold the mouse button as you drag the cursor to the opposite corner of the block.
- Release the mouse button when the cursor is in the cell at the opposite corner of the block.

or

- Select a cell at one corner of the block of cells.
- Move the cursor to the opposite corner of the block.
- Hold down the <Shift> key and click the mouse button.

If the cells are not in a contiguous block:

- Use either procedure above to select some of the cells.
- Select additional cells by holding down the <Control> key while clicking-and-dragging.
- Continue selecting rows, columns, or blocks until you have selected all the cells you want.

Copying Cell Contents

Copy the contents of a cell or of multiple cells by selecting the cell or cells and using either the Edit | Copy command or the keyboard shortcut +c or <Control>+c.

Cutting Cell Contents

Cutting is similar to copying except that copying leaves the original cell(s) unchanged, whereas cutting deletes the contents of the cut cell(s) once they have been pasted into another cell. The Cut command is Edit | Cut under the Edit menu; the shortcut is +x or <Control>+x.

Pasting into a Cell

Paste information that you copied or cut from one cell into another cell by executing the Edit | Paste command or the keyboard shortcut +v or <Control>+v.

Cell Addresses

Every cell has an **address**, consisting of its column letter and row number. The top-left cell's address is A1; two cells to the right is cell C1; two cells down from C1 is cell C3 (see Figure 2). When you carry out spreadsheet operations, such as finding the sum of two cells or the mean of a column of cells, you must tell the program the addresses of the cells to operate upon. You use addresses rather than entering the **values** to operate upon, because this allows you use a principal advantage of spreadsheet programs: their ability to update calculations when you change cell contents.

You can type single cell addresses—A1, C3, etc.—or you can type a range of cell addresses in the form **A1:C3**. The latter designates a contiguous block of cells with its top-left corner at cell A1 and its bottom-right corner at cell C3. You can designate any contiguous block of cells by entering the addresses of any two opposite corners, separated by a colon. A block may also consist of a single column (e.g., **A1:A10**) or single row (e.g., **B3:B20**). Other spreadsheet programs may use different symbols than the colon, so consult your spreadsheet user's manual if the colon doesn't work.

Entering Literals

The titles, headings, notes, and other pieces of text (or numbers) that you want to appear on your spreadsheet are called **literals** because the program does not interpret them, but represents them literally (i.e., exactly as you type them). To enter a literal, select the cell in which you want the text to appear, and type.

Press the <Return> (or <Enter>) key *only* when you have finished entering text. The <Return> key ends text entry; it does not give you a second line of text. If you want a label of more than one line, one way is to type the first line, press <Return> or the down arrow key, place the cursor in the cell below (if it's not already there), and type the second line. Another way is to type all the text into a single cell and then format the cell to turn on text wrapping (see p. 13 for how to format cells).

As you type text or numbers into a cell, what you type will appear in the cell and in the **formula bar** above the spreadsheet column headings (Figure 3). If you make a mistake, use your mouse to place the cursor on the mistake either in the cell or in the formula bar. Then use the backspace or delete key to erase the mistake, or highlight the mistake using click-and-drag, and retype. The text will appear in the selected cell after you press <Return>. If you discover an error later, you can simply select the cell again and correct your mistake as above.

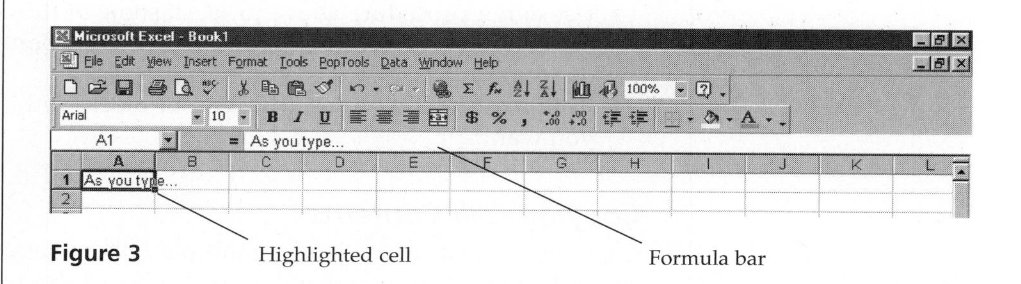

Figure 3

Sometimes strange things happen when you enter a literal, depending on your program and how it is set up. For instance, if you enter **5-10** (meaning a range of values from 5 to 10), the cell may show **May 10**. This is because the program interprets some entries as dates. To force the program to treat your entry as a literal, precede it with an apostrophe, **'5-10**, or open Format | Cells | General.

Another potentially confusing aspect of entering literals is spill-over. If the text you enter is too long to fit into a single cell, it may spill over into adjacent cells if they are empty, as does the text "Spreadsheet Hints and Tips" in cell A1 of Figure 4. The entire text is actually in cell A1, although it appears to occupy cell B1 as well, because cell B1 is empty. If the adjacent cell holds information, the text is truncated rather than spilling over. Note that the same text is present in cell A2 (as you can see in the formula bar), but because cell B2 holds the text "Example," the text in cell A2 is truncated.

A2 = Spreadsheet Hints and Tips

Example1.xls

	A	B	C
1	Spreadsheet Hints and Tips		
2	Spreadsheet	Example	
3			
4	Individual	Height (cm)	Weight (g)
5	1	10.0	35
6	2	4.0	15
7	3	6.0	22
8	4	6.0	20
9	5	7.0	26
10			
11	Sample Size	5.0	5.0
12	Sum	33.0	118.0
13	Mean	6.6	23.6
14	Minimum	4.0	15.0
15	Maximum	10.0	35.0
16	Range	6.0	20.0
17	Standard Dev	2.2	7.5

Sheet1 Sheet2 Sheet3

Figure 4

Entering Formulae

A very important part of spreadsheet programming is entering formulae. A **formula** tells the spreadsheet to carry out some operation(s) on the contents of one or more cells, and to place the result into the cell where the formula is. A formula usually contains one or more cell addresses and operations to be performed on the contents of the referenced cells. A formula must begin with a symbol to alert the spreadsheet that it is a formula rather than a literal. In Excel, the symbol is typically the equal sign (=), but other symbols (such as +) may work in this or other spreadsheet programs.

Two useful tips to remember regarding formulas:

- The formula appears in the formula bar as you type it, and it will appear there again if you select the cell later. But once you press <Return>, only the *result* of the formula appears in the cell itself.
- A formula may not refer to the cell in which it resides; therefore, e.g., do not enter the formula **=2*B2** into cell B2. This will generate an error message complaining about a "circular reference."

In Figure 5 we wanted the range of height values (the maximum value minus the minimum value) to appear in cell B16, so we entered **=B15-B14** into cell B16. Although the result (6.0) is shown in the cell, the formula bar shows the formula.

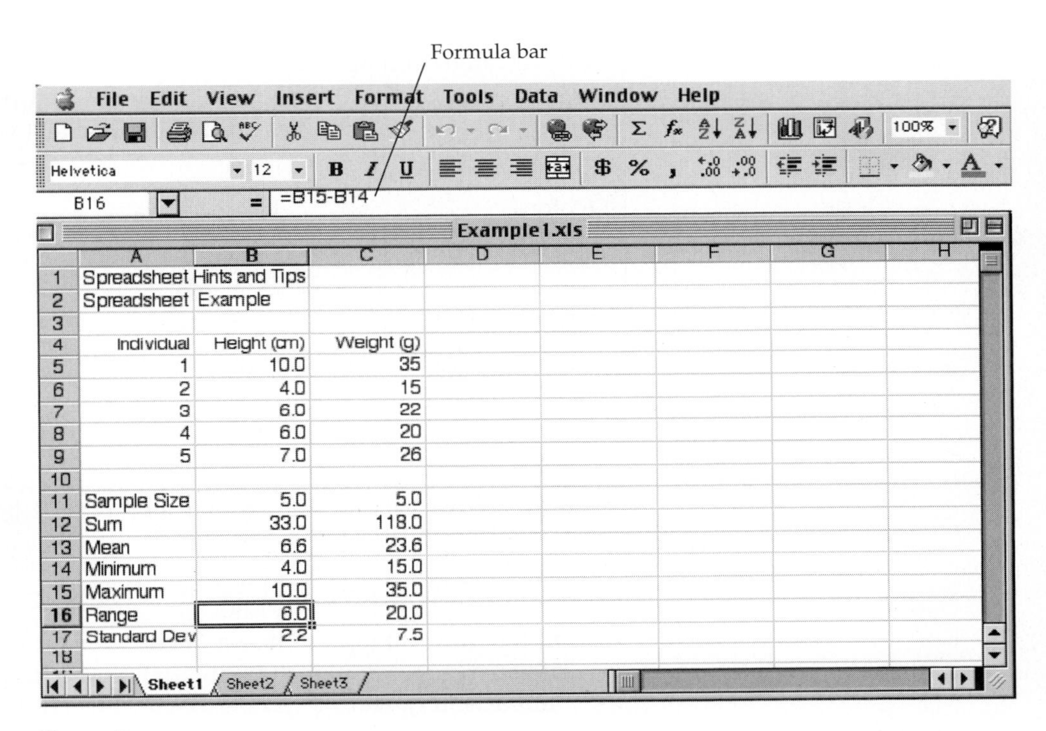

Figure 5

Calculation Operators in Formulae

Spreadsheet operators are keyboard entries that specify the type of calculation that you want to perform on the elements of a formula. Microsoft Excel has four different types of calculation operators: arithmetic, comparison, concatenation, and reference. These are listed in Table 2.

- **Arithmetic operators** perform basic operations such as addition, subtraction, or multiplication; combine numbers; and produces **numeric results**. The asterisk (*) is used to specify multiplication; the forward slash (/) represents division;

Table 2. Calculation Operators in Microsoft Excel Formulae

Operator	Meaning	Example
Arithmetic operators		
+ (plus sign)	Add	3+3
- (hyphen)	Subtract	3-1
- (hyphen	Negation (negative value)	-1
* (asterisk)	Multiply	3*3
/ (forward slash)	Divide	3/3
% (percent sign)	Percent	20%
^ (caret)	Exponentiation	10^3 (10 to the third power, or 1,000)
Comparison operators		
= (equal sign)	Equal to*	A1=B1
> (right angle)	Greater than	A1>B1
< (left angle)	Less than	A1<B1
>=	Greater than or equal to	A1>=B1
<=	Less than or equal to	A1<=B1
<>	Not equal to	A1<>B1
Text concatenation operator		
& (ampersand)	Join two values to produce one continuous text value	"A1"&"A2" becomes "A1A2"
Reference operators		
: (colon)	Range operator	B5:B15 (Produces one reference to all the cells between B5 and B15, including those two cells)
, (comma)	Union operator	SUM(B5:B15,D5:D15) (Combines multiple references into one reference)

*Recall that the equal sign is also a "start signal" that tells Excel to consider what follows as a formula, as in =A1+B1.

and the carat (^) represents exponentiation (raising to a power). Other arithemetic operators include the standard + and -.

- **Comparison operators** compare two values (for example, whether two values are equal, or one is greater than the other) and return a **logical value**—either true or false—for specified calculations.
- The ampersand (&) is the **text concatenation operator**. It joins, or "concatenates" two strings of text to produce a continues text string.
- **Reference operators** are the colon (:) and the comma (,). These operators combine ranges of cells for calculations.

If you *combine* several operations in a single formula, Microsoft Excel performs the operations in the order shown in Table 3. If a formula contains multiple operators with the same precedence (i.e., if a formula contains both a multiplication and a division operator), the program evaluates the operators from left to right. You can change the order of evaluation by enclosing the part of the formula to be calculated first in parentheses.

Table 3. Order of Operation in Microsoft Excel Formula

Precedence of calculation	Description	Operator
1	Reference operators	: ,
2	Negation	-
3	Percent	%
4	Exponentiation	^
5	Multiplication and division	* /
6	Addition and subtraction	+ –
7	Concatenation	%
8	Comparison	= < > <= >= <>

Entering Functions

A **function** is similar to a formula, but it usually carries out a more complex operation or set of operations, and it has been prewritten for you by the spreadsheet programmers. We use functions extensively; many of the exercises in this book rely on them. Excel has over 100 functions, and you will probably not remember them all. Fortunately, most spreadsheet packages provide a simple means of entering functions so that you don't need to memorize them.

Functions are entered by pasting them into the formula bar. You can use the "Paste Function" button on the toolbar, f_x (indicated by an arrow in Figure 6), or you can open Insert | Function to guide you through entering a function. Either way, the dialog box headed Paste Function will appear (Figure 6).

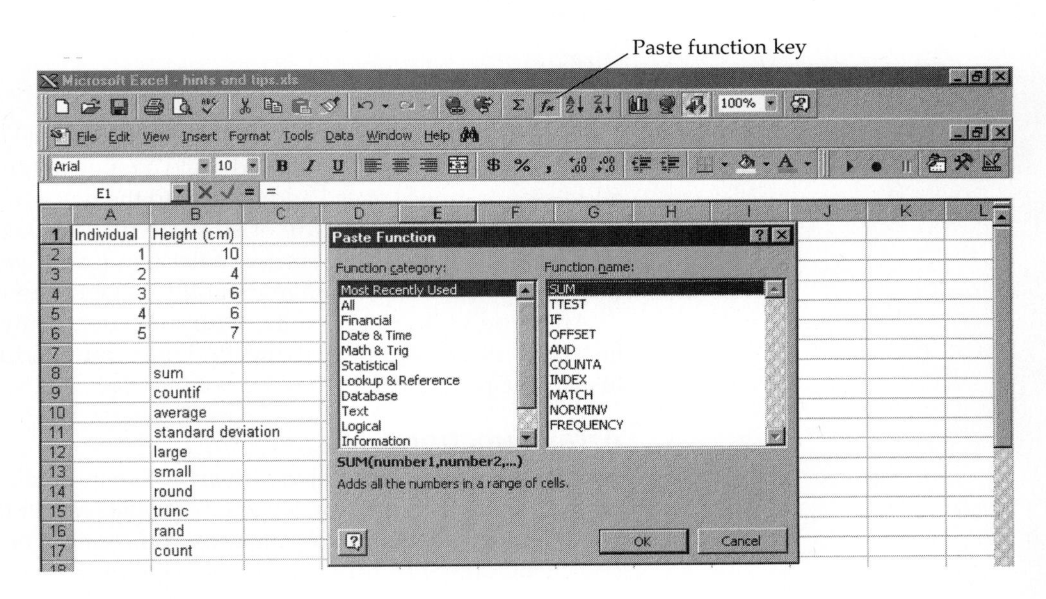

Figure 6

Look at the column on the left side of the dialog box, labelled Function category. It asks what kinds of functions you want to examine. In the figure, the Most Recently Used category was selected, so a list of the most recently used functions appears in the right side of the dialog box. Note that the function **SUM** is selected, and the program displays a brief description of the **SUM** function at the bottom of the window. If you choose the Function category All, you'll see every function available, listed in alphabetical order.

Use your mouse to select the function you want, and a brief description of the function will appear. Click OK when you've got the function you want. When you select a function, a new dialog box will appear (Figure 7). In Figure 7, we selected the **SUM** function. Excel asks you to specify the cells you want to sum. There are two handy features in this dialog box. First, notice the small figure with the arrow pointing upward and leftward (located to the right of the blank space labeled Number 1). If you click on this arrow, the dialog box will shrink, exposing your spreadsheet so that you can use your mouse to select the range of cells you want to sum. After you've selected the cells you want to sum (in this case, cells **B2:B6**), click on the arrow again and the **SUM** dialog box will reappear. Click OK and Excel will return the calculated value.

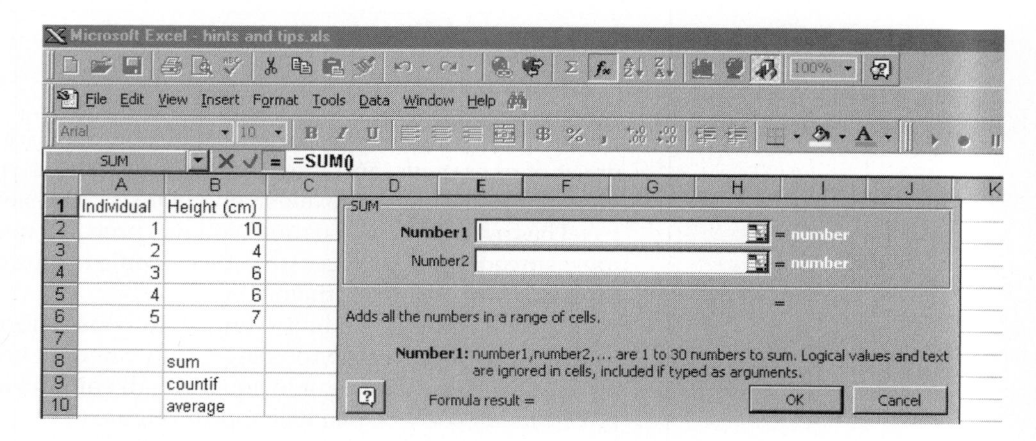

Figure 7

Note that although the box is labeled Number 1, it is not limited to a single cell address, but can (and often should) hold a range of cell addresses. You can also type cell addresses or ranges of cell addresses into the boxes, if that's easier.

The second handy feature of all paste function dialog boxes is the question mark located at the bottom-left corner of the window. If you don't know how the function works, click on the question mark and Excel will provide more information.

After you've become familiar with some frequently used functions, you may find it faster to type them into a cell directly. Like formulae, functions begin with an equal sign to alert the program that they are not literals.

Array Functions

In some exercises, you will use an array function rather than a standard function. An **array function** acts on *two or more sets of values* rather than on a single value. These sets of values are called **array arguments**. You create array formulae in the same way that you create other formulae, with this major exception: Instead of selecting a single cell to enter a formula, you need to select a series of cells, then enter a formula, and then press <Control>+<Shift>+<Enter> (Windows) or +Return (Macs)to enter the formula for all of the cells you have selected.

For example, the **FREQUENCY()** function is an array function that calculates how often values occur within a range of values, and then returns a vertical array of numbers. Suppose you want to construct a frequency distribution for the weights (in grams) of 10 individuals (Figure 8).

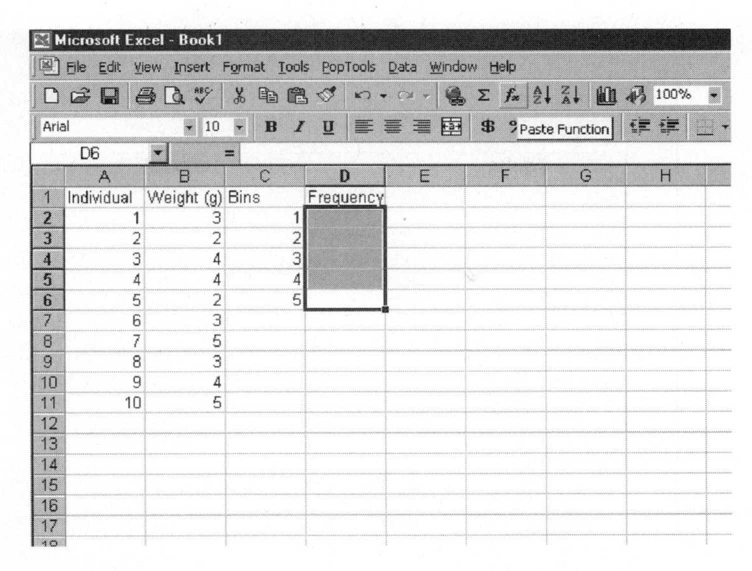

Figure 8

In Figure 8, the column labeled "Bins" tells Excel how you want your data grouped. You can think of a "bin" as a bucket in which specific numbers go. The bins may be very small (hold only a few numbers) or very large (hold a large set of numbers). For example, suppose you want to count the number of individuals that are 1 g, 2 g, 3 g, 4 g, and 5 g. The numbers 1 through 5 represent the five bins. If we want Excel to return the number of individuals of given weights in cells D2–D6, then we need to first select those cells (rather than a single cell) *before* using the paste function key to summon the frequency procedure. The dialog box in Figure 9 will appear.

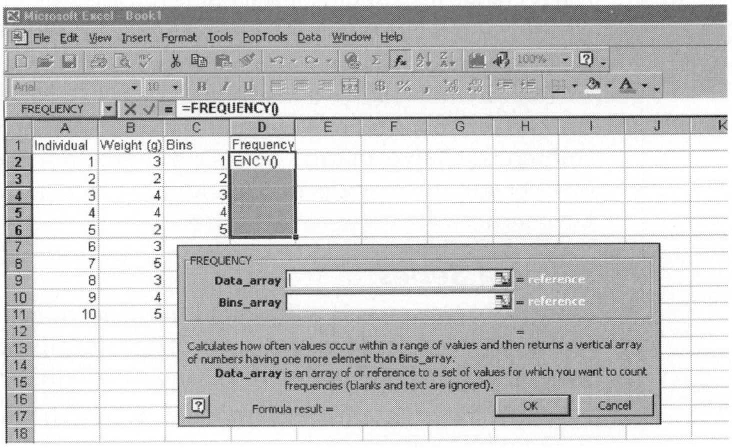

Figure 9

The **Data_array** is simply the data you want to summarize, given in cells B2:B11. The **Bins_array** is cells C2–C5. Instead of clicking OK, press <Control>+<Shift>+<Enter> on Windows machines; Excel will return your frequencies. On Macs, type the formula in by hand, then press +Return. After you've obtained your results, examine the formulas in cells D2 through D6 (Figure 10). Every cell will have a formula that looks like this: **{=FREQUENCY(B2:B11,C2:C6)}**. The { } symbols indicate that the formula is part of an array, rather than a standard formula.

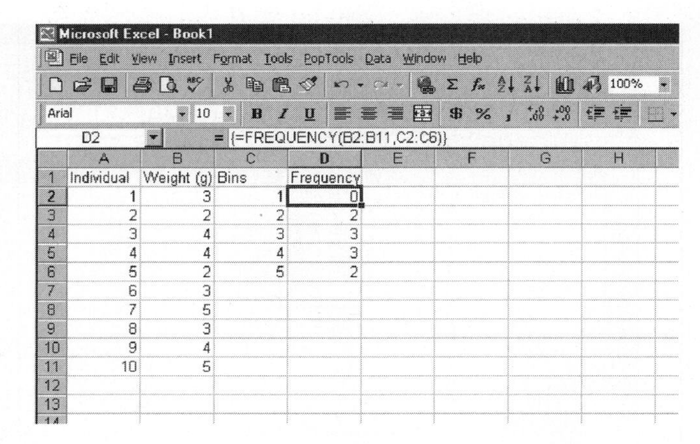

Figure 10

Relative and Absolute Cell Addresses

Cell addresses are said to be either "relative" or "absolute." It's critical that you know the difference between these two kinds of addresses. A **relative address** refers to the position of a cell *relative to the position of the currently selected cell.* For example, if you enter the formula **=2*B2** into cell C3, the cell address B2 does not really refer to cell B2; it refers to a cell one column to the left and one row up from the cell you're typing into (cell C3). If you copy this formula into cell D5, the program will automatically change the formula into **=2*C4**, which is one column to the left and one row up from cell D5.

In Excel, the dollar sign (**$**) indicates an **absolute address**. An absolute address *always refers to the same cell*, even if you copy or move the formula to a new cell. For example, if you enter the formula **=2*$B$2** into cell C3, the cell address **$B$2** really does refer to cell B2 regardless of which cell holds the formula. If you copy *this* formula into cell D5, it will still read **=2*$B$2**. Addresses without dollar signs are relative addresses. Other programs may use symbols other than $ to indicate an absolute address.

You can mix relative and absolute references in one address. In the address **$B2**, the column reference is absolute, and the row reference is relative. In the address **B$2**, the column is relative and the row is absolute. (In the Windows version of Excel, you can quickly add dollar signs to cell addresses by pressing the F4 button at the top of your keyboard.)

Filling a Series

In many exercises, you will be told to create, or **fill**, a series of values, usually in a column. What we mean is to create a sequence of numbers, like the one shown in column A, Cells A5–A9 of Figure 11. You can do this in either of two ways. The first is:

- Give the program an example of what you want (e.g., enter 1 into cell A5 and 2 into cell A6).
- Tell the program to extend this series by selecting the example cells (A3 and A4), then placing the cursor at the bottom-right corner of the last cell in the example (cell A6).
- The cursor will turn into a bold cross. Click and hold the mouse button while dragging down the column to cell A9.
- The program will extend the series down the column, showing you the current value in a small box as it goes.
- When the series reaches the maximum desired value, release the mouse button.

The alternatetive way to fill a series is:

- Enter the first value of the series in the first cell (enter 1 into cell A5).

- Enter a formula to calculate the next value in the series into cell A6 (=A3+1).
- Copy the formula in cell A6 (select the cell and press <control>+c or +c).
- Select the cells to hold the rest of the series (select cells A7:A9).
- Paste the formula into the selected cells (<control>+v or +v).

You can also just click on the bottom-right hand corner of cell A6 (the cursor will change to a bold cross) and then "drag" the formula down to cell A9. Any of these procedures will work with series in rows as well as in columns.

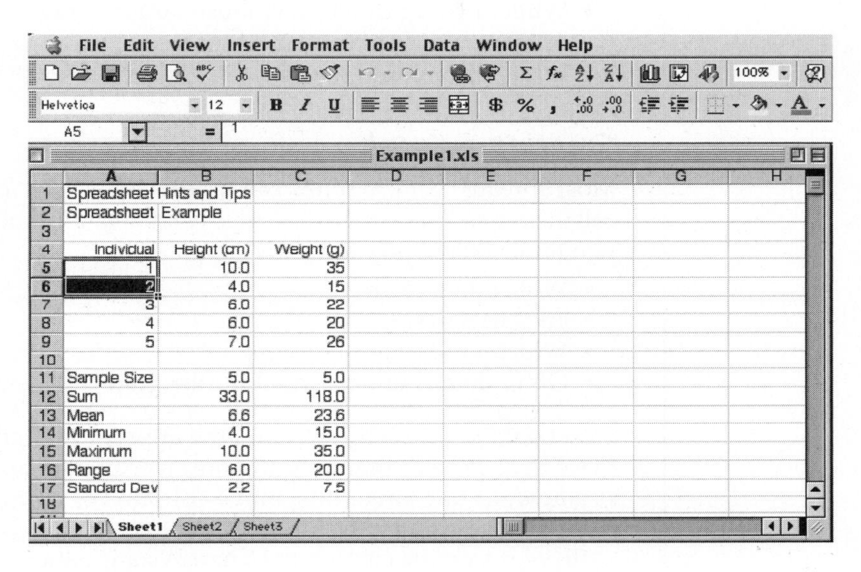

	A	B	C
1	Spreadsheet Hints and Tips		
2	Spreadsheet Example		
3			
4	Individual	Height (cm)	Weight (g)
5	1	10.0	35
6	2	4.0	15
7	3	6.0	22
8	4	6.0	20
9	5	7.0	26
10			
11	Sample Size	5.0	5.0
12	Sum	33.0	118.0
13	Mean	6.6	23.6
14	Minimum	4.0	15.0
15	Maximum	10.0	35.0
16	Range	6.0	20.0
17	Standard Dev	2.2	7.5
18			

Figure 11

Formatting Cells

The appearance of a cell's contents depends on how the cell is formatted. To access all the options for formatting a cell or range of cells, select the cell(s) and then open Format | Cells. You can also use toolbar shortcuts to format font, size, alignment, number of decimal places, borders, shading, or color.

With some exceptions (an important one, is formatting column width), formatting cells is a matter of taste. Our guiding principles have been to keep fancy formatting to a minimum, and to format cells to enhance readability. In the exercises in this book, you will see cells with borders, shading, bold type, and other formats. Unless otherwise noted, you need not reproduce these unless you wish to.

However, some aspects of formatting cells are *not* just a matter of appearance. If a number is too large to fit in the space provided by a cell, it will be represented by hashmarks (########). To see the number, you must either reduce the number of decimal places (which may not be applicable or desirable), or expand the column width to accommodate the number. There are several ways to format column width. All begin with the same first step:

- Select the column to be formatted either by clicking in a cell in the column or by clicking on the column letter at the top of the column.

You can then follow *one* of three procedures. The first procedure is:

- Open Format | Column | Width.
- Type a number in the dialog box.
- The relationship of the number to the column width is obscure (i.e., we don't understand it), so you'll have to experiment until you get the result you want.

The above steps can be used to adjust several columns to a uniform width. A second procedure is:

- Open Format | Column | AutoFit Selection. Excel will adjust the column width to permit display of the widest element in the selected block or column.

A third alternative:

- Place the cursor at the right-hand edge of the space around the letter at the top of the column to be adjusted. The cursor will change to a vertical bar with arrows pointing to the right and the left.
- Click and hold down the mouse button.
- While holding down the mouse button, drag to the right to widen the column or to the left to narrow it.
- When the column width is appropriate, release the mouse button.

Creating a Graph

Most spreadsheet programs call graphs "charts." We will follow scientific usage and call them graphs. In these exercises, you'll make lots of graphs. To create a graph (chart), you must tell the program:

- Which data to graph
- To start a graph
- Which kind of graph to use
- Other details of how to set up the graph

Select data to graph by selecting the appropriate cells (see p. 4–5). Excel will always place the leftmost column or topmost row of data on the horizontal axis of the graph. If you want to change this, move columns or rows using the cut-and-paste procedures described on page 5.

To start a graph, click on the Chart Wizard button (the little bar graph in the toolbar; Figure 11) or open Insert | Chart. You will be presented with a series of dialog boxes that take you through the process of creating a graph. After finishing each dialog box, move to the next by clicking on the OK button.

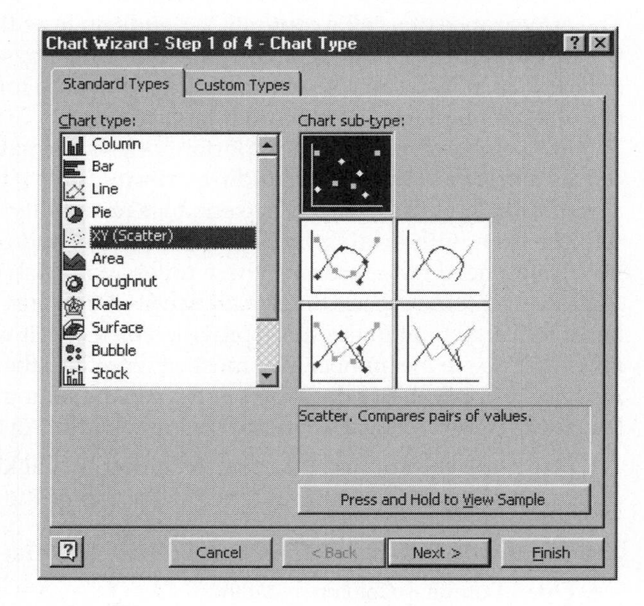

Figure 12

In the first dialog box (Chart Type), click on the kind of graph you want to create (Figure 12). You will frequently choose an X-Y axis scatterplot, XY (Scatter), or sometimes a line graph (Line) or a vertical bar graph (Column), or other.

We strongly advise you to avoid "chart junk." Three-dimensional graphs, lots of colors, and bizarre chart-types usually detract from the readability of a graph. Keep in mind that your purpose is to communicate clearly and immediately, not to impress with fancy graphics.

In the second dialog box (Chart Source Data), you will be given some choices about the data to be graphed (Figure 13). Most often, the default settings will work, but sometimes you may have to tell the program that your data are arranged in rows rather than columns, or vice versa. The Series tab provides additional options. This window enables you to name a series of values (such as weight) and to specify the x and y values to be used in the chart if the default values are not appropriate.

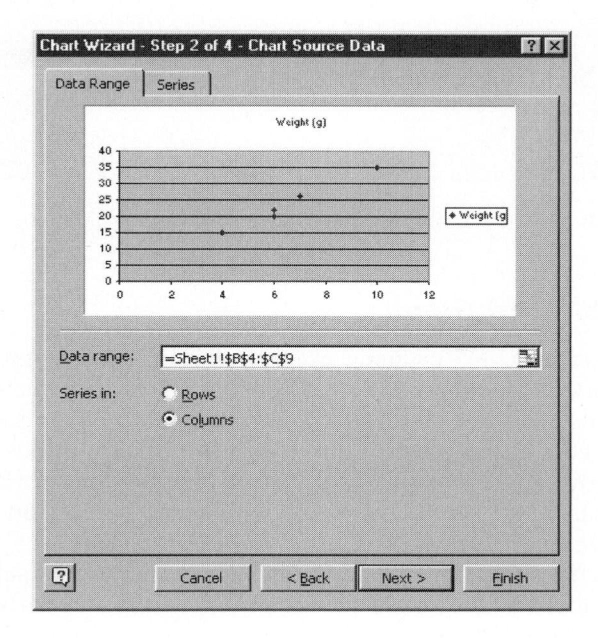

Figure 13

In the third dialog box (Chart Options), you will be presented with a variety of choices for formatting your graph (Figure 14). This dialog box is very important because it is your opportunity to label the graph, its axes, and legend. It is extremely important to label your graphs thoroughly, including units when appropriate.

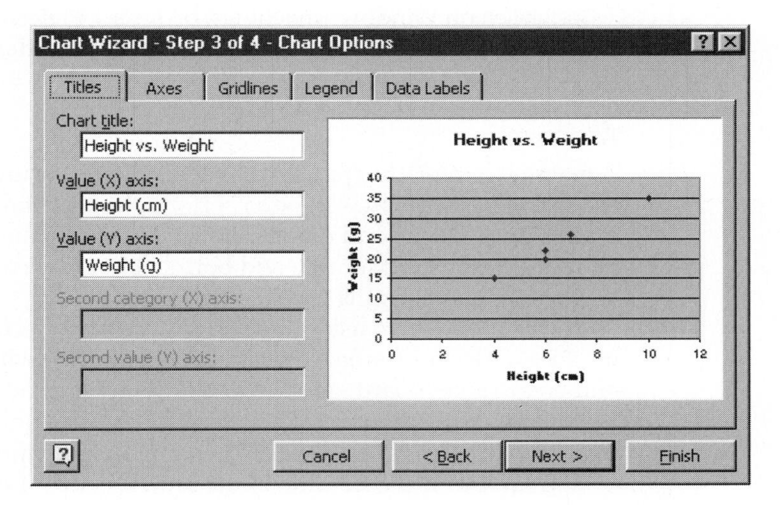

Figure 14

In the final dialog box (Chart Location; Figure 15), you will be asked to specify where to save the graph (Figure 15). Most commonly (and by default) we choose to save the graph on the spreadsheet, but in some circumstances you may want to save it on a separate sheet. Click on the Finish button and your chart will appear on your spreadsheet.

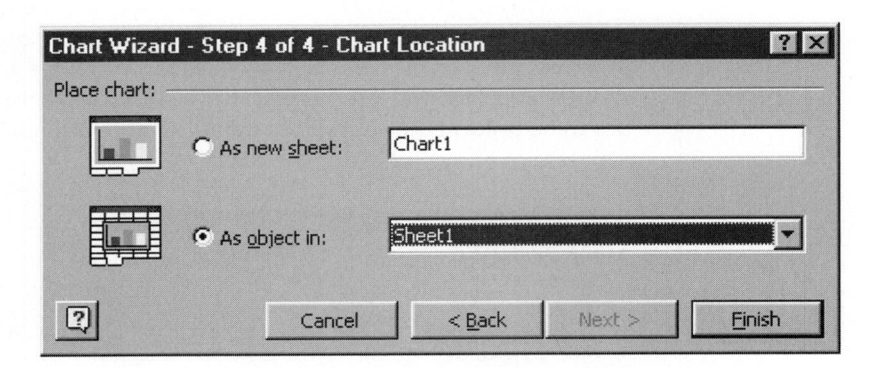

Figure 15

Editing a Graph

After you have created a graph, you can change its appearance by editing it in various ways. To begin, select the graph by clicking anywhere in it. To change a feature of the graph, double-click (two mouse clicks in rapid succession) on the feature you want to change, and choose the desired options from those offered in the resulting dialog box(es). When you have finished changing that feature, click on OK. For example, to change an axis to a logarithmic scale, double-click on the axis, click in the box for logarithmic scale, and click OK.

Alternatively, you may open the Chart menu after selecting the graph. The submenus within the Chart menu will allow you to modify nearly any feature of the graph to suit your needs.

Automatic and Manual Calculation

By default, the spreadsheet program re-calculates all formulae and functions every time you press the <Return> or <Enter> key (or perform certain other actions). This is called **automatic calculation**. In some circumstances, you will want to prevent this, and take direct control of when calculations are updated. This is called **manual calculation**. You can choose whether calculation is automatic or manual by opening Tools | Options | Calculation on Windows machines, or Tools | Preferences | Calculation on Macs.

After you set calculation to manual, you can update all formulae and functions by pressing the recalculate key: F9 on Windows, or += on Macs.

Macros

A **macro** is a miniature program that you create to run a sequence of Excel actions. For example, suppose you wanted to perform the same fairly long, tedious series of actions many times. Typing and mouse-clicking your way through them over and over would not only be time-consuming and boring, but also error-prone. A macro allows you to achieve the same results with a single command.

You create a macro using Excel's built-in macro recorder. Start the recorder by opening Tools | Macro | Record New Macro. The program will prompt you to name the macro and create a keyboard shortcut. Then, a small window will appear with the macro recorder controls (Figure 17). If this button does not appear, go to View | Toolbars | Stop Recording, and the Stop Recording figure will appear.

The square on the left side of the button is the **Stop Recording** button (Figure 17). When you press this square, you will stop recording your macro The button on the right

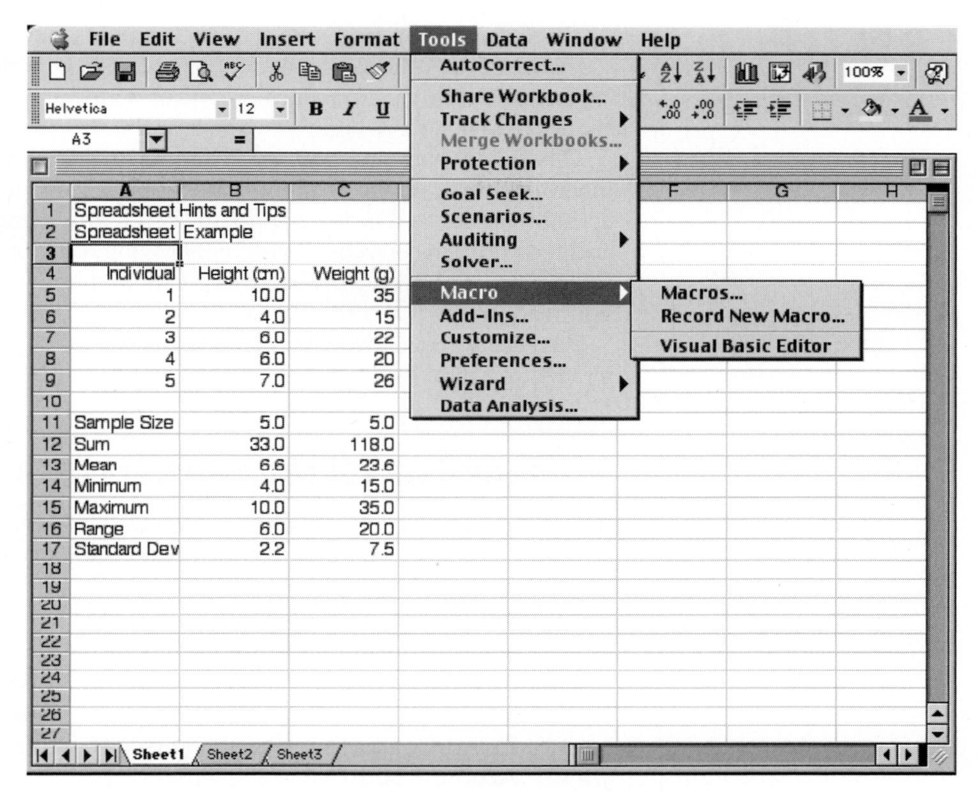

Figure 16

is the **relative reference button**. By default this button is not selected so that your macro recorder assumes that the cell references you make in the course of developing your macro are absolute. In other words, if you select cell A1 as part of a macro, Excel will interpret your keystroke as cell **$A$1**. There are cases (for example, the survival analysis exercise) in which you will want to select the relative reference button as you create your macro.

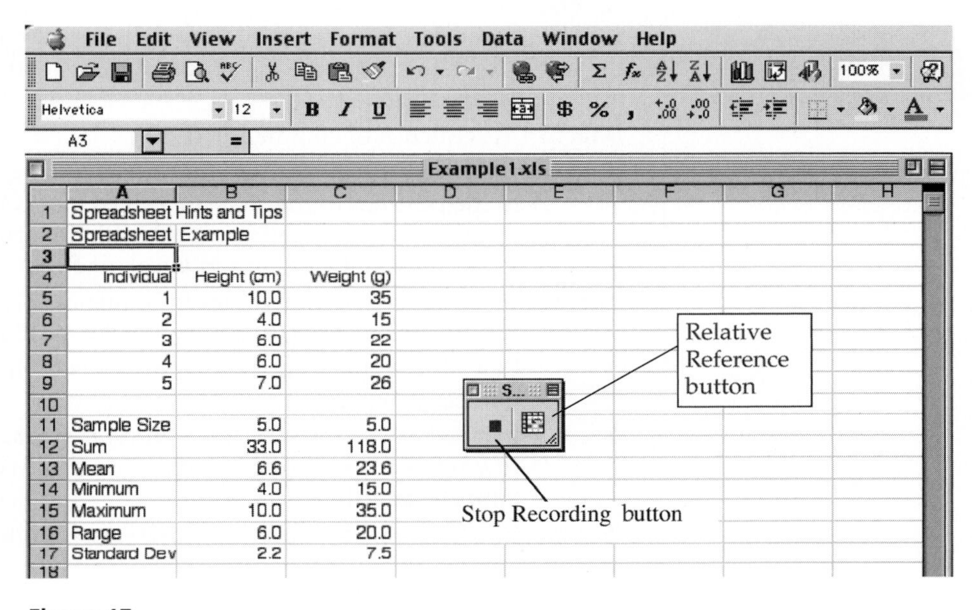

Figure 17

From this point on, Excel will record every action you take. Carry out the entire sequence of operations you want the spreadsheet to do, and then press the Stop Recording button in the macro recorder control window. The program will mimic that entire sequence of actions whenever you press the shortcut key or issue the macro command.

Obviously, planning pays off when recording a macro. If you're creating your own macro, go through the sequence of actions at least once in preparation to make sure it actually achieves the desired result. Write down each action, so that you can repeat and record them correctly. If you're following our instructions to create a macro, be careful to execute each step precisely as given. Remember, the computer doesn't know what you want to do; it records everything faithfully, mistakes and all.

Exercise 2, "Spreadsheet Functions and Macros," provides exercises to help you master creating macros.

GLOSSARY OF TERMS AND SYMBOLS

Absolute address A cell address (see Cell address) that refers to a specific location in the spreadsheet, regardless of its position relative to the selected cell (see p. 12). An absolute address does not change if copied to a new location. In Excel, an absolute address is indicated by preceding the column letter or row number (or both) by a dollar sign ($).

Cell address The location of a cell in the spreadsheet. The cell address consists of a letter representing the column and a number representing the row (see p. 5). Addresses may be relative (see Relative address) or absolute (see Absolute address).

Formula A symbolic representation of a set of operations to be carried out by the spreadsheet (see p. 7). Usually, a formula contains one or more cell addresses and one or more mathematical operations to be carried out on the contents of those cells. The result of the operation(s) appears in the cell in which the formula is entered. In Excel, formulae begin with the equal sign (=).

Function A prewritten formula or set of formulae (see p. 9). Enter a function by typing it in, by opening Insert | Function and choosing from the list, or by clicking the Paste Function button (f_x) and choosing from the list. In Excel, functions begin with the equal sign (=).

Literal Text or a number that is not interpreted or manipulated by the spreadsheet program (see p. 5). Row labels, column labels, and model constants are literals. To force the program to treat an entry as a literal, begin it with an apostrophe (').

Macro A sequence of commands to be executed automatically (see p. 16).

Relative address A cell address that refers to a location in the spreadsheet relative to the position of the selected cell (see p. 12). A relative address changes if copied to a new location, preserving the original relationship. Cell addresses are relative by default in Excel, and require no special symbol.

Series A column or row of values in sequence. Most frequently these will be a simple linear series (0, 1, 2, 3, ...). See p. 12 for shortcuts to enter a series.

***** In a *formula*, the asterisk (*) represents multiplication. In *text*, it represents a wildcard: a stand-in for any letter or digit.

**$** In a cell address, the dollar sign ($) indicates that the following column or row reference is absolute rather than relative. See Cell address, Absolute address, and Relative address.

^ In a formula, the carat (^) represents exponentiation. That is, **3^2** is equivalent to 3^2.

1 MATHEMATICAL FUNCTIONS AND GRAPHS

Objectives

- Learn how to enter formulae and create and edit graphs.
- Familiarize yourself with three classes of functions: linear, exponential, and power.
- Explore effects of logarithmic plots on graphs of each kind of function.

INTRODUCTION

This exercise serves two main purposes: to allow you to practice some of the procedures outlined in "Spreadsheet Hints and Tips," and to acquaint you with three classes of mathematical functions. Biology, like all sciences, uses mathematical relations to describe natural phenomena. In many cases, the mathematics is only implied, as in any graph of one variable against another. In other cases, it is made explicit in the form of an equation. Such relationships take a variety of forms, but you will encounter three classes of relationships with some regularity in textbooks and journal articles: linear functions, exponential functions, and power functions.

For example, the number of lizard species in a given area of desert habitat rises *linearly* with the length of the growing season; a bacterial population introduced into an empty vial of nutrient broth will grow *exponentially* (at least for a time); and the number of species on an island is a *power* function of the island's area.

A mathematical **function** relates one **variable** to another. For example, we may say that the death rate in a population is a function of population density, meaning that death rate and population density (both numbers that change from population to population, and even within a population—i.e., numbers that are "variable") are related in some way. By writing an equation, we can specify precisely how these variables relate to one another.

For convenience, we usually refer to one variable as the **independent variable** and the other as the **dependent variable**, and we speak of the dependent variable "depending on" the independent variable. For example, we may say that death rate depends on population density. If one variable is clearly a cause of the other, we take the cause as the independent variable and the effect as the dependent variable. But in many cases, cause and effect relationships are not clear, or each variable may in a sense cause the other and be an effect of the other. Population density and death rate offer an example of such a mutual cause-effect relationship. In such cases, our choice of which variable to treat as independent and which to treat as dependent is a matter of convenience or convention.

As a matter of convention, we denote the independent variable as x and plot it on the horizontal axis of a graph, and we denote the dependent variable as y and plot it on the vertical axis.

More strictly speaking, a function is a rule that produces one and only one value of y for any given value of x. Some equations, such as $y = \sqrt{x}$, are not functions because they produce more than one value of y for a given value of x. We can often treat such equations as functions by imposing some additional rule; in this case, we might restrict ourselves to positive square roots.

Functions take a variety of forms, but to begin with, we will concern ourselves with the three broad categories of functions mentioned earlier: linear, exponential, and power. **Linear functions** take the form

$$y = a + bx$$

where a is called the **y-intercept** and b is called the **slope**. The reasons for these terms will become clear in the course of this exercise. **Exponential functions** take the form

$$y = a + q^x$$

Power functions take the form

$$y = a + kx^p$$

Note the difference between exponential functions and power functions. Exponential functions have a constant base (q) raised to a variable power (x); power functions have a variable base (x) raised to a constant power (p). The base is multiplied by a constant (k) after raising it to the power (p).

PROCEDURES

The left-hand column of instructions gives rather generic directions; the right-hand column gives a step-by-step breakdown of these and explanatory comments or annotations. If you are not familiar with an operation called for in these instructions, refer to "Spreadsheet Hints and Tips."

Try to think through and carry out the instructions in the left-hand column before referring to the right-hand column for confirmation. This way, you will learn more about using the spreadsheet, rather than simply following directions. We hope that, with practice, you will gain enough skill in using the spreadsheet that you will be able to modify our models, or create your own from scratch, to suit your own uses.

Your goals in this exercise are to learn how to use a spreadsheet program to calculate and graph these functions and to see how these graphs look with linear and logarithmic axes. In achieving these goals, you will learn about the behavior of these classes of functions, how to use formulae, how to make graphs, and the utility of logarithmic plots. Save your work frequently to disk!

INSTRUCTIONS	ANNOTATION

A. Set up the spreadsheet.

1. Enter titles and headings through Row 9, as shown in Figure 1. You need not enter the text shown in Rows 2 through 6, but if you don't enter the text, leave these rows blank so that the cell addresses in your formulae will match the ones given in these instructions.

These are all literals, so select each cell by clicking in it with the mouse, then type in each title or heading. Use the delete (backspace) key or highlight and overtype to correct errors.

	A	B	C	D	E	F
1	Functions and Graphs					
2	The first part of this exercise will familiarize you with several kinds of mathematical					
3		functions, entering formulae, and graphing in Excel.				
4	The second part will compare functions.					
5						
6	Part 1: Kinds of Functions					
7	Independent					
8	variable		Linear functions			
9	(x)	y=5+1x	y=0+5x	y=10+5x	y=60-5x	
10	0	5	0	10	60	
11	1	6	5	15	55	
12	2	7	10	20	50	
13	3	8	15	25	45	

Figure 1

Linear Functions

2. Set up a linear series from 0 to 9 in cells A10–A19. This will provide values for the independent variable x.

Enter the value 0 as a literal in cell A10.
In cell A11, enter the formula **=A10+1**. Copy the formula in cell A11.
Select cells A12–A19. Paste.

3. In cell B10, enter a spreadsheet formula that expresses the equation shown in cell B9.

In cell B10, type the formula **=5+1*A10**.
We could omit the 1 in the equation and in the formula, but we keep it for consistency with the others.

4. Copy the formula in cell B10 down the column through cell B19.

Copy the contents of cell B10.
Select cells B11–B19. Paste.

5. Enter formulae for the equations shown in cells C10, D10, and E10 into cells C11, D11, and E11, respectively.

These should be:

Cell C10: **=0+5*A10**
Cell D10: **=10+5*A10**
Cell E10: **= 60-5*A10**

6. Copy these formulae down their respective columns.

Select cells C10–E10. Copy.
Select cells C11–E19. Paste.

7. Adjust the widths of columns to accommodate text and numbers.

Select the column(s) to be modified.
You can either open Format | Column | AutoFit Selection, or click and drag column boundaries at the top of the page to achieve the desired widths.

Exponential Functions

8. Enter titles and headings in Rows 21 and 22.

These are all literals, so enter them as before (see Step 1).

	A	B	C	D	E
21		Exponential functions			
22	x	y=0+1.1^x	y=0+1.5^x	y=0+1.5^-x	
23	0	1.00	1.00	1.00	
24	1	1.10	1.50	0.67	
25	2	1.21	2.25	0.44	

Figure 2

9. Set up a linear series from 0 to 9 in cells A23–A32. This will provide values for the independent variable *x*.

Enter the number 0 as a literal in cell A23.
In cell A24, enter the formula **=A23+1**.
Copy the formula in cell A24.
Select cells A25–A32. Paste.

10. In cells B23–D23, enter spreadsheet formulae that express the equations shown in cells B22–D22.

These should be

Cell B23: **=0+1.1^A23**
Cell C23: **=0+1.5^A23**
Cell D23: **=0+1.5^-A23**

We could omit the zeros in the equations and in the formulae, but we keep them for consistency with the others.

11. Copy the formulae in cells B23–D23 into cells B24–D32.

Select cells B23–D23. Copy.
Select cells B24–D32. Paste.
At this point, your spreadsheet should contain the values shown above.

12. If needed, adjust column widths to accommodate text and numbers.

See Step 7.

Power Functions

13. Enter titles and headings in Rows 34 and 35.

These are all literals, so enter them as before (see Step 1).

	A	B	C	D	E
34		Power functions			
35	x	y=0+x^2	y=0+x^0.5	y=0+x^-0.5	
36	1	1.00	1.00	1.00	
37	2	4.00	1.41	0.71	
38	3	9.00	1.73	0.58	

Figure 3

14. Set up a linear series from 1 to 10 in cells A36–A45. This will provide values for the independent variable *x*.

Enter the number 1 as a literal in cell A36.
In cell A37, enter the formula **=A36+1**.
Copy the formula in cell A37.
Select cells A38–A45. Paste.
Note that this differs from previous examples by starting at 1 rather than 0. We will explain why later.

15. In cells B36–D36, enter spreadsheet formulae that express the equations shown in cells B35–D35.

These should be
Cell B36: **=0+A36^2**
Cell C36: **=0+A36^0.5**
Cell D36: **=0+A36^-0.5**
Again, we could omit the zeros in the equations and in the formulae, but we keep them for consistency with the others.

16. Copy the formulae in cells B36–D36 into cells B37–D45.

Select cells B36–D36. Copy.
Select cells B37–D45. Paste.
At this point, your spreadsheet should contain the values shown above.

Comparing Functions

17. Enter titles and headings in Rows 47–55. Also enter the values shown for the parameters (constants).

These are all literals, so enter them as before (see Step 1).

	A	B	C	D	E
47	Part 2:Comparing Functions				
48		Parameters (constants)			
49		y-Intercept (a)	0		
50		Slope (b)	1		
51		Base (q)	2		
52		Power (p)	3		
53					
54		Linear	Exponential	Power	
55	x	y=a+bx	y=a+q^x	y=a+x^p	
56	1	1	2	1	
57	2	2	4	8	
58	3	3	8	27	

Figure 4

18. Set up a linear series from 1 to 10 in cells A56–A65.

Enter the number 1 as a literal in cell A56.
In cell A57, enter the formula **=A56+1**.
Copy the formula in cell A57.
Select cells A58–A65. Paste.

19. Enter formulae into cells B56–D56 to calculate the functions in cells B55–D55.

The formulae should read:
Cell B56: **=$C$49+$C$50*A56**
Cell C56: **=$C$49+$C$51^A56**
Cell D56: **=$C$49+A56^$C$52**

20. Copy the formulae down their columns.

Select cells B56–D56. Copy
Select cells B57–D65. Paste.
At this point, your spreadsheet should contain the values shown in Figure 4.

Your spreadsheet is complete. Save your work!

B. Create graphs.

Linear Functions

1. Graph all four linear functions on the same graph.

Select the contiguous block of cells from cell A9 through cell E19. Note that you should select the column headings as well as the data to be graphed. This lets the program label the graph legend correctly.
Click on the Chart Wizard icon or open Insert | Chart.
In the Chart Type dialog box, select XY (Scatter). Then, from the chart subtypes shown, choose the one at bottom left, which has data points connected with straight lines.

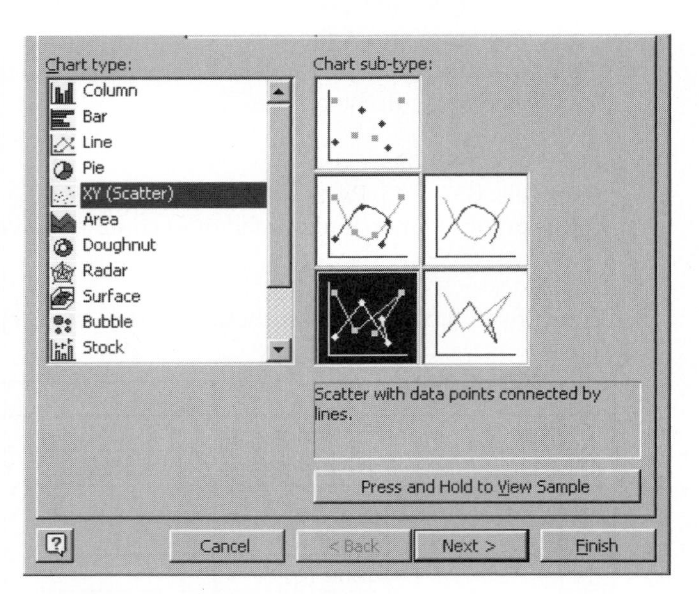

Figure 5

Click the Next button.

In the Chart Source Data dialog box, select Series in Columns. This will probably already be selected for you, in which case you need only click on the Next button.

In the Chart Options dialog box, enter a figure title and axis labels as shown in Figure 6.

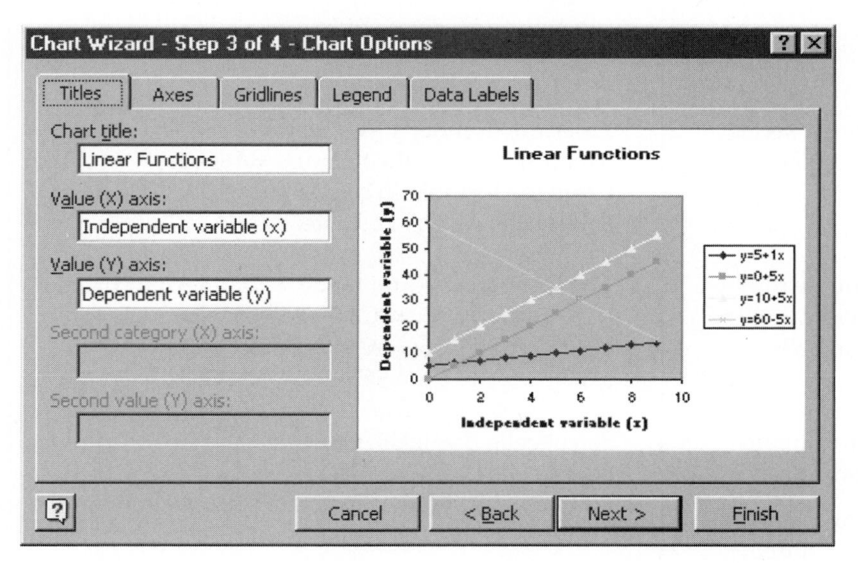

Figure 6

Note the tabs across the top of the dialog box. Clicking on one of these will take you to another page of chart options. We usually go to the gridlines page and remove the horizontal gridlines that appear by default because we find them distracting. This has already been done in Figure 6.
Click the Next button.

In the Chart Location dialog box, select Place Chart: As Object In: Sheet 1 and click on the Finish button.

2. Edit your graph to improve readability. Change to an uncolored background.

Often, the shaded background and default colors of data markers and lines are difficult to see and print poorly, especially on black-and-white printers. To change to an uncolored (clear) background, double-click inside the graph axes, away from any lines or data markers, and you should see the dialog box shown below.
Click on the buttons labeled None for Border and Area, as shown in Figure 7.

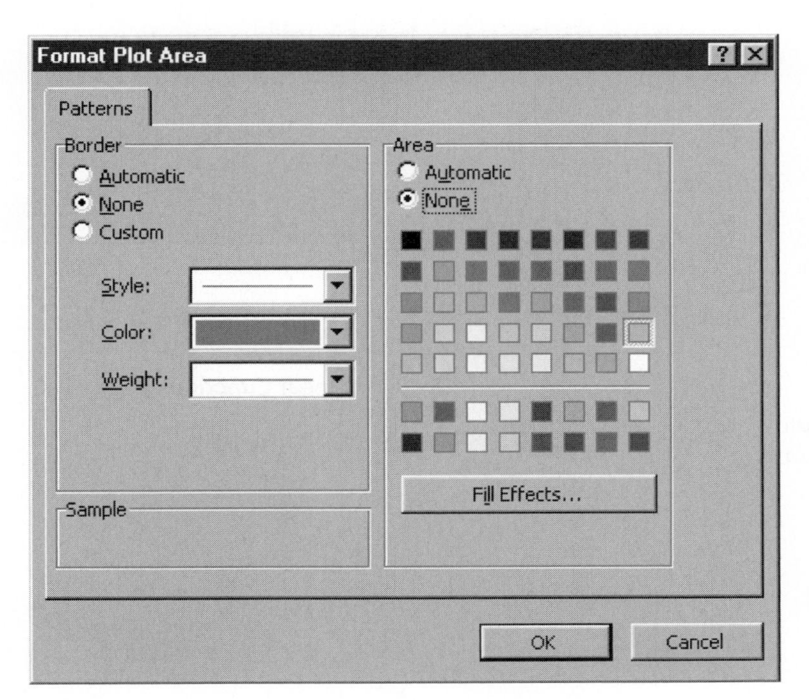

Figure 7

3. Make all data lines and markers black and give each function an easily distinguished marker or line type.

Double-click on a data point marker, and you should see the dialog box in Figure 8.

The left-hand section offers several options for formatting the line connecting data points. Click and hold on the arrow in the box labeled Color and a color palette will pop up. Still holding down the mouse button, select Black.
You can change the style of the line (solid, dashed, dotted, etc.) and its weight (thickness) similarly. In general, you should *not* use the smoothed line option.

The right-hand section offers options for formatting data markers. Change the foreground and background colors to black as you did for line color. You can use the Style pop-up menu to choose the shape of the data marker. To make hollow markers, choose No Color from the color palette for background color.
Edit each data series similarly, making all black and choosing easily distinguished markers or line-styles.

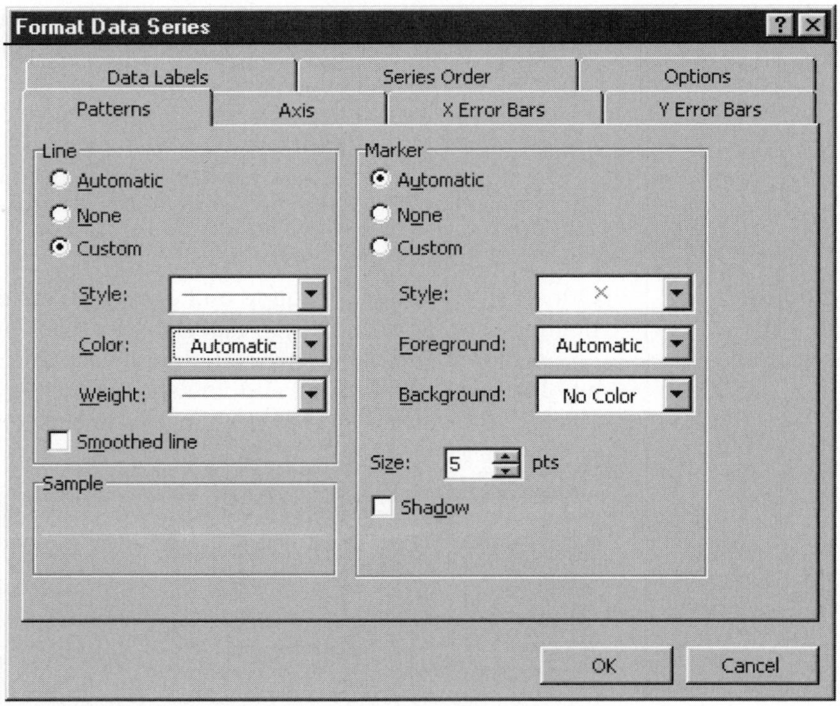

Format Data Series
Data Labels
Series Order
Options
Patterns
Axis
X Error Bars
Y Error Bars
Line
Automatic
None
Custom
Style:
Color:
Automatic
Weight:
Smoothed line
Sample
Marker
Automatic
None
Custom
Style:
Foreground:
Automatic
Background:
No Color
Size:
5
pts
Shadow
OK
Cancel

Figure 8

4. Your graph should now resemble the one in Figure 9.

Linear Functions
Dependent variable (y)
Independent variable (x)
0
10
20
30
40
50
60
70
0
5
10
y=5+1x
y=0+5x
y=10+5x
y=60-5x

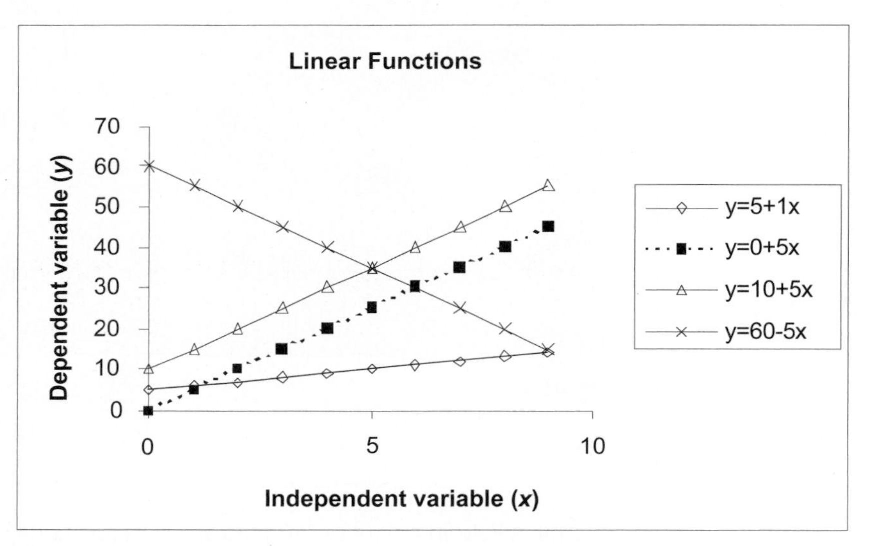

Figure 9

5. If the graph obscures cells A19–E9 of your spreadsheet, drag it to the right so that those cells are visible.

Click once inside the box around the graph, but outside the graph axes. The graph box should now have small, square "handles" at the middle of each side. If it does not, try clicking in a different place inside the graph box.
Press and hold the mouse button while dragging the graph to the desired location.
If only part of the graph moves, rather than the entire graph moving as a unit, open Edit | Undo Move and try again.

Exponential Functions

6. Graph all three exponential functions on a new graph.

Select the contiguous block of cells A22–D32. Note that you should select the column headings as well as the data to be graphed. This lets the program label the graph legend correctly.

Click on the Chart Wizard icon or open Insert | Chart.
Follow the steps for graphing linear functions given in Section B1.

7. Edit your graph to improve readability.

Follow the steps given in Section B2 on linear functions: Remove gridlines and label the graph and its axes. Remove background color and change all lines and data markers to black. Choose markers and line types so that different functions are clearly labeled. When you are done, your graph should look something like the graph in Figure 10.

Exponential Functions
Dependent variable (y)
45
40
35
30
25
20
15
10
5
0
0
2
4
6
8
10
Independent variable (x)
y=0+1.1^x
y=0+1.5^x
y=0+1.5^-x

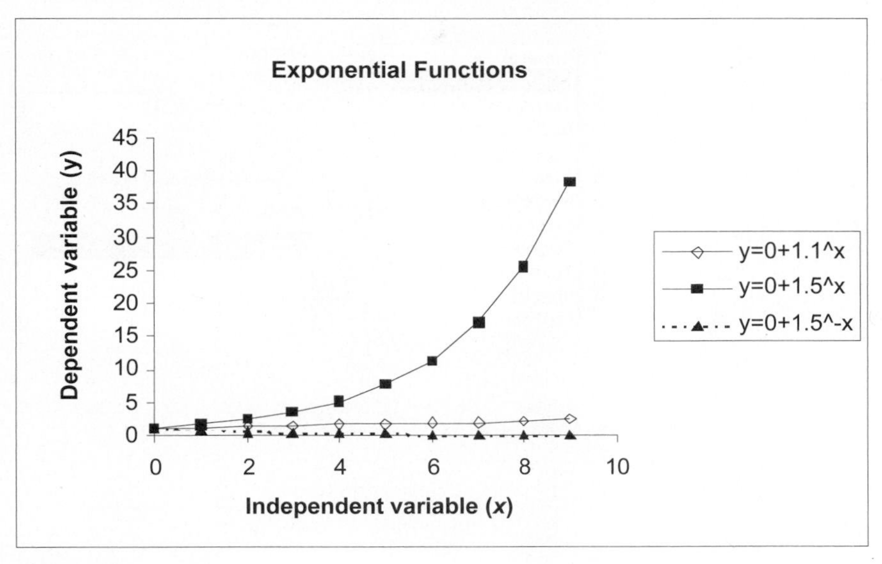

Figure 10

8. Change the vertical axis to a logarithmic scale.

Double-click on the vertical axis. A dialog box will appear. Click on the tab labeled Scale. The page shown in Figure 11 will appear. Click in the box labeled Logarithmic Scale. Do ***not*** click on the OK button yet.

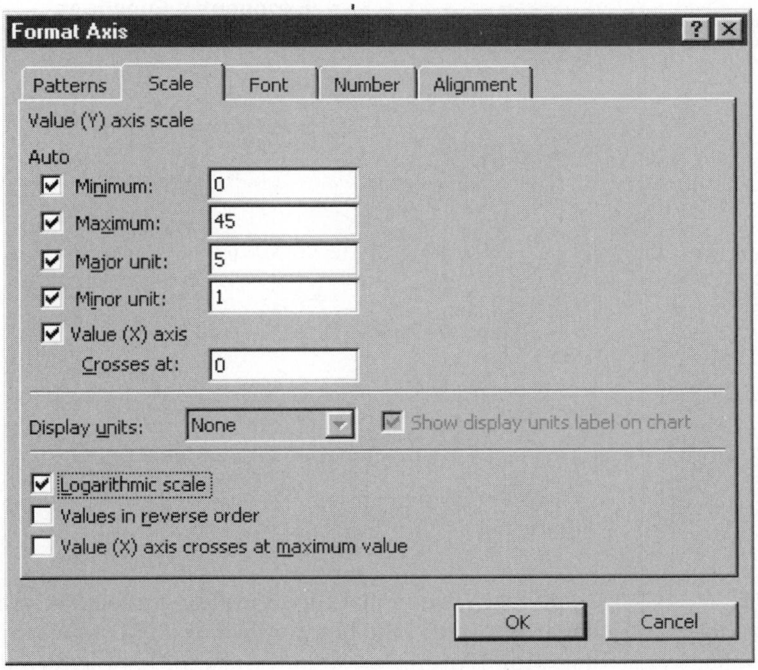

Format Axis
Patterns
Scale
Font
Number
Alignment
Value (Y) axis scale
Auto
Minimum: 0
Maximum: 45
Major unit: 5
Minor unit: 1
Value (X) axis
Crosses at: 0
Display units: None
Show display units label on chart
Logarithmic scale
Values in reverse order
Value (X) axis crosses at maximum value
OK
Cancel

Figure 11

9. Change the numbers on the vertical axis to display two decimal places.

Click on the tab labeled Number. The page shown in Figure 12 will appear. Select Number from the category list on the left. Use the little arrows next to the Decimal Places box to select 2 decimal places.

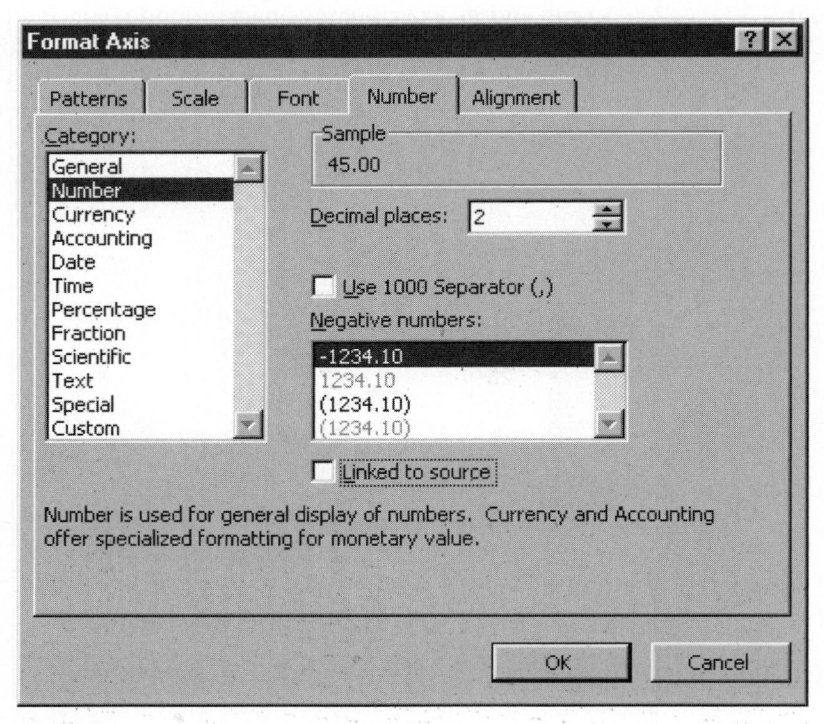

Figure 12

Now click on the OK button.

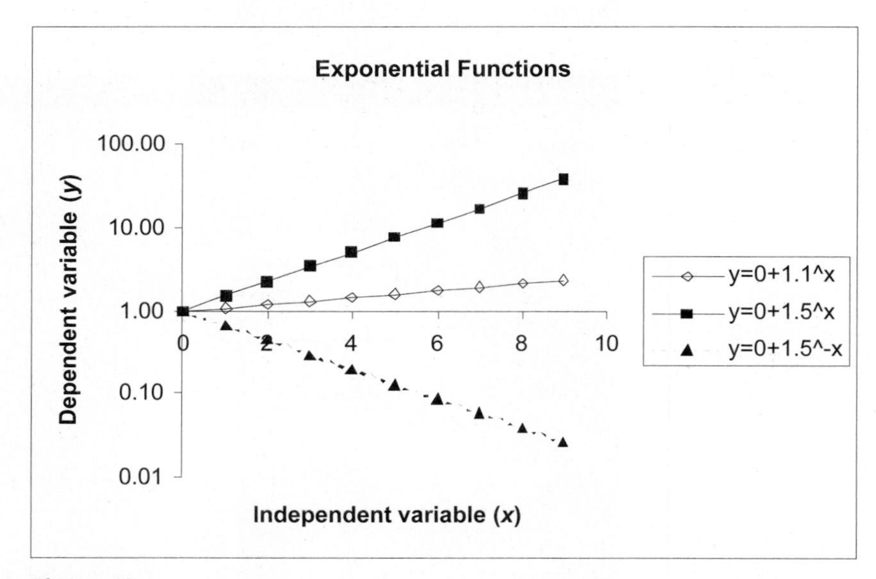

Figure 13

10. Your graph should now resemble the one in Figure 13.

Note that exponential functions are graphed as straight lines when the vertical axis is logarithmic and the horizontal axis is linear. A graph with such axes is called a **semi-log plot**. Plotting variables on a semi-log plot is a good way to test for an exponential relationship.

Power Functions

11. Graph all three power functions on a new graph.

Select the contiguous block of cells A35–D45. Note that you should select the column headings as well as the data to be graphed. This lets the program label the graph legend correctly.
Click on the Chart Wizard icon or open Insert | Chart.
Follow the steps given in Section B1 on linear functions.

12. Edit your graph to improve readability. Your graph should resemble the one in Figure 14.
Graphing each function separately reveals the shapes of their graphs.

Follow the steps given in Section B2 on linear functions: Remove gridlines and label the graph and its axes. Remove background color and change all lines and data markers to black. Choose markers and line types so that different functions are clearly labeled.

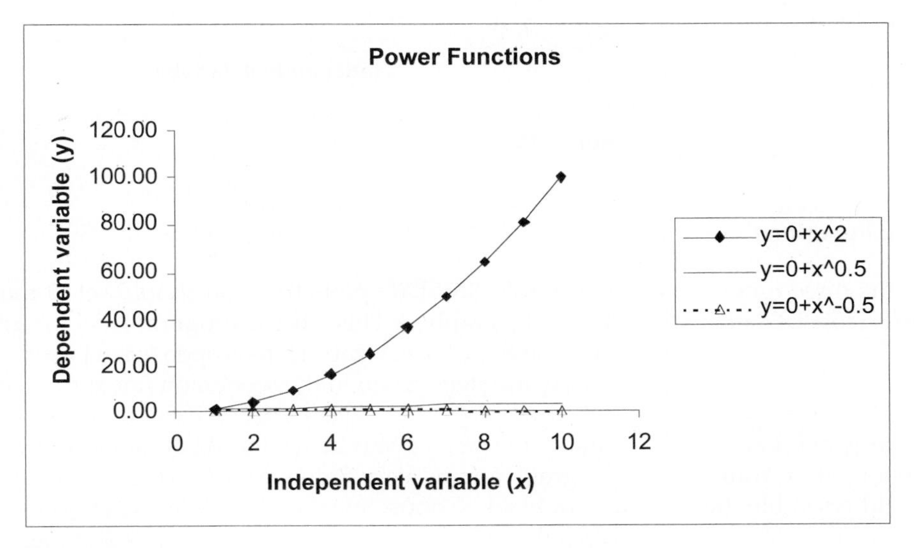

Figure 14

The graph of $y = x^2$ resembles an exponential function but, as we will show shortly, it is not. The other functions lie almost on top of the x-axis.

13. Change the vertical axis to a logarithmic scale.

Double-click on the vertical axis.
In the dialog box, click on the Scale tab and select Logarithmic Scale. Do not click OK yet.

14. Change the numbers on the vertical axis to display one decimal place.

Click the Number tab, and use the Decimal Places box to select 1 decimal place.
Now click OK.

Note that none of the functions appears as a straight line; this tells you that they are not exponential functions.

15. Change the horizontal axis to a logarithmic scale. Your graph should now resemble the one in Figure 15.

Follow the same procedure that you used in changing the vertical axis to a logarithmic scale.

Note that all these power functions are graphed as straight lines when both axes are logarithmic. A graph with such axes is called a **log-log plot**. Plotting variables on a log-log plot is a good way to test for a power relationship.

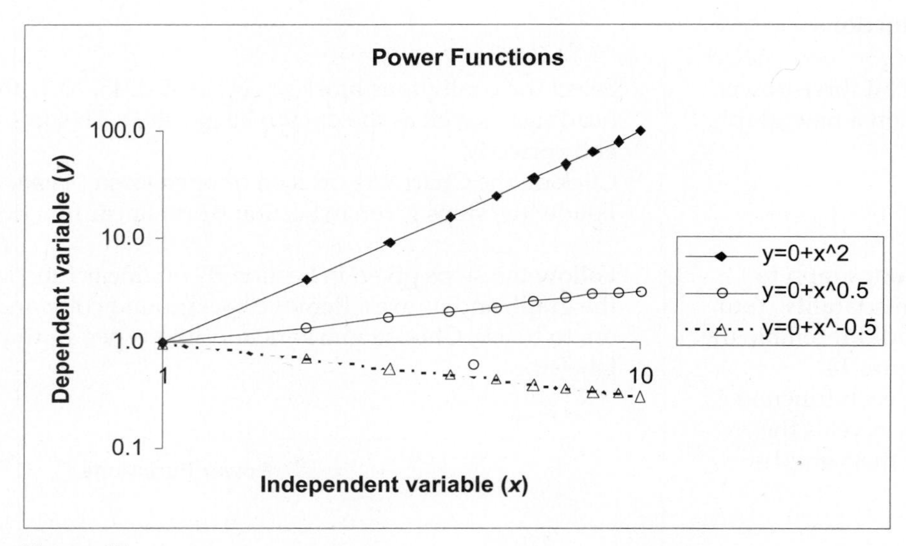

Figure 15

Comparing Functions

16. Graph the three functions in cells A55–D65 on a new graph.

Select cells A55–D65. Note that you should select the column headings as well as the data to be graphed. This lets the program label the graph legend correctly.
Click on the Chart Wizard icon or open Insert | Chart.
Follow the steps given in the section on linear functions.

17. Edit your graph to improve readability. Your graph should resemble the one in Figure 16.

Follow the steps given in Section B2 on linear functions: Remove gridlines and label the graph and its axes. Remove background color and change all lines and data markers to black. Choose markers and line types so that different functions are clearly labeled.

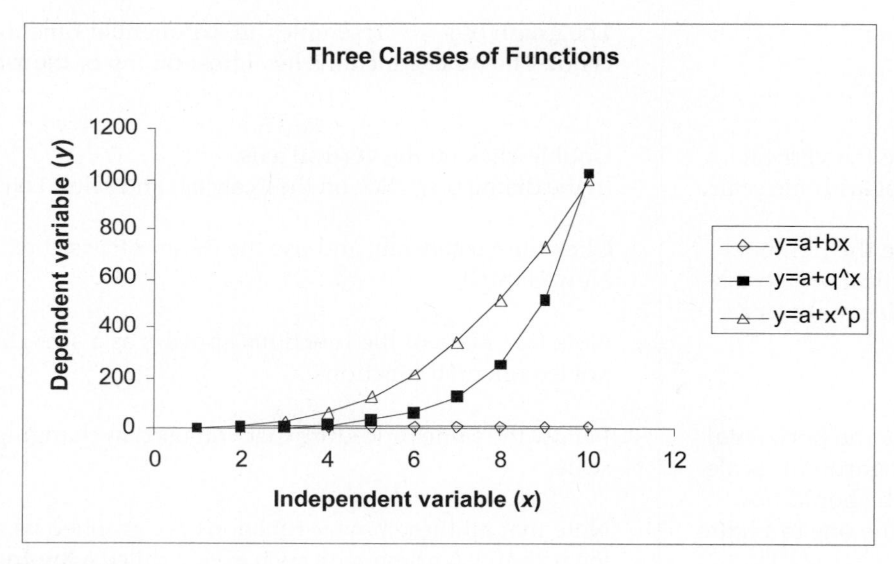

Figure 16

18. Experiment with different combinations of logarithmic and linear axes.

Try:

- Both axes linear
- Logarithmic x-axis, linear y-axis(semi-log)
- Both axes logarithmic (log-log)

See instructions above for details of changing axis scaling.

19. Experiment with different values of y-intercept, slope, base, and power, and observe the effects on the graph.

Simply enter new values in the cells labeled "Parameters" (constants)—cells C49 through C53. You do not need to edit the formulae.

QUESTIONS

1. How does changing the value of the y-intercept (a) affect each of the kinds of functions? Enter different values in cell C49 and observe the effects on your graph of three kinds of functions. The effects may be difficult to see at first, because the spreadsheet automatically rescales the y-axis to accommodate values to be graphed. Be sure to note the values along the y-axis in your comparisons. Also compare the four linear functions you graphed in step B1.
2. How does changing the value of the slope (b) in cell C50 affect the linear function? Try values greater and less than zero. Also compare the four linear functions you graphed in step B1.
3. How does the exponential function look if you enter different values for the base (q) in cell C51? Try values greater than one, equal to one, less than one, and less than zero. You will have to reformat the axes of your graph to see some of these effects. Also compare the three exponential functions you graphed in step B6.
4. How does the power function look if you enter different values for the power (p) in cell C52? Try values greater than one, equal to one, less than one, and less than zero. You will have to reformat the axes of your graph to see some of these effects. Also compare the three power functions you graphed in step B11.
5. Find examples of all three kinds of functions in your textbook or in other books or papers about ecology or biology. Look for explicit equations and for graphs that imply these functions by their axis formats (both axes linear, y-axis logarithmic, or both axes logarithmic).

2

SPREADSHEET FUNCTIONS AND MACROS

Objectives

- Learn how to use the Paste Function menu on your spreadsheet to carry out a set of mathematical operations.
- Become familiar with three types of spreadsheet functions: standard functions, nested functions, and array functions.
- Practice using a variety of common spreadsheet functions.
- Develop and run a macro.

INTRODUCTION

Mathematical functions describe natural phenomena in the form of an equation, relating one variable to another. In Exercise 1, you learned about linear, exponential, and power mathematical functions.

In this exercise, the "function" under discussion is quite different. **Spreadsheet functions** are formulae that have been written by a computer programmer to perform mathematical and other operations (see pp. 9–12). Your spreadsheet package likely has over 100 functions available for your use. These functions can make modeling easier for you, and you will use them extensively throughout this book.

Standard Functions

As an introduction to spreadsheet functions, let's suppose that there are eight people in an elevator. The names of the eight individuals and their weights are given in Figure 1.

	A	B
1	**Individual**	**Weight (lbs)**
2	Tim	180
3	Anne	135
4	Pat	200
5	Donna	140
6	Kathleen	142
7	Joe	190
8	Mike	176
9	Tansy	135
10	**SUM =>**	

Figure 1

Imagine that the elevator can hold a maximum of 1,500 pounds, and that a ninth person would like to get on. Would the addition of a ninth person exceed the 1,500-pound safety limit? To answer this question, we need to know how much the eight people in the elevator collectively weigh, and the weight of the ninth person. We could add cells B2–B9 to determine how much the eight people weigh. If we entered a mathematical formula in cell B10 to compute this, the formula reads **=B2+B3+B4+B5+B6+B7+B8+B9**. The result is 1,298 pounds. The more complicated a formula becomes, however, the more likely it is that you will make a mistake in entering it. This is where spreadsheet functions come into play. Instead of entering **=B2+B3+B4+B5+B6+B7+B8+B9** in cell B10, we can use the **SUM** spreadsheet function and have the spreadsheet do the work.

To enter a spreadsheet function, first select the cell in which you want the function to be computed (in this case, cell B10). Then you can do either of one of two things. You can use the Paste Function button f_x on your toolbar (indicated in Figure 2), or you can open Insert | Function to guide you through entering a function. Either way, the dialog box will appear as shown in Figure 2.

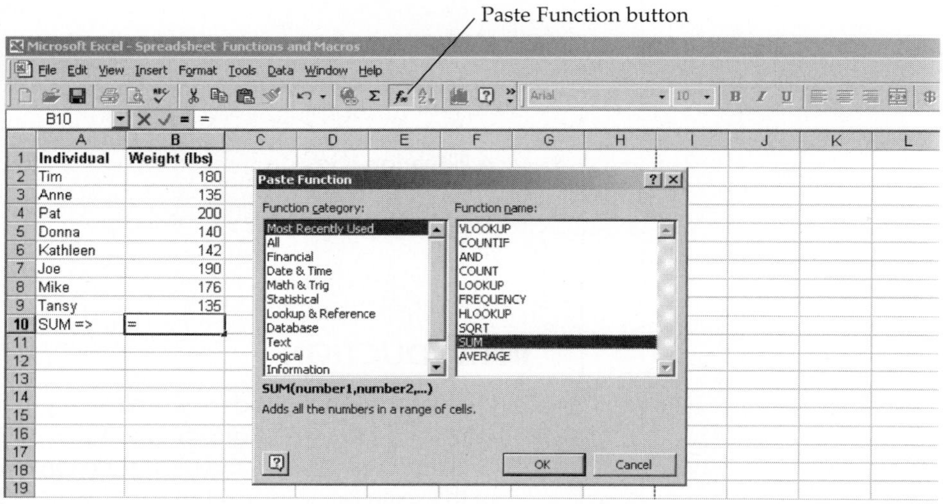

Paste Function button

Figure 2

Look at the column on the left side of the dialog box. It asks what kinds of function category you want to examine. You could choose to look at the most recently used functions, or you can look at all the available functions, or you can check out the functions in a specific category, such as financial functions, statistical functions, and so on. If you choose All as a Function category, you'll see every function available in your spreadsheet package, listed in alphabetical order.

In Figure 2, we selected the Most Recently Used function category, so a list of the most recently used functions appears in the right side of the dialog box. Note that the function **SUM** is selected, and the program displays a brief description of the function at the bottom of the box: "Adds all the numbers in a range of cells." Click OK when you've got the function you want (in this case, the **SUM** function). Another box will then appear, called the **formula palette** (Figure 3). Each function has its own formula palette. You are asked to enter the addresses of the cells you wish to sum in the **SUM** formula palette. You can enter cell B2 as Number 1, cell B3 as Number 2, cell B4 as Number 3, and so on. Or you can type in the range **B2:B9** as Number 1 and the spreadsheet will recognize that the entire range of cells is to be added. When you are finished, click the OK box, or click on the green check-mark button to the left of the formula bar. If you

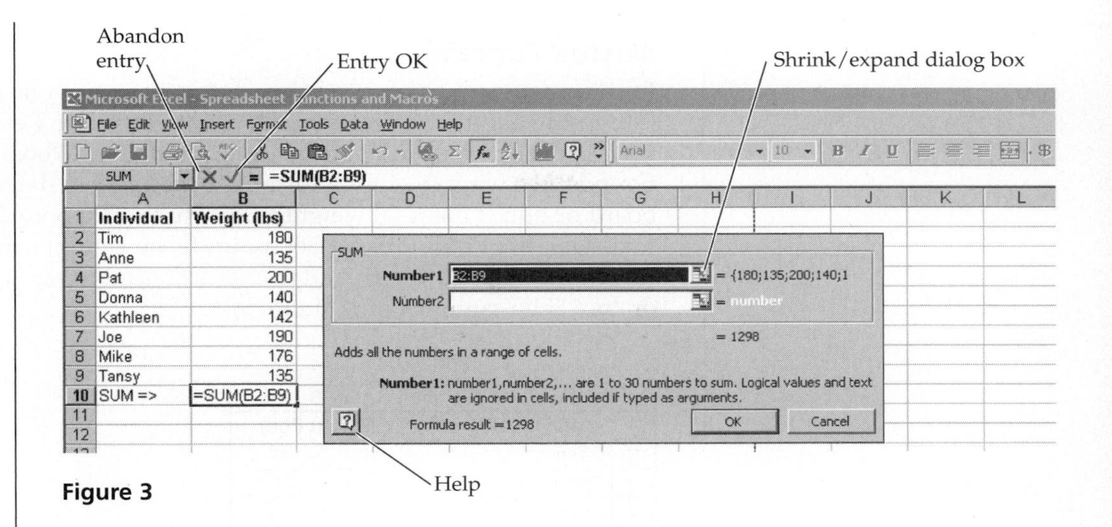

Figure 3

change your mind and decide to abandon the formula entry, click on the red × button to the left of the formula bar.

There are two handy features in a formula palette that you should note.

- First, notice the small figure with a red arrow pointing upward and leftward (located to the right of the blank space labeled Number 1). If you click on this arrow, the dialog box will shrink, exposing your spreadsheet so that you can use your mouse to select the range of cells you want to add. This is handy because you don't have to type in the cell references—just point and click on the appropriate cells. After you've selected the cells you want to add (in this case, use your mouse to highlight cells B2–B9), click on the arrow again and the SUM dialog box will reappear.
- The second handy feature of all Paste Function dialog boxes is "Help" information, accessed by clicking on the question mark located at the bottom-left corner of the window. If you don't know how the function works, clicking on the question mark will provide additional information.

Once you have entered all the necessary data and pressed OK, the spreadsheet will return the answer in cell B10. Although the spreadsheet displays the answer (1,298) in cell B10, the formula bar shows that the cell really contains the function **=SUM(B2:B9)**. Note that the spreadsheet automatically inserted an equal sign before the function name, alerting the spreadsheet that a function is being used (Figure 4).

Microsoft Excel - Spreadsheet Functions and Macros

B10 = =SUM(B2:B9)

	A	B
1	**Individual**	**Weight (lbs)**
2	Tim	180
3	Anne	135
4	Pat	200
5	Donna	140
6	Kathleen	142
7	Joe	190
8	Mike	176
9	Tansy	135
10	SUM =>	1298
11		

Figure 4

Nested Functions

In some cases, you may need to perform more than one function, "nesting" one function inside another to give you the result you want. Returning to our elevator example, suppose that a ninth person, Peter, would like to board the elevator. He weighs 200 pounds. We want to enter a formula in cell B13 to determine whether he can safely board or not. If the total weight is less than 1,500 pounds, he can safely board. If the total weight is more than 1,500 pounds, he cannot safely board. We can use an **IF** function in cell B13 to carry out the operation and return the word "yes" if he can board or "no" if he cannot board (Figure 5).

	A	B
1	**Individual**	**Weight (lbs)**
2	Tim	180
3	Anne	135
4	Pat	200
5	Donna	140
6	Kathleen	142
7	Joe	190
8	Mike	176
9	Tansy	135
10	**SUM =>**	1298
11		
12	Peter	200
13	**SAFELY BOARD?**	

Figure 5

As with the **SUM** function, you can use the Paste Function menu and then search for and select the **IF** function (Figure 6). You will notice at the bottom of the dialog box the words **IF(logical_test,value_if_true,value_if_false)**. This is the syntax for the **IF** formula, and it provides the "rules" for entering an **IF** function. You should also see a brief description of the function that tells you the function "returns one value if a condition you specify evaluates to TRUE and another value if it evaluates to FALSE."

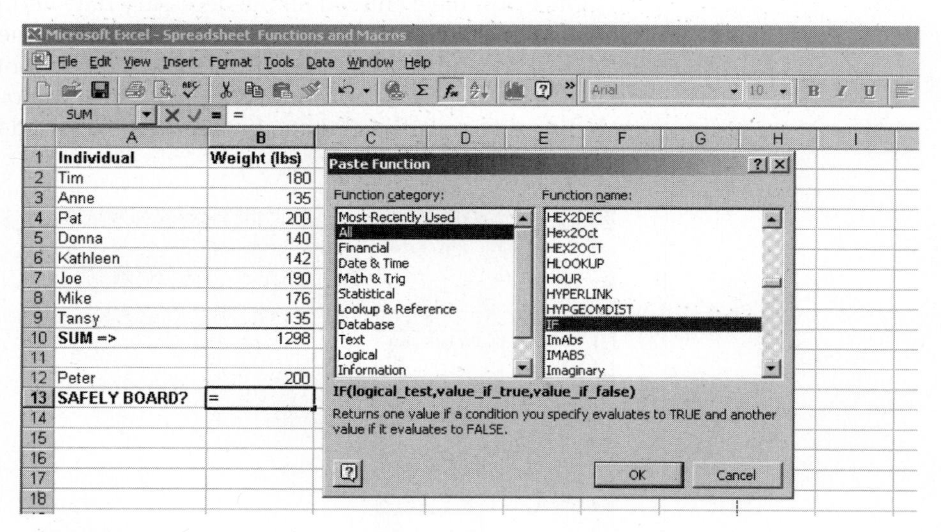

Figure 6

For our example, we want to determine whether the total weight is less than or equal to 1,500 pounds. This is the logical test. If the logical test is TRUE, we want the word YES to be returned (he can safely board). If the logical test is FALSE, we want the word NO to be returned (he should not board). The formula palette for the **IF** function is shown in Figure 7.

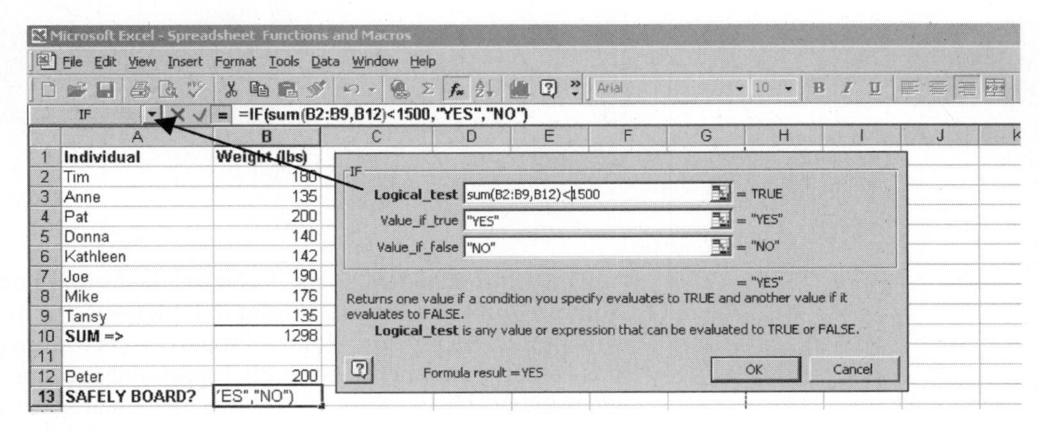

Figure 7

The logical test requires that we sum the weights of the original eight individuals in cells B2–B9 and the weight of the ninth individual (cell B12) and determine whether the sum is less than 1,500. Because the logical test (**IF** function) contains the **SUM** function, it is called a **nested function**. To nest the **SUM** function within the **IF** function, place your cursor within the Logical_test box. Then select the down arrow to the left of the formula bar. A list of functions appears. Search for the **SUM** function and click on it, and the **SUM** function palette will appear as shown in Figure 3. Enter the cell range **B2:B9** as Number 1, and cell **B12** as number 2. Instead of clicking OK when you are finished with the **SUM** function, click on the word **IF** on the formula bar; you will be returned to the **IF** formula palette and can complete the **IF** function entries.

Notice that the formula palette in Figure 7 displays the result of the logical test (TRUE) and the formula result (YES), indicating that Peter can board the elevator safely. The final function in cell B13 reads **=IF(SUM(B2:B9,B12)<1500,"YES","NO")**. When functions are nested within other functions, the spreadsheet will compute the answer to the "nested" functions (in this case, **SUM**) first and then will complete the outer functions.

Array Formulae

Functions such as **SUM** perform a calculation and generate a result in a single cell. An **array formula**, on the other hand, can perform multiple calculations, returning either a single result or multiple results. Array formulae act on two or more sets of values known as "array arguments."

You create array formulae in the same way you create other formulae, with a few major exceptions. First, instead of selecting a single cell to enter a formula, you need to select a *series* of cells, then enter an array formula. And second, instead of pressing OK after you have completed the entries in the function palette, you press <Control>+<Shift>+<Enter> (on Windows-based machines) or <>+<Return> (on Macs) to enter the formula for all of the cells you have selected.

Let's consider a new example. Suppose you want to construct a frequency distribution from the data in Figure 8. The weights (in grams) for 10 individuals are given in column B. Suppose you want to count the number of individuals that are 1 gram, 2 grams, 3 grams, 4 grams, and 5 grams. You could use the **FREQUENCY** function, which is an array formula to generate frequency data quickly.

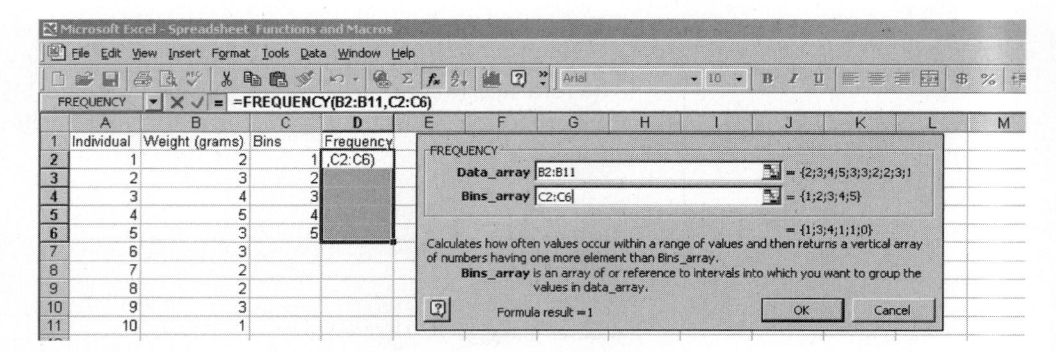

Figure 8

The column labeled "Bins" in Figure 8 tells Excel how you want your data grouped. You can think of a bin as a bucket in which specific numbers go. The bins may be very small (hold a single or a few numbers) or very large (hold a large set of numbers). In this case, the numbers 1 through 5 represent the bins, and each bin "holds" just a single number. The task now is to have the spreadsheet count the number of individuals in each bin and return the answer in cells D2–D6. Because the frequency function is an array function, we need to select cells D2–D6 (rather than a single cell) *before* using the f_x button to summon the **FREQUENCY** formula.

The **FREQUENCY** formula palette will appear (Figure 8) and will guide you through the entries. The Data_array is simply the data you want to summarize, given in cells **B2:B11**. The Bins_array is cells **C2:C6**. Instead of clicking OK, press <Control>+<Shift>+<Enter> on Windows machines, or <>+<Return> on Macs, and the spreadsheet will return your frequencies.If we examine the formulae in cells D2–D6, every cell will have the formula **{=FREQUENCY(B2:B11,C2:C6)}**. The **{ }** symbols indicate that the formula is part of an array.

Typically, frequency data are depicted graphically as shown in Figure 9. If you change the data set in some way, the spreadsheet will automatically update the frequencies. If for some reason you get "stuck" in an array formula, just hit the Escape key and start again.

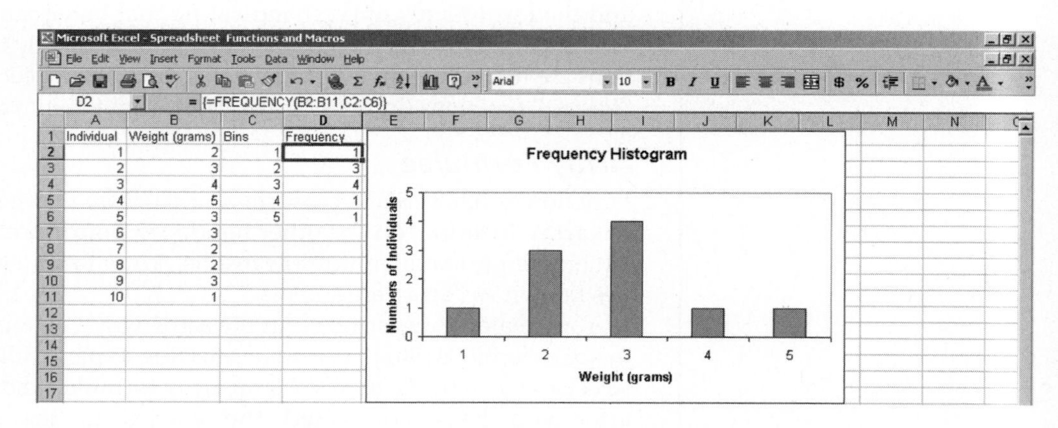

Figure 9

MACROS

As noted in the Introduction (p. 16), a **macro** is a miniature program that you build for yourself in order to run a sequence of spreadsheet actions. Typing and mouse-clicking your way through a long series of commands over and over is time-consuming, boring, and error-prone. A macro allows you to achieve the same results with a single command.

You record a macro using Excel's built-in macro recorder. Start the recorder by opening Tools | Macro | Record New Macro (Figure 10).

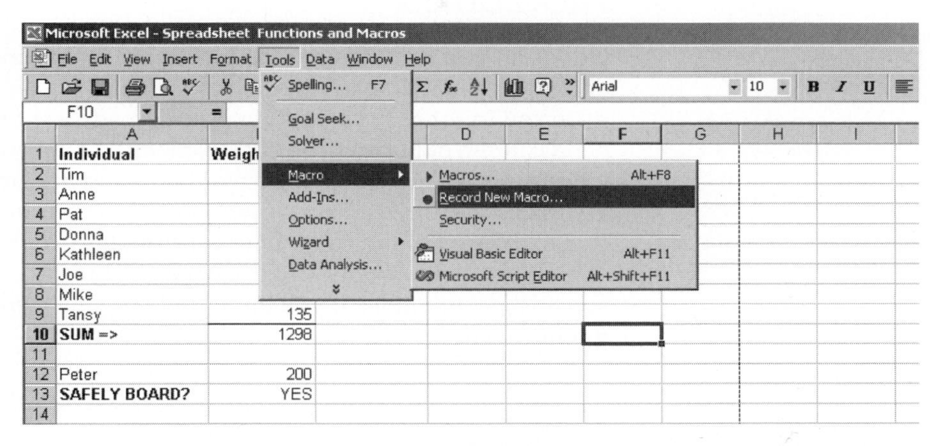

Figure 10

The program will prompt you to name the macro and create a keyboard shortcut. Then a small window will appear with the macro recorder controls (Figure 11). If this button does not appear, go to View | Toolbars | Stop Recording, and the Stop Recording figure will appear. The square on the left side of the button is the Stop Recording button. When you press this square, you will stop recording your macro. The button on the right is the Relative Reference button. By default this button is not selected so that your macro recorder assumes that the cell references you make in the course of developing your macro are absolute. In other words, if you select cell A1 as part of a macro, Excel will interpret your keystroke as cell **$A$1**. There are cases (for example, the Survival Analysis exercise) in which you will want to select the relative reference button as you record your macro.

Once you have entered the macro name and shortcut key, the spreadsheet will record every action you take. Carry out the entire sequence of operations you want the macro

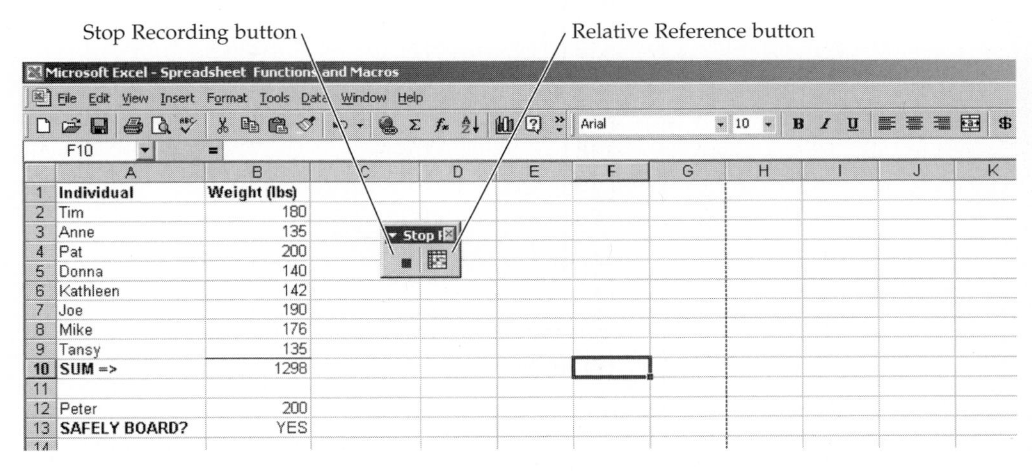

Figure 11

to do, and then press the Stop Recording button in the macro recorder control window. From this point on, Excel will mimic that entire sequence of actions whenever you press the keyboard shortcut or issue the macro command.

PROCEDURES

Now that you have been introduced to simple functions, nested functions, arrays, and macros, it's time to put them into practice. The following instructions will introduce you to some 20 commonly used spreadsheet functions. As in Exercise 1, the left-hand column of instructions gives rather generic directions, and the right-hand column gives a step-by-step breakdown of these and explanatory comments or annotations. Try to think through and carry out the instructions in the left-hand column before referring to the right-hand column for confirmation. It's tempting to jump to the right hand column for the answers and explanation, but you will learn a lot more about using spreadsheet functions if you attempt it on your own. As always, **save your work frequently to disk.**

INSTRUCTIONS

ANNOTATION

A. Set up the spreadsheet.

1. Open a new spreadsheet and enter headings as shown in Figure 12.

	A	B	C
1	*Spreadsheet Functions and Macros*		
2			
3	Individual	Height (cm)	
4	1	12	
5	2	2	
6	3	8	
7	4	20	
8	5	3	
9	6	5	
10	7	12	
11	8	6	
12	9	4	
13	10	9	
14	11	7	
15	12	4	
16	13	1	
17	14	7	
18	15	7	
19	16	10	
20	17	1	
21	18	3	
22	19	2	
23	20	4	

Figure 12

2. Set up a linear series from 1 to 20 in cells A4–A23.

We will consider a sample of 20 individuals and their heights.
Enter 1 in cell A4.
Enter **=1+A4** in cell A5.
Select cell A5 and copy it down to cell A23.

3. Enter the heights for the 20 individuals in cells B4–B23 as shown.

These are the actual data, so just type in the numbers as shown in Figure 12.

B. Compute simple functions.

In this section, you'll use 11 standard spreadsheet functions to compute various things, like the average height of the 20 individuals. For all functions, use the Paste Function menu (the Paste Function button, f_x, or open Insert | Function) to locate the appropriate function, review the function's formula palette, and complete the entries. You can double-check your results with ours at the end of the section.

1. Set up new headings as shown in Figure 13.

	D	E
3	**Simple functions**	
4		
5	***Count***	
6	***Sum***	
7	***Average***	
8	***Median***	
9	***Mode***	
10	***Min***	
11	***Max***	
12	***Stdev***	
13	***4th large***	
14	***Rand***	
15	***Randbetween***	

Figure 13

2. In cell E5, use the **COUNT** spreadsheet function to count the total number of individuals in the sample.

The **COUNT** function counts the number of cells that contain numbers. In this case, you want to count the number of times that a number is contained in cells B4–B23. Select the **COUNT** function from the Paste Function menu and compute this result. After you are finished, cell E5 should display the number 20, and its formula should be **=COUNT(B4:B23)**.

3. In cells E6–E12, use the spreadsheet functions **SUM, AVERAGE, MEDIAN, MODE, MIN, MAX,** and **STDEV** to compute basic descriptive statistics for the population.

For each formula, use the Paste Function menu and read through the information on the formula palette carefully. If you are unsure of the kind of information a statistic provides, click on the question mark on the bottom-left corner of the formula palette. After you have finished, the formulae in your spreadsheet should look like Figure 14, except that instead of seeing the formula in cells E5–E12, the *answers* to each formula will be displayed.

	D	E
3	**Simple functions**	
4		
5	***Count***	=COUNT(B4:B23)
6	***Sum***	=SUM(B4:B23)
7	***Average***	=AVERAGE(B4:B23)
8	***Median***	=MEDIAN(B4:B23)
9	***Mode***	=MODE(B4:B23)
10	***Min***	=MIN(B4:B23)
11	***Max***	=MAX(B4:B23)
12	***Stdev***	=STDEV(B4:B23)

Figure 14

4. In cell E13, use the **LARGE** function to compute the fourth largest height.

The **LARGE** function returns the kth largest value in a range of cells. In this case, the range of cells is B4–B23 (Figure 12), and $k = 4$. Your formula should read **=LARGE(B4:B23,4)**, and the answer should be 10.

5. In cell E14, use the **RAND** formula to generate a random number between 0 and 1.

You will use the **RAND** function in many of the exercises in this book. This function has the form **=RAND()**. The **(** and **)** are open and closed parentheses; you do not need to put anything inside them.

6. In cell E15, use the **RANDBETWEEN** function to generate a random number between 1 and 20.

The **RANDBETWEEN** function generates a random integer between two specified values. The bottom value is the lowermost integer that can be randomly selected (1), and the top value is the uppermost integer that can be randomly selected (20). This function could be used to randomly select an individual from the population. The formula in cell E15 should read **=RANDBETWEEN(A4,A23)** or **=RANDBETWEEN(1,20)**.

Note: If your spreadsheet doesn't have the **RANDBETWEEN** function, you can enter the nested functions **=ROUNDUP(RAND()*20,0)**. This will generate a random number between 0 and 1, multiply it by 20, and round it up to the nearest zero decimal places (i.e., to the nearest integer).

7. Press F9, the Calculate key, to generate new random numbers in cells E14 and E15.

The Calculate key in Windows is the F9 key, located at the top of your keyboard.* When this button is pushed, the spreadsheet will recalculate all of the formulae in the spreadsheet. For random numbers, such as those generated by the **RAND** or **RANDBETWEEN** functions, a new random number will be generated when the spreadsheet is calculated.
Verify this by examining the results in cells E14–E15 each time you press F9.

8. Save your work.

C. Compute multistep and nested functions.

Now we will turn to nested functions and multi-step functions. Multi-step functions are actually standard functions like **SUM**, **MIN**, and **MAX**, but there are more entries involved in the formula palette. A function is *nested* if it uses more than one function to complete the calculations.

*The F9 function key will work on Macintosh machines provided the Hot Function Key option in the Keyboard Control dialog box is turned OFF. If the F9 key does not work on your Mac, use the alternative, ⌘+=.

1. Set up new headings as shown in Figure 15.

	F	G
3	**Multi-step and**	
4	**nested functions**	
5	***Countif***	
6	***And***	
7	***Or***	
8	***Concatenate***	
9	***Vlookup***	
10	***Norminv***	
11	***Round***	
12	***If***	
13	***Random height***	

Figure 15

2. In cell G5, use the **COUNTIF** formula to count the number of times the modal value (given in cell E9) occurs.

We use the **COUNTIF** formula extensively. It counts the number of times a specific value occurs within a range of cells. Your formula should read **=COUNTIF(B4:B23,E9)** in cell G5, and your result should be 3, indicating that 3 individuals are 4 cm. in height.

3. In cell G6, use the **AND** function to determine if the value in cell B4 = 12 and the value in cell B5 = 2.

The **AND** function returns the word TRUE or FALSE. It returns the word TRUE if all of the arguments in the formula are true (cell B4 = 12 *and* cell B5 = 2). If either condition is not true, the spreadsheet returns the word FALSE. Your result should be TRUE.

4. In cell G7, use the **OR** function to determine if the value in cell B5 is either 1 or 2.

The **OR** function is similar to the **AND** function in that it returns the word TRUE or FALSE. It returns the word TRUE if any of the arguments in the formula are true (cell B5 = 1 *or* cell B5 = 2). Your result should be TRUE.

5. In cell G8, use the **CONCATENATE** function to join the text in cell F6 with the text in cell F7.

The **CONCATENATE** function joins several text strings into a single text string. The formula =**CONCATENATE(F6,F7)** should return the word "AndOr." This doesn't mean anything, but serves to illustrate the function. We will use this function in many of the genetics exercises. (The formula =**F6&F7** would generate the same result.)

6. In cell G9, use the **VLOOKUP** function to return the height of individual 1.

The **VLOOKUP** function searches in the first column of a table for a value that you specify and returns the value of the corresponding cell in a different column. The **VLOOKUP** function needs three pieces of information: the value you want to find in the first column of the table, the cells that define the table (the upper-left and lower-right cells of the table), and the number of the column in the table that holds the information you want the formula to return. The formula =**VLOOKUP(1,A4:B23,2)** looks for the number 1 in the first column of the table defined by cells A4–B23, and it returns the value of the cell from the same row in the second column. In our spreadsheet, this formula returns the height of individual 1.

7. In cell G10, use the **NORMINV** function to draw a random data point from a distribution whose mean is given in cell E7, and whose standard deviation is given in cell E12.

The **NORMINV** function is used extensively throughout the book, and is described more fully in Exercise 3, "Statistical Distributions." Since here you will use the **RAND** function within the **NORMINV** function, this is a *nested* formula. Generally speaking, for a set of normally distributed data, the function will generate a data value if you specify a probability associated with a normal curve. The function in cell G10 should read **=NORMINV(RAND(),E7,E12)**. In this case, we will first generate a random probability between 0 and 1. This probability will be applied to a normal distribution whose mean is given in cell E7 and whose standard deviation is given in cell E12. The spreadsheet will then return the data value associated with that probability. Note when you press F9, the Calculate key, a new random number is computed, and thus a new random data point from the normal distribution is drawn. Also note that occasionally a negative number will appear. This is because the mean is close to 0 (6.35) and the standard deviation is quite large (4.65), so some of the data points within this distribution are below 0.

8. In cell G11, use the **ROUND** function to round cell G10 to 0 decimal places.

Your formula should read **=ROUND(G10,0)**. Once you are familiar with this function, you may find yourself typing it in by hand.

9. In cell G12, use an **IF** function to return the number 0 if cell G11 is a negative number.

Your formula should read, **=IF(G11<0,0,G11)**. This tells the spreadsheet to evaluate the value in cell G11; if the number is < 0, return a 0; otherwise, return the number given in cell G11. This formula will prevent the spreadsheet from generating negative heights.

10. In cell G13, use the **VLOOKUP** function to look up the height of the randomly selected individual listed in cell E15.

Your formula should read **=VLOOKUP(E15,A4:B23,2)**.

11. Save your work.

D. Utilize an array function.

1. Set up new headings as shown in Figure 16.

	H	I
3		
4	Array function	
5	*"Bin"*	*Frequency*
6	5	
7	10	
8	15	
9	>15	
10		

Figure 16

2. Select cells I6–I9; then use the **FREQUENCY** function to generate frequency data of heights in the population. Use the bins in cells H6–H9.

Remember that the **FREQUENCY** function is an array function. For this example, each bin "holds" several numbers. The bin labeled 5 holds heights that are up to and including 5 cm. The bin labeled 10 holds heights that are 6, 7, 8, 9, and 10 cm. Don't forget that to enter an array function such as the **FREQUENCY** function, you must press <Control>+<Shift>+<Enter> to generate a proper result. Cells I6–I9 should have the formula **{=FREQUENCY(B4:B23,H6:H9)}**.

3. Create a frequency histogram of the data in cells I6–I9. Label your axes fully (Figure 17).

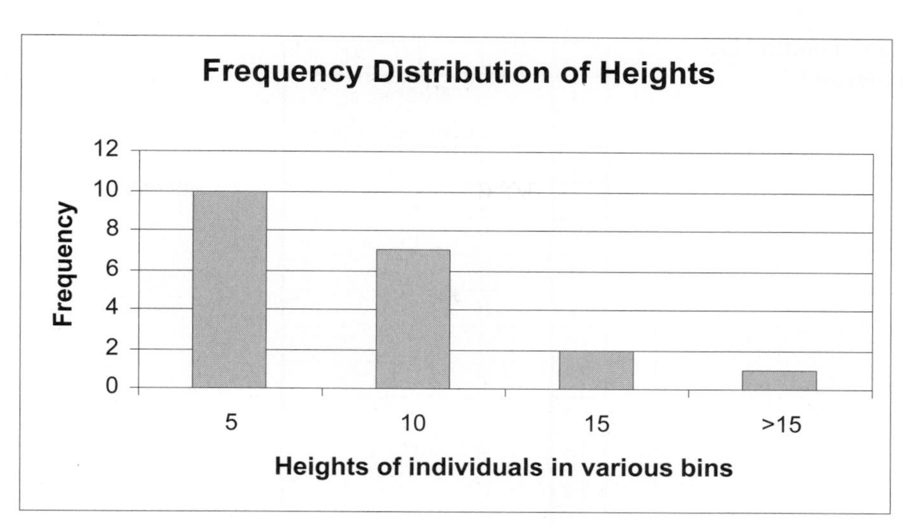

Figure 17

Your spreadsheet should now look as shown in Figure 18. Note that you will likely have different values in cells E14–E15, G10–G11, and G13 because random numbers are used to generate the results shown.

4. Double-check your results.

	D	E	F	G	H	I
3	Simple functions		Multi-step and			
4			nested functions		Array function	
5	***Count***	20	***Countif***	3	***"Bin"***	***Frequency***
6	***Sum***	127	***And***	TRUE	5	10
7	***Average***	6.35	***Or***	TRUE	10	7
8	***Median***	5.5	***Concatenate***	AndOr	15	2
9	***Mode***	4	***Vlookup***	12	>15	1
10	***Min***	1	***Norminv***	8.418740538		20
11	***Max***	20	***Round***	8		
12	***Stdev***	4.648429276	***If***	8		
13	***4th large***	10	***Random height***	4		
14	***Rand***	0.498679379				
15	***Randbetween***	20				

Figure 18

5. Save your work.

E. Write a macro to randomly select individuals from the population.

Now we will write a macro to randomly select an individual from the population, and we will record its height in column K. We will do this for 20 samples. Remember that you generated a random number between 1 and 20 in cell E15. You also looked up this randomly selected individual's height with the **VLOOKUP** function in cell G13. In our macro, we will press F9 to generate a new randomly selected individual, then we will copy the value in cell G13 into cell K5. We will repeat the process for the second sample, but we will record the height of the randomly selected individual in cell K6 (and so on).

1. Set up new headings as shown in Figure 19.

	J	K
3	Macro	
4	Sample	Height
5	1	
6	2	
7	3	
8	4	
9	5	
10	6	
11	7	
12	8	
13	9	
14	10	
15	11	
16	12	
17	13	
18	14	
19	15	
20	16	
21	17	
22	18	
23	19	
24	20	
25	average	

Figure 19

2. Write a macro to record the heights of 20 randomly sampled individuals from the population.

There are many ways you can construct a macro to complete the task; here is one suggestion.

- Open Tools | Options | Calculation, and set the Calculation key to manual.
- Open Tools | Macro | Record New Macro. A dialog box will appear.
- Enter in a macro name (such as Sample) and a shortcut key (such as <Control>+<t>).
- If the Stop Recording button does not appear, open View | Toolbars | Stop Recording. You should now see the Stop Recording toolbar on your spreadsheet. The filled square on the left is the Stop Recording button. Press this button with your mouse when you are finished recording your macro.
- Press F9, the calculate key, to generate a new randomly selected individual.
- Select cell G13, the height of the randomly selected individual, and open Edit | Copy.
- Select cell K4, the top row of the height column.
- Open Edit | Find. A dialog box will appear (Figure 21).

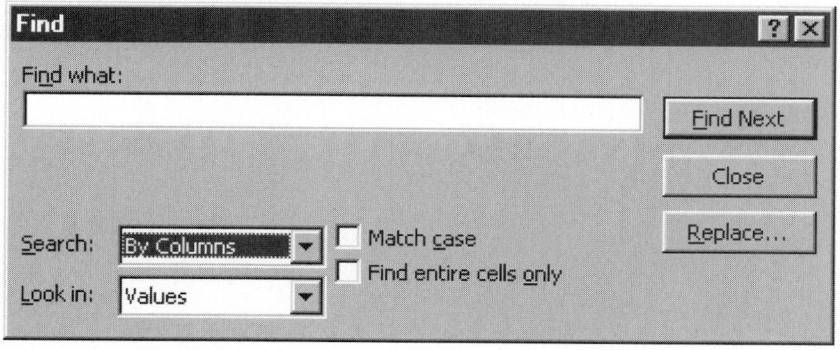

Figure 20

- Leave the box labeled Find What blank, and select the Search By Columns option. Click on Find Next, then on Close. Your cursor should have moved down to the next empty cell on your spreadsheet.
- Open Edit | Paste Special, and select the Paste Values option. Click OK.
- Click on the Stop Recording button.

That's all there is to it. Now when you press the shortcut key, <Control>+<t>, the spreadsheet will repeat the steps in the macro automatically. Run your macro until you have obtained the heights of 20 randomly sampled individuals. (Note that with this process, the same individuals can be sampled more than once.)

You can view the code that the spreadsheet "wrote" as a result of your keystrokes by going to Tools | Macros | Macro. Select the macro name of interest, and click on Edit. The Visual Basic code will be revealed. When you are finished, click on the x button in the upper-right hand corner of the spreadsheet to close the Visual Basic code. You will be returned to your original spreadsheet.

3. Save your work.

You may want to switch your calculation key back to automatic; otherwise, you must press F9 any time you want your spreadsheet to calculate values.

QUESTIONS

1. Explore the formulae used in the exercise by changing some of the heights of the individuals. For example, change cell B5 from 2 to 1. How does this change affect the outcome of the **AND** and **OR** functions? Change other values in the data set as well. How do your changes affect the frequency distribution of the data?

2. Click on the Paste Function button, f_x, and select the function category ALL. A list of all functions is displayed on the right-hand side of the Paste Function dialog box. Click on a function that looks interesting, and notice the description of the function that appears in the lower portion of the dialog box. Select three functions that were *not* used in this exercise and explore how each function works. Choose functions that are likely to be relevant to the data set in the exercise.

3 STATISTICAL DISTRIBUTIONS

Objectives

- Become familiar with properties of the normal distribution.
- Construct a frequency histogram of a trait for a population.
- Become familiar with properties of the binomial distribution.
- Become familiar with properties of the Poisson distribution.

INTRODUCTION

In your studies of ecology and evolution, you will very likely come across a variety of statistical distributions and their uses. If you haven't taken a course on statistics, learning about these distributions may seem like learning a foreign language. However, since they are so widely used in the sciences, it is important that you become familiar with the most common statistical distributions used in ecology and evolution. In this exercise, you will learn about three distributions: the **normal** (or **Gaussian**) distribution, the **binomial distribution**, and the **Poisson distribution**.

Normal Distribution

Let's start with some very basic concepts before introducing the normal distribution. In the biological sense, a **population** is a group of organisms that occupy a certain space and that can potentially interact with one another. In statistics the term population has a slightly different meaning. A statistical population is *the totality of individual observations about which inferences are made, existing anywhere in the world or at least within a specified sampling area limited in space and time* (Sokal and Rohlf 1981). Suppose you want to make a statement about the average height of humans on earth. Your statistical population would then include all of the individuals that currently occupy the planet earth. Usually, statistical populations are smaller than that. For example, if you want to make a statement about the size of a certain fish species in a local stream or pond, your statistical population consists of all of the fish currently occurring within the boundaries of a stream or pond. Other examples of statistical populations include a population of business firms, of record cards kept in a filing system, of trees, or of motor vehicles. By convention, Greek letters are used to describe the nature of a population. For example, the average height of humans on earth would be denoted with the Greek letter μ, and the variance in height would be denoted with the Greek letter σ^2, and the standard deviation would be denoted as σ. (We will define these terms shortly.)

In practice, it would be very difficult to measure the heights of all the individuals on Earth or even to measure *all* the fish in a local pond. So, we **sample** from the population. A sample is a subset of the population that we can deal with and measure. The goal of sampling is to make scientific statements about the greater population from the information we obtain in the sample. Quantities gathered from samples are called **statistics**. Statistics are denoted by letters from the Latin alphabet (i.e., from the same alphabet we use for writing English). For example, the mean of our sampled population would be denoted by the Latin letter $\bar{x}$, the variance is denoted by S^2, and the standard deviation is denoted by S.

The most important pictorial representation of a set of data that make up a sample is called a **frequency distribution**. If we sampled plants in an area of interest and recorded their biomasses in grams, we could then construct a frequency distribution such as Figure 1 and examine the shape of our data. Biomass would go on the x-axis (on the bottom), and numbers of individuals of a certain biomass would go on the y-axis (the vertical axis).

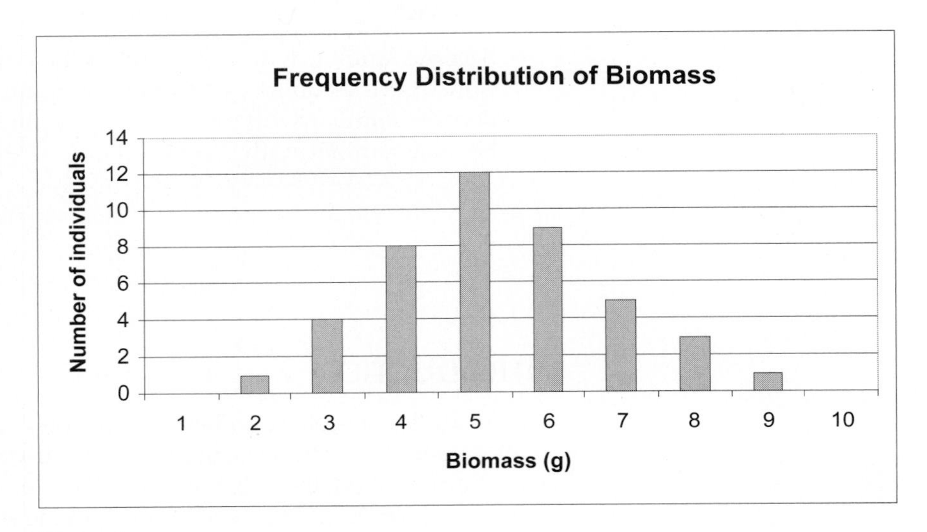

Figure 1

In published papers, you rarely see frequency distributions because they take up too much space in print, and they usually provide more information than a reader needs. Instead, ecologists and evolutionary biologists often report two kinds of summary statistics: (1) measures of **central tendency** (average value, middleness), and (2) measures of **dispersion** (how spread or dispersed the raw data are). Examine Figure 1. How would you characterize the "average plant" in terms of biomass? There are three common measures of central tendency: the mean, the mode, and the median. The **mean**, denoted by $\bar{x}$, is simply the arithmetic average: sum up the total biomass and divide by the number of individuals in the sample.

$$\bar{x} = \frac{\sum_{i=1}^{N} x}{N} \qquad \text{Equation 1}$$

If our sample consisted of the values 4, 6, 10, and 12, those values represent the little x's in equation 1, and $N = 4$ since there are four values in the sample. The average is (4 + 6 + 10 + 12) divided by 4. In Figure 1, the average is 4.3 grams of biomass. The **mode** is the most frequently occurring value. It is the high point of the frequency distribution. In our example, 5 is the mode since this value occurs 12 times. The **median** is the middle number in a data set when the samples are ordered. For example, if our sample consisted of the values 1, 3, 4, 6, and 10, the median would be 4 because it is the middle value. If the data set consisted of an even number of observations, then the median is the average of the two middlemost numbers.

Now let's consider the spread of the data in Figure 1. How can we characterize this spread? One way is to record the range of values the data assume. The lowest observed biomass was 2 grams, and the highest observed biomass was 9 grams. The range of biomass for our sample then is 9 – 2, or 7 grams. The data points at the extremes really affect the range, so it is not a very stable estimate of variability. A second method, called **average error**, describes how far each data point is, on average, from the mean. It is calculated as

$$\frac{\sum(x-\bar{x})}{N} \qquad \text{Equation 2}$$

However, because some scores will fall above the mean, and others will fall below it, this sum will always be 0! How can we overcome this problem? By squaring the deviations from the mean, and by subtracting 1 from the total sample size, we end up a definition of **variance**, or S^2:

$$S^2 = \frac{\sum(x-\bar{x})^2}{N-1} \qquad \text{Equation 3}$$

Thus, variance can be defined as (almost) the average squared deviation of scores from the mean. This is a very useful way of describing the spread of data in a given data set. However, all of the units have now been squared (e.g., biomass^2). To get rid of the squaring, we take the square root of both sides and arrive at the equation for computing the **standard deviation** of a sample, or S.

$$S = \sqrt{\frac{\sum(x-\bar{x})^2}{N-1}} \qquad \text{Equation 4}$$

With this background, we can now proceed to talk about the **normal distribution**. This distribution is one of the most familiar in statistics. Let us first return to a statistical population, rather than a sample. For a normally distributed trait, the frequency of distribution takes on a bell-shape that is completely symmetrical and has tails that approach the x-axis. The shape and position of the normal curve is determined by both the mean (μ) and the standard deviation (σ): μ sets the position of the curve along the x axis, while σ determines the spread of the curve. Two normal curves are shown in Figure 2. They have different μ but the same σ; thus they are similar in shape but are positioned in different locations along the x-axis.

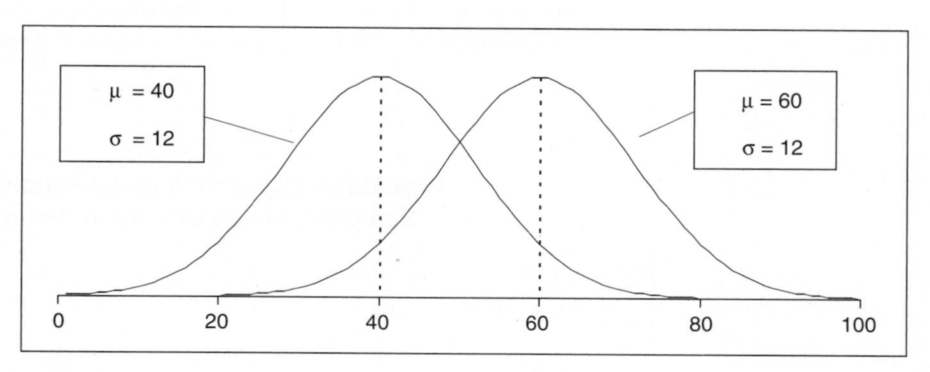

Figure 2

The standard deviation determines the spread of the normal curve. Figure 3 shows two normal curves with the same μ, 40, but different σ. Note that when σ is small, most of the data are distributed close to the mean, and when σ is large, the curve "flattens out" because the data vary more from the mean.

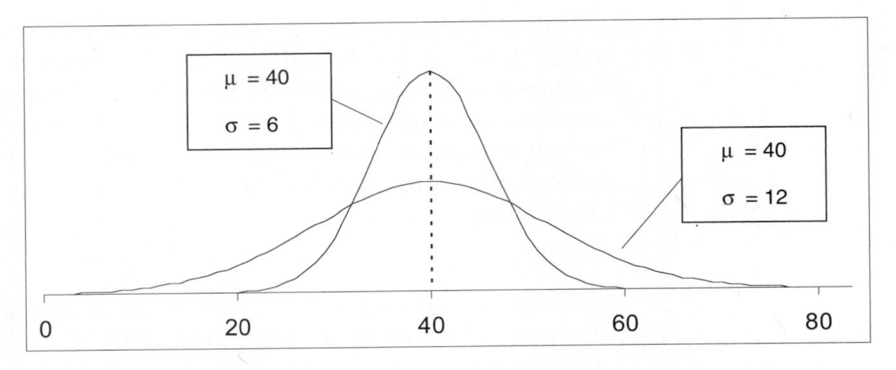

Figure 3

A property of normal curves is that the total area under the curve is equal to 1. (This is true of all probability models or models of frequency distributions.) Another property is that the most of the data fall in the middle of the curve around the mean. Normal distributions are completely symmetrical about the mean, and the mean equals the median and the mode. For normal distributions, approximately 68% of the observations will fall between the mean and plus or minus 1 standard deviation. If we assume, for example, that the mean length for a population of seeds is 10 mm and that *S* is 1.0, and if we assume that seed length is normally distributed, then 68% of the seed length values will fall between 9 and 11 mm (i.e., the mean, 10 mm, plus or minus 1.0, which is 1 standard deviation). And approximately 95% of the observations will fall between the mean and plus or minus 2 standard deviations. These properties make it possible to compute the specific probability that, for example, a seed of 8 mm length will be sampled from the population.

Figure 4 shows that, for a population with a mean of 10 and a standard deviation of 1.0, this probability is 0.054. This probability was computed in Excel with the NORMDIST function. The probability of sampling a seed of 10 mm length is 0.4. The **cumulative probability** gives the probability of sampling a seed of a certain size or less. For example, the probability of sampling a seed of *at least* 10 mm is 0.5. As you can see, with the parameters given, the cumulative probability is 1 when the seed length is 13 mm. This means that there is a 100% chance of sampling a seed of 13 mm or less, given that the population has a mean length of 10 mm and a standard deviation of 1.

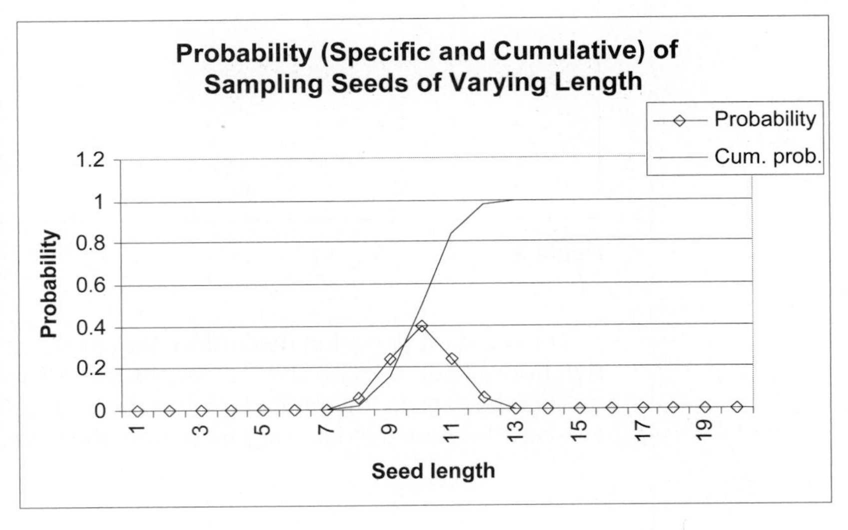

Figure 4

If we change the standard deviation to 3 mm, and keep the mean at 10 mm, the probabilities will be different (Figure 5).

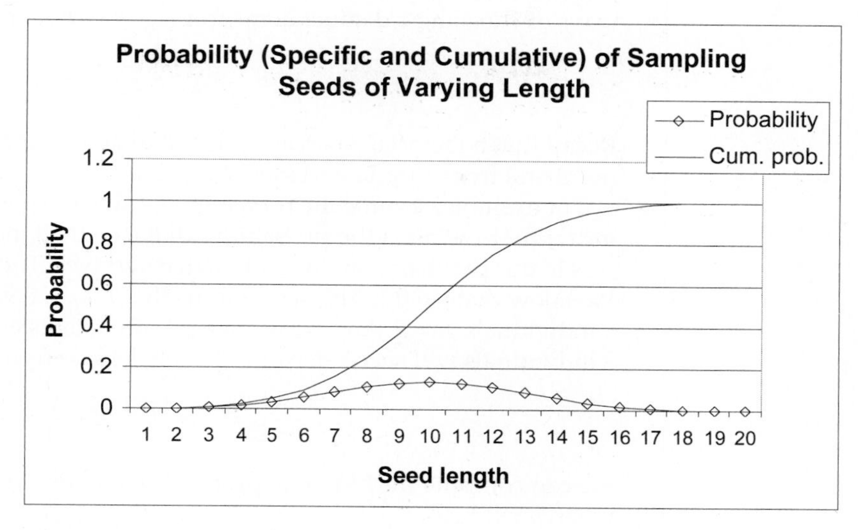

Figure 5

Knowledge about the normal distribution is important to ecologists because many statistical procedures, such as a *t* test, assume that the sampled data are normally distributed. These properties are handy from a modeling perspective; in many of the exercises in this book, we will "draw" samples from a normal distribution whose mean and standard deviation are specified.

Binomial Distribution

Some situations in ecology are binary: There are only two possible outcomes. For example, suppose we are tracking the fates of individuals over time and are interested in the number of deaths. During this period, there are only two outcomes: death or survival. In this situation, a binomial distribution can be used to describe the relative number of times that a particular event will occur (death) among groups of observations. Another example may be the relative numbers of trees in flower among a series of samples of a particular size. The **binomial distribution** is used when a researcher is interested in the *occurrence* of an event, not in its magnitude. The binomial distribution describes, for instance, the relative numbers of individuals that flower, not how well they flower.

The binomial distribution is specified by the number of observations, n, and the probability of occurrence, which is denoted by p. Here are some things to keep in mind when using the binomial distribution:

- Each outcome must be classified as a "success" (the type of outcome that we're interested in) or as a "failure."
- Since we're dealing with a count of successes, this probability distribution is discrete. (The x-axis is the number of successes, and it cannot be a fraction).
- Each trial is independent. The probability of success (p) and the probability of failure ($1 - p$) is the same for each trial. Thus, if one tree in your sample has fruits, you don't know anything about the next sample, other than it has a probability p of having fruit.

The formula for calculating the probability of x successes out of n trials of a binomial experiment, where the probability of success on an individual trial is p, is

$$f(x) = {}_nC_x \times p^x \times (1 - p)^{n-x} \qquad \text{Equation 5}$$

In this equation, p is the probability of success and its exponent, x, is the number of successes. The probability of failure is $1 - p$, and its exponent, $n - x$, is the number of failures. The term ${}_nC_x$ means "out of n samples, let x succeed." This gives the number of ways of choosing x distinct items from a set of n items, and it is calculated as

$$ {}_nC_x = \frac{n!}{x! \times (n-x)!} \qquad \text{Equation 6} $$

Recall that a factorial, such as $n!$, is calculated by multiplying all the integers (whole numbers) from 1 up to and including n.

For example, assume the probability of surviving is 0.1. If we have a population of 5 individuals, what is the probability that exactly 3 individuals will survive? The success in this problem is an individual that survives. The failure is an individual that dies. We know that $p = 0.1$. This also tells us that $1 - p = 0.9$. Since our population consists of 5 individuals, $n = 5$. And we are specifically interested in knowing the probability that 3 individuals will succeed, so $x = 3$. First, let's compute ${}_5C_3$. It is

$$ \frac{5!}{3! \times 2!} = \frac{5 \times 4 \times 3 \times 2 \times 1}{(3 \times 2 \times 1)(2 \times 1)} = \frac{20}{2} = 10 $$

We can compute the binomial probability that exactly 3 of 5 individuals will survive when $p = 0.1$ as

$$ \begin{aligned} f(3) &= {}_5C_3 \times (0.1)^3 \times (0.9)^2 \\ &= (10) \times (0.001) \times (0.81) \\ &= 0.0081 \end{aligned} $$

The probability that exactly 3 of these 5 individuals survive is 0.0081. Similarly, the probability that 0, 1, 2, 4, and 5 individuals survive could be calculated (rather easily with the **BINOMDIST** function). We can graph these binomial probabilities as shown in Figure 6.

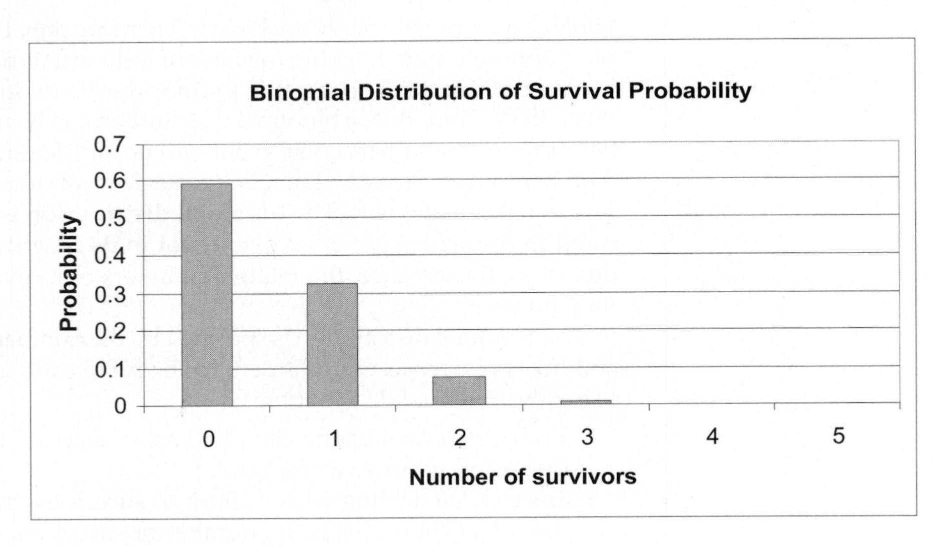

Figure 6

If we change our survival probability to 0.7, our binomial distribution of probabilities will differ, as shown in Figure 7. As with the normal distribution, we could also plot the *cumulative* probability that *at least* x number of individuals survive.

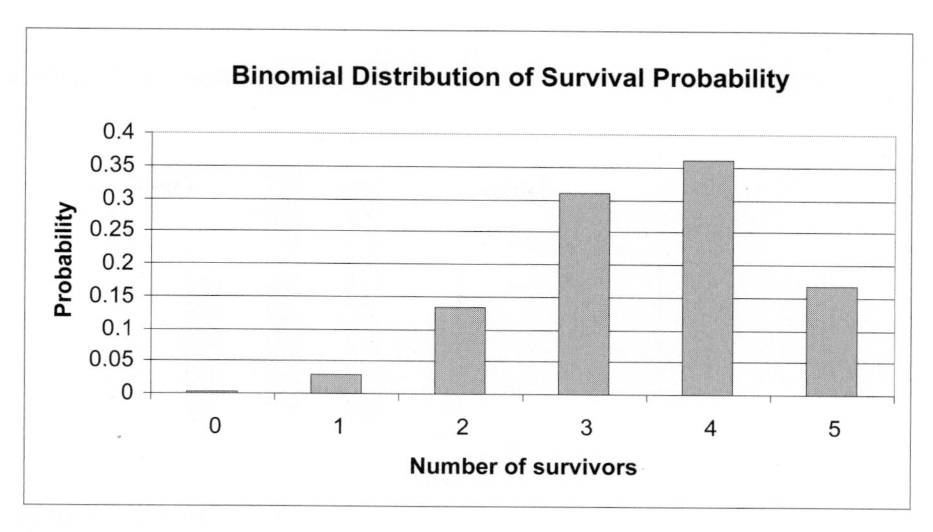

Figure 7

Poisson Distribution

The **Poisson distribution** is similar to the binomial distribution in that the number of events is counted. However, the events are not limited to two outcomes. For example, ecologists may be interested in the number of birth events in a given period of time. The Poisson distribution is a mathematical rule that assigns probabilities to the number occurrences. The French mathematician Poisson derived this distribution in 1837, and evidently its first application was the description of the number of deaths in the Prussian army due to horse kicking (Bortkiewicz 1898). The only thing we need to know to specify the Poisson distribution is the mean number of occurrences, such as the mean number of births. Contrast this to the binomial distribution, in which both the probability that an event will occur and the total number of individuals in the population must be known. For example, in the binomial distribution all individuals are studied to see whether they had survived or not, whereas using the Poisson distribution only the individuals that survived are studied.

The formula for calculating the Poisson probability is

$$f(x) = \frac{\lambda^x \times e^{-\lambda}}{x!} \qquad \text{Equation 7}$$

where λ is the mean number of successes in a given period of time, x is the number of successes we are interested in, and e is the natural logarithm constant (approximately 2.718). As an example, suppose the average number of offspring produced per individual in a population is 2.1; what is the probability that an individual will have exactly 4 offspring? The probability would be calculated as

$$f(4) = \frac{2.1^4 \times e^{-2.1}}{4 \times 3 \times 2 \times 1} = 0.0992 \qquad \text{Equation 6}$$

We could calculate the probability that exactly 0, 1, 2, 3, 5, 6, 7, … offspring were produced, given the average, with the **POISSON** spreadsheet function. Our Poisson distribution is shown in Figure 8.

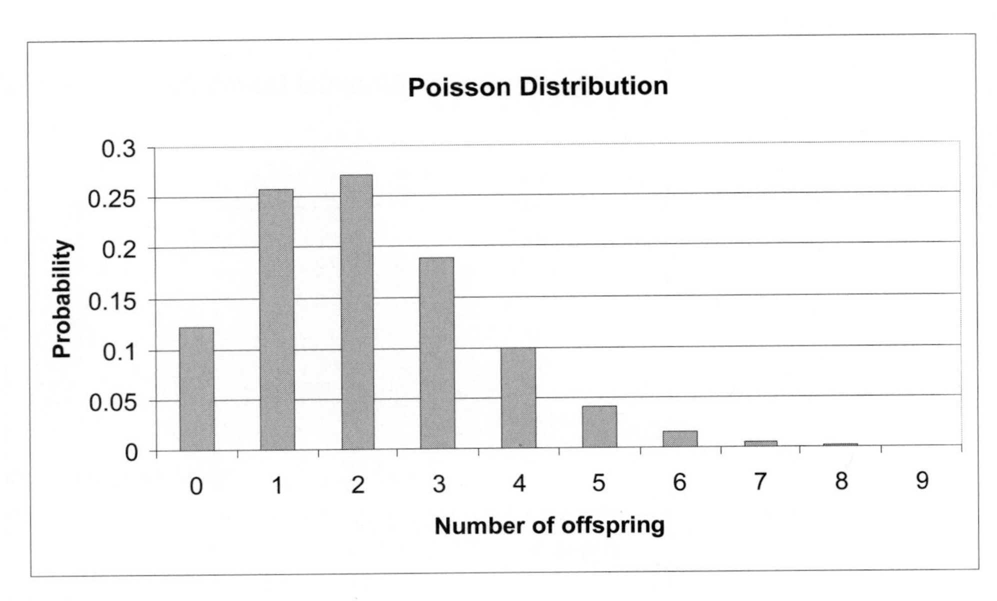

Figure 8

In this exercise, you'll use a spreadsheet to explore properties of the normal, binomial, and Poisson distributions. As always, save your work frequently to disk.

INSTRUCTIONS

ANNOTATIONS

A. Set up the spreadsheet for normal distribution.

We will start our exercise by investigating properties of the normal distribution, and we will compare a trait (height, for example) between two different populations, each consisting of 30 individuals.

1. Open a new spreadsheet and set up column headings as shown in Figure 9.

	A	B	C	D	E	F	G
1	***Normal Distribution***						
2							
3		**Mean**	**Std**				
4	Population 1	50	5				
5	Population 2	30	5				
6							
7	Individual	Pop 1	Pop 2		**Frequency distribution**		
8	1				Bins	Pop 1	Pop 2

Figure 9

2. Set up a linear series from 1 to 30 in cells A8–A37.

Enter 1 in cell A8.
Enter **=1+A8** in cell A9. Copy this formula down to cell A37.

Next we will assign a height to each of the 30 individuals in population 1, drawn from a normal distribution with a mean given in cell B4 and a standard deviation given in cell C4. We don't really have individuals to measure, of course, but the **NORMINV** function allows us to simulate this. The **NORMINV** function consists of three parts, each separated by a comma. It has the form **NORMINV(probability,mean,standard_dev)** where **probability** is a probability (from 0 to 1) corresponding to the cumulative normal distribution, **mean** is the arithmetic mean of the normal distribution, and **standard_dev** is the standard deviation of the distribution.

3. In cell B8, enter a **NORMINV** formula to generate a random height for an individual in population 1. Copy your formula down to cell B37.

In cell B8, enter the formula **=ROUND(NORMINV(RAND(),$B$4,$C$4),1)**. Copy this formula down to cell B37.

The formula =**NORMINV(RAND(),$B$4,$C$4)** tells the spreadsheet to draw a random cumulative probability between 0 and 1 (the **RAND()** portion of the formula) from a normal distribution that has a mean given in cell B4 and a standard deviation given in cell C4. The formula returns the inverse of this probability; it changes the cumulative probability into an actual number from the distribution. Excel will return a value, which is the height of the individual. You'll note that this formula is embedded within a **ROUND** formula, which consists of two parts that are separated by a comma. The first part is the number that should be rounded **(NORMINV(RAND(),$B$4,$C$4)**, and the second part is the number of decimal places to which the number should be rounded. Note that when you press F9, the calculate key, the spreadsheet will generate a new random number, and hence will generate a new cumulative probability and height for individual 1 in Population 1.

4. In cells C8–C37, enter a formula to generate a random height for an individual in Population 2.

Enter the formula **=ROUND(NORMINV(RAND(),$B$5,$C$5),1)** in cell 8. Copy your formula down to cell C37.
Note that the references to the mean and standard deviation are absolute cell references (indicated by the dollar signs), so that when you copy the formula down to cell C37 the heights will be drawn from a distribution whose mean and standard deviation are fixed in cells B5 and C5.

5. Save your work.

B. Construct the frequency distribution.

The most common way to depict a population's values is as a frequency distribution. A frequency distribution is a plot of the raw data, in this case height, against the frequency that each value appears in the population.

We will use the **FREQUENCY** function to generate a frequency distribution of heights for Population 1 and Population 2. This formula is a bit tricky, so pay attention to these instructions. The **FREQUENCY** function calculates how often values occur within a range of values.

1. In cell E9, enter the number 5. In cell E10, enter **=5+E9**. Copy this formula down to cell E28.

Use the **FREQUENCY** function to count the number of heights that fall 5 mm or lower, within 6 and 10 mm, within 11 and 15 mm, and so on. These groupings are called "bins." The bins may be very small (hold only a few numbers) or very large (hold a large set of numbers). Our bins will cluster heights in groups of 5 mm. The bin labeled 5 (cell E9) will "hold" heights up to and including 5 mm (0, 1, 2, 3, 4, and 5 mm). The bin labeled 10 (cell (E10) will "hold" heights from 6 to 10 mm, and so on.

2. Use the **FREQUENCY** function in cells F9–F28 to compute frequencies of heights for Population 1.

The **FREQUENCY** returns an array of values (in our case the values will be in cells F9–F28), it must be entered as an array formula, which is a bit different than the normal formula entries. It has the syntax **FREQUENCY(data_array,bins_array)**, where **data_array** is the set of values for which you want to count frequencies (heights), and **bins_array** is the array of intervals into which you want to group the values in data_array. You can think of a bin as a bucket in which specific numbers go.

The **FREQUENCY** formula works best when you use the f_x button and follow the cues for entering a formula. Since you will be entering this formula for an array of cells, the mechanics of entering this formula are a bit different than the typical formula entry. *Instead of selecting a single cell to enter a formula, you need to select a series of cells, enter a* *formula, and then press <Control>+<Shift>+<Enter> (Windows) to enter the formula for all of the cells you have selected.*

Let's try it to determine the frequencies of heights for Population 1. Select cells F9–F28 with your mouse, then use your f_x button and select the **FREQUENCY** function. (If it doesn't show up in the list of most recently used functions, you will have to view the list of all functions.) To define the data array, use your mouse to highlight the heights of all 30 individuals in Population 1 (cells B8–B37). To define the bins array, select cells E9–E28. Next, instead of clicking "OK," press <Control>+<Shift>+<Enter> to return your height frequencies. After you've obtained your results, examine the formula in cells F9–F28. Your formula should look like this:

{ =FREQUENCY(B8:B37,E9:E28)}

The { } symbols indicate that the formula is part of an array. If for some reason you get "stuck" in an array formula, press the Escape key and start over.

3. Use the **FREQUENCY** function in cells G9–G28 to compute frequencies of heights for Population 2.

Follow steps 1 and 2. Your formula should be **{=FREQUENCY(C8:C37,E9:E28)}** in cells G9–G28.

4. Graph the frequencies of Population 1 and Population 2.

Use the column graph option and label your axes fully. Your graph should resemble Figure 10.

5. Save your work.

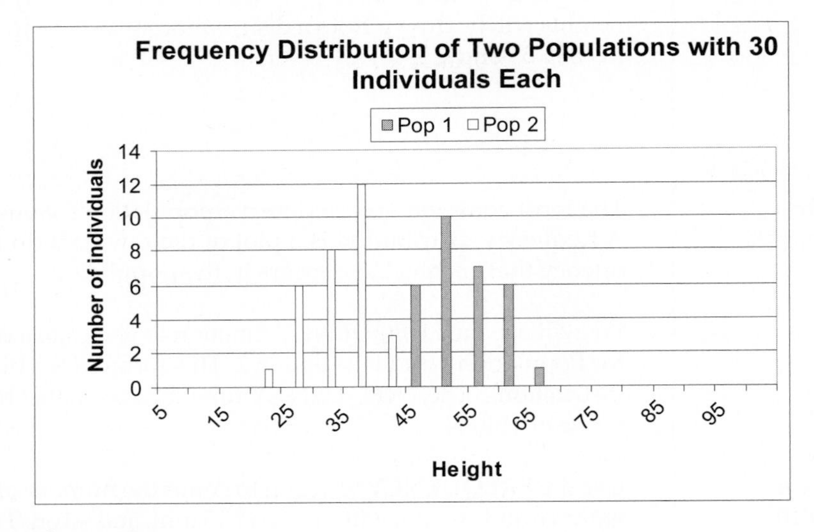

Figure 10

C. Compute statistics.

1. Set up new spreadsheet headings as shown in Figure 11.

	I	J	K
7		***Pop 1***	***Pop 2***
8	Mean		
9	Median		
10	Mode		
11	Standard deviation		
12	Minimum		
13	Maximum		
14	Range		
15	Count		

Figure 11

2. Enter formulae to compute measures of central tendency: the mean, median, and mode height for Populations 1 and 2 in cells J8–K10.

Use the f_x button to guide you through the formulae. Your results should be

- J8 **=AVERAGE(B8:B37)**
- J9 **=MEDIAN(B8:B37)**
- J10 **=MODE(B8:B37)**
- K8 **=AVERAGE(C8:C37)**
- K9 **=MEDIAN(C8:C37)**
- K10 **=MODE(C8:C37)**

If Excel cannot find a most commonly occurring number (i.e., if there is no mode), it will return **#N/A**.

3. Enter formulae in cells J11–K14 to compute measures of dispersion: standard deviation, minimum, maximum, and range.

Use the f_x button to guide you through the formulae. Your results should be:

- J11 **=STDEV(B8:B37)**
- J12 **=MIN(B8:B37)**
- J13 **=MAX(B8:B37)**
- J14 **=J13-J12**
- K11 **=STDEV(C8:C37)**
- K12 **=MIN(C8:C37)**
- K13 **=MAX(C8:C37)**
- K14 **=K13-K12**

4. Enter a formula in cells J15–K15 to count the sample size of each population.

Enter the formulae:

- J15 **=COUNT(B8:B37)**
- K15 **=COUNT(C8:C37)**

5. Save your work, and answer Questions 1–4 at the end of the exercise. Your spreadsheet should now resemble Figure 12, although your numbers will be different. Each time you press the F9 key to generate new heights, the statistics for each population will be automatically updated.

	I	J	K
7		***Pop 1***	***Pop 2***
8	Mean	50.4	29.4
9	Median	50.0	30.2
10	Mode	52.6	31.6
11	Standard deviation	5.3	4.7
12	Minimum	40.1	18.8
13	Maximum	61.2	38.1
14	Range	21.1	19.3
15	Count	30.0	30.0

Figure 12

D. Set up the binomial distribution spreadsheet.

1. Click on Sheet 2 and set up new headings as shown in Figure 13.

	A	B	C	D	E	F	G
1	***Binomial and Poisson Distributions***						
2							
3	Probability of survival =		0.5				
4	Mean number of offspring =		20				
5	Number of individuals =		30				
6							
7		**Binomial**				**Poisson**	
8	# Survivors	Probability	Cum. prob.		# Offspring	Probabity	Cum. prob.

Figure 13

2. Set up a linear series from 0 to 30 in cells A9–A39.

First, we will consider the number of survivors over a period of time in a population that again consists of 30 individuals. There are only two outcomes for each individual (survive or die), which makes survival probabilities an appropriate use of the binomial distribution. We will consider the probability that 0, 1, 2, ..., 30 individuals will survive the period with a binomial distribution, given that the survival probability is 0.5 (cell C3) and that there are 30 individuals (cell C5).

Enter **0** in cell A9.
Enter **=1+A9** in cell A10. Copy this formula down to cell A39.

3. In cells B9–B39, enter a formula to calculate the probability that the exact number of individuals given in cell A9 will survive.

In cell B9, enter the formula **=BINOMDIST(A9,$C$5,$C$3,FALSE)**. Copy this formula down to cell B39.

The **BINOMDIST** function returns the probability of success (survival) from the binomial distribution, given the number of trials (the number of individuals in the population) and the probability of success (the probability of survival). This function consists of four parts, each separated by a comma. The first part is the number of individuals in the population, the second part is the number of survivors in the population, the third part is the probability of survival for the whole population, and the fourth part tells the spreadsheet whether you want the binomial probability to be a cumulative probability (e.g, the probability that there will be *up to but not more than* 15 survivors), or simply the probability that a given number of individuals will survive (e.g., the probability that 4 out of 30 individuals in the population will survive). The word "True" returns the cumulative probability, while the word "False" returns the specific probability.

For example, the formula in cell B9 returns the binomial probability that there will be 0 survivors (cell A9) when the population consists of 30 individuals (given in cell C5) and the average survival probability is 0.5 (given in cell C3). The **FALSE** part of the formula indicates that the program should return the probability for the exact number of survivors, not the cumulative probability.

4. Enter a formula in cell C9 to calculate the cumulative probability that no more than the number of individuals given in cell A9 will survive.

In cell C9, enter the formula **=BINOMDIST(A9,$C$5,$C$3,TRUE)**. Copy this formula down to cell C39.

This formula is identical to the one just entered in cells B9–B39, except that the last part of the formula is **TRUE**, indicating that the program should return the cumulative probability, or the probability that there will be *up to* a certain number of survivors.

5. Graph the probability of survival against the number of survivors (cells B9–B39).

Use the column graph option and label your axes fully. You could also use the Scatterplot graph option if you prefer.

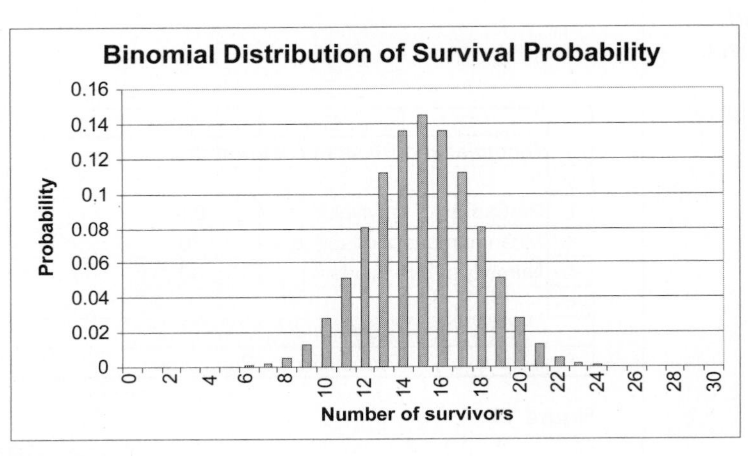

Figure 14

6. Graph the probability of survival, and the cumulative probability of survival, against the number of survivors (cells B9–C39).

7. Save your work.

Use the column graph option and label your axes fully. Your graph should resemble Figure 15.

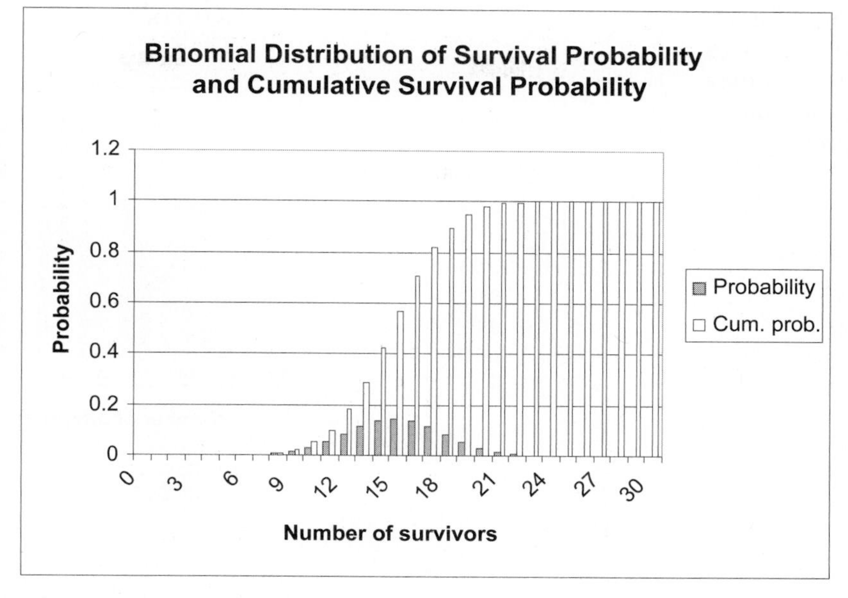

Figure 15

E. Set up the Poisson distribution spreadsheet.

Now we will consider the number of births over a period of time in a population that once again consists of 30 individuals. For this exercise, we will assume that there are between 0 and 30 births possible. Because there are several discrete numbers of births possible, this analysis is an appropriate use of the Poisson distribution. We will consider the probability that 0, 1, 2, ..., 30 individuals will be born during a time period of interest, given that the average number of offspring for the population is 20 (cell C4).

1. Set up a linear series from 0 to 30 in cells E9–E39.

Enter 0 in cell E9.
Enter **=1+E9** in cell E10. Copy this formula down to cell E39.

2. In cell F9, enter a formula to calulate the probability that the exact number of young given in cell E9 will be born. Copy this formula down to cells F10–F39.

In cell F9, enter the formula **=POISSON(E9,$C$4,FALSE)**. Copy this formula down to cell F39.

Cell F9 uses the **POISSON** function to give the probability that a certain number of young will be born, given the average number of young born per period of time. This function has three parts, each separated by a comma. This first part gives the number of young born (e.g., 0 young, cell E9). The second part gives the expected number of young born (cell C4). The third part, like the **BINOMIAL** function, indicates whether you want the cumulative probability (e.g., the probability that up to 8 young will be born) or the probability that a specific number of young are born (e.g., the probability that exactly 10 young will be born). **FALSE** returns the exact probability, whereas **TRUE** returns the cumulative probability.

3. In cell G9, enter a formula to calulate the probability that no more than the number of young given in cell E9 will be born. Copy this formula down to cells G10–G39.

In cell G9, enter the formula **=POISSON(E9,$C$4,TRUE)**. Copy this formula down to cell G39.

4. Graph the number of offspring and the Poisson probability (exact, cells E9–F39). Use the column graph and label your axes fully (Figure 16). You may also use the Scattergraph option if you prefer.

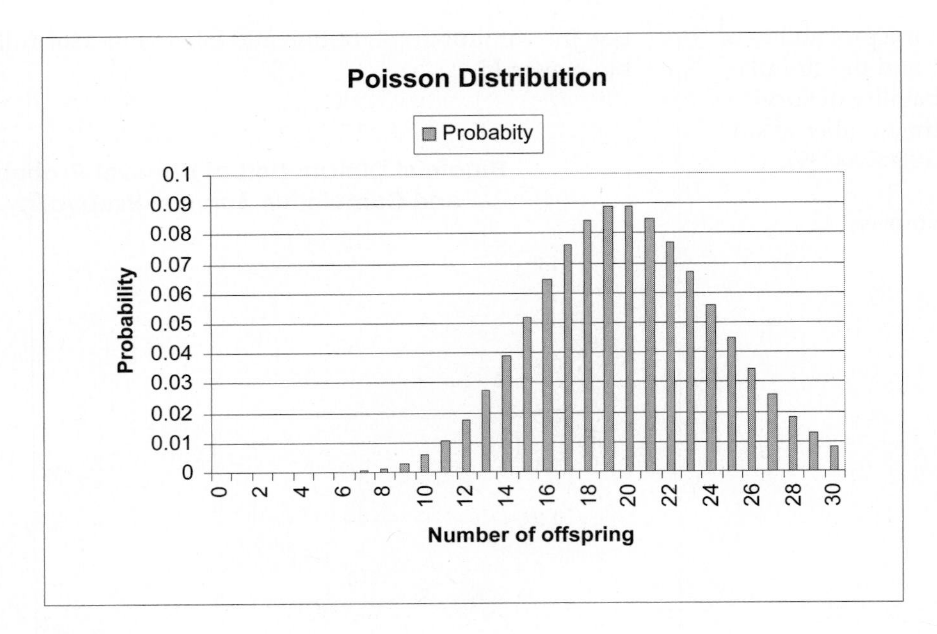

Figure 16

5. Graph the number of offspring and the cumulative Poisson probability (cells G9–G39). Use the column graph and label your axes fully (Figure 17).

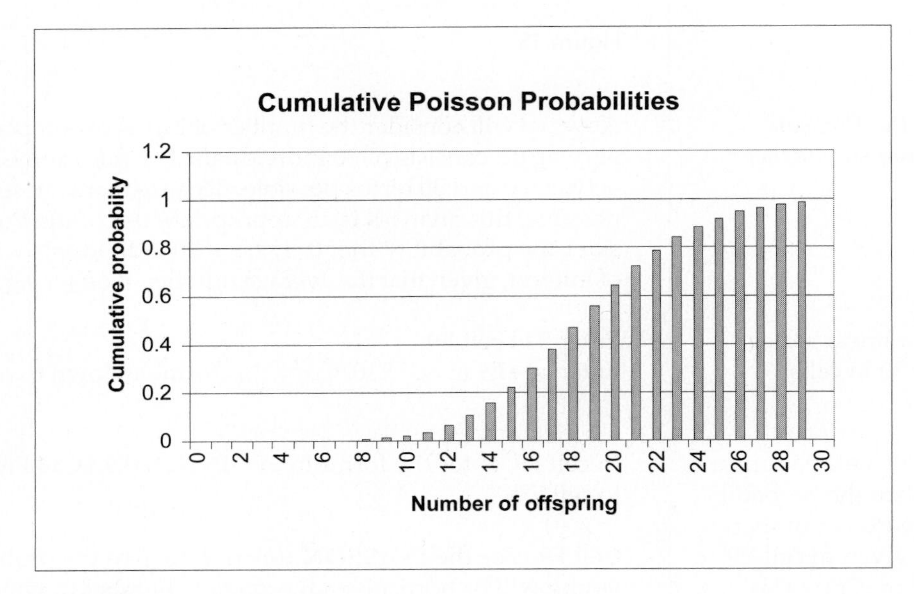

Figure 17

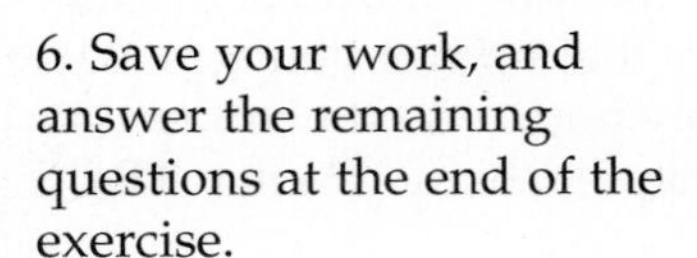

6. Save your work, and answer the remaining questions at the end of the exercise.

QUESTIONS

1. What parameter controls the location along the x-axis of the data in your frequency distribution? Change the value in cell B4 (Population 1, try several values) and examine your distribution.
2. What parameter controls the spread of the data in your frequency distribution? Change the value in cell C4 (Population 1, try several values) and examine your distribution. What happens when this value is almost 0 (0.0001)?
3. One of the properties of the normal distribution is that the mean, mode, and median are equal. Why might this not be the case in your spreadsheet? How could you increase the chances that the mean, mode, and median would be equal?
4. Assume that instead of heights, we are comparing the annual salaries (in thousands of dollars) of 30, randomly selected individuals. Set up cell values as shown:

	A	B	C
3		**Mean**	**Std**
4	Population 1	50	5
5	Population 2	50	5

   Furthermore, assume that Bill Gates is part of our sample in Population 1, and his salary is entered in cell B8. Enter 1000 in cell B8 (overwrite the formula in that cell). Assess which measure of "middleness" is the most appropriate descriptor of average salaries.

5. Assume you are a biologist working on a mark-recapture study of a population of salmon, and you have tagged 20 salmon. You estimate that 50% of the salmon will survive to the time set for recapture. What is the probability that *exactly* 10 of the marked salmon are still alive when it is time to recapture? What is the probability that *up to* 10 of the marked salmon are still alive?
6. How do your answers from Question 5 change if the survival estimate is 30%?
7. Set cell C3 to 0.5. Change the value in cell C5, starting with 0, and increase by twos up to 20. How does changing cell C5 (n) affect the location and shape of the binomial distribution?
8. How does changing cell C4 (λ) affect the location and shape of the Poisson distribution? Change the value in cell C4, from 0 to 10, in increments of 1. As λ increases, what kind of shape does the Poisson distribution take?

LITERATURE CITED

Sokal, R. R. and F. J. Rohlf. 1981. *Biometry*, 2nd Ed. W. H. Freeman and Company, New York.

4 CENTRAL LIMIT THEOREM

Objectives

- Set up a spreadsheet model to examine the properties of the central limit theorem.
- Develop frequency distributions and sampling distributions, and differentiate between the two.
- Develop a bootstrap analysis of the mean for various sample sizes.
- Evaluate the relationship between standard error and sample size, and standard deviation and sample size.

Suggested Preliminary Exercise: Statistical Distribution

INTRODUCTION

You have probably come across the term "population" in your studies of biology. In the biological sense, the term "population" refers to a group of organisms that occupy a defined space and that can potentially interact with one another. The Hardy-Weinberg equilibrium principle is an example of a population-level study. In statistics the term population has a slightly different meaning. A statistical population is *the totality of individual observations about which inferences are made, existing anywhere in the world, or at least within a specified sampling area limited in space and time* (Sokal and Rohlf 1995).

Suppose you want to make a statement about the average height of humans on earth. Your statistical population would include all the individuals that currently occupy the planet earth. Usually, statistical populations are smaller than that, and the researcher determines the size of the statistical population. For example, if you want to make a statement about the length of dandelion stems in your hometown, your statistical population consists of all of the dandelions currently occurring within the boundaries of your hometown. Other examples of statistical populations include a population of all the record cards kept in a filing system, of trees in a county park, or motor vehicles in the state of Vermont.

In practice, it would be very difficult to measure the heights of *all* the individuals on earth, or even to measure *all* the dandelions in your hometown. So we take a sample from the population. A **sample** is a subset of the population that we can deal with and measure. The goal of sampling is to make scientific statements about the greater population based on the information we obtain in the sample. Quantities gathered from samples are called **statistics**.

"How many samples should I take?" and "How should I choose my samples?" are very important questions that any investigator should ask before starting a scientific study. In this exercise, we'll consider **simple random sampling**. If you sample 10 dandelions in your hometown with the intent of making scientific statements about all of the dandelions that occupy your town, then each and every individual in the population must have the same chance of being selected as part of the sample. In other words, a simple random sample is a sample selected by a process that gives every possible sample (of that size from that population) the same chance of being selected.

Let's imagine that you use a simple random sampling scheme to sample the stem lengths of 10 dandelions in your hometown. And let's further imagine that the *actual* average stem length of the dandelion population in your hometown is $\mu = 10$ mm; you are trying to estimate this parameter through sampling. You carefully measure the stem length of each of the 10 sampled dandelions, and then calculate and record the mean of the sample on your computer spreadsheet. The mean you have calculated is called an **estimator**, usually designated as $\bar{x}$, which estimates the true population mean, μ (which in this case is 10 mm). If you plot your raw data on a graph, your graph is called a **frequency distribution**. This is a pictorial description of how frequent or common different values (in this case stem lengths) appear in the population. A frequency distribution reveals many things about the nature of your samples, including the sample size, the mean, the shape of the distribution (normal, skewed, etc.), the range of values, and modality of the data (Figure 1).

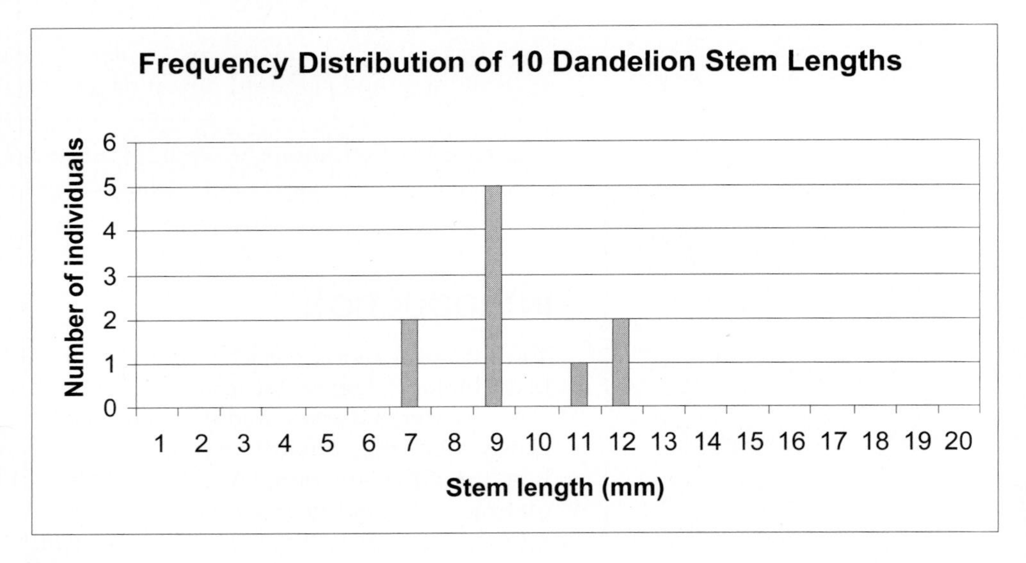

Figure 1

In the example in Figure 1, our sample of 10 dandelions had a mean value of 9.4 mm. How do you know how close your estimator is to the true mean, μ, if you can't actually measure μ? The central paradox of sampling is that it is impossible to know, based on a single sample, how well the sample represents μ. If you obtain another sample of 10 dandelions, and calculate a mean, you will now have two estimates of the population mean, μ. What if they are different? How will you know which is the "best" estimator?

Here is where the central limit theorem comes into play. If you repeat this sampling process and obtain a set of estimators (say, for example, 10 estimators in total, each based on a sample size of 10 dandelions), you now have a **sampling distribution** of the sample average (note the difference between the sampling distribution and the frequency distribution). The sampling distribution shows the possible values that the estimator can take and the frequency with which they occur. The standard deviation of a sampling distribution is called the standard error.

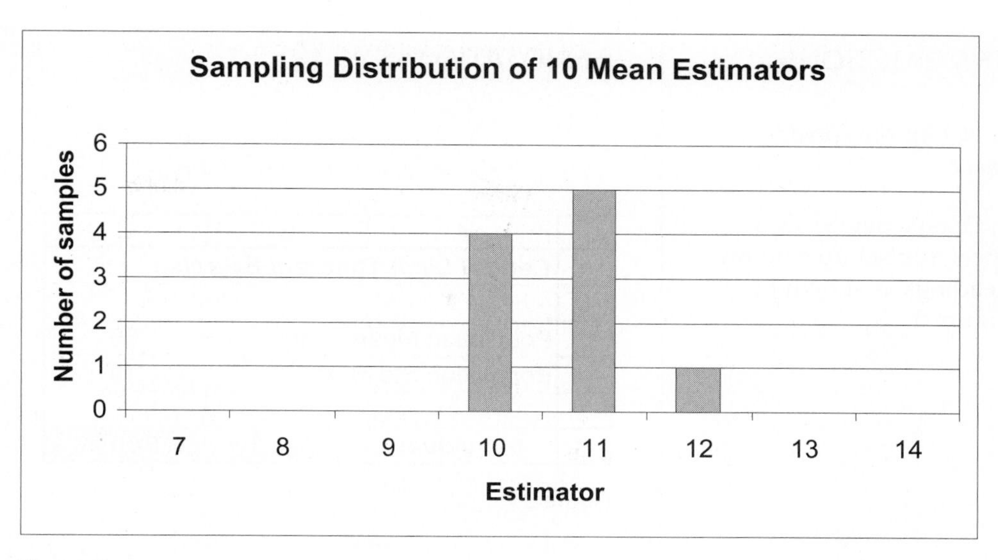

Figure 2

The central limit theorem, one of the most important statistical concepts you will encounter, states that in a finite population with a mean μ and variance σ^2, the *sampling distribution of the means* approaches a *normal distribution* with a sampling mean μ and a sampling variance σ^2/N as N (N = number of individuals in the sample) increases. In Figure 2, 4 of our 10 samples had a mean of 10 mm, 5 samples had a mean of 11 mm, and 1 sample had a mean of 12 mm. The central limit theorem says that this sampling distribution will become more and more "normal" (a bell-shaped curve on a graph) as the sample size increases. It also says that the mean of the sampling distribution is an unbiased estimator of μ, and that the variance of the estimators is σ^2/N.

In this exercise, you will set up two populations that have the same mean, μ, of 50 mm. You will try to estimate this parameter through sampling. Both populations contain 500 individuals. The mean stem lengths of Population 1 follow a normal distribution. Population 2 has a somewhat funky, **bimodal distribution** in which individuals have stem lengths of either 0 or 100. We will obtain samples from each population, from which we will estimate the mean of each population.

The method by which we will sample is called the **bootstrap method**, a very common sampling method in statistics (Efron 1982). The bootstrap involves *repeated reestimation of a parameter (such as a mean) using random samples with replacement from the original data*. Because the sampling is with replacement, some items in the data set are selected two or more times and other are not selected at all. We will do a bootstrap analysis of the mean when sample sizes of 5, 10, 15, and 20 are drawn (with replacement) from each population. When the procedure is repeated a hundred or a thousand times, we get "pseudosamples" that behave similarly to the underlying distribution of the data. In turn, you can evaluate how biased your estimator is (whether your estimator gives a good estimate of μ or not), the confidence intervals of the estimator, and the bootstrap standard error of your estimator. All of this will become more clear as you work through the exercise.

As always, save your work frequently to disk.

INSTRUCTIONS | **ANNOTATION**

A. Set up the spreadsheet.

1. Open a new spreadsheet and set up column headings as shown in Figure 3.

	A	B	C
1	***Central Limit Theorem Exercise***		
2			
3	Population Mean => μ		50
4	Population Std => σ		10
5			
6	**Individual**	**Pop 1**	**Pop 2**

Figure 3

2. In cells A7–A506, assign a number to each individual in the populations, starting with 1 in cell A7 and ending with 500 in cell A506.

Enter 1 in cell A7.
Enter **=A7+1** in cell A8.
Copy this formula down to cell A506 to designate the 500 individuals.

3. Enter a population mean of 50 in cell C3.

We will compare two populations of dandelions (actual statistical populations), each consisting of 500 individuals. Both populations, Population 1 and Population 2, have an actual mean stem length (μ) of 50 mm, which is designated in cell C3.

4. Enter the standard deviation for Population 1 in cell C4.

Population 1 will consist of 500 individuals that have a mean, μ, of 50 mm and a standard deviation of 10 mm. We'll assume that Population 1 is normally distributed. Thus, the raw data are distributed in a bell-shaped curve that is completely symmetrical and has tails that approach but never touch the *x*-axis. The shape and position of the normal curve is determined by μ and σ: μ sets the position of the curve while σ determines the spread of the curve. Figure 4 shows two normal curves. They have different means (μ) but have the same σ, thus they are similar in shape but are positioned in different locations along the *x*-axis.

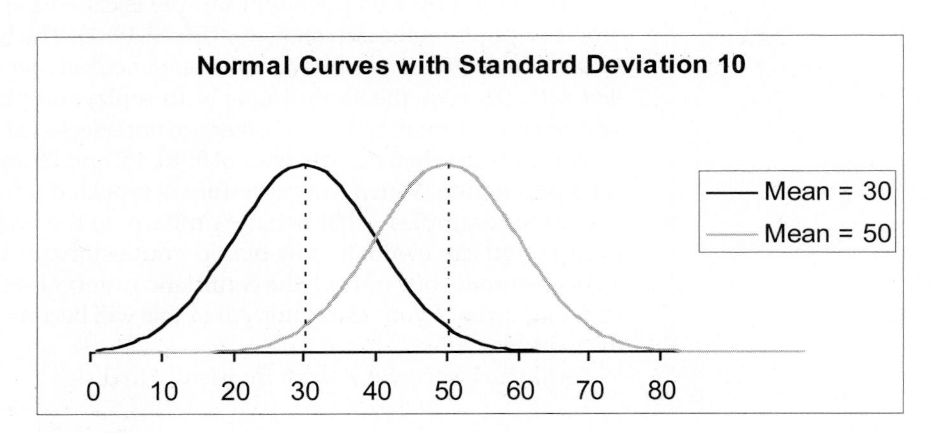

Figure 4

A property of normal curves is that the total area under the curve is equal to 1. (This is true of all probability models or models of frequency distributions). Another property is that the most of the data fall in the middle of the curve around the mean. For normal distributions, approximately 68% of the observations will fall between the mean and ±1 standard deviation. In our dandelion population, this means that 68% of the individuals in the population will have a stem length between 40 mm and 60 mm (which is the mean, 50 mm, ±10, which is 1 standard deviation). About 95% of the observations will fall between the mean and ±2 standard deviations. Since our dandelion Population 1 is normally distributed, approximately 95% of the individuals will have stem lengths between 30 mm and 70 mm (2 standard deviations, or 20 mm, from the mean in either direction).

5. In cell B7, use the **NORMINV** function to obtain a stem length for Individual 1 in Population 1, whose mean and standard deviation are given in cell C3 and C4. Copy this formula down to obtain stem lengths for the remaining 499 individuals in Population 1.

We used the formula **=NORMINV(RAND(),$C$3,$C$4)**. This formula allows us to draw a random probability from a normal distribution whose mean is 50 and standard deviation is 10, and convert it to a data point from the same distribution. In this way we can assign stem lengths to each individual in Population 1 and end up with a population that has (approximately) the desired mean and standard deviation.

Let's look at the formula carefully. The NORMINV function consists of three parts, each separated by a comma. It has the form **NORMINV(probability, mean, standard_dev)**, where probability corresponds to the cumulative probability from the normal distribution, mean is the arithmetic mean of the distribution, and standard_dev is the standard deviation of the distribution. For example, the formula **=NORMINV(RAND(), $C$3,$C$4)** tells Excel to draw a random cumulative probability between 0 and 1 (the **RAND()** portion of the formula) from a normal distribution that has a mean given in cell C3 and a standard deviation given in cell C4. The formula returns the inverse of this probability; it changes the cumulative probability into an actual number from the distribution. Excel will return a value, which is the stem length of the individual.

6. Copy cells B7–B506 and paste their values in place of the formulae.

Now we need to "fix" the stem lengths for Population 1 in cells B7–B506. (Otherwise, Excel will generate new stem lengths for Population 1 every time the spreadsheet recalculates its formulae).

Copy cells B7–B506.
Select cell B7.
Go to Edit | Paste Special | Paste Values. The **NORMINV** formula will be overwritten and the values will occupy the cells.

7. Enter 0 in cell C7, and fill this value down to cell C256. In cell C257, enter 100 and fill this value down to cell C506.

Population 2 also has a mean stem length, μ, of 50 mm. Stem lengths in this population are highly variable, where individuals either have a very long stem of 100 mm or no stem at all (0 mm).

8. Label cell A507 as "Mean" and cell A509 as "Std" as shown in Figure 5.

	A	B	C
507	**Mean =**		
508	**Std =**		

Figure 5

9. Calculate the mean stem lengths and standard deviation for the two populations in cells B507–C508.

10. Save your work.

We used the following formulae:

- Cell B507 =**AVERAGE(B7:B506)**
- Cell B508 =**STDEV(B7:B506)**
- Cell C507 =**AVERAGE(C7:C506)**
- Cell C508 =**STDEV(C7:C506)**

Note that both populations have approximately the same mean, but are very different in terms of how stem lengths are distributed in the population.

B. Construct a frequency distribution of the raw data.

1. Set up new column headings as shown in Figure 6. Enter values in cells F7–G16.

	F	G	H	I
3	**Frequencies of Values in Populations**			
4				
5				
6	**"Bin"**	**Stem lengths**	**Pop 1**	**Pop 2**
7	9	<10	0	250
8	19	<20	0	0
9	29	<30	6	0
10	39	<40	64	0
11	49	<50	153	0
12	59	<60	186	0
13	69	<70	79	0
14	79	<80	12	0
15	89	<90	0	0
16		<100	0	250

Figure 6

2. Use the **FREQUENCY** formula to generate the frequencies of the various stem lengths in Population 1. For example, in cell H7, count the number of individuals in Population 1 whose stem lengths are <10 mm. In cell H8, count the number of individuals whose stem lengths are within 10 and 19 mm, and so on.

The most common way to depict a population's values is through a **frequency distribution**. A frequency distribution is a plot of the raw data, which we can generate using Excel's **FREQUENCY** function. This is an array formula (see pp xxx) and is a bit tricky, so proceed carefully.

The **FREQUENCY** function calculates how often values occur within a range of values, and then returns an array (or series) of numbers. For example, you will use it to count the number of stems that fall within 0 and 9 mm, 10 and 19 mm, and all of the other potential categories listed in Figure 6. Because **FREQUENCY** returns an array, it must be entered as an array formula. The function has the syntax **FREQUENCY(data_array, bins_array)**, where data_array is a set of values for which you want to count frequencies, and bins_array is a reference to intervals into which you want to group the values. You can think of a "bin" as a bucket in which specific numbers go. The bins may be very small (hold only a few numbers) or very large (hold a large set of numbers). In our example, we used bins that hold 10 numbers each. For example, a bin labeled 9 holds numbers 0, 1, 2, 3, 4, 5, 6, 7, 8, and 9. The bin labeled 19 holds numbers 10, 11, 12, 13, 14, 15, 16, 17, 18, and 19. The bin labeled 89 holds numbers 80, 81, 82, 83, 84, 85, 86, 87, 88, and 89. Any data points greater than 89 go into a final "default" bin, which is not technically listed as a bin.

The **FREQUENCY** function works best when you use the f_x button and follow the cues for entering a formula. Since you will be entering this formula for an array of cells, the mechanics of entering this formula is different than the typical formula entry. Instead of selecting a single cell to enter a formula, you need to select a series of cells, then enter a formula, and then press <Control>+<Shift>+<Enter> (Windows) to simultaneously enter the formula for all of the cells you have selected. (Press the <Control>, <Shift>, and <Enter> keys in that order, making sure to hold the <Control> and <Shift> keys—or the key if you use a Mac—down until the <Enter> key is pressed.

OK, let's try it. Select cells H7–H16 (where we are building the frequency distribution for Population 1) with your mouse, then press the f_x button and select the **FREQUENCY** function. Click on the button just to the right of the Data_array box (the button with the little arrow pointing up and left; see Figure 9 on p. 11); this will allow you to indicate the cells with the appropriate data by selecting them with your mouse. Select all of the individuals in Population 1 (i.e., cells B7–B506 of your data array) and click again on the button just to the right of the box again to return to the Frequency dialogue box. Then use the button next to the Bins_array box to select cells F7–F15 for your bins. Instead of clicking OK, press <Control>+<Shift>+<Enter>, and Excel will return your frequencies for Population 1. After you've obtained your results, examine the formulas in cells H7–H16. Your formula should look like this: **{=FREQUENCY(B7:B506,F7:F15)}**. This formula will be identical in all of the cells. The { } symbols indicate that the formula is an array formula.

3. Obtain the frequencies for Population 2 in I7–I16.

Your formulae should be **{=FREQUENCY(C7:C506,F7:F15)}** in cells I7–I16.

4. Construct a frequency histogram of the two populations. Select the data in G6–I16.
The data in the G column will form the x-axis, and the data in the H and I columns will make up the frequencies. Make sure you label your axes fully.

5. Save your work.

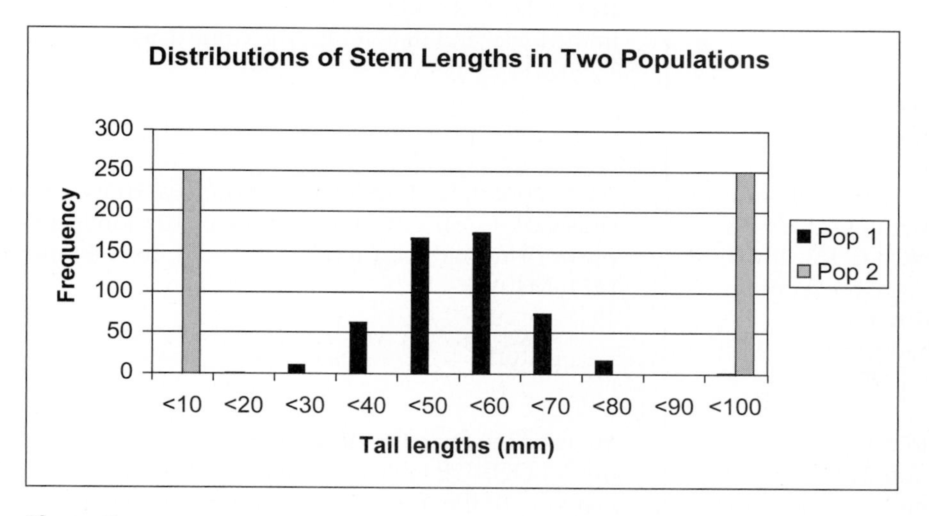

Figure 7

Based on Figure 7, it's easy to see that both populations have a mean around 50 mm, although their variances are quite different.

C. Obtain random samples from each population.

1. Set up new column headings as shown in Figure 8, but extend the series in row F to cell F40.

	F	G	H	I	J	K	L	M	N
19		n = 5		n = 10		n = 15		n = 20	
20	**Individual #**	**Pop 1**	**Pop 2**	**Pop 1**	**Pop 2**	**Pop 1**	**Pop 2**	**Pop 1**	**Pop 2**
21	1								
22	2								
23	3								
24	4								
25	5								

Figure 8

Remember that our goal is not to measure all 500 individuals in each population, but to sample from each population and estimate μ with a statistic. We will now randomly sample individuals from the population (with replacement), and estimate μ. We will do this for sample sizes of 5, 10, 15, and 20 individuals.

2. In cell F18, generate a random number between 1 and 500.

The random number will select which individuals from the population will be part of a sample. For example, if the random number is 324, then individual number 324 will be selected as part of the sample. Two formulae can be used to generate a random number between 1 and 500: **=RANDBETWEEN(1,500)** and **=ROUNDUP(RAND()*500,0)**

Press F9, the calculate key, several times to obtain new random numbers in cell F18.

3. Enter a formula in cell G21 to return the stem length of a random individual in Population 1.

Now we will draw a random sample from Population 1, and output the individual's stem length in cell G1. We'll use the **VLOOKUP** formula, combined with the **RANDBETWEEN** (or **ROUNDUP(RAND()** formula) above, to accomplish this task. The **VLOOKUP** formula searches for a value in the leftmost column of a table you specify (in this case, the table consists of cells A7–B506; the leftmost column is column A, which gives the individual's number). The function finds the individuals number, then returns a value associated with that individual from a different column in the table (in this case, the stem length associated with the randomly drawn individual).

Enter one of the following formulae (depending on whether or not you have the **RANDBETWEEN** function) in cell G21: **=VLOOKUP(RANDBETWEEN(1,500),$A$7:$B$506,2)** *or* **=VLOOKUP((ROUNDUP(RAND()*500,0),$A$7:$B$506,2)**. This formula tells Excel to generate a random number between 1 and 500 (the **RANDBETWEEN** or **ROUNDUP(RAND)** portion of the formula), find that number in the left-hand column in the table, and then return the value listed in the second column of the table.

4. Copy cell G21 into cells I21, K21, and M21.

5. Copy the formula in G21 down to G25. Copy the formula in I21 down to I30. Copy the formula in K21 down to K35. Copy the formula in M21 down to M40.

At this point, for Population 1, you have drawn a random sample of 5 individuals (in cells G21–G25), a random sample of 10 individuals (in cells I21–I30), a random sample of 15 individuals (in cells K21–K35), and a random sample of 20 individuals (in cells M21–M40).

6. Obtain samples from Population 2 and output stem lengths in the appropriate cells.

We used the formula **=VLOOKUP(RANDBETWEEN(1,500),$A$7:$C$506,3)**. Note that our VLOOKUP table now includes columns A through C, and returns the value associated with the third column of data (stem lengths from Population 2).

Your spreadsheet should now look like Figure 9 (the values in the cells will be different).

	F	G	H	I	J	K	L	M	N
19		n = 5		n = 10		n = 15		n = 20	
20	Individual #	Pop 1	Pop 2	Pop 1	Pop 2	Pop 1	Pop 2	Pop 1	Pop 2
21	1	49	0	37	0	51	100	34	100
22	2	46	0	54	0	62	0	69	100
23	3	43	0	32	0	69	100	58	100
24	4	52	0	51	0	49	0	62	100
25	5	46	100	52	100	62	100	28	0
26	6	4		8	0	48	100	45	0
27	7	2		6	100	58	0	54	100
28	8	4		9	0	33	100	56	0
29	9	5		8	0	31	0	32	100
30	10			62	100	54	0	39	100
31	11					46	100	42	100
32	12					63	0	62	0
33	13					46	100	44	0
34	14					54	0	58	100
35	15					44	0	41	100
36	16							45	100
37	17							59	0
38	18							45	0
39	19							69	100
40	20							50	0

Figure 9

7. Calculate the mean for each population and each sample size in cells G41–N41.

Enter **=AVERAGE(G21:G40)** in cell G41. Copy this formula over to cell N41. Now you have an estimator of the mean for each population when various sample sizes (*N*) are taken.

8. Save your work.

D. Set up the bootstrap.

The central limit theorem says that if we repeat this process many times and construct a graph of the frequency distribution of our *sampling means*—or estimates—the average of that sampling distribution will in fact be close to μ, the actual mean stem length of the population. So far, you've run one "trial." To make a sampling distribution of the means, you'll want to run several trials with a bootstrap analysis. We'll do 25 trials in this exercise, which should be just enough to show you the general principles of the central limit theorem. (You can do more trials if you'd like.)

1. Set up new column headings as shown, but extend the trials to 25 in cell F69.

	F	G	H	I	J	K	L	M	N
43		*n* = 5		*n* = 10		*n* = 15		*n* = 20	
44		**Pop 1**	**Pop 2**	**Pop 1**	**Pop 2**	**Pop 1**	**Pop 2**	**Pop 1**	**Pop 2**
45	**Trial 1**								
46	**Trial 2**								
47	**Trial 3**								
48	**Trial 4**								
49	**Trial 5**								

Figure 10

2. Develop a bootstrap macro.

The following steps will create a bootstrap macro:

- Open Tools | Options | Calculation and set the calculation key to manual.
- Open Tools | Macro | Record New Macro. A dialog box will appear. Type in a name (bootstrap) and a shortcut key (<Control>+b).
- Press F9, the calculate key, to generate a new set of random samples from both populations.
- Select cells G41–N41, the estimators of μ for various sample sizes.
- Open Edit | Copy. Select cell G44.
- Open Edit | Find. A dialog box will appear. Leave the Find What box completely blank. Search by columns and look in values, then select Find Next and then Close. Your cursor should move down to cell G45 (the next blank cell in that column).

- Open Edit | Paste Special | Paste Values. Select OK.
- Open Macros | Stop Recording (or, if the Stop Recording menu is visible, press the Stop Recording button).
- Open Tools | Options | Calculation and return your calculation to automatic.

3. Save your work.

Your bootstrap macro is finished. When you press <Control>+b 24 more times, you will have resampled your population and computed new means for 25 different trials. This is the bootstrap analysis.

E. Construct a Sampling Distribution of the Means.

1. Set up column headings as shown in Figure 11.

	E	F	G	H	I
72		**Frequency of Estimated Mean**			
73	"Bin"	n = 5	n = 10	n = 15	n = 20
74	40				
75	41				
76	42				
77	43				
78	44				
79	45				
80	46				
81	47				
82	48				
83	49				
84	50				
85	51				
86	52				
87	53				
88	54				
89	55				
90	56				
91	57				
92	58				
93	59				
94	60				
95	>60				

Figure 11

2. Use the **FREQUENCY** function to count the frequency in which certain values (estimators) were obtained for Population 1 for various sample sizes.

We used the following formulae:

- **F74–F95 {=FREQUENCY(G45:G69,E74:E94)}**
- **G74–G95 {=FREQUENCY(I45:I69,E74:E94)}**
- **H74–H95{=FREQUENCY(K45:K69,E74:E94)}**
- **I74–I95 {=FREQUENCY(M45:M69,E74:E94)}**

3. Construct a sampling distribution of the *means* (Figure 12) by plotting the results from the previous step.

For clarity, we have graphed only the cases $N = 5$ and $N = 20$. Your own graph will look different.

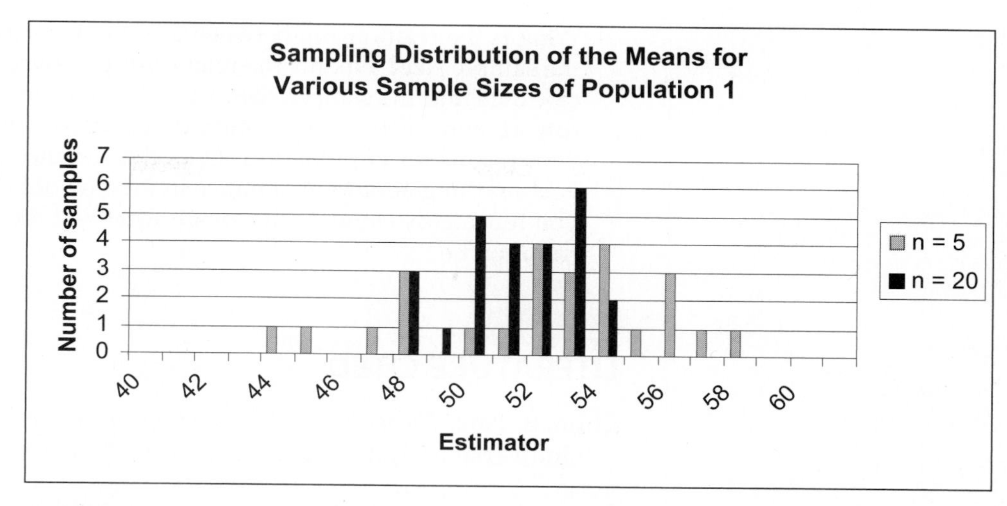

Figure 12

QUESTIONS

1. Examine your graph from Part E, Step 3. How does N, the sample size, affect the sampling distribution's mean and variance?

2. Repeat Part E for Population 2. Set up column headings and bins as shown in Figure 13. Explain why different bins are necessary for this population. Population 2 has a very strong bimodal distribution. Does the sampling distribution at $N = 20$ also have a bimodal shape? How does the shape of the sampling distribution change as sample size changes?

	J	K	L	M	N
71			Population 2		
72		Frequency of Estimated Mean			
73	"Bin"	n = 5	n = 10	n = 15	n = 20
74	0				
75	10				
76	20				
77	30				
78	40				
79	50				
80	60				
81	70				
82	80				
83	90				
84	100				

Figure 13

3. Review the definition of the central limit theorem (given at the top of Page 67). How close was the average of your bootstrap analyses to μ? How did sample size affect this? Did the two populations show similar results? Why or why not?

4. What is the relationship between the standard error of the sample means and the sample size? What is the relationship between the standard deviation of the raw data and the sample size? Calculate the standard deviation of samples in row 42, and calculate the standard deviation of your 25 trials in row 70. Plot your results for Population 1. Does the variance in the sampling distribution tell you anything about the variance in the raw data? If your sample size is 1 and you repeatedly estimate the mean, what will the variance of your sampling distribution be?

LITERATURE CITED

Efron, B. 1982. *The Jackknife, the Bootstrap, and Other Resampling Plans*. Society for Industrial and Applied Mathematics, Philadelphia.

Sokal, R. R., and F. J. Rohlf. 1995. *Biometry*. 3rd Edition. W. H. Freeman & Co., New York.

5 HYPOTHESIS TESTING: ALPHA, BETA, AND POWER

Objectives

- Understand the concepts of statistical errors, sample variability, and effect size.
- Explore the interplay among alpha, beta, effect size, sample variability, and the power of test.

Suggested Preliminary Exercise: Statistical Distributions

INTRODUCTION

Much research in ecology involves making statistical tests of one kind or another. We frequently want to know if two or more populations differ from one another with respect to some parameter that they share. For example, are trees from one forest "significantly" larger than those from another? Do older rabbits have thicker coats than younger rabbits? Is the species diversity of the restored prairie different from that of the degraded one? These comparisons generally involve estimating the value of a parameter in each population using data obtained through sampling. Typically, these estimates are compared using a statistical test to identify a difference or lack thereof.

Sampling and Uncertainty

Because sampling always involves some uncertainty (with sampling we are never entirely sure that we have properly estimated the true value of the parameter for the population), we have to consider the possibility that any difference that we see between two estimated parameters could result from sampling flukes. That is, the populations we sampled don't actually differ, but we drew unrepresentative samples by chance that give the incorrect appearance of a difference. This is a **Type I** statistical error. The probability of committing a Type I error is called **alpha (α)**.

Alternatively, the populations we are interested in may actually be different, but from some fluke in sampling we drew two samples that showed no differences. This is a **Type II** error. The probability of committing a Type II error is known as **beta (β)**. Type II errors may occur because the actual difference between the populations (the "effect size") is small and the variability in our samples obscures the difference and prevents us from detecting it. We are obviously more likely to detect a difference between populations the more precisely we have estimated the parameters in each of them (perhaps because we sampled each population well) or where the difference is substantial enough to detect despite the variability in samples.

Thus, the major challenge in performing a statistical test is simple: Ensure that you don't commit a type I or II error and thereby confidently detect any differences that might exist (Sokal and Rohlf, 1981). You can guard against committing a Type I error by using an appropriately stringent α level, say, 0.05 or lower. Guarding against a Type II error can be more problematic. A test that will detect differences if they exist, regardless of the sample variability and the effect size, is said to be have high statistical **power**. Power = $1 - \beta$, so a low β (probability of missing an important difference) equates to a high power of the test. Statistical power of the test is an important concept because ensur-

Summary of Type I and Type II Errors, and Power

Suppose we sample coat thickness of two populations of rabbits. The **null hypothesis (H_0)** is that the groups do not differ in coat thickness. We hope to gather evidence to reject the null hypothesis at a given probability level (α). If H_0 is in fact true and the populations do not actually differ in coat thickness, but you reject H_0 and conclude that the populations are different, you have committed a Type I error. If H_0 is false and the populations have different coat thicknesses but you fail to reject the H_0, you have committed a Type II error; your sampling lacked power to detect actual differences. **Power** is the probability of rejecting H_0 when it is in fact false.

	Reject H_0	**Fail to reject H_0**
H_0 is true:	Type I error (α)	Correct decision. Other ideas?
H_0 is false:	Correct decision. Nobel Prize!	Type II error (β)

ing that a given test has high power means that it will accomplish what you hope it will: that is, it will detect differences should they exist. All too often we regard a lack of difference, as indicated by a nonsignificant result on a test, to reflect no real difference between populations, when it may actually be the result of a poorly designed study (too variable or too small a sample to detect a subtle difference that nonetheless exists).

Choosing acceptable α and β values is worth additional consideration. Standard biological literature generally sets α to 0.05 and β at 0.2. In many cases, it may make sense to use other values. If the goal is to detect important differences, perhaps doing so at the risk of an increased level of false detections, then designing a test using high α and a low β (high power) would be advisable. This might be the case, for example, in looking for trends in a population of an endangered species. You want to quickly detect any declines in the species so you can step in and do something about them, but you are comfortable exploring some false reports of declines should they occur. On the other hand, if wrongly detecting a difference is very costly, then you might want to use a low α to guard against committing a Type I error. The important message is that "statistical significance" is only relative to the levels of α and β that you consider to be reasonable and that you set in advance.

The purpose of this exercise is to enable you to explore the interplay among α, β, effect size, sample variability, and the power of test. If you clearly understand the trade-offs among these parameters, you will greatly enhance your ability to design appropriate sampling schemes for detecting differences, should they exist, among populations. As always, save your work frequently to disk.

INSTRUCTIONS

ANNOTATION

A. Set up and sample two model populations.

1. Open a new spreadsheet and set up column headings as shown in Figure 1.

Generate the α symbol by typing an "a." Select the letter in the formula bar and change the font to symbol font.

	A	B	C	D	E	F	G
1							
2		TYPE I ERRORS			TYPE II ERRORS and POWER		
3							
4		mean	sd			mean	sd
5	Pop 1 =>	50	5		Pop 1 =>	45	5
6	Pop 2 =>	50	5		Pop 2 =>	50	5
7	Effect size =	0			Effect size =	5	
8							
9	Individual	Pop 1	Pop 2		Individual	Pop 1	Pop 2
10	1				1		
11	2				2		
12	3				3		
13	4				4		
14	5				5		
15	6				6		
16	7				7		
17	8				8		
18	9				9		
19	10				10		
20							
21	mean				mean		
22	std				std		
23							
24	α=	t-test:	significant?		α=	t-test:	significant?
25	0.05				0.05		
26							
27							
28	trial	t-test	significant?		trial	t-test	significant?

Figure 1

2. Enter the values shown in cells B5–C6.

We'll start by exploring Type I errors in columns A, B, and C. We'll make a statistical comparison of two populations (columns B and C) that have identical means and variances. Enter 50 in cells B5 and B6 to indicate a mean value of the population, say, height. Enter a standard deviation of 5 in cells C5 and C6. Thus, both populations have the same mean (μ) and standard deviation (σ^2) in height (of course, you don't really know these are the true means and variances of the populations; you will sample individuals to estimate these parameters).

3. In cell B7, calculate the **effect size** as the difference between the means of the two populations. Save your work.

Enter the formula **=ABS(B5–B6)** in cell B7. In this case, the effect size is 0.

4. In cell B10, enter the formula **=NORMINV (RAND(),$B$5, $C$5)**. Copy the formula down to cell B19.

Now we will "sample" 10 individuals from population 1 by generating random measurements as if they came from a population with a normal height distribution. We can use the **NORMINV** function and **RAND** function to do this. The **NORMINV** function returns the inverse of the normal cumulative distribution for the specified mean and standard deviation, and has the form **NORMINV(probability,mean,standard_dev)**. The B10 formula tells Excel to draw a random probability (the **RAND()** portion of the formula) from a normal distribution with a mean height given in cell B5 and a standard deviation given in cell C5; Excel will convert that random probability into a value (height) from that distribution.

5. In cells C10–C19, obtain 10 samples from population 2.

Obtain heights of individuals for 10 individuals drawn at random from Population 2. We used the formula **=NORMINV(RAND(),$B$6,$C$6)** in cells C10–C19 (following the procedure in Step 4).

6. In cells B21 and C21, enter a formula to calculate the mean of your sample for populations 1 and 2, respectively.

In cell B21 we used the formula **=AVERAGE(B10:B19)**.
In cell C21 we used the formula **=AVERAGE(C10:C19)**.

7. In cells B22 and C22, enter a formula to calculate the standard deviation of your sample for population 1 and 2, respectively. Save your work.

Enter the formula **=STDEV(B10:B19)** in cell B22.
Enter the formula **=STDEV(C10:C19)** in cell C22.

B. Conduct a* t*-test to determine if samples from populations 1 and 2 differ in height.

1. Enter 0.05 in cell A25.

In cell A25, you need to specify what α will be. By convention, $\alpha = 0.05$ is used. Remember that α is the probability of committing a Type I error—rejecting the null hypothesis when the null hypothesis is in fact true. In the next step you will generate a *t*-test statistic and a probability associated with that test statistic. If the test statistic has a probability that is less than or equal to the α level you have selected, you would conclude that the two populations are different. If the test statistic has a probability that is greater than the α level you have selected, you would conclude that the populations are not statistically different. You can set α to any level you like (although $\alpha > 0.15$ will raise eyebrows). For now, we will use the conventional $\alpha = 0.05$, and will change α levels later in the exercise.

2. In cell B25, use the **TTEST** function to conduct a *t*-test on the two population sample means.

Enter the formula **=TTEST(B10:B19,C10:C19,2,2) in cell B25.**
Now that you have determined what kind of Type I error rate you can live with, you're ready to perform a *t*-test to compare the sample means of the two populations. The **TTEST** formula returns the *probability* associated with a Student's *t*-Test (it does not return the value of the test statistic itself). You will use **TTEST** to determine whether the two samples are likely to have come from two underlying populations that have the same mean. The **TTEST** formula has the form **TTEST(array1,array2,tails,type).** Array1 is the first data set (or the 10 individuals sampled from population 1), Array 2 is the second data set (or the 10 individuals sampled from population 2), tails refers to whether you want to conduct a one- or two-tailed test (choose 2), and type is the kind of *t*-test to perform (for now, choose two-sample equal variance).

3. In cell C25, enter an **IF** formula to return a 0 if your *t*-test statistic is greater than alpha, and a 1 if your *t*-test statistic is less than alpha.

Enter the formula **=IF($B$25>$A$25,0,1)** in cell C25.
Now that you have a test statistic probability, you need to compare it to the α level you've chosen. If the probability of the test statistic is <0.05 (your α level), you would conclude that the two populations are different. If the test statistic probability is >0.05 you would conclude the populations are not different (or, more correctly, the samples failed to show differences). The **IF** formula returns one value if a condition you specify is true, and another value if the condition you specify is false. It has the syntax **IF(logical_test,value_if_true,value_if_false)**. A score of 1 indicates that the two populations are statistically different; a score of 0 indicates they are not statistically different. Based on your test, what conclusions can you make about the two populations?

C. Run 100 sampling trials.

1. Set up a linear series from 1 to 100 in cells A29–A128.

Enter **1** in cell A29.
Enter the formula **=1+A29** in cell A30. Copy this formula down to cell A128.
A value of $\alpha = 0.05$ means that if you ran your *t*-test on samples (new samples) over and over again, about 5 times in 100 you would conclude that the two populations are different when in fact they are identical. We'll prove that to ourselves by running a number of trials in which we randomly draw 10 individuals from each population, calculate their means, run a *t*-test, and determine if the two populations are statistically different or not.

2. Under Trial 1 in cells B29–C29, re-enter by hand the results you obtained in cells B25 and C25.

Now that you've run your first trial and recorded your results, you are ready to run 99 more trials.

3. Switch to Manual Calculation.

Under Tools | Options | Calculation, select Manual Calculation.

4. Write a macro to run 99 more trials and record results in cells B30–C128.

Open the macro program and assign a shortcut key (refer to Exercise 2 for details on building macros). In Record mode, perform the following tasks:

- Select Tools | Macro | Record New Macro. Name your macro and assign it a shortcut key. For example, you might name your macro Type_I and assign it the shortcut "control t". Every keystroke you now make will be recorded as part of the macro.
- Press F9, the calculate key, to obtain new random samples from Population 1 and Population 2.
- Use your mouse to highlight cells B25 and C25, the new *t*-test statistic probability and significance result, and open Edit | Copy.
- Highlight cell B28, then go to Edit | Find. A dialog box will appear. You want to leave the Find What box completely blank, and search by columns. Click the Find Next button, then Close. Excel will move your cursor to the next blank cell in column B.
- Select Edit | Paste Special | Paste Values.
- You're finished. Select Tools | Macro | Stop Recording. Now when you press your shortcut key 99 times, your new results will automatically fill into the appropriate cells. Run your macro until you have results from 100 trials.

5. Save your work.

Our first five results looked like Figure 2; yours will very likely look different.

	A	B	C
28	**trial**	**t-test**	**significant?**
29	1	0.602499	0
30	2	0.910298643	0
31	3	0.70263163	0
32	4	0.810947176	0
33	5	0.099062869	0

Figure 2

D. Calculate Type I error rate.

1. Set up new headings as shown in Figure 3:

	A	B	C
130	**number of tests showing**		
131		**significant differences = >**	
132		**probability of Type I error = >**	

Figure 3

Switch back to automatic calculation, and visually inspect the t-test probabilities you obtained in your trials. Most of the results should indicate that the two populations are not statistically different from each other. Occasionally, however—about 5 times in 100—you will conclude that the two populations are different even though they have exactly the same mean height (μ) and standard deviation (σ^2). These are Type I errors. By a sampling fluke, you concluded the populations were different when in fact they are not.

2. In cell C131, use the **SUM** function to count the number of Type I errors committed.

We used the formula **=SUM(C29:C128)**.

3. In cell C132, calculate the Type I error rate as the number of Type 1 errors divided by 100 trials.

We used the formula = **C131/100**.
Your answer should be somewhat close to 0.05 because you established a Type I error rate of 0.05 in cell A25.

4. Save your work, and answer Question 1 at the end of the exercise.

E. Type II errors and power.

1. Enter values shown in cells F5–G6 (see Figure 1).

Now let's switch gears and think about Type II errors, which we'll deal with in Columns E, F, and G. Let's assume that the two populations really have different underlying distributions in terms of height. In cell F5, enter 45 to indicate that population 1 has an average height (μ) of 45 mm and a standard deviation (σ^2) of 5 mm (entered in cell G5). In cell F6, enter 50 to indicate that population 2 has an average height (μ) of 50 mm and a standard deviation (σ^2) of 5 mm (entered in cell G6). The effect size is entered in cell F7 as **=ABS(F5-F6)**. Although the effect size may seem small, these differences in height might be biologically meaningful, and you'd like to know this.

2. Calculate the effect size in cell F7.

3. Enter 0.05 in cell E25.

Set $\alpha = 0.05$ in cell E25.

4. Obtain samples from your population, and run 100 trials as you did earlier. You will need to create a new macro to keep track of results from 100 trials in cells F29–G128.

You'll sample from these populations, calculate a t-test, determine if you conclude the two populations are statistically different or not, and run 100 trials in total. Your spreadsheet columns E, F, and G should look like columns A, B, and C in appearance, although you will be sampling from different populations. In case you get stuck, the formulae we used are given at the top of the next page:

- F10 – F19 **=NORMINV(RAND(),$F$5,$G$5)**
- G10 – G19 **=NORMINV(RAND(),$F$6,$G$6)**
- F21 **=AVERAGE(F10:F19)**
- G21 **=AVERAGE(G10:G19)**
- F22 **=STDEV(F10:F19)**
- G22 **=STDEV(G10:G19)**
- F25 **=TTEST(F10:F19,G10:G19,2,2)**
- G25 **=IF($F$25>$E$25,0,1)**

5. Set up headings as shown in Figure 4.

	E	F	G
130	number of tests NOT showing		
131		significant differences =>	
132	probability of Type II error = β=>		
133		Power = 1 - β =>	

Figure 4

6. In cell G131, use the **COUNTIF** formula to count the number of tests *not* showing a significant difference.

Remember that the populations really are different biologically, and we're trying to determine if they are different based on our samples. The **COUNTIF** formula counts the number of cells within a range that meet a given criterion. It has the syntax **COUNTIF(range,criteria)**. We used the formula **=COUNTIF(G29:G128,0)** to count the number of times our *t*-test was not significant. These are the Type II errors. By a sampling fluke, you concluded that the populations are not different when in fact they are.

7. Calculate the probability of a Type II error (β) in cell G132.

Remember that a Type II error is falsely concluding that the two populations are similar when in fact they are different. Enter the formula **=G131/100** in cell G132. Is your Type II error rate acceptable, or is it too high for your liking?

8. Calculate power as 1– β in cell G133.

Enter the formula **=1– G132** in cell G133.
Scientists usually calculate the **power** of their design to detect differences assuming that they really exist, rather than reporting the probability of a Type II error. Remember that power is simply 1– β.

9. Save your work, and answer Questions 2–6.

QUESTIONS

1. If you change α in cell A25 to 0.1, approximately how many Type I errors are you likely to make if you run 100 trials again? How many Type I errors are you likely to commit if you set α to 0.01?
2. How does decreasing the standard deviation of the two populations affect Type II error rates and power? Enter 1 in cells G5 and G6. Press F9, the calculate key, 20 times and examine the significance of your 20 *t*-tests in cells F25 and G25. Keep track of the number of Type II errors out of 20 trials.
3. How does increasing the standard deviation of the two populations affect Type II error rates and power? Enter 10 in cells G5 and G6. Press F9 20 times and keep track of the number of Type II errors out of 20 trials.
4. How does effect size influence Type II error rates? Enter 45 in cell F5 and enter 55 in cell F6 (effect size = 10). Enter 5 in cells G5 and G6. Press the F9 key 20 times and keep track of the number of Type I and Type II errors out of 20 trials.
5. Does changing the α level in cell E25 affect β or power? Clear your macro results in cells F29–F128 and run 100 trials with varying α levels. Interpret your results.

6. How does sample size affect Type I and Type II error rates? Set cells B5–B6 and cells F5–G6 back to their original values. Then, develop a new model with population sizes of 1000 individuals, and compare the Type I and Type II error rates for populations of size 10 (currently modeled) with your new populations.

LITERATURE CITED AND FURTHER READINGS

Johnson, D. H. 1999. The insignificance of statistical significance testing. *Journal of Wildlife Management* 63(3): 763–772.

Sokal, R. R. and F. J. Rohlf. 1981. *Biometry*, 2nd Edition. W. H. Freeman, New York.

Taylor, B. L. and T. Gerrodette. 1993. The uses of statistical power in conservation biology: The vaquita and northern spotted owl. *Conservation Biology* 7: 489–500.

6 SAMPLING SPECIES RICHNESS

Objectives

- Simulate a population of 1000 individuals composed of various species.
- Calculate species richness by sampling.
- Determine how community composition affects species richness estimates.
- Develop a bootstrap analysis of how sample size affects species richness estimates.

INTRODUCTION

Imagine you are a conservation biologist conducting surveys of insect species in previously unstudied areas. Your mission is to estimate the number of species occurring in different habitat types across a large region. The number of species that occurs in a particular area is called its **species richness**, and it is just one of many measures of biodiversity. A practice known as a *rapid biodiversity assessment* is currently being used by many conservation organizations to survey the biodiversity of plants and animals before pristine habitats are altered and developed (see, for example, http://www.conservation.org/RAP/Default.htm). Assume there are 10 locations that must be sampled in a short period of time. How many samples should you take at each site to estimate the number of insect species in a location before moving onto the next location? Time and funding are short and you will not be able to do a complete survey of the insect biota.

A basic problem is that it is nearly impossible to count every single species in a community. If funding and time were unlimited, you might conduct a complete census and enumerate all of the species in the community. However, this is not often the case; instead you must settle for *sampling* the community and estimating its species richness based on this sample of individuals. Estimating species richness by sampling presents some major challenges. First, you are likely to miss some species. And second, although the more you sample in a particular area the more likely you are to find new, previously unsampled species, there is a point of diminishing returns that must be considered in your sampling efforts.

For example, consider a community that consists of 1000 insect species, and you sample insects by sweeping the vegetation with a net. In your first sweep, you capture 25 species. In your second sweep, you capture 30 species, but 20 of these were already captured in the first sweep. Thus, with 2 samples your total species richness is 35 (25 new species recorded with the first sweep, and 10 new species recorded with the second sweep). With each sweep (sample), the chances of adding a new, previously unsampled species decreases. At some point it becomes cost-effective to move

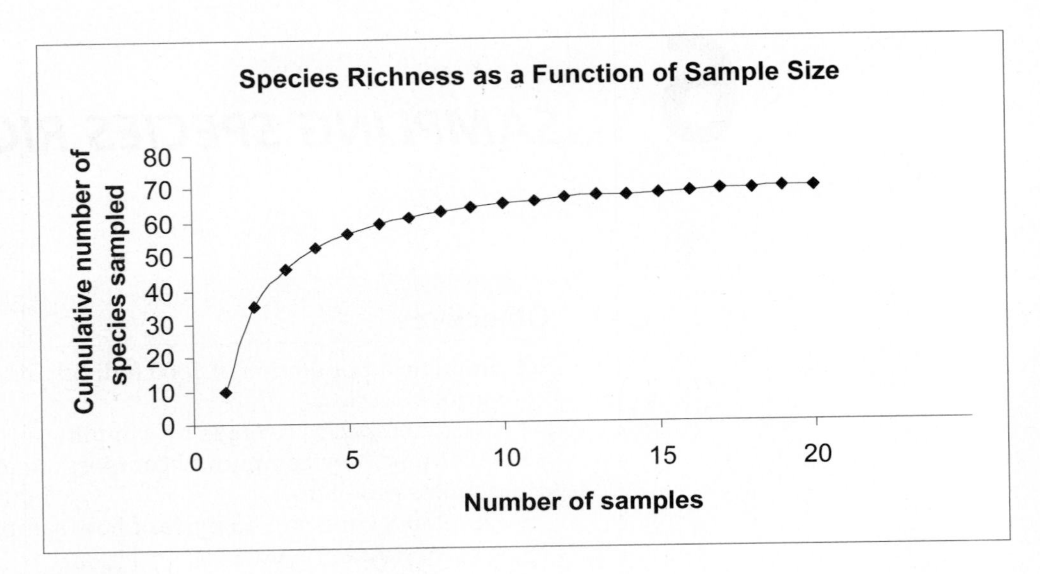

Figure 1

to the next location and start sampling anew. In the example shown in Figure 1, taking 15 samples will yield more or less the same species richness estimate as taking 18 or 20 samples.

What factors will determine the shape of a sampling curve such as Figure 1? One factor is the distribution of the individuals within the community. If the community consists of 100 species, but 90% of the total individuals are from species 1, most of our samples will consist of species 1, and we may have to take many samples to encounter one of the rarer species. In contrast, if the numbers of individuals in the community are more or less evenly distributed across 100 species, so that no single species dominates the community, you may not have to sample as much because all species are equally abundant.

Another general problem with sampling is that you will never *really* know how well your species richness estimate measured the *true* species richness in a community. After all, this is what you are trying to estimate with your sampling. With advances in computing, however, it is now possible to ask the question, *"If we take a different, random sample from a community with a known number of species, how does the species richness estimate change as sample size changes?"* The difference between the actual species richness of the community and the estimated species richness based on sampling is called **bias**.

One method for analyzing bias is a **bootstrap analysis**, which involves taking random samples of the data (with replacement so that the same individuals can be sampled more than once), calculating the parameter of interest (in this case, species richness), repeating the process for 1,000 or more trials for a given sample size, and then estimating the mean and standard deviation of species richness from the replicate bootstrap estimates. As discussed in Exercise 4, this process is relatively straightforward with spreadsheets.

Since the number of species in the community in your bootstrap analysis is known *a priori* (known beforehand), the bootstrap analysis gives you an indication of how sample size, as well as community composition, biases your estimate of species richness. The purpose of this exercise is to introduce you to sampling and bootstrap methods as they pertain to species richness. As always, save your work frequently to disk.

INSTRUCTIONS

ANNOTATION

A. Set up the model community.

1. Open a new spreadsheet and set up column headings as shown in Figure 2.

	A	B	C	D	E	F
1	***Sampling Species Richness***					
2						
3			**Tally**			
4	**Species**	**# in pop**	0			
5	1	100				
6	2	100				
7	3	100				
8	4	100				
9	5	100				
10	6	100				
11	7	100				
12	8	100				
13	9	100				
14	10	100		**<-- This number must equal 1000.**		
15						
16	**Total =**	1000				

Figure 2

We will consider a community in which there are 1000 total individuals and up to 10 different species. The species identification is given in cells A5–A14. The numbers of individuals of each species are given in cells B5–B14.

2. Enter the values shown in cells B5–B14.

To begin, let's consider a community that is evenly distributed with 100 individuals of each species. Later in the exercise, you will be able to change the composition of the community by altering the values in cells B5–B14.

3. In cell B16, enter a formula to sum the total number of individuals in the community.

Enter the equation **=SUM(B5:B14)** in cell B16. Your result should be 1000.

4. Graph the distribution of the 1000 individuals among the 10 species. Use a column graph, and label your axes fully (Figure 3).

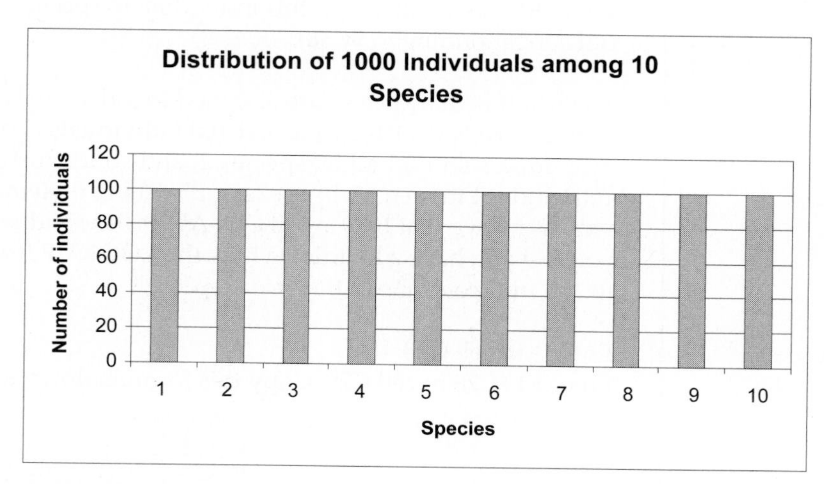

Figure 3

5. Compute a "running tally" of individuals in C4–C14.

Enter 0 in cell C4.
Enter the formula **=B5+C4** in cell C5 and copy this formula down to cell C14.
The formula in cell C5 gives the tally of individuals when only the first species, species 1, has been considered. Copying the formula down the column keeps a running tally of the number of individuals in the community as more species are observed. The result in cell C14 should be 1000, to account for all of the individuals present in the community. This "tally" will allow you to assign a species identification to individuals in a later step.

6. Save your work.

B. Sample from the community and compute species richness.

Now we are ready to sample from this community (one individual at a time) and estimate species richness. Since there are 10 species present (each with 100 individuals), species richness is 10. You will try to estimate this parameter by randomly sampling the population and computing richness.

1. Set up new spreadsheet headings as shown in Figure 4.

	A	B	C	D	E	F
26				Random sample		
27	Individual	Species	Sample size	Individual	Species	Richness

Figure 4

2. Set up a linear series from 0 to 999 in cells A28–A1027.

Enter 0 in cell A28.
Enter **=1+A28** in cell A29. Copy this formula down to cell A1027.
This series will represent the 1000 individuals in the community.

3. In cell B28, use the **LOOKUP** function to assign a species to the individual in cell A28. Copy this formula down to cell B1027.

Now we will identify which species each individual belongs to, based on the species identification (1–10) given in column A and the tally given in cells C4–C14. In cell B28, enter the formula **=LOOKUP(A28,$C$4:$C$14,$A$5:$A$14)**. The **LOOKUP** function looks up a value (the value in cell A28) in a vector that you specify (**$C$4:$C$14**), and returns a value from a corresponding vector (**$A$5:$A$14**). (A *vector* is a single row or column of values). In this case, it compares the value in cell A28 (which is 1) to the values in cells C4–C14; it finds that A28 is equal to 0 (the value in **$C$4**), so it returns the value in **$A$5**, which is 1. In other words, it assigns individual 1 to species 1. (Note that with this formula, the value in the tally and the species assignments are offset by one row.)

The **LOOKUP** function is handy for assigning species to individuals because if the function can't find the exact lookup value, it matches the largest value in the lookup vector (cells C4–C14) that is less than or equal to the lookup value. For example, when it looks for individual 449 in **$C$4:$C$14**, the largest value it can find that is less than 449 is 400, so it will assign this individual to species 5 (the value in **$A$9**, which is the cell corresponding to **$C$8**).

The result is that species are assigned to individuals with the distribution you determined in cells B5–B14. Your first 100 individuals should all be species 1, the next 100 individuals should all be species 2, and so forth. To test the function, set cell B6 to 1000 and set the remaining cells in B5–B14 to 0. Remember that the final tally of individuals must equal 1000 in cell C14. All 1000 individuals should now be species 2. When you feel you have a handle on how the **LOOKUP** function works, return cells B5–B14 to 100, and continue to the next step.

4. Set up a linear series from 1 to 1000 in cells C28–C1027.

Enter 1 in cell C28.
Enter **=1+C28** in cell C29. Copy this formula down to cell C1027.

5. In cells D28–D1027, generate a random number between 0 and 999 to designate a randomly sampled individual in the population.

Enter the formula **=ROUND(RAND()*1000,0)** in cell D28. Copy this formula down to cell D1027.

Cell D28 represents the first individual sampled, cell D29 represents the second individual sampled, and so on. Note that an individual can be sampled more than once if the same random number is drawn.

The **RAND()** function generates a random number between 0 and 1. When the random number is multiplied by 1000 and then rounded to 0 decimal places with the **ROUND** function, the result is a randomly sampled individual from the population. (If your program has the **RANDBETWEEN** function, the formula **=RANDBETWEEN(1,1000)** will do the same thing.)

6. In cell E28, enter a **LOOKUP** formula to identify the species of the randomly chosen individual in cell D28. Copy this formula down to cell E1028.

Enter the formula **=LOOKUP(D28,$A$28:$A$1027,$B$28:$B$1027)** in cell E28. Copy it down to cell E1028. Column E returns the species of each randomly selected individual. It uses another **LOOKUP** function to do this. The formula in cell E28 tells Excel to lookup the value in cell D28 (the randomly selected individual) in the vector of cells A28–A1027 and return this individual's species identification, given in cells B28–B1027.

7. Enter the number 1 in cell F28.

Finally we are ready to compute species richness—the total number of species—as our sampling progresses. Cell F28 is the first sample, so species richness will be equal to 1.

8. In cell F29, enter a nested **IF(COUNTIF()** formula to calculate the species richness, and copy this formula down to cell F1028.

With our second sample, we need to evaluate whether species richness is 1 (i.e., we sampled the same species in sample 2 as we did in sample 1) or 2 (i.e., we sampled a new species in sample 2). Enter the formula **=IF(COUNTIF($E$28:E28,E29)>0, F28,F28+1)** in cell F29. This is an **IF** formula with a **COUNTIF** formula nested within it. An **IF** formula has 3 parts to it, each separated by a comma. The first part is called the *criterion*. In this case, our criterion is **COUNTIF($E$28:E28,E29)>0.** The **COUNTIF** formula counts the number of times a certain value appears in a range of cells. Our formula tells the spreadsheet to examine cell E29 and count the number of times this value appears in the range of cells E28–E28. If this number is greater than 0 (the second sample was also recorded in the first sample), the program carries out the second part of the **IF** statement; if this number is not greater than 0, it carries out the third part of the **IF** statement. Thus, our example will look at the second species sampled (cell E29), and if this species number has appeared in the previous samples (E28–E28), the species richness value will remain at the previous number (cell F28); otherwise the richness will be increased by 1 (cell F28+1).

9. Graph species richness as a function of sample size. Use the scatter graph option, and label your axes fully (Figure 5).

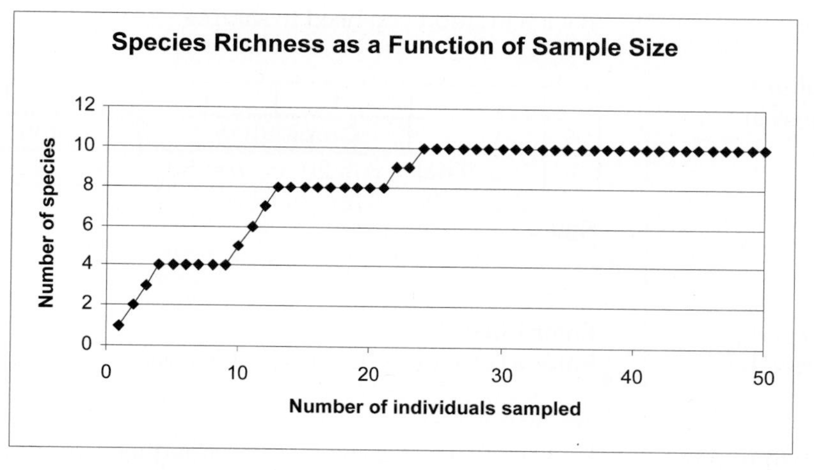

Figure 5

Your graph will look different than ours because your random samples likely differed than ours. Keep in mind that the actual species richness of the community is 10 species. In our example, 24 individuals needed to be sampled to arrive at this number.

10. Press F9, the calculate key, a number of times to generate new samples.

Pressing F9 will generate new random numbers, and hence a new set of individuals that are sampled. With each simulation, you will notice that your species richness estimates change as samples accumulate. For example, a new simulation required over 40 individuals to be sampled to generate an unbiased estimate of species richness (Figure 6).

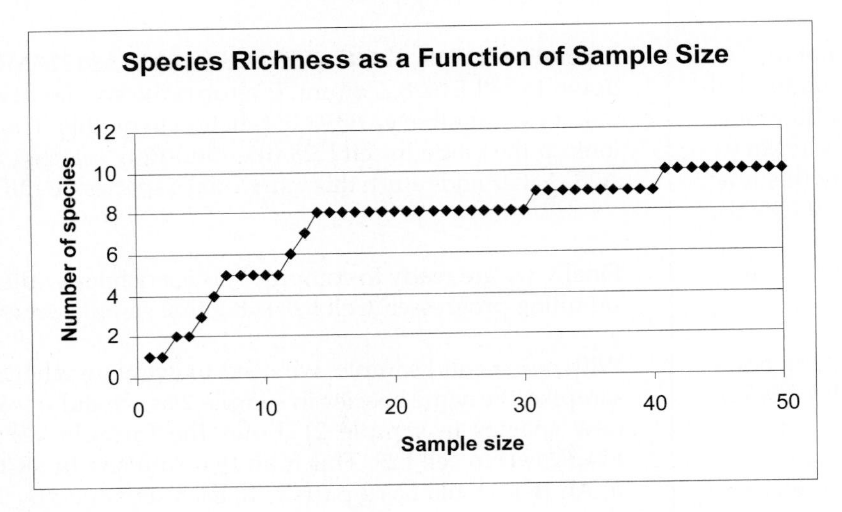

Figure 6

11. Save your work.

C. Set up the bootstrap.

The fact that each sampling simulation generates new and different results suggests the need for a bootstrap analysis. For example, if we took only 20 samples, how would our species richness estimate change from simulation to simulation? By "bootstrapping"—conducting many "replicate" sampling simulations—we can characterize the nature (mean and standard deviation) of our sampling with respect to species richness. We will do this for two of sample sizes (n = 20 and n = 50). We will run 1000 trials for each sample size, recording our species richness estimate with each simulation. This will provide useful information for deciding how many samples would be adequate at each location you need to sample.

1. Set up new column headings as shown in Figure 7.

	G	H	I	J	K
4		Community 1		Community 2	
5	Trial	n = 20	n = 50	n = 20	n = 50

Figure 7

2. Set up a linear series from 1 to 1000 in cells G6–G1005.

Enter 1 in cell G6.
Enter **=1+G6** in cell G7. Copy this formula down to cell G1005.

3. Create a macro to record species richness for sample size of 20 for 1000 trials.

First go to Tools | Options | Calculation and set your calculation key to Manual. Then put your Macro function in the "Record Macro" mode and assign a name and shortcut key. This macro provides one way to keep track of the species richness estimates when the sample size consists of 20 individuals. These estimates will be output into cells H6–H1005.

Record the following steps:

- Press F9, the calculate key, to generate a new set of random numbers, and hence a new set of randomly selected individuals.
- Select cell F47, the species richness estimate associated with a sample size of 20.
- Select Edit | Copy.
- Select cell H5, and then go to Edit | Find (Figure 8). Leave the Find What box completely blank; choose By Columns in the Search box and Values in the Look In box. Click Find Next and Close. Your cursor should move down to the next blank cell (trial 1).
- Go to Edit | Paste Special, and paste in Values, which is the species richness estimate for that trial.
- Select Tools | Macro | Stop Recording.

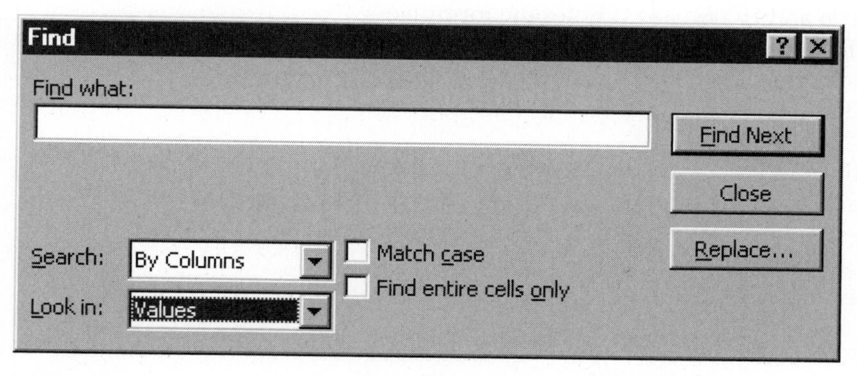

Figure 8

Now when you press your shortcut key, the macro will automatically conduct a new replicate sample and record the species richness values in the appropriate place. Run the macro 1000 times to complete your bootstrap analysis. This may take a while. If you like shortcuts, you can edit your macro's Visual Basic code by inserting two lines of code in the Visual Basic program, as follows:

- Open Tools | Macro | Macros.
- Click the Edit button to edit your macro called Trials. You should now see the Visual Basic code (Figure 9).

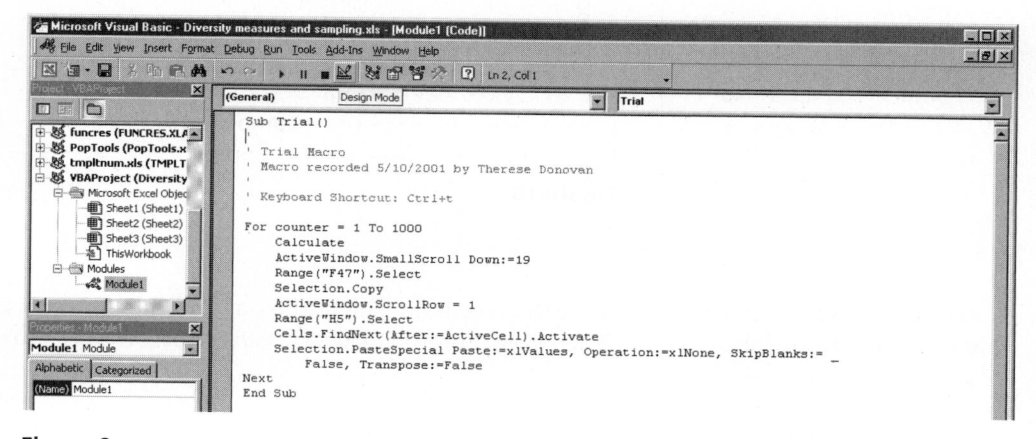

Figure 9

- Below line 4 (**Keyboard Shortcut**), enter a new line and type in the words **For counter = 1 to 1000** as shown in Figure 10.
- Above the last line (**End Sub**), enter a new line and type in the word **Next**.
- Exit the Visual Basic editor by clicking the close box in the upper right hand corner of the spreadsheet. You will be returned to your spreadsheet. Now when you press <Control>t, Excel will run 1000 trials for you.

4. Conduct a bootstrap analysis for a sample size of 50, and record the results of each bootstrap trial in column I.

You can record brand new macros, or edit the Visual Basic code in your existing macro. For the sample size of 50, you would highlight cell F77 (which is the species richness for a sample size of 50), and select cell I5 to record the results in the appropriate column. These slight adjustments can be made in the existing visual basic code. After you are finished, switch back to Automatic Calculation.

5. In cells H1006 and I1006, enter a formula to compute the mean species richness from the 1000 trials.

Enter the formulae

- H1006 **=AVERAGE(H6:H1005)**
- I1006 **=AVERAGE(I6:I1005)**

6. In cells H1007 and I1007, enter a formula to compute the standard deviation of species richness from the 100 trials.

Enter the formulae

- H1007 **=STDEV(H6:H1005)**
- I1007 **=STDEV(I6:I1005)**

7. In cells H1008 and I1008, enter a formula to divide the standard deviations by 2.

This step is necessary for graphing the standard deviations in the next step. Enter the formulae

- H1008 **=H1007/2**
- I1008 **=I1007/2**

8. Graph the mean species richness for the 1000 trials. Use a column graph and label your axes fully. Your graph should resemble Figure 10.

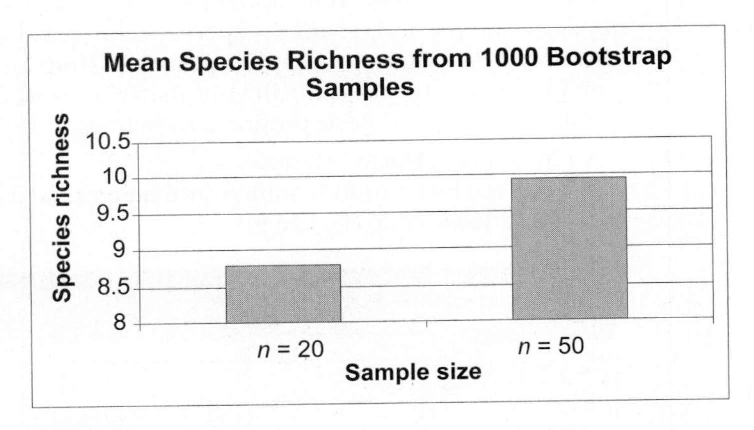

Figure 10

9. Add the standard deviation bars to your graph.

To add error bars, select the bars on the chart by clicking once on one of the bars. Then go to Format | Selected Data Series. A dialog box will appear (Figure 11).

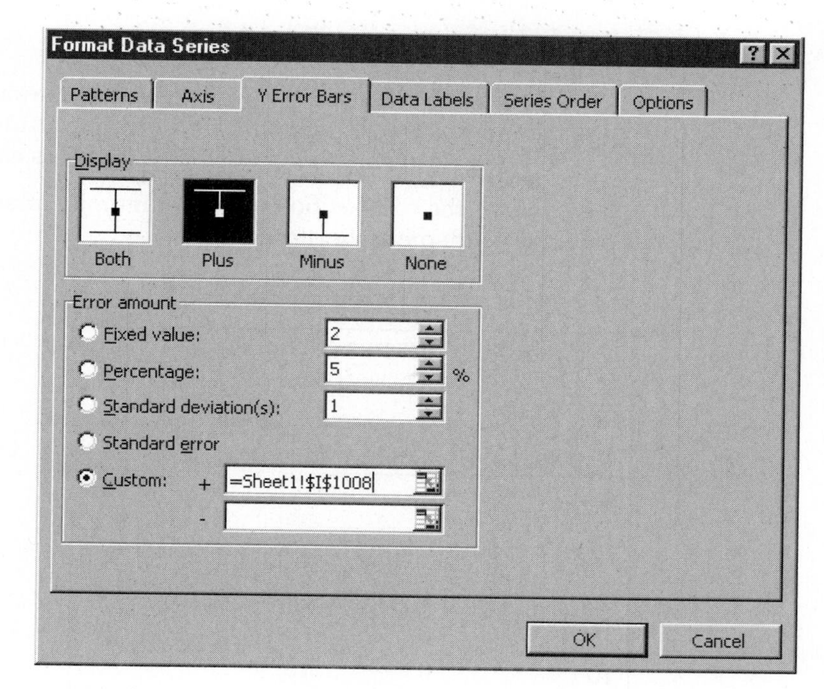

Figure 11

If you want to show only the top half of the errors, click on the Plus display, and then choose the Custom button. Then, in the window to the right of the + symbol, click on the little red arrow to shrink the box, use your mouse to select cell H1008, type in a comma, and use your mouse to highlight select cell I1008. Click again on the red arrow to bring the dialog box up again. Press OK and your graph should be updated (Figure 12). You should notice instantly that the larger sample size has a much smaller standard deviation than the smaller sample size, and that the larger sample provides a less biased estimate of species richness than the smaller sample. You must now consider the trade-offs between sampling a site intensively (n = 50 or more) at the expense of sampling a large number of sites.

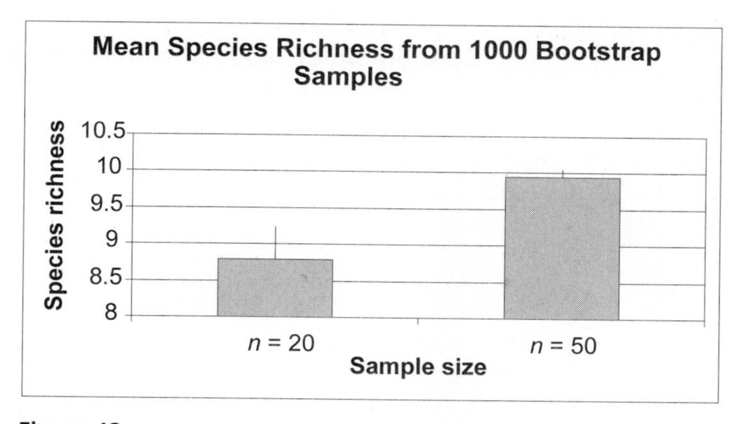

Figure 12

10. Save your work.

QUESTIONS

1. Fully interpret the last graph you created, the results of the bootstrap analysis for sample sizes of 20 and 50. Based on your results, is it worth sampling 50 individuals to ensure that your species richness estimate is unbiased?

2. How does the composition of the community affect species richness estimates? Set up your spreadsheet as follows:

	A	B	C
1	*Sampling Species Richness*		
2			
3			**Tally**
4	**Species**	**# in pop**	0
5	1	900	900
6	2	20	920
7	3	10	930
8	4	10	940
9	5	10	950
10	6	10	960
11	7	10	970
12	8	10	980
13	9	10	990
14	10	10	1000
15			
16	**Total =**	1000	

The new frequency distribution for species in this community should look like Figure 13. Develop a new macro, and sample from this new community with sample sizes of 20 and 50. Record your output under community 2 (columns J and K), and compare the bootstrap analysis for community 1 and community 2. Use graphs to explain your answer.

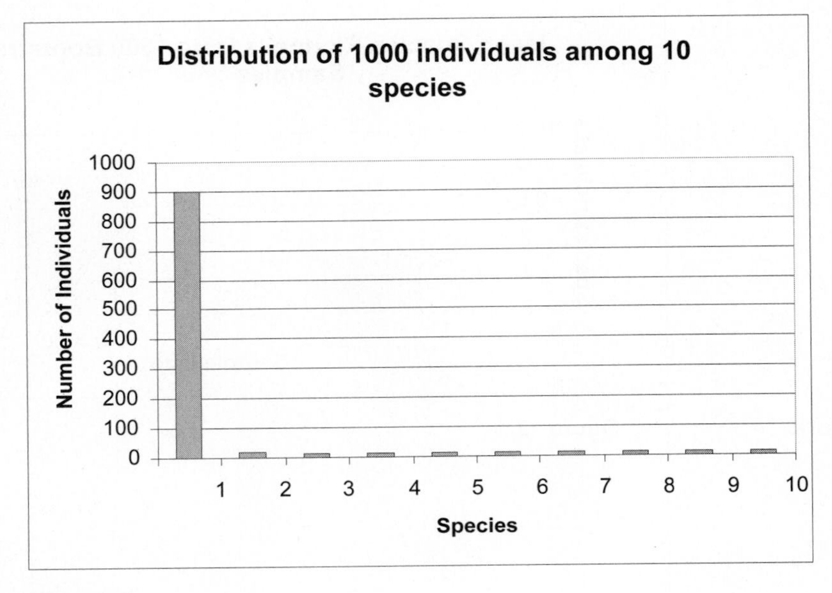

Figure 13

3. Species richness is only one measure of biodiversity for a community, but it is frequently used. Can you think of any shortcomings or assumptions of assigning conservation priorities to various locations based on species richness estimates?

LITERATURE CITED AND ADDITIONAL READINGS

Krebs, C. 1999. *Ecological Methodology*. 2nd Ed. Addison-Wesley Educational Publishers, Inc. Menlo Park, CA.

Moguel, P. and V. M. Toledo. 1998. Biodiversity conservation in traditional coffee systems of Mexico. *Conservation Biology* 13: 11–21.

Soberon, M. and J. B. Llorente. 1993. The use of species accumulation functions for the prediction of species richness. *Conservation Biology* 7: 480–488.

7 GEOMETRIC AND EXPONENTIAL POPULATION MODELS

Objectives

- Understand the demographic processes that affect population size, including raw birth and death rates, per capita birth and death rates, and rates of immigration and emigration.
- Explore the derivations of of geometric (discrete-time) and exponential (continuous-time) models of populations.
- Investigate the relationship between geometric and exponential models.
- Set up spreadsheet models of geometric and exponential population growth and graph the results.

INTRODUCTION

The study of population dynamics has been and continues to be an important area of investigation in ecology. A **population** is a group of individual organisms belonging to the same species living in the same area at the same time. Members of a population are often considered to be actually or potentially interbreeding or exchanging genes.

The term **population dynamics** means change in population size (number of individuals) or population density (number of individuals per unit area) over time. In general, population dynamics are influenced by four fundamental **demographic processes**: birth, death, immigration (individuals moving into the population), and emigration (individuals moving out of the population).

In this exercise, we will ignore immigration and emigration so that we may concentrate on births and deaths. For many populations (e.g., the human population of the earth) this is a realistic simplification. Other populations (e.g., the human population of the United States) are more open, however, and immigration and emigration must be considered. Fortunately, the addition of immigration and emigration does not complicate the models very much.

We will begin by developing a model in discrete time. That is, we will treat time as if it moved in steps, rather than continuously. This allows us to use difference equations rather than differential equations, and thereby avoid the calculus. It is also a natural way to work in spreadsheets, and is realistic for many populations that have seasonal, synchronous reproduction. Strictly speaking, the discrete-time model represents **geometric population growth**. Later in the exercise, we will develop a continuous-time model, properly called an **exponential** model.

Many textbooks present only the continuous-time exponential model. The discrete-time geometric model developed in this exercise behaves very much like its continuous-time exponential counterpart, but there are some interesting differences, which we will explore at the end of the exercise.

Model Development

To begin, we can write a very simple equation expressing the relationship between population size and the four demographic processes. Let

N_t represent the size or density of the population at some arbitrary time t (we will ignore the distinction between population size and population density)
N_{t+1} represent population size one arbitrary time-unit later
B_t represent the total number of births in the interval from time t to time $t + 1$
D_t represent the total number of deaths in the same time interval
I_t represent the total number of immigrants in the same time interval
E_t represent the total number of emigrants in the same time interval

Then we can write

$$N_{t+1} = N_t + B_t - D_t + I_t - E_t$$

For simplicity, this exercise ignores immigration and emigration. Our equation becomes

$$N_{t+1} = N_t + B_t - D_t$$

This equation is easy to understand but inconvenient for modeling. The problem lies in the use of "raw" birth and death rates (B_t and D_t). We have no obvious, biologically reasonable starting assumptions about these numbers. However, if we switch from *raw* birth and death rates to *per capita* birth and death rates, we can do some fruitful modeling.

Geometric (Discrete-Time) Model of Population Growth

A **per capita** rate is a rate per individual; that is, the per capita birth rate is the number of births per individual in the population per unit time, and the per capita death rate is the number of deaths per individual in the population per unit time. Per capita birth rate is easy to understand, and seems a reasonable thing to model because reproduction (giving birth) is something individuals rather than whole populations do. Per capita death rate may seem strange at first; after all, an individual can die only once. But remember, this rate is calculated per unit time. You can think of per capita birth and death rates as *each individual's probability of giving birth or dying in a given unit of time.*

Keeping in mind that per capita rates are per individual rates, we can translate the raw rates B_t and D_t into per capita rates, which we will represent with lower-case letters (b_t and d_t) to distinguish them from the raw numbers. To calculate per capita rates, we divide the raw numbers by the population size. Thus,

$$b_t = B_t/N_t \quad \text{and} \quad d_t = D_t/N_t$$

Conversely,

$$B_t = b_tN_t \quad \text{and} \quad D_t = d_tN_t$$

Now we can rewrite our model in terms of per capita rates:

$$N_{t+1} = N_t + b_tN_t - d_tN_t$$

Perhaps this seems to have gotten us nowhere, but it turns out to be a very informative model if we make one further assumption. Let us assume, just to see what happens, that per capita rates of birth and death remain constant over time. In other words, let us assume that average number of births per unit time per individual in the population and the average risk of dying per unit time remain unchanged over some period of time. What will happen to population size?

Because we assume constant per capita birth and death rates, we can make one further, minor modification to our equation by leaving off the time subscripts on b and d:

$$N_{t+1} = N_t + bN_t - dN_t \qquad \text{Equation 1}$$

At this point, you're probably thinking that this assumption is unrealistic—that per capita rates of birth and death *are* likely to change over time for a variety of reasons.* You are quite correct, but the model is still useful for three reasons:

- It provides a starting point for a more complex and realistic model in which per capita rates of birth and death do change over time. (You will build such a model in the "Logistic Population Models" exercise.)
- It is a good *heuristic model*—that is, it can lead to insights and learning despite its lack of realism.
- Many populations *do* in fact grow as predicted by this model, under certain conditions and for limited periods of time.

Because per capita birth and death rates do not change in response to the size (or density) of the population, this model is said to be **density-independent**.

We can further simplify Equation 1 by factoring N_t out of the birth and death terms:

$$N_{t+1} = N_t + (b - d)N_t$$

The term $(b - d)$ is so important in population biology that it is given its own symbol, R. Thus $\mathbf{R = b - d}$, and is called the **geometric rate of increase**. Substituting R for $(b - d)$ gives us

$$N_{t+1} = N_t + RN_t \qquad \text{Equation 2}$$

To further define R, we can calculate the *rate of change* in population size, ΔN_t, by subtracting N_t from both sides of Equation 2:

$$\Delta N_t = N_{t+1} - N_t = RN_t$$

Because $\Delta N_t = N_{t+1} - N_t$, we can simply write

$$\Delta N_t = RN_t \qquad \text{Equation 3}$$

In words, the rate of change in population size is proportional to the population size, and the constant of proportionality is R.

We can convert this to per capita rate of change in population size if we divide both sides by N_t:

$$\frac{\Delta N_t}{N_t} = R \qquad \text{Equation 4}$$

In other words, the parameter R represents the (discrete-time) per capita rate of change in the size of the population.

---

*You may also wonder why we use this complex model (Equation 1) rather than the simpler forms of the geometric and exponential models presented in most textbooks (and developed in this exercise beginning with Equation 2). We prefer Equation 1 for three reasons:

- It emphasizes the roles of per capita birth and death rates rather than the more abstract quantities R or r (explained later).
- It allows you to manipulate per capita birth and death rates directly and separately, and discover that neither alone, but rather the difference between them, determines population growth rate.
- It allows you to discover that the per capita rate of population growth ($\Delta N_t/N_t$) is a constant, which you can then relate to R (and r if desired).

Moving on, we can simplify Equation 2 ($N_{t+1} = N_t + RN_t$) even further by factoring N_t out of the terms on the right-hand side, to get

$$N_{t+1} = (1 + R)N_t$$

The quantity $(1 + R)$ is often given its own symbol, λ (lambda), and its own name: the **finite rate of increase**. Substituting λ, we can write

$$N_{t+1} = \lambda N_t \qquad \text{Equation 5}$$

The quantity λ can be very useful in analyzing real population data. Some additional algebra will show us how.

If we divide both sides of Equation 5 by N_t, we get

$$\frac{N_{t+1}}{N_t} = \lambda \qquad \text{Equation 6}$$

In words, λ is the ratio of the population size at one time to its size one time-unit earlier. We can calculate λ from population counts at successive times, even if we do not know per capita rates of birth and death. You will use this tool to analyze human population data in Question 10 at the end of this exercise.

In Equations 2 and 5, we showed how to calculate the size of the population one time unit into the future. What if you wanted to know how big the population will be at some distant future time? You could carry out the one-time-step calculations many times, until you arrived at the desired answer, and you will do this in the spreadsheet. But there is also a shortcut. Let us start with Equation 5:

$$N_{t+1} = \lambda N_t$$

Starting at time 0, we can carry this calculation through a few times to calculate population sizes at time 1, time 2, and time 3. The population size at time 0 can be written N_0. Thus the populations at times 1, 2, and 3 would be

$$N_1 = \lambda N_0$$

$$N_2 = \lambda N_1 = \lambda(\lambda N_0)$$

$$N_3 = \lambda N_2 = \lambda[\lambda(\lambda N_0)]$$

Do you see a pattern here? Population size at time 1 is $\lambda^1 N_0$, at time 2 it is $\lambda^2 N_0$, and at time 3 it is $\lambda^3 N_0$. In general, we can write

$$N_t = \lambda^t N_0 \qquad \text{Equation 7}$$

This expression may strike you as rather abstract. One way to understand its impact is to use Equation 7 to calculate doubling time (t_{double})—that is, the time required for the population to double in size.* If we plug the doubling time into Equation 7, we get

$$N_{t_{double}} = \lambda^{t_{double}} N_0$$

We can derive doubling time by exploiting the fact that the population at time t_{double} is, by definition, twice the population at time 0:

$$N_{t_{double}} = 2N_0$$

Substituting $2N_0$ for $N_{t_{double}}$ gives us

*This derivation follows Gotelli (2001).

$$2N_0 = \lambda^{t_{\text{double}}} N_0$$

If we divide both sides by N_0, we get

$$2 = \lambda^{t_{\text{double}}}$$

Taking the logarithm of both sides gives us

$$\ln 2 = t_{\text{double}} \ln\lambda$$

Dividing both sides by $\ln\lambda$, we get

$$\frac{\ln 2}{\ln \lambda} = t_{\text{double}} \qquad \text{Equation 8}$$

What does this mean? Suppose $R = 0.1$ individuals/individual/year. Therefore, $\lambda = 1 + R = 1.1$. This implies that the population increases by 10% per year, which doesn't sound like much. But, if you plug this value of λ into Equation 8, you'll find that the population doubles in about 7.27 years, which seems more impressive.

You may be wondering how a population that grows in discrete intervals of a year can double in a non-integer number of years. It can't, of course. This calculation really means that the population will not quite double in 7 years, and will more than double in 8 years.

Exponential (Continuous-Time) Model of Population Growth

Population growth can also be modeled in continuous time, which is more realistic for populations that reproduce continuously, rather than seasonally. Continuous-time models also allow use of the calculus, which provides many powerful analytical tools. In this exercise, we will eschew the calculus, and simply present some results.

Most textbooks begin with the continuous-time analog of Equation 3:

$$dN/dt = rN \qquad \text{Equation 9}$$

The left-hand side of Equation 9 represents the **instantaneous** rate of change in population size, which is different from the rate of change over some discrete time interval, $\Delta N_t/N_t$, that we looked at in Equation 7. Therefore, we use a lowercase r to distinguish the continuous-time exponential model from the discrete-time geometric model. The symbol r is called the **instantaneous rate of increase** or the **intrinsic rate of increase**. The parameters r and R are *not* equal, although they are related, as we will show below.

As we did with the discrete-time model, we can calculate the per capita rate of population growth by dividing both sides of Equation 9 by N:

$$\frac{(dN/dt)}{N} = r \qquad \text{Equation 10}$$

You can use the calculus to operate on Equation 10 and calculate the size of the population at any time. We will spare you the derivation, but the resulting equation is

$$N_t = N_0 e^{rt} \qquad \text{Equation 11}$$

where e is the root of the natural logarithms ($e \cong 2.71828$).

You can derive the relationship between r and R as follows. Suppose we start two populations with the same initial number of individuals, N_0, and both grow at the same rate. However, one grows in continuous time and the other grows in discrete time. Because they grow at the same rate, at some later time, t, they will have reached the same size, N_t. If we write the discrete-time population on the left and the continuous-time population on the right we can derive as follows:

$$N_t = N_t$$

$$N_0 \lambda^t = N_0 e^{rt}$$

$$\lambda^t = e^{rt}$$

$$\ln(\lambda^t) = \ln(e^{rt})$$

$$t \ln\lambda = rt \ln e$$

$$\ln\lambda = r \qquad \text{Equation 12}$$

$$\lambda = e^r \qquad \text{Equation 13}$$

So we can convert back and forth between continuous-and-discrete time models. Remember that $\lambda = 1 + R$.

Suppose we have a population growing in continuous time with some value of r, and a population growing in discrete time with the same value of R, i.e., $r = R$. Which will grow faster? As we did with the geometric model, we can derive the doubling time for the exponential model (Gotelli 2001). We begin with Equation 11, and plug in t_{double}:

$$N_{t_{double}} = N_0 e^{rt_{double}}$$

Substituting $2N_0$ for Nt_{double}, we get

$$2N_0 = N_0 e^{rt_{double}}$$

Dividing both sides by N_0 gives us

$$2 = e^{rt_{double}}$$

and taking the natural logarithm of both sides yields

$$\ln 2 = rt_{double}$$

Finally, we divide both sides by r, and rearrange, to get

$$t_{double} = \frac{\ln 2}{r}$$

Parallel to our earlier example, let us suppose $r = 0.1$ individuals/individual/year. As before, this implies a 10% annual increase in the population, but now this increase occurs continuously rather than in discrete time intervals. How long does it take for this population to double? Plugging in the value 0.1 for r yields a doubling time of 6.93 years, somewhat faster than indicated by the geometric model.

PROCEDURES

The following exercises will set up spreadsheets and allow you to graph both the geometric and exponential growth of populations. As always, save your work frequently to disk.

INSTRUCTIONS	ANNOTATION
A. Geometric (discrete-time) model.	
1. Open a new spreadsheet and set up titles and column headings as shown in Figure 1.	Enter only the text items for now. These are all literals, so just select the appropriate cells and type them in.

	A	B	C	D	E	F	G	H	I
1	Geometric Model of Population Growth								
2	Assumes constant per capita rates of birth and death.								
3		Variables					Constants		
4	t	Nt	Total births	Total deaths	ΔN_t	$(\Delta N_t)/N_t$	b	d	R
5	0						1.25	0.50	0.75
6	1								
7	2								

Figure 1

INSTRUCTIONS	ANNOTATION
2. Set up a linear time series from 0 to 20 in column A.	In cell A5, enter the number 0. In cell A6, enter the formula **=A5+1**. Copy cell A6. Select cells A7–A25. Paste.
3. Enter the values shown for per capita birth and death rates, b and d.	In cell G5, enter the number 1.25. In cell H5, enter the number 0.50.
4. Enter a formula to calculate R in cell I5.	In cell I5, enter the formula **=G5-H5**.
5. Enter an initial population size of 100.	In cell B5, enter the number 100.
6. Enter the formulae for total births (bN_t) and deaths (dN_t) into cells C5 and D5.	In cell C5, enter the formula **=$G$5*B5**. In cell D5, enter the formula **=$H$5*B5**. Note that references to per capita birth rate (**$G$5**) and per capita death rate (**$H$5**) use absolute addresses, but the references to current population size (**B5**) use a relative address. This is because you will later copy these formulae down their columns, and you want them to refer, respectively, to constants—per capita birth and death rates—and to a variable—the population size at time t.
7. Enter the formula for N_{t+1} into cell B6.	In cell B6, enter the formula **=B5+C5-D5**. Note that this formula uses the total births and deaths you have already calculated. This mimics the chain of biological cause and effect: per capita rates of birth and death, in conjunction with the number of individuals in the population, determine the total number of births and deaths, which in turn determine the size of the population at the next time.
8. Copy the formulae for total births and deaths into cells C6 and D6.	Select cells C5 and D5. Copy. Select cells C6 and D6. Paste.

9. Copy the formulae for N_t, total birth and total deaths, down their columns.

See annotation at Step 8 for the commands involved.

10. In cell E5, enter a formula to calculate the change in population size (ΔN_t) from time 0 to time 1.

In cell E5, enter the formula **=B6-B5**.

Note that this change in population size is calculated for the coming time interval. You could do it differently, but this way gives an interesting result, seen in the next step.

11. In cell F5, enter a formula for the per capita change in population size ($\Delta N_t/N_t$) from time 0 to time 1.

In cell F5, enter the formula **=E5/B5**.

Like all per capita rates, this one is calculated by dividing the change in population size by the current population size. How does the value of ($\Delta N_t/N_t$) compare to the value of *R*?

12. Copy the formulae for ΔN_t and $\Delta N_t/N_t$ down their columns.

See step 8 for the commands involved.
Your model is now complete and you are ready to create graphs.
Save your work.

13. Graph N_t, total births, total deaths, ΔN_t, and $\Delta N_t/N_t$ against time.

Select cells A4–F24. Note that you should include column headings in your selection, so that the legend will be labeled properly. Do not include row 25 because ΔN_t, and $\Delta N_t/N_t$ are undefined there.

Click on the Chart Wizard button or open Insert | Chart. (Details are given in the Introduction, "Spreadsheet Hints and Tips," and in Exercise 1, "Mathematical Functions and Graphs.") Follow the prompts in the resulting dialog boxes to set up an XY chart (Scatterplot) with time on the *x*-axis. *Do not use a line chart.*

14. Edit your graph for readability. The result should resemble Figure 2.

Put $\Delta N_t/N_t$ on the secondary *y*-axis and scale that axis from 0 to 1. (Again, refer to the Introduction and to Exercise 1; or just try clicking on things in the graph, and see what happens.)

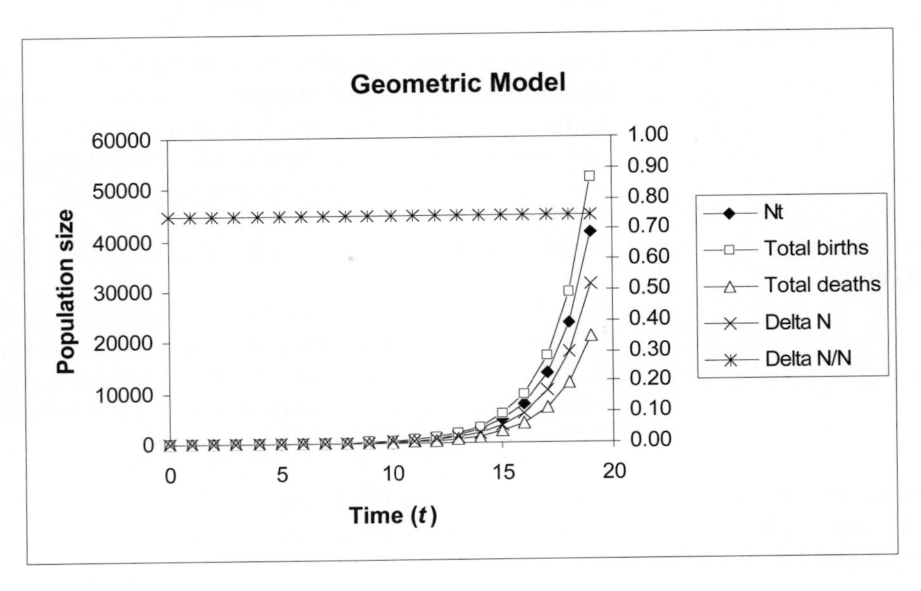

Figure 2

Strictly speaking, the graph in Figure 2 is inaccurate, because it implies that population size increases smoothly and continuously between time steps. Actually, population size remains unchanged from one time (t) to the next ($t + 1$), and then instantaneously takes its new value. Thus, the graph should look like a flight of stairs that gets steeper exponentially. However, such a graph is difficult to produce in Excel, so we will have to settle for this one and bear this inaccuracy in mind.

15. Graph total births, total deaths, and ΔN_t on the vertical axis against population size on the horizontal axis.

Select cells B4–E24. Note that this differs from your previous graph in that you do *not* include time (column A). Include column headings in your selection so that the legend will be labeled properly.

Click on the Chart Wizard or open Insert | Chart. Follow the prompts in the resulting dialog boxes to set up an XY chart (Scatterplot) with N_t on the x-axis. *Do not use a line chart.*

16. Edit your graph for readability. The result should resemble Figure 3.

See Step 2 above.

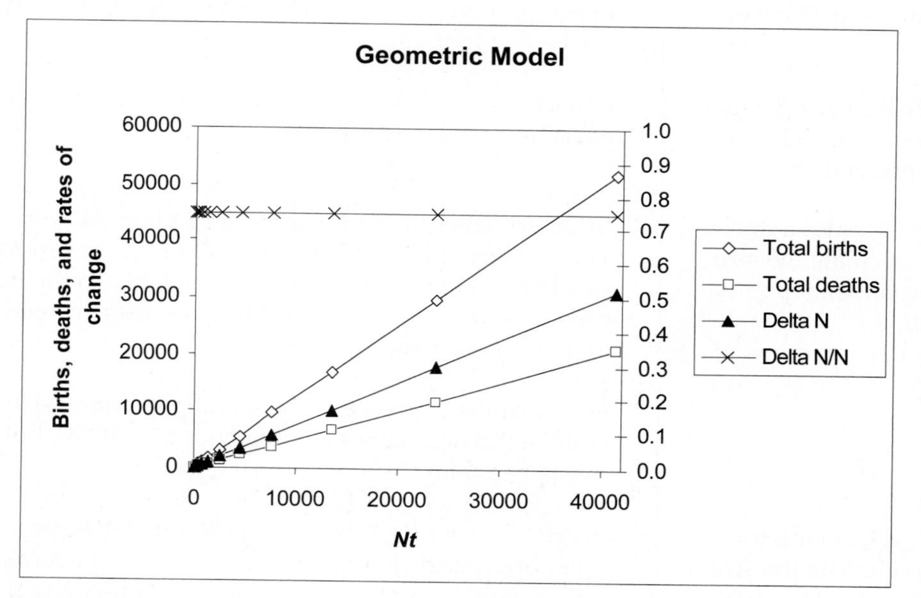

Figure 3

B. Exponential (continuous-time) model.

1. Open a new spreadsheet and set up titles and column headings as shown in Figure 4. Enter the values shown for r and R.

These are all literals, so just select the appropriate cells and type them in. We will set up an exponential (continuous-time) model and a geometric (discrete-time) model side-by-side for comparison.

	A	B	C	D	E
1	Comparison of Exponential and Geometric Models				
2		Constants			
3		r =	0.25		
4		R =	0.25		
5				Calculated values	
6		*Nt*	*Nt*	*r*	*R*
7	Time(*t*)	Exponential	Geometric	Exponential	Geometric
8	0				
9	1				
10	2				

Figure 4

2. Set up a linear time series from 0 to 20 in column A.

Enter the value 0 in cell A8.
In cell A9, enter the formula **=1+A8**. Copy cell A9 and paste into cells A10–A28.

3. In cells B8 and C8, enter initial population sizes for the two populations.

Enter the value 1.00 into cells B8 and C8. Later, you can change these values to see the effect on population growth.

4. In cell B9, enter a formula to calculate the size of the exponential population at time 1.

In cell B9, enter the formula **=$B$8*EXP($C$4*A9).**
This corresponds to Equation 11, $N_t = N_0e^{rt}$. The function **EXP($C$4*A9)** is the spreadsheet version of e^{rt}. Note that the reference to the initial population size (a constant) uses an absolute cell address (**$B$8**), as does the reference to r (**$C$4**), but the reference to time (**A9**) is relative (a variable).

Note that the reference to the initial population size (a constant) uses an absolute cell address (**$B$8**), as does the reference to r (**$C$4**), but the reference to time (**A9**) is relative (a variable).

5. In cell C9, enter a formula to calculate the size of the geometric population at time 1.

In cell C9, enter the formula **=(1+$C$5)^A9*$C$8.**
This corresponds to Equation 7: $N_t = \lambda^t N_0$. The term **(1+$C$5)** calculates λ, (which is 1 + R, remember) and the expression **^A9** raises λ to the power t. Note that the reference to the initial population size uses an absolute cell address (**$C$8**), as does the reference to R (**$C$5**), but the reference to time (**A9**) is relative.

6. Copy the formulae in cells B9 and C9 down their columns.

Select cells B9 and C9. Copy.
Select cells B10–C28. Paste.

7. Enter a formula in cell D8 to calculate r from the population sizes in cells B8 and B9.

In cell D8, enter the formula **=LN(B9/B8).**
This formula calculates λ from the population sizes at times 0 and 1, as if the population were growing in discrete time, and then converts λ to the continuous-time r by taking the natural logarithm of λ. Review Equation 12 for the derivation of this relationship.

We use this roundabout method to set the stage for analyzing real population data, as you will do in answering Question 10 at the end of this exercise. In some cases, we may know population sizes at different times, but not per capita rates of birth and death. Using this method allows us to determine r from population sizes, and predict population dynamics without knowing per capita birth and death rates.

8. Enter a formula in cell E8 to calculate R from the population sizes in cells C8 and C9.

In cell E8, enter the formula **=C9/C8-1**.
Remember that $\lambda = 1 + R$, so $R = \lambda - 1$. The rationale for this calculation is the same as for our calculation of r in step 7.

9. Copy the formulae in cells D8 and E8 down their columns to row 27.

Do not copy the formulae into cells D28 and E28 because they become undefined there.

10. Save your work.

11. Graph population size against time for exponential and geometric models on the same graph.

See "Spreadsheet Hints and Tips" and Exercise 2, "Spreadsheet Functions and Graphs," for detailed instructions. Your finished graph should resemble Figure 5.

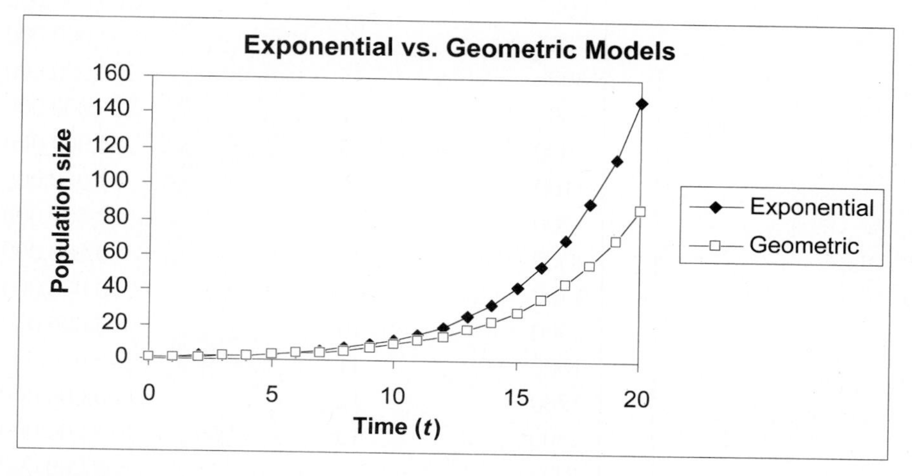

Figure 5

QUESTIONS

1. Under the assumptions $b > d$ and both b and d constant, how does the population grow? How can you verify your answer?
2. How does population size change over time if $b < d$? Before you start plugging values into the model, sketch what you think the graph of N_t against time will look like.
3. How does population size change over time if $b = d$?
4. Which of the following determine the rate of population growth (ΔN_t)?
   - per capita birth rate
   - per capita death rate
   - the product of the two
   - the ratio of the two
   - the difference between the two
5. How does the rate of population growth (ΔN_t) change over time?
6. How do total births, total deaths, and ΔN_t relate to population size?
7. How does per capita rate of population growth ($\Delta N_t/N_t$) relate to population size (N_t)?

8. Which grows faster, the continuous-time population or the discrete-time population? Why?

9. How much larger than r must R be in order to produce equal population growth rates?

10. How has the human population grown over the past 12 centuries or so? Analyze the following data from the U.S. Census Bureau website (http://www.census.gov):

Date (year C.E.)	Time (years elapsed since 500 C.E.)	Estimated population size
500	0	190,000,000
600	1	200,000,000
700	2	207,000,000
800	3	220,000,000
900	4	226,000,000
1000	5	254,000,000
1100	6	301,000,000
1200	7	360,000,000
1300	8	360,000,000
1400	9	350,000,000
1500	10	425,000,000
1600	11	545,000,000
1700	12	600,000,000
1800	13	813,000,000
1900	14	1,550,000,000

LITERATURE CITED

Gotelli, N. J. 2001. *A Primer of Ecology*, 3rd Edition. Sinauer Associates, Sunderland, MA.

8 LOGISTIC POPULATION MODELS

Objectives

- Explore various aspects of logistic population growth models, such as per capita rates of birth and death, population growth rate, and carrying capacity.
- Understand the concepts of density dependence and density independence.
- Set up spreadsheet models and graphs of logistic population growth.
- Compare the model to real populations.

Suggested Preliminary Exercise: Geometric and Exponential Population Models

INTRODUCTION

This exercise builds on the models developed in Exercise 7, "Geometric and Exponential Population Models." If you have not already done that exercise, you should do it first, or at least read its introduction.

As in the earlier exercise, we begin with a model of population dynamics in discrete time, with explicit parameters for per capita rates of birth and death. We choose the discrete-time model as the starting point for several reasons:

- It emphasizes the roles of per capita birth and death rates (b and d) rather than the more abstract quantities r (or R) and K (explained later).
- It allows you to manipulate, directly and separately, per capita birth and death rates and density-dependent rates of change in per capita birth and death rates. You will discover that none of these alone, but rather the relationship between them, determines logistic population growth and whether the population eventually stabilizes.
- It allows you to discover that for a population to stabilize, per capita birth and death rates must change as the population grows, and they must become equal at some equilibrium population size.
- It drives home, in ways that algebraically simpler models cannot, the meaning of density dependence: change in per capita birth and death rates in response to change in population size.

The logistic model with explicit birth and death rates, presented first here, lies at the heart of this particular exercise. For the sake of compatibility with a variety of textbooks, and to provide background for other exercises in this book, we present two other logistic models: a more commonly encountered discrete-time version incorporating carrying capacity (K), and a continuous-time version. We see no need to build all three versions; which one you do will depend on your instructor's aims. To instructors, we strongly suggest that the first version, with explicit per capita birth and death rates, is the best learning tool, for the reasons given above. In our experience, students have no difficulty switching to the R-K version for later exercises.

Model Development: Logistic Model with Explicit Birth and Death Rates

In Exercise 7, we developed the following geometric model of population dynamics:

$$N_{t+1} = N_t + bN_t - dN_t \qquad \text{Equation 1}$$

where

N_t = population size at time t
N_{t+1} = population size one time unit later
b = per capita birth rate
d = per capita death rate

As you discovered in the earlier exercise, this model produces geometric population growth (the discrete-time analog of exponential growth) if b and d are held constant and $b > d$. However, the assumption that per capita rates of birth and death remain constant is unrealistic, so in this exercise you will develop a model in which these rates change.

Birth and death rates may change for many reasons, such as changes in climate conditions, food supply, or populations of natural enemies (competitors, predators, parasites, and pathogens). To keep our model manageable, in this exercise we will consider only one cause of changes in per capita birth and death rates: the size of the population itself. In other words, we will assume that environmental conditions, food supply, and so on remain constant; only the size of the population itself changes. Because per capita rates of birth and death *do* change in response to population size or density, logistic models are **density-dependent**, in contrast to geometric and exponential models, which are density-independent. As the population grows, less food and water, fewer nesting and hiding sites, and fewer resources in general are available to each individual, affecting both an individual's rate of reproduction and its risk of death. Our model will thus include **intraspecific competition** (competition among members of the same species) for resources. Later exercises will develop models of interspecific (between two species) competition and predator-prey dynamics.

We now add two new terms to our model to represent changes in per capita rates of birth and death:

b' = the amount by which the per capita birth rate changes in response to the addition of one individual to the population
d' = the amount by which the per capita death rate changes in response to the addition of one individual to the population

We can now add these terms to our geometric model to produce a discrete-time logistic model:

$$N_{t+1} = N_t + (b + b'N_t)N_t - (d + d'N_t)N_t \qquad \text{Equation 2}$$

This model replaces the simple per capita birth rate b with the more complex expression $(b + b'N_t)$, and it replaces d with $(d + d'N_t)$. The symbols b and d now represent per capita rates of birth and death when the population is very small. The terms $(b + b'N_t)N_t$ and $(d + d'N_t)N_t$ represent total births and total deaths, respectively. Thus, our model still represents the fundamental insight that

$$N_{t+1} = N_t + \text{Births} - \text{Deaths} \qquad \text{Equation 3}$$

Most textbooks that use this model use a slightly different form, in which the birth term is written $(b - b'N_t)N_t$, because per capita birth rate normally decreases as population size increases. We prefer to add $b'N_t$ rather than subtract it, because our way forces you to use a negative number for b', reinforcing the idea of decreasing per capita births. It also allows you to experiment with the model to see what happens if per capita birth rate *increases* with population size.

All four parameters (b, b' d, and d') are assumed to remain constant, as you can tell from the absence of time subscripts. Let's try to visualize what happens to per capita rates of birth and death as the population grows according to this model. When the population is small, there are plenty of resources for each individual, so per capita birth rate should be high, per capita death rate should be low, and the population will grow larger. As new individuals are added, available resources will be divided among more individuals, and each individual will get less. We would expect per capita birth rate to decline (so b' should be less than zero) in proportion to the number of individuals in the population (so we multiply b' by N_t). We would also expect per capita death rate to increase (so d' should be greater than zero), also in proportion to population size (so we multiply d' by N_t as well).

As simple as it is, this model has proven useful in several contexts. Many populations grow as predicted by this model, and (in the form of Equation 7, below) it was one of the origins of chaos theory. Logistic models are used in studying interspecific as well as intraspecific competition and predator-prey relationships. These models also inform practical decisions in the management of fisheries and game animal populations and are used to predict the growth of the human population.

The rate of population growth is not easy to visualize from this equation, so you will explore its behavior using the spreadsheet. However, we can see informally that when the population is very small, it will grow almost geometrically (exponentially), because the parameters b' and d' are multiplied by a small number (N_t is small), and thus the model reduces (almost) to a geometric model. As the population grows larger, however, the influence of b' and d' increases, and population growth slows. What will be the endpoint of this slowing rate of growth? Will the population stabilize, will it continue to grow at an ever-decreasing rate, or will it decrease in size?

We can show formally that there is an **equilibrium population size** in this model. In other words, appropriate values of b, d, b', and d' will produce a model population that grows until it reaches a stable size. To prove that such an equilibrium exists, we try a commonly used tactic: We will *assume* that the equilibrium population size exists, and try to calculate its value. If the equilibrium does not exist, this procedure will lead us to a logical contradiction. If the equilibrium exists, we will find its value.

Let us begin with Equation 2:

$$N_{t+1} = N_t + (b + b'N_t)N_t - (d + d'N_t)N_t$$

Assume that an equilibrium population size exists, and call it N_{eq}. If N_{eq} exists, then plugging it into Equation 2 in place of N_t should produce no change in population size. Therefore N_{t+1} will also equal N_{eq}. If we substitute N_{eq} for N_t and N_{t+1}, we get

$$N_{eq} = N_{eq} + (b + b'N_{eq})N_{eq} - (d + d'N_{eq})N_{eq}$$

Subtracting N_{eq} from both sides gives us

$$0 = (b + b'N_{eq})N_{eq} - (d + d'N_{eq})N_{eq}$$

Adding $(d + d'N_{eq})N_{eq}$ to both sides, we get

$$(d + d'N_{eq})N_{eq} = (b + b'N_{eq})N_{eq}$$

In words, the population is at equilibrium when *total* deaths equal *total* births (compare to Equation 3 above). This seems a sensible result. Let us continue by dividing both sides by N_{eq}, to get

$$d + d'N_{eq} = b + b'N_{eq}$$

This tells us that the population is at equilibrium when *per capita* rates of birth and death are equal, which also makes sense.
Subtracting d and $b'N_{eq}$ from both sides gives us

$$d'N_{eq} - b'N_{eq} = b - d$$

Factoring N_{eq} out of the left-hand side produces

$$(d' - b')N_{eq} = b - d$$

and dividing both sides by $(d' - b')$ gives us

$$N_{eq} = \frac{b-d}{d'-b'} \qquad \text{Equation 4}$$

Note that the numerator on the right-hand side of Equation 4 is the geometric growth factor R, as defined in Exercise 7, "Geometric and Exponential Population Growth."

Equation 4 gives us our equilibrium population size. The derivation shows that values of b, d, b', and d' exist that will produce a stable population. Be aware, however, that it does not show that *any* values of these parameters will do so—that is, there also may exist values of these parameters that will produce population growth that does *not* reach equilibrium. It also shows that the equilibrium population depends on all four parameters, in the particular way shown in Equation 4.

Logistic Model with Explicit Carrying Capacity

Because the equilibrium defined in Equation 4 is so important in population biology, it is given its own name—the **carrying capacity**. The carrying capacity is defined as the largest population that can be supported indefinitely, given the resources available in the environment. Most logistic models presented in textbooks represent this carrying capacity with its own parameter, K, and build it into the model explicitly. We develop this model below.

Most textbooks present logistic population growth in terms of a differential equation in continuous time:

$$\frac{dN}{dt} = rN\left(\frac{K-N}{K}\right) \qquad \text{Equation 5}$$

The discrete-time analog of this equation is

$$\Delta N_t = RN_t\left(\frac{K-N_t}{K}\right) \qquad \text{Equation 6}$$

In Equation 6, ΔN_t represents the difference between the population size at time $t + 1$ and at time t. We can therefore write $\Delta N_t = N_{t+1} - N_t$ and substitute that into Equation 6. This gives us

$$N_{t+1} - N_t = RN_t\left(\frac{K-N_t}{K}\right)$$

Adding N_t to both sides gives us our discrete-time model of logistic population growth; we get

$$N_{t+1} = N_t + RN_t\left(\frac{K-N_t}{K}\right) \qquad \text{Equation 7}$$

Because this model has fewer parameters, it is more convenient to use in studying interspecific competition, predator-prey relationships, and harvesting populations.

The behavior of Equation 7 is not difficult to visualize. If we begin with a very small population, the term $(K - N_t)/K$ is very nearly equal to K/K, or 1. The model will then behave like a geometric model, and the population will grow, provided $R > 1$. The population will grow slowly at first, because the parameter R is also being multiplied by a number (N_t) that is nearly equal to zero, but it will grow faster and faster, at least for a while. At some point, however, population growth will begin to slow because the term $(K - N_t)/K$ is getting smaller and smaller as N_t gets larger and closer to K.

At the other extreme, imagine a population that starts out at a size very close to its carrying capacity, K. The term $(K - N_t)/K$ becomes nearly equal to zero, and population growth is extremely slow. When $N_t = K$, the population stops growing altogether.

The actual dynamics of this model can be much more complex, as you will see when you build the spreadsheet model and play around with its parameters. With some values of b, d, b', and d', or of R and K, the population can temporarily overshoot its carrying capacity, oscillate around it, or become chaotic.

The two discrete-time models (expressed in Equations 2 and 7) are mathematically equivalent. This is not obvious from the equations, and the proof is not directly relevant to our modeling concerns, but if you're curious you can read the proof at the end of the exercise (pp. 121–122).

Continuous-Time Logistic Model

As we said above, most textbooks begin with the model given by Equation 5:

$$\frac{dN}{dt} = rN\left(\frac{K-N}{K}\right)$$

As stated, this tells you only the rate of change in population size, not the population size at any time t. To derive the equation for population size requires the calculus, so we will simply give the result (Roughgarden 1998):

$$N_t = \frac{K}{1+\left[(K-N_0)/N_0\right]e^{-rt}} \qquad \text{Equation 8}$$

This model behaves as described for the discrete-time version. An important difference, however, is that the continuous-time model always grows smoothly to its carrying capacity and stabilizes there. The discrete-time model can display more interesting behavior.

PROCEDURES

Your instructor may assign all of the following three parts, or only one or two. As always, save your work frequently to disk.

INSTRUCTIONS	ANNOTATIONS
Part 1. Discrete-Time Logistic Model with Explicit Birth and Death Rates	
A. Set up the spreadsheet.	
1. Open a new spreadsheet and set up titles and column headings as shown in Figure 1.	Enter only the text items for now. These are all literals, so just select the appropriate cells and type them in.

	A	B	C	D	E	F	G	H	I	J	K
1	Logistic Model of Population Growth										
2	Includes explicit terms for per capita rates of birth and death, and for changes in these rates										
3					Variables						
4			Per capita	Total	Per capita	Total			Constants		
5	Time (t)	N_t	birth rate	births	death rate	deaths	ΔN_t	$(\Delta N_t)/N_t$	b	d	R
6	0								1.2500	0.50	0.75
7	1								b'	d'	K
8	2								-0.010	0.005	50.00
9	3										
10	4										

Figure 1

2. Set up a linear time series from 0 to 20 in column A.

In cell A6, enter the number **0**.
In cell A7, enter the formula **=A6+1**.
Copy cell A7.
Select cells A8–A26. Paste.

3. Enter the values shown in Figure 1 for per capita birth and death rates, b and d, and per capita rates of change in b and d, b' and d'.

Be sure to enter a negative number for b'. This indicates that per capita birth rate *decreases* as each new member is added to the population. A positive value of d' indicates that per capita death rate *increases* as each new member is added to the population.

4. Enter formulae to calculate R and K.

In cell K6, enter the formula **=I6-J6**. Remember, by definition, $R = b - d$.
In cell K8, enter the formula **=(I6-J6)/(J8-I8)**. This is the spreadsheet version of Equation 4 in the Introduction. It represents the largest population that can be sustained indefinitely on the resources available.

5. Enter an initial population size of 1.00.

In cell B6, enter the value 1.00.

6. Enter formulae for per capita birth and death rates (b and d).

In cell C6, enter the formula **=$I$6+$I$8*B6**.
In cell E6, enter the formula **=$J$6+$J$8*B6**.
These formulae correspond to the per capita birth and death rates, $(b + b'N_1)$ and $(d + d'N_1)N_1$, in Equation 2.
We calculate per capita births and deaths explicitly because it is important to understand how these rates respond to changes in population size. You will graph these quantities later in the exercise.

7. Enter formulae for total births and total deaths.

In cell D6, enter the formula **=C6*B6**.
In cell F6, enter the formula **=E6*B6**.
These formulae correspond to the total births, $(b + b'N_t)N_t$, and total deaths, $(d + d'N_t)N_t$, in Equation 2. We calculate total births and deaths as an intermediate step in calculating N_{t+1} (see next step).

8. Enter a formula to calculate the size of the population at time 1.

In cell B7, enter the formula **=B6+D6-F6**.
This corresponds to Equation 3, $N_{t+1} = N_t$ + Births – Deaths.

9. Copy the formulae in cells C6–F6 into cells C7–F7.

See "Spreadsheet Hints and Tips" for instructions on copying and pasting.

10. Copy the formulae from cells B7–F7 into cells B8–F26.

11. Enter a formula for ΔN_t.

In cell G6, enter the formula **=B7-B6**.
Note that we calculate ΔN_t over the coming time interval, as we did in Exercise 7, "Geometric and Exponential Population Models."

12. Enter a formula for $\Delta N_t / N_t$.

In cell H6, enter the formula **=G6/B6**.
This is the *per capita* rate of change in population size over the interval from time 0 to time 1. Like ΔN_t, it is calculated over the coming time interval.

13. Copy the formulae in cells G6 and H6 into cells G7–H25.

Note that we do *not* paste these formulae into cells G26 and H26. This is because ΔN_t and $\Delta N_t / N_t$ are calculated over the coming time interval, and would therefore be undefined for the last population size calculated (cell B26).

B. Create graphs.

1. Graph N_t against time, and edit your graph for readability.

Select cells A5–B26. Note that you should include column headings in your selection, so that the legend will be labeled properly.

Click on the Chart Wizard or open Insert | Chart. Details of the steps involved are given in "Spreadsheet Hints and Tips" and in Exercise 2, "Mathematical Functions and Graphs." Follow the prompts in the dialog boxes to set up an XY chart (Scatter graph) with time on the x-axis. Do not use a line chart.

Your graph should resemble Figure 2.

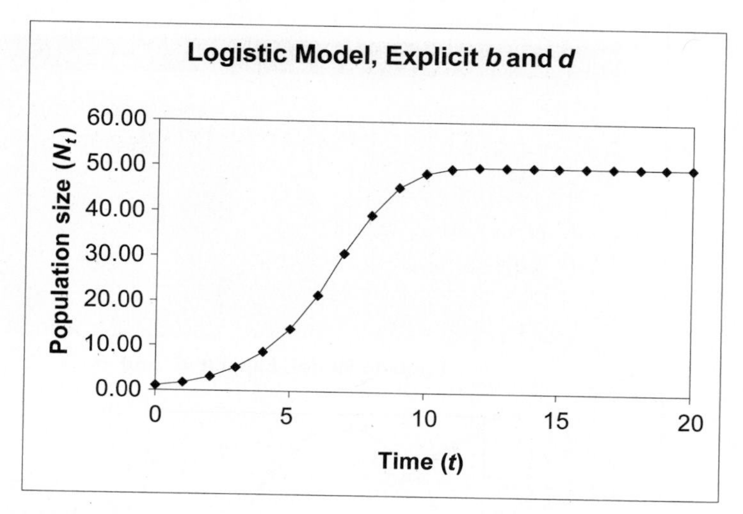

Figure 2

2. Graph per capita birth and death rates against N_t, and edit your graph for readability.

Note that you are *not* graphing against time. Select cells B5–C26.
Hold down the <Control> or key and select cells E5–E26.
Make an XY chart (Scatter graph) (see previous step).

Your graph should resemble Figure 3.

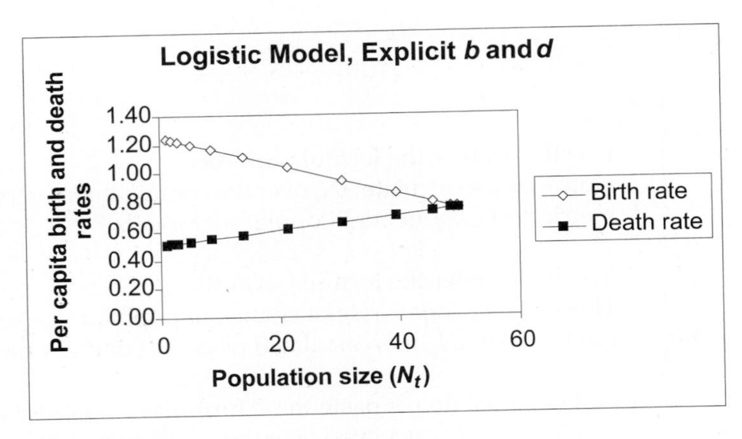

Figure 3

3. Graph ΔN_t and $\Delta N_t/N_t$ against N_t.

Select cells B5–B25. Note that you should *not* include cell B26 in your selection. Hold down the <Control> or key and select cells G5–H25.
Make an XY chart, per the previous step.
Because the ranges of values taken by ΔN_t is so much larger than the range of $\Delta N_t/N_t$, the latter gets squashed down against the x-axis. You will fix this in the next step.

4. Graph ΔN_t on a second y-axis of the same graph.

Select the curve for ΔN_t in your graph by double-clicking on the line or on one of the data points.
In the dialog box that pops up, select the Axis tab, and then click the button for Secondary axis, as shown in Figure 4. Click on the OK button.

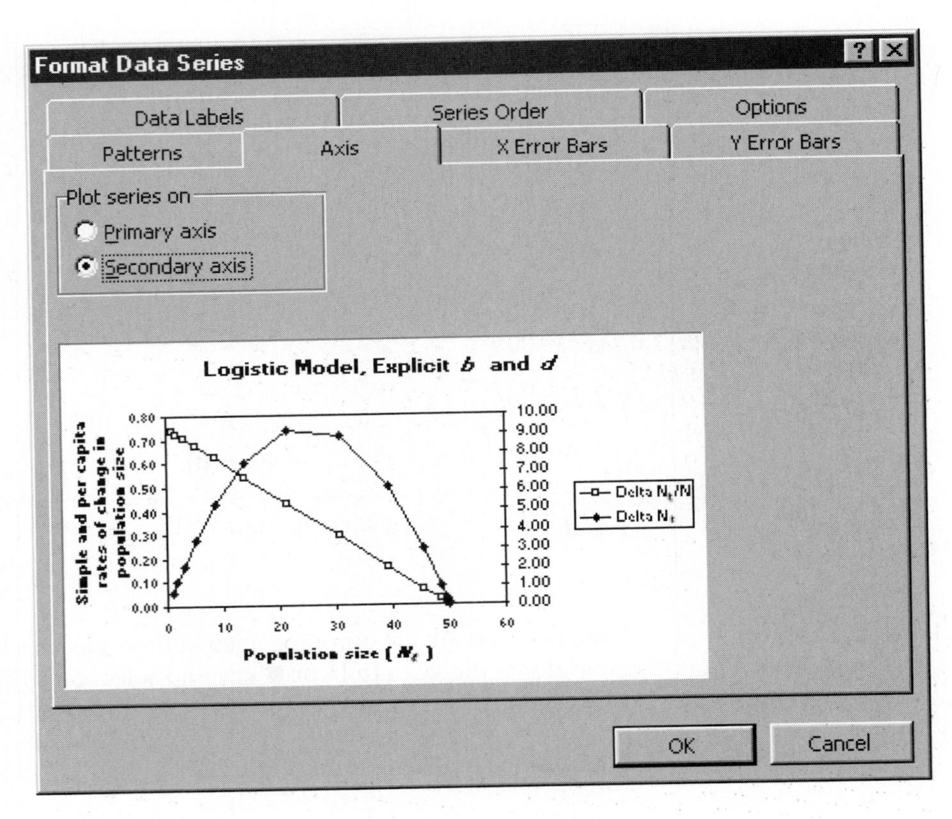

Figure 4

5. Edit your graph for readability.

Set the minimum of the left-hand y-axis to zero: Double-click on the left-hand y-axis. In the dialog box that pops up, click the tab for Scale, and enter the value 0 in the Minimum box. This will make no difference now, but it will prevent graphing errors later in the exercise.

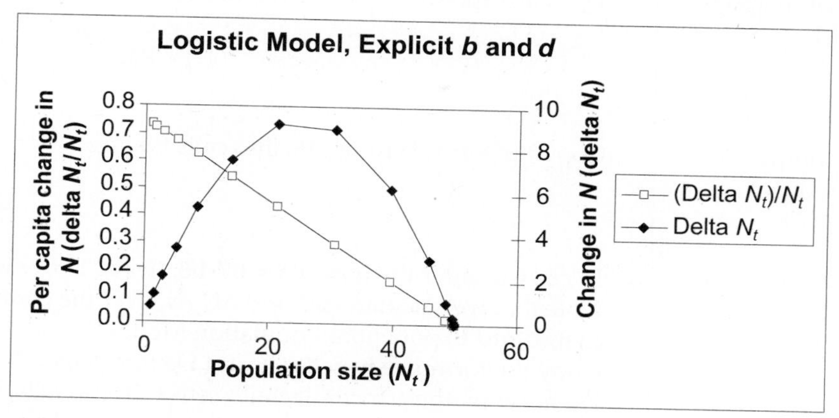

Figure 5

Your graph should resemble Figure 5. To label the right-hand y-axis, select the whole chart by clicking once inside it. Then open Chart|Chart Options|Titles. Enter the label for the right-hand y-axis in the text box for Second value (Y) axis.

Part 2. Discrete-Time Logistic Model with Explicit Carrying Capacity

C. Set up the spreadsheet.

1. Open a new spreadsheet and set up titles and column headings as shown in Figure 6.

	A	B	C	D	E	F	G
1	Logistic Model of Population Growth						
2	Assumes density-dependent changing per capita rates of birth and death.						
3							
4			Variables		Constants		
5	Time (t)	N_t	ΔN_t	$(\Delta N_t)/N_t$	R	K	
6	0				0.75	50.00	
7	1						
8	2						
9	3						

Figure 6

Enter only the text items for now. These are all literals, so just select the appropriate cells and type them in.

2. Set up a linear time series from 0 to 20 in column A.

In cell A6, enter the value 0.
In cell A7, enter the formula **=A6+1**.
Copy the formula in cell A7 into cells A8–A26.

3. Enter the values shown for initial population size, R, and K.

In cell B6, enter the value 1.00.
In cell E6, enter the value 0.75.
In cell F6, enter the value 50.

4. Enter a formula to calculate the size of the population at time 1.

In cell B7, enter the formula **=B6+$E$6*B6*($F$6-B6)/$F$6.**
This corresponds to the right-hand side of Equation 7:

$$N_t + RN_t\left(\frac{K - N_t}{K}\right)$$

5. Extend the population-size calculation down its column.

Copy the formula in cell B7 into cells B8–B26.

6. Enter formulae to calculate ΔN_t and $\Delta N_t/N_t$, and copy them down their columns.

In cell C6, enter the formula = **B7-B6**. In cell D6, enter the formula **=C6/B6.**
Note that we calculate ΔN_t and $\Delta N_t/N_t$ over the coming time interval, as we did "Geometric and Exponential Population Models."
Copy the formulae in cells C6 and D6 into cells C7–D25.
Do not copy these formulae into cells C26 and D26, because they would be undefined for the last population size calculated.

D. Create graphs.

1. Graph N_t against time and edit your graph for readability.

1. Select cells A5–B26. Create an XY graph. Your graph should resemble Figure 7.

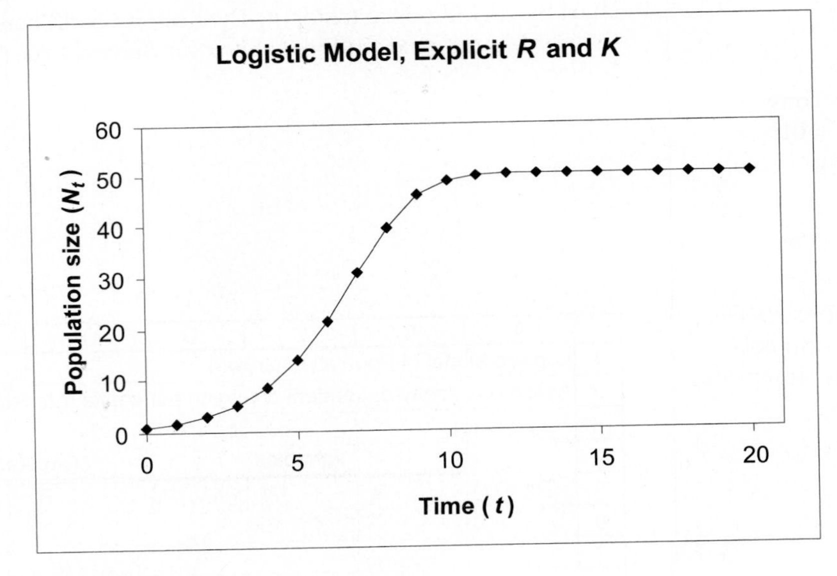

Figure 7

2. Graph ΔN_t and $\Delta N_t/N_t$ against N_t.

Select cells B5–D25 and make an XY graph.
Because the range of values taken by ΔN_t is so much larger than the range of $\Delta N_t/N_t$, the latter gets squashed down against the x-axis. You will fix this in the next step.

3. Graph ΔN_t on a second y-axis of the same graph.

Select the curve for ΔN_t in your graph, by double-clicking on the line or on one of the data points.
In the dialog box that pops up, select the Axis tab, and then click the button for Secondary axis (see Figure 4). Click on the OK button.

4. Edit your graph for readability.

Set the minimum of the left-hand y-axis to zero. Double-click on the left-hand y-axis. In the dialog box that pops up, click the tab for Scale, and enter the value 0 in the Minimum box. This will make no difference now, but it will prevent graphing errors later in the exercise. Your graph should resemble Figure 8.

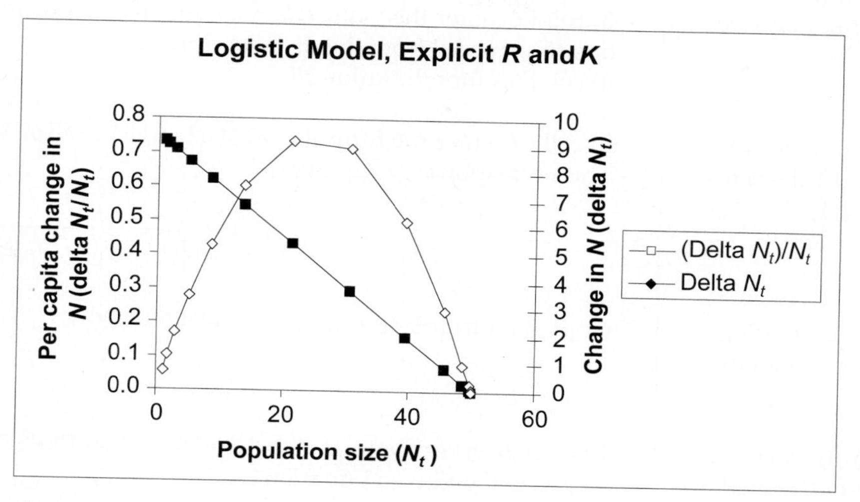

Figure 8

To label the right-hand y-axis, select the whole chart by clicking once inside it. Then open Chart|Chart Options|Titles. Enter the label for the right-hand y-axis in the text box for Second value (Y) axis.

Part 3. Continuous-Time Logistic Model

E. Set up the spreadsheet.

1. Open a new spreadsheet and set up titles and column headings as shown in Figure 9.

Enter only the text items for now. These are all literals, so just select the appropriate cells and type them in. Note that we use the differential notation dN/dt and $(dN/dt)/N$ instead of the difference notation ΔN_t and $\Delta N_t/N_t$, and r in place of R.

	A	B	C	D	E	F
1	Logistic Model of Population Growth					
2	Continuous-time version					
3						
4			Variables		Constants	
5	Time (t)	N_t	dN/dt	$(dN/dt)/N$	r	K
6	0				0.75	50.00
7	1					
8	2					
9	3					

Figure 9

2. Set up a linear time series from 0 to 20 in column A.

In cell A6, enter the value 0.
In cell A7, enter the formula **=A6+1**.
Copy the formula in cell A7 into cells A8–A26.

3. Enter the values shown for initial population size, *r*, and *K*.

In cell B6, enter the value 1.00.
In cell E6, enter the value 0.5. Note the use of lowercase *r*, to distinguish the continuous-time model from the discrete-time version.
In cell F6, enter the value 50.

4. Enter a formula to calculate the size of the population at time 1.

In cell B7, enter the formula **=$F$6/(1+(($F$6-$B$6)/$B$6)*EXP(-1*$E$6*A7))**.
This corresponds to Equation 8:

$$N_t = \frac{K}{1+\left[(K-N_0)/N_0\right]e^{-rt}}$$

5. Extend the population-size calculation down its column.

Copy the formula in cell B7 into cells B8–B26.

6. Enter formulae to calculate *dN/dt* and *(dN/dt)/N* and copy them down their columns. (Hint: Refer to Equation 5.)

In cell C6, enter the formula **=$E$6*B6*($F$6-B6)/$F$6**.
This corresponds to Equation 5:

$$\frac{dN}{dt} = rN\left(\frac{K-N}{K}\right)$$

In cell D6, enter the formula **=C6/B6**.
Copy the formulae in cells C6 and D6 into cells C7–D26. In this case, we *do* copy these formulae into cells C26 and D26, because we are calculating them instantaneously from the current N_t. This is an important difference between this model and the two previous discrete-time logistic models. If we had used the same difference method to calculate these rates of change as we used before, we would get different values.

F. Create graphs.

1. Graph N_t against time and edit your graph for readability.

Select cells A5–B26. Make an XY graph and edit for readability. Your graph should resemble Figure 10.

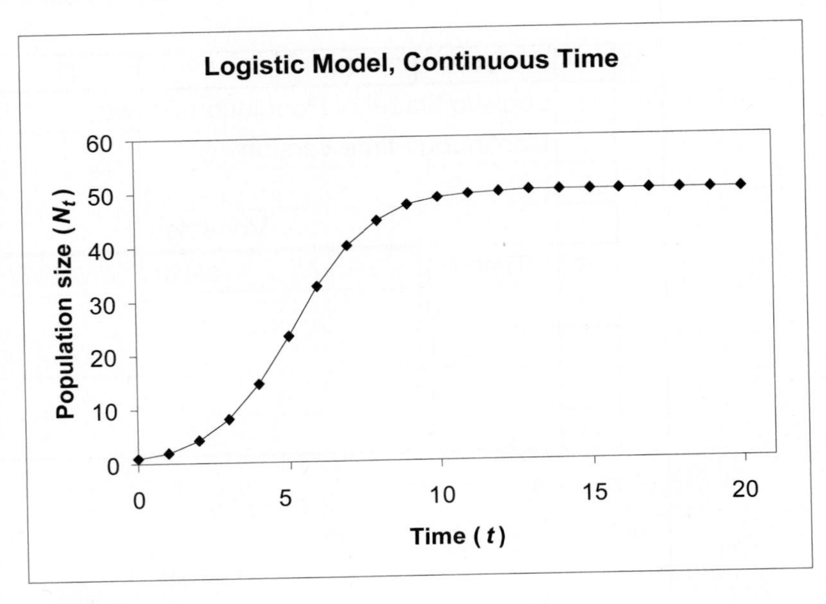

Figure 10

2. Graph dN/dt and $(dN/dt)/N$ against N_t.

Select cells B5–D26 and make an XY graph.
Because the ranges of values taken by dN/dt is so much larger than the range of $(dN/dt)/N$, the latter gets squashed down against the x-axis. You will fix this in the next step.

3. Graph dN/dt on a second y-axis of the same graph.

Select the curve for dN/dt in your graph by double-clicking on the line or on one of the data points.
In the dialog box that pops up, select the Axis tab, and then click the button for Secondary axis (see Figure 4). Then click on the OK button.

4. Edit your graph for readability.

Set the minimum of the left-hand y-axis to zero. Double-click on the left-hand y-axis. In the dialog box that pops up, click the tab for Scale, and enter the value 0 in the Minimum box. Your graph should resemble Figure 11.

This will make no difference now but will prevent graphing errors later in the exercise.

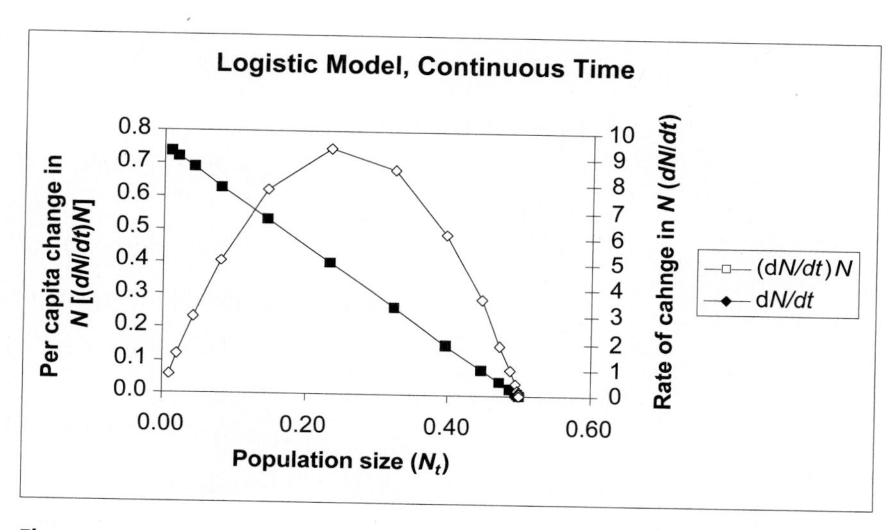

Figure 11

To label the right-hand y-axis, select the whole chart by clicking once inside it. Then open Chart|Chart Options|Titles. Enter the label for the right-hand y-axis in the text box for Second value (Y) axis.

Proof That the Two Discrete-Time Models Are Equivalent

The following proof* demonstrates that the two discrete-time models (Equation 2 and Equation 7) are in fact equivalent. Begin with Equation 7:

$$N_{t+1} = N_t + RN_t\left(\frac{K - N_t}{K}\right) \qquad \text{Equation 7}$$

Rewrite the term in parentheses:

$$N_{t+1} = N_t + RN_t\left(\frac{K}{K} - \frac{N_t}{K}\right)$$

Because $K/K = 1$, we can write

$$N_{t+1} = N_t + RN_t\left(1 - \frac{N_t}{K}\right)$$

We showed early in the exercise that $K = (b - d)/d' - b')$, so we can substitute

$$N_{t+1} = N_t + RN_t\left(1 - \frac{N_t}{\left(\frac{b-d}{d'-b'}\right)}\right)$$

Rearranging gives us

$$N_{t+1} = N_t + RN_t\left(1 - \frac{(d'-b')N_t}{b-d}\right)$$

Because $R = b - d$ by definition, we can substitute

$$N_{t+1} = N_t + (b-d)N_t\left(1 - \frac{(d'-b')N_t}{b-d}\right)$$

and carry out the multiplication across the parentheses:

$$N_{t+1} = N_t + (b-d)N_t - \frac{(b-d)(d'-b')N_t^2}{(b-d)}$$

Canceling terms gives us

$$N_{t+1} = N_t + (b-d)N_t - (d'-b')N_t^2$$

Factoring out N_t, we get

$$N_{t+1} = N_t + N_t[(b-d) - (d'-b')N_t]$$

Carrying out the multiplication inside the square brackets, we get

$$N_{t+1} = N_t + N_t[b - d - d'N_t + b'N_t]$$

Rearranging terms gives us

$$N_{t+1} = N_t + N_t[(b + b'N_t) - (d + d'N_t)]$$

Multiplying through by N_t gives us Equation 2:

$$N_{t+1} = N_t + (b + b'N_t)N_t - (d + d'N_t)N_t \qquad \text{Equation 2}$$

Unfortunately, the graph does not indicate which axis relates to which curve. You must look at the values in the spreadsheet to see that the left-hand *y*-axis relates to *dN/dt/N* because that ratio takes values from 0.74 to 0 (see column D). Likewise, the right-hand axis relates to *dN/dt* because that difference takes values from 9.33 to 0 (see column C).

QUESTIONS

1. How does the behavior of the logistic model differ from that of the geometric and exponential models in the previous exercise?
2. Why does the population stabilize at the carrying capacity?
3. How do ΔN_t and $\Delta N_t/N_t$, or *dN/dt* and *(dN/dt)/N*, change as the population grows? How does the behavior of these quantities differ from the geometric and exponential models?

---

*Thanks to Shannon Cleary, a student in Charles Welden's Community and Population Ecology class at Southern Oregon University, for deriving this proof.

4. What is the y-intercept of the $\Delta N_t/N_t$ line in the graph of $\Delta N_t/N_t$ against N_t? What is the x-intercept? If you used the continuous-time version, ask the same questions about the $(dN/dt)/N$ line. Answering this question will lead you to a powerful tool for analyzing real populations for density-dependence, and for estimating R (or r) and K.

5. What happens if the population overshoots its carrying capacity? This might happen, for example, if resources decreased dramatically from one year to the next, causing the carrying capacity to decrease. If population were at its old carrying capacity, it would suddenly find itself above its new carrying capacity. What would happen?

6. Is the carrying capacity a *stable equilibrium* or an *unstable equilibrium*? If an equilibrium is stable, the system (the population, in this case) will tend to return to equilibrium after a disturbance. If an equilibrium is unstable, the system will show no tendency to return to equilibrium after disturbance.

7. We have assumed so far that as the population grows, per capita births decrease and per capita deaths increase. However, that need not be the case. Per capita births may increase as the population grows if, for example, mates become easier to find. Per capita deaths may decrease if, for example, a bigger herd is safer from predators.

   What happens if per capita birth rate increases with increasing N_t, or if per capita death rate decreases with increasing N_t?

8. What happens if the per capita birth rate and per capita death rate change equally (so that the difference between them remains constant) as the population grows?

9. What happens if the difference between per capita birth and death rates increases as the population grows?

10. So far, we have kept the population growth rate relatively slow, and population size has changed smoothly and predictably. What happens if the population grows more rapidly?

11. Has the human population grown exponentially or logistically since 1963? Can you estimate r and K for the human population? Estimating K is especially important because it amounts to a prediction of the size of our population when (and if) it stabilizes. Estimating r will allow you to predict when the population may stabilize.

LITERATURE CITED

Roughgarden, Jonathan. 1998. *Primer of Ecological Theory*. Prentice Hall, Upper Saddle River, NJ.

U.S. Census Bureau Web site *http://www.census.gov/*

9 INTERSPECIFIC COMPETITION AND COMPETITIVE EXCLUSION

Objectives

- Program the Lotka-Volterra model of interspecific competition in a spreadsheet.
- Understand the competitive exclusion principle and how it relates to the model.
- Use the model to explore competitive exclusion and coexistence.
- Determine under what conditions two competing species can coexist, in terms of their competition coefficients, carrying capacities, and intrinsic rates of increase.

Suggested Preliminary Exercise: Logistic Population Models

INTRODUCTION

Our previous models of population dynamics considered only one population. As informative as those models were, it should be obvious that real populations do not exist in isolation, but share habitats with populations of other species. In many cases, coexisting species will interact by interspecific competition, predation, parasitism, mutualism, or other ecological interactions. More realistic models must take such interactions into account. In the 1920s, Vito Volterra and Alfred Lotka (1932) independently developed models of **interspecific competition** (competition between two species), and investigated the conditions that would permit competing species to coexist indefinitely. In this exercise, you will build a discrete-time version of their continuous-time models.

An important ecological generalization, the **competitive exclusion principle**, has grown out of the Lotka-Volterra model and from other sources. This principle states that *two species cannot coexist unless their niches are sufficiently different that each limits its own population growth more than it limits that of the other*. In other words, if there is too much niche overlap, one species will competitively exclude the other. In reality, whether two species coexist depends not only on their competitive interactions with each other, but also on their interactions with the abiotic environment and with other species not included in this simple model. Nevertheless, as with other models in this book, the competitive exclusion principle has proven fruitful in stimulating research and understanding ecological interactions in the natural world.

Model Development

To review, the geometric model of population growth, $N_{t+1} = N_t + RN_t$, includes no effect of competition. The population increases by RN_t in every time interval, without any limitations such as might be imposed by finite resources.

The logistic model of population growth includes **intraspecific competition** (competition between individuals of the same species). To keep things (relatively) simple, we will develop our model of interspecific competition beginning with this form of the logistic model:

$$N_{t+1} = N_t + RN_t \frac{K - N_t}{K} \qquad \text{Equation 1}$$

where K is the carrying capacity, or largest sustainable population. The value of K is set by available resources and by each individual's resource demand. This version of the logistic model has intraspecific competition built into it in the term $(K - N_t)/K$. This term reduces the population growth rate in response to the addition of each new member of the population, representing the reduction in per capita birth rate, and increase in per capita death rate, caused by competition for limited resources. You can review Exercise 8, "Logistic Population Models," for more information about this model.

The Lotka-Volterra model of interspecific competition builds on the logistic model of a single population. It begins with a separate logistic model of the population of each of the two competing species.

$$\text{Population 1:} \quad N_{1,t+1} = N_{1,t} + R_1 N_{1,t} \frac{K_1 - N_{1,t}}{K_1}$$

$$\text{Population 2:} \quad N_{2,t+1} = N_{2,t} + R_2 N_{2,t} \frac{K_2 - N_{2,t}}{K_2}$$

Note the use of subscripts 1 and 2 to denote which species' population is being modeled. Each population has its own rate of increase R and carrying capacity K, and these may differ between the two species.

Next we build interspecific competition into each of these equations. In the model of population 1 above, we assume that each new member of population 1 reduces resources available to each member of population 1, and thus reduces population growth rate. In the two-species model, new members of population 2 will also reduce resources available to members of population 1—this is, after all, the meaning of interspecific competiton.

The simplest way to model this would be to modify the $(K_1 - N_{1,t})/K$ term into $(K_1 - N_{1,t} - N_{2,t})/K_1$. However, this assumes that each additional member of population 2 will affect population 1 exactly as much as an additional member of population 1. That is not necessarily the case, so we multiply $N_{2,t}$ in this term by a **competition coefficient**, α_{12} to express how much effect each additional member of population 2 has on population 1, relative to the effect of a new member of population 1. We modify the model for population 2 in a parallel way. The resulting Lotka-Volterra model of two-species competition is:

$$\text{Population 1:} \quad N_{1,t+1} = N_{1,t} + R_1 N_{1,t} \frac{K_1 - N_{1,t} - \alpha_{12} N_{2,t}}{K_1} \qquad \text{Equation 2}$$

$$\text{Population 2:} \quad N_{2,t+1} = N_{2,t} + R_2 N_{2,t} \frac{K_2 - N_{2,t} - \alpha_{21} N_{1,t}}{K_2} \qquad \text{Equation 3}$$

Note the subscripts on the competition coefficients: α_{12} expresses the effect of one member of population 2 on the growth rate of population 1; α_{21} expresses the effect of one member of population 1 on the growth rate of population 2.

In broad terms, the question Lotka and Volterra asked was, What will happen to the population dynamics of these two populations, given various values of the model parameters? Are there parameter values that will produce a winner and a loser,—one population that persists while the other goes extinct? This would be competitive exclusion. Will other values result in coexistence, in which both competing populations persist indefinitely? You will look for answers to these questions both analytically (algebraically) and graphically (using the spreadsheet).

Equilibrium Solutions

One approach to answering the questions posed above is to look for equilibrium solutions to Equations 2 and 3. If population 1 is at equilibrium, then $N_{1,t+1} = N_{1,t}$ and we can substitute $N_{1,t}$ for $N_{1,t+1}$:

$$N_{1,t} = N_{1,t} + R_1 N_{1,t} \frac{K_1 - N_{1,t} - \alpha_{12} N_{2,t}}{K_1}$$

Subtracting $N_{1,t}$ from both sides of the equation gives us

$$0 = R_1 N_{1,t} \frac{K_1 - N_{1,t} - \alpha_{12} N_{2,t}}{K_1}$$

In words, this equation says the population stops growing when it is at equilibrium, which should come as no surprise. This equation is satisfied if $N_{1,t} = 0$ or if $R_1 = 0$, but these solutions are trivial.

The equation is also satisfied by the more interesting case of

$$K_1 - N_{1,t} - \alpha_{12} N_{2,t} = 0$$

If we add $N_{1,t}$ to both sides and rearrange the terms, we get

$$N_{1,t} = K_1 - \alpha_{12} N_{2,t} \qquad \text{Equation 4}$$

Notice that this equation is in the general form of a linear equation, $y = a + bx$, and is therefore a straight line. We call this line a **zero net growth isocline**, or **ZNGI**, because anywhere along it, population 1 has zero net growth. In other words, this is an equilibrium solution for population 1.

Just as x and y in the general linear equation $y = a + bx$ can be used as coordinates for graphing, so we can use $N_{1,t}$ and $N_{2,t}$ as coordinates to graph Equation 4. We can graph this isocline by finding any two points along it and connecting them with a straight line. Two convenient points are where $N_{2,t} = 0$ and where $N_{1,t} = 0$.

If $N_{2,t} = 0$, then we solve for $N_{1,t}$. Equation 4 becomes

$$N_{1,t} = K_1 - \alpha_{12} 0$$

which reduces to

$$N_{1,t} = K_1$$

In words, if there are no members of population 2 in the habitat, population 1 will stabilize at its own carrying capacity, K_1. This seems a reasonable solution.

If we set $N_{1,t} = 0$, and then solve for $N_{2,t}$. Equation 4 becomes

$$0 = K_1 - \alpha_{12} N_{2,t}$$

and adding $a_{12} N_{2,t}$ to both sides gives us

$$\alpha_{12} N_{2,t} = K_1$$

Dividing both sides by α_{12} gives us

$$N_{2,t} = K_1 / \alpha_{12}$$

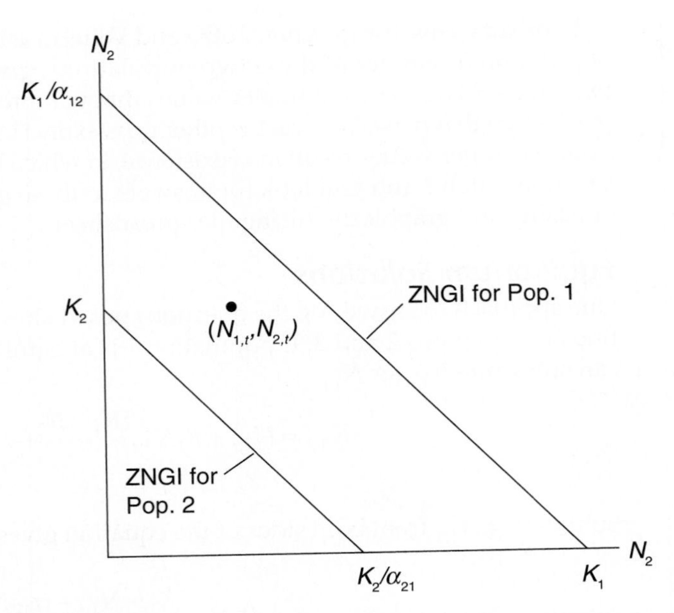

Figure 1 Zero net growth isoclines (ZNGIs) generated by the Lotka-Volterra model of two-species competition. The point $(N_{1,t}, N_{2,t})$ represents the two populations at time t.

In words, if there are K_1/α_{12} members of population 2 in the habitat, there will be no resources left over for population 1, and its numbers will go to zero.

We can find a ZNGI and two points on it for population 2 in the same manner.

$$N_{2,t} = K_2 - \alpha_{21}N_{1,t}$$

$$\text{If } N_{1,t} = 0, \text{ then } N_{2,t} = K_2$$

$$\text{If } N_{2,t} = 0, \text{ then } N_{1,t} = K_2/\alpha_{21}$$

We can draw these isoclines on a linear graph of the two populations as shown in Figure 1. If we plot N_1 on the horizontal axis and N_2 on the vertical, then the solution points found become the intercepts of the isoclines on the axes.

We can graph the populations of the two species at any time by a point on a graph. If the point falls below and/or to the left of a species' isocline, that population will continue to increase. If the point falls above and/or to the right of a species' isocline, that population will decrease. In the case of the point shown in Figure 1, population 1 will increase and population 2 will decrease. As time passes, the point will move downward (population 2 decreases) and to the right (population 1 increases), and the point describing the two populations will trace some trajectory across the graph.

Notice that time does not appear on either axis of this graph. Figure 1 is called a **phase diagram**, and the space bounded by its axes is called **phase space**. You will plot the trajectory of two changing populations through the phase space and from that determine whether one species excludes the other, or if they coexist. The isoclines need not be arranged as shown in Figure 1; their arrangement will depend on the values of K_1, K_2, α_{12}, and α_{21}.

PROCEDURES

The questions Lotka and Volterra asked, and which you will answer in this exercise, are: What values of these parameters will cause population 1 to exclude population 2, and vice versa? What parameter values will allow the two populations to coexist indefinitely? What do these outcomes, and their associated parameter values, mean in ecological terms?

As always, save your work frequently to disk.

INSTRUCTIONS

ANNOTATION

A. Set up the spreadsheet.

1. Open a new spreadsheet and set up titles and column headings as shown in Figure 2.

	A	B	C	D	E	F
1	***Lotka-Volterra Model of Interspecific Competition***					
2						
3		*N*1	*N*2		Parameters	
4		End points			*R*1 -->	1.00
5	*N*1 = 0 -->	0	1000	<-- *N*2 = *K*2	*K*1 -->	1200
6	*N*1 = *K*2/*a*21 -->	2000	0	<-- *N*2 = 0	*a*12 -->	0.75
7	*N*1 = *K*1 -->	1200	0	<-- *N*2 = 0	*R*2 -->	1.00
8	*N*1 = 0 -->	0	1600	<-- *N*2 = *K*1/*a*12	*K*2 -->	1000
9					*a*21 -->	0.50
10	Time (*t*)					
11	0					
12	1					
13	2					

Figure 2

Enter only the text items for now. These are all literals, so just select the appropriate cells and type them in.

You must leave cells B10 and C10 empty for your graphs to come out properly.

The values in cells B5 through C8 are the coordinates of the endpoints of the ZNGIs for the two species. How we got these values will be explained in subsequent steps.

2. Set up a linear time series from 0 to 50 in cells A11 through A61.

See the exercise "Spreadsheet Hints and Tips" for details.

3. Enter the values shown for the parameters.

These are in cells F4 through F9. Do not enter anything in cells B5 through C8 yet.

4. Enter zeros in cells B5, C6, C7, and B8.

These are ZNGI endpoints where each population is itself at zero. Cells B5 through C8 hold coordinates for the endpoints of the two ZNGIs. You must lay out these endpoint cells as shown for your graphs to work properly.

5. In cells B7 and C5, enter formulae to echo the carrying capacities of populations 1 and 2, respectively.

In cell B7, enter the formula **=F5**.

In cell C5, enter the formula **=F8**.

These are ZNGI endpoints where the competing population is at zero. When you change carrying capacities later in the exercise, your changes will automatically be carried over to the ZNGI endpoints.

6. Enter formulae to calculate the other ZNGI endpoints.

In cell B6, enter the formula **=F8/F9**. This corresponds to $N_{1,t} = K_2/\alpha_{21}$.

In cell C8, enter the formula **=F5/F6**. This corresponds to $N_{2,t} = K_1/\alpha_{12}$.

7. Enter initial population sizes ($N_{1,0}$ and $N_{2,0}$).

In cell B11, enter the value 100. In cell C11, enter the value 50. You will change these values later.

8. Enter formulae to calculate populations sizes at times $t = 0$ through $t = 50$.

In cell B12, enter the formula **=B11+$F$4*B11*($F$5-B11-$F$6*C11)/$F$5**. This corresponds to Equation 2:

$$N_{1,t+1} = N_{1,t} + R_1 N_{1,t} \frac{K_1 - N_{1,t} - \alpha_{12} N_{2,t}}{K_1}$$

In cell C12, enter the formula **=C11+$F$7*C11*($F$8-C11-$F$9*B11)/$F$8**. This corresponds to Equation 3:

$$N_{2,t+1} = N_{2,t} + R_2 N_{2,t} \frac{K_2 - N_{2,t} - \alpha_{21} N_{1,t}}{K_2}$$

Be sure to use absolute and relative addresses as shown.

9. Copy and paste the formulae in cells B12 and C12 down their columns through row 51.

See "Spreadsheet Hints and Tips" for details on copying and pasting.

B. Create graphs.

1. Graph N_1 and N_2 (vertical axis) against time (horizontal axis).

Use an XY graph (scatterplot). Include only cells A11 through C51 in the block of data to graph. Leave out the ZNGI endpoints (cells B5 through C8).
Use the *second* Chart Wizard dialog box to name your series so that they will be labeled properly in the legend.

In the dialog box (Figure 3), click the Series tab. Select Series1 and type "Pop 1" in the box to the right. Then select Series 2 and type "Pop 2" in the box. Your finished graph should resemble Figure 4.

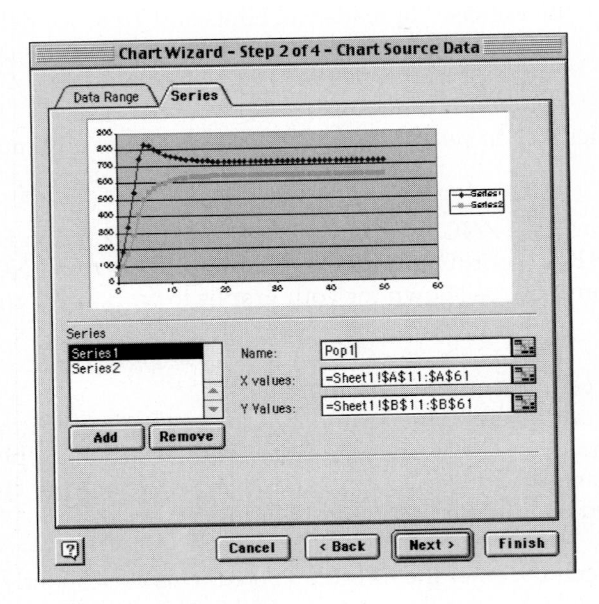

Figure 3

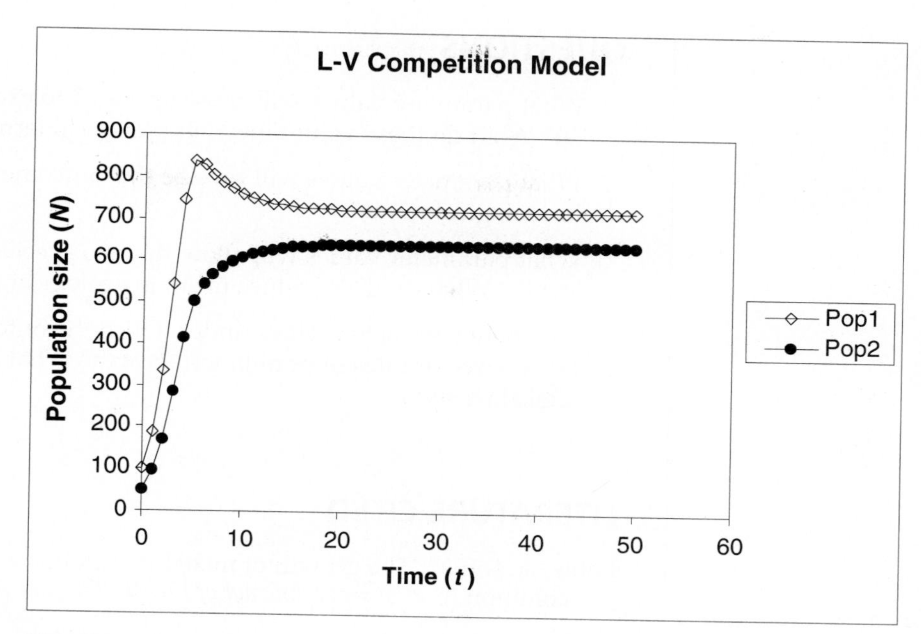

Figure 4

2. Graph N_2 (vertical axis) against N_1 (horizontal axis).

Include cells B5 through C61 in the block to graph—in other words, this time *include* the ZNGI endpoints, but *leave out* "Time" (column A). Use an XY graph (scatterplot). Your graph should resemble Figure 5.

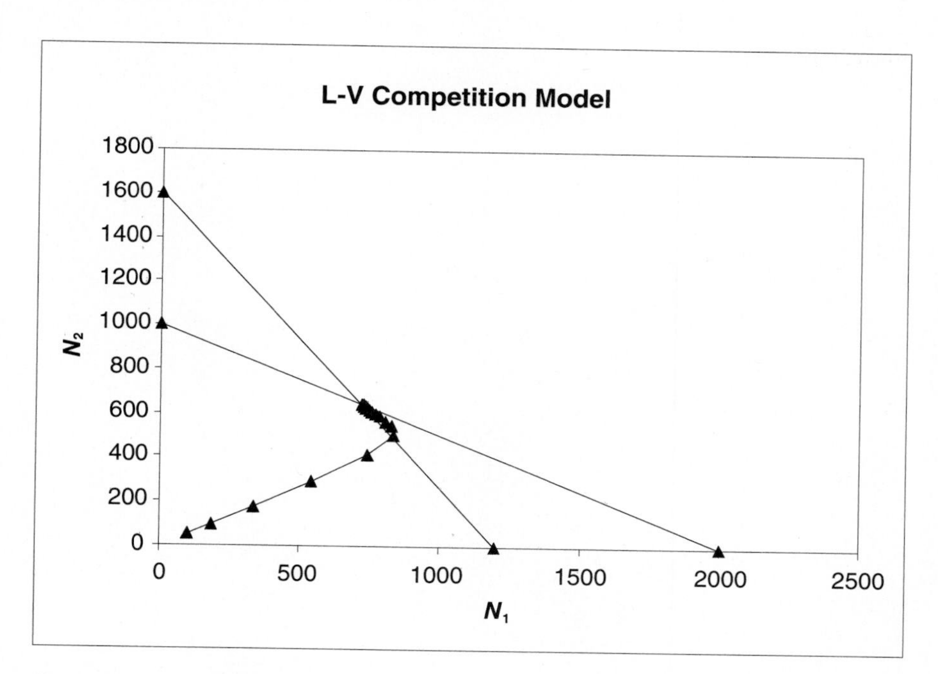

Figure 5

Unfortunately, the program does not label the ZNGI endpoints for you. You will have to identify each endpoint by its coordinates in the spreadsheet. In Figure 5, the top-left endpoint is $(0, K_1/\alpha_{12})$; the lower-left endpoint is $(0, K_2)$; the bottom-right endpoint is $(K_2/\alpha_{21}, 0)$; and the bottom-left endpoint is $(K_1, 0)$.

QUESTIONS

1. What parameter values will cause species 1 to exclude species 2 from the habitat? What do these values mean in ecological terms?
2. What parameter values will reverse this outcome? What do these values mean in ecological terms?
3. What parameter values will allow the two species to coexist indefinitely and stably? What do these values mean in ecological terms?
4. Are there parameter values under which the outcome depends on initial population sizes or rates of population growth? What do these values mean in ecological terms?

LITERATURE CITED

Lotka, A. J. 1932. The growth of mixed populations: two species competing for a common food supply. *Journal of the Washington Academy of Sciences* 22: 461–469.

10 PREDATOR-PREY DYNAMICS

Objectives

- Set up a spreadsheet model of interacting predator and prey populations.
- Modify the model to include an explicit carrying capacity for the prey population, independent of the effect of predation.
- Explore the effects of different prey reproductive rates on the dynamics of both models.
- Explore the effects of different predator attack rates and reproductive efficiencies on the dynamics of both models.
- Evaluate the stability of these models.
- Evaluate these models in comparison to real predator and prey populations.

Suggested Preliminary Exercises: Geometric and Exponential Population Models; Logistic Population Models

INTRODUCTION

In this exercise, you will set up a spreadsheet model of interacting predator and prey populations. You will begin with the classic Lotka-Volterra predator-prey model (Rosenzweig and MacArthur 1963), which treats each population as if it were growing exponentially. After exploring the predictions of this model, you will modify it to include refuges for the prey and see how this changes the behavior of the model.

Next, you will modify the model of the prey population to include an explicit carrying capacity. This reflects the idea that the prey population may be limited by available resources in addition to any limitation by the effects of predation.

Finally, you may modify the predator model to include an explicit carrying capacity. This would represent some limitation on the predator population other than the availability of prey. Such limitation might arise from other required resources or from direct interference among predators.

Model Development

This exercise departs somewhat from the format of others in this book, because we want to follow the progression of increasingly complex and realistic models outlined above. You will build the simplest model first, make some graphs, and answer some questions about the model and its ecological meaning. Then you will return to the spreadsheet to modify the model, reexamine the same questions, and repeat this process a third time.

In the models that follow, we will use the symbols explained in Table 1.

TABLE 1 Symbols used in predator-prey models

Symbol	Name	Description
C_t	Predator population	Think "Consumer"
V_t	Prey population	Think "Victim"
R	Prey population growth	Per capita growth rate of prey population
K_c	Predator carrying capacity	Maximum sustainable predator population
K_v	Prey carrying capacity	Maximum sustainable prey population
q	Predator starvation rate	Per capita rate of mortality of predators due to starvation
a	Attack rate	The ability of a predator to find and consume prey
f	Conversion efficiency	The efficiency with which a predator converts consumed prey into predator offspring

First Model: A Classical Lotka-Volterra Predator-Prey Model

To begin, we will build a discrete-time version of the continuous-time model developed by Alfred Lotka and Vito Volterra. In this model, neither prey population nor predator population has an explicit carrying capacity. Be aware, however, that either or both may have an implicit carrying capacity imposed by the interaction between the two populations.

To model the prey population, we begin with a basic geometric model for the prey population

$$V_{t+1} = V_t + RV_t$$

and subtract the number of prey individuals killed by predators in the interval from t to $t + 1$. This number killed will depend on the number of predators: the more predators, the more prey they will kill. It will also depend on the number of prey available: the more prey, the more successful the predators. Finally, it will depend on the **attack rate**: the ability of a predator to find and consume prey. The number of prey killed in one time interval will be the product of these, or using the symbols given above, aC_tV_t. The equation for the prey population thus becomes

$$V_{t+1} = V_t + RV_t - aC_tV_t \qquad \text{Equation 1}$$

In words, the prey population grows according to its per capita growth rate minus losses to predators. Losses are determined by attack rate, predator population, and prey population.

To model the predator population, we also begin with an exponential model, in concept. However, there is a wrinkle in this model, because we cannot assume a constant per capita rate of population growth. There is no simple R for the predator population because its growth rate will depend on how many prey are caught. As in the prey model, the number of prey caught will be aC_tV_t. The growth of the predator population will depend on this number, and on the efficiency with which predators convert consumed prey into predator offspring. We will represent this conversion efficiency with the parameter f, so the per capita population growth of predators will be afV_tC_t. We should reduce this predator population growth by some quantity to represent the starvation rate of predators who fail to consume prey. This will be the product of the per capita starvation rate times the predator population: qC_t. Taking all this into account, we can write an equation for the predator population:

$$C_{t+1} = C_t + afV_tC_t - qC_t \qquad \text{Equation 2}$$

In words, the predator population grows according to the attack rate, conversion efficiency, and prey population, minus losses to starvation. Note that the product afV_t acts as the predator's R.

Having created these models, we can ask several questions about the interaction they portray, such as

- Under what conditions (i.e., parameter values) will the predator population drive the prey to extinction?
- Under what conditions will the predator population die off, leaving the prey population to expand unhindered?
- Under what conditions will predator and prey populations both persist indefinitely? What will be their population dynamics while they coexist? In other words, will one or both populations stabilize, or will they continue to change over time?

Equilibrium Solutions

As we did in the Interspecific Competition exercise, we will begin to answer these questions by seeking equilibrium solutions to Equations 1 and 2. For the prey population, we want to find values of predator and prey population sizes at which the prey population remains stable. In other words, we want to solve for $\Delta V_t = 0$.

Beginning with Equation 1

$$V_{t+1} = V_t + RV_t - aC_tV_t$$

we subtract V_t from both sides, and get

$$V_{t+1} - V_t = RV_t - aC_tV_t$$

Because $V_{t+1} - V_t = \Delta V_t$ we can substitute into the equation and get

$$\Delta V_t = RV_t - aC_tV_t$$

We are looking for a solution when $\Delta V_t = 0$, so we substitute again:

$$0 = RV_t - aC_tV_t$$

Adding aC_tV_t to both sides gives us

$$aC_tV_t = RV_t$$

Dividing both sides by V_t, we get

$$aC_t = R$$

Dividing both sides by a gives us our solution:

$$C_t = R/a \qquad \text{Equation 3}$$

In words, the prey population reaches equilibrium when the predator population equals the prey's per capita growth rate divided by the predator's attack rate. Note that this is a constant. Strangely, the equilibrium size of the prey population is not determined by this solution, which says, in effect, that the prey population can be stable at any size as long as the predator population is at the specified size.

For the predator population, we follow the same strategy, and solve for $\Delta C = 0$.

Beginning with Equation 2,

$$C_{t+1} = C_t + afV_tC_t - qC_t$$

we subtract C_t from both sides, and get

$$C_{t+1} - C_t = afV_tC_t - qC_t$$

Because $C_{t+1} - C_t = \Delta C_t$, we can substitute into the equation and get

$$\Delta C_t = afV_tC_t - qC_t$$

We are looking for a solution when $\Delta C_t = 0$, so we substitute again:

$$0 = afV_tC_t - qC_t$$

Adding qC_t to both sides gives us

$$qC_t = afV_tC_t$$

Dividing both sides by C_t, we get

$$q = afV_t$$

Dividing both sides by af gives us our solution:

$$q/af = V_t \qquad \text{Equation 4}$$

In words, the predator population reaches equilibrium when the prey population equals the predator's starvation rate over the product of attack rate times conversion efficiency. Note that this is also a constant, and like the solution for the prey population, it does not specify the equilibrium size of the predator population, only the size of the prey population at which the predators are at equilibrium.

As we did in the model of interspecific competition, we can plot the population sizes of the two interacting populations on the two axes of a graph (Figure 1). The equilibrium solutions (Equations 3 and 4) then become straight-line zero net growth isoclines

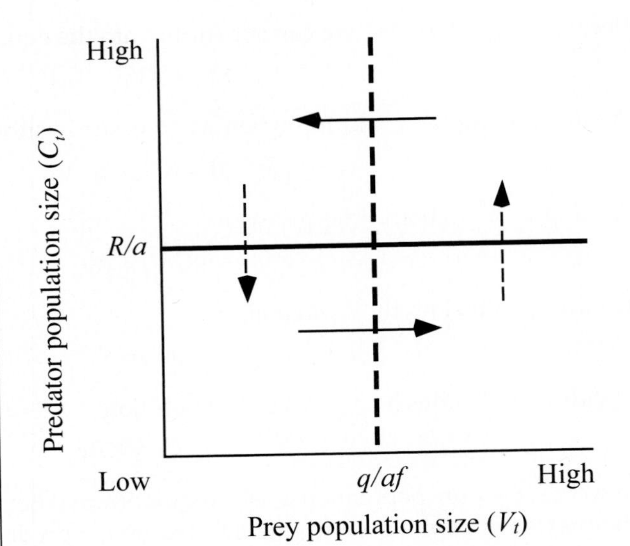

Figure 1 Graph of prey and predator zero net growth isoclines (ZNGIs), according to the Lotka-Volterra model of predator-prey dynamics. The horizontal line is the ZNGI for the prey population, and horizontal arrows show areas of population increase or decrease for the prey population. The vertical line is the ZNGI for the predator population, and vertical arrows show areas of increase or decrease for the predator population.

(ZNGIs), as they did in the interspecific competition model. On this graph, the ZNGI for the prey population is a horizontal line at $C_t = R/a$ (the solid line in Figure 1), below which the prey population increases, and above which it decreases (solid arrows). The ZNGI for the predator population is a vertical line at $V_t = q/af$ (dashed line), to the left of which the predator population decreases, and to the right of which it increases (dashed arrows). Where the two lines cross—at the point $[(q/af), (R/a)]$—the two populations are at equilibrium. As in the Interspecific Competition exercise, the two populations are represented by a point on this phase diagram, and that point will trace out a trajectory through phase space as the populations change in size.

As discussed in most ecology texts, the continuous-time Lotka-Volterra model predicts that the point representing the two populations will cycle endlessly around the point where the two ZNGIs cross. The discrete-time model, however, behaves rather differently, as you will discover.

PROCEDURES

We will use the spreadsheet to explore the behavior of the model developed so far before we introduce the models with explicit prey and predator carrying capacities.

As always, save your work frequently to disk.

INSTRUCTIONS	ANNOTATION

Part 1. Discrete-Time Version of the Lotka-Volterra Model

A. Set up the spreadsheet.

1. Open a new spreadsheet and set up titles and column headings as shown in Figure 2.

Enter only the text items for now. These are all literals, so just select the appropriate cells and type them in. Note that cells B12 through C13 must be empty.

	A	B	C	D	E	F	G	H
1	***Predator-Prey Dynamics***							
2	Uses an exponentially-growing prey population, with an additional term for losses to predators.							
3	Uses an exponentially-growing predator population with per capita pop growth rate determined							
4		by prey capture and conversion efficiency.						
5								
6		**Zero net growth isoclines**			**Prey parameters**		**Predator parameters**	
7		3649.232	25.000		*R*	0.250	Starvation rate (*q*)	0.100
8		0.000	25.000				Conversion efficiency (*f*)	0.008
9		0.000	0.000				Attack rate (*a*)	0.010
10		1250.000	0.000					
11		1250.000	41.999					
12								
13	Time							
14	0	1000.000	20.000					
15	1	1050.000	19.600					
16	2	1106.700	19.286					

Figure 2

2. Set up a linear series from 0 to 100 in column A (cells A14–A114).

Enter the value 0 in cell A14.
Enter the formula **=A14+1** in cell A15. Copy this formula down to cell A114.

3. Enter the values shown for the parameters R, q, f, and a.

Type the values shown into cells F7, H7, H8, and H9.
Cells F8 and H10 remain empty for now.

4. Enter the initial population sizes (V_0 and C_0).

Enter the value 1000 into cell B14.
Enter the value 20 into cell C14.
Leave cells B12 through C13 empty.

5. Enter formulae and values into cells B7 through C11 to define the prey and predator ZNGIs.

This will force the spreadsheet to plot the ZNGIs on the graph, as shown in Figure 1.

In cell B7, enter the formula **=MAX(B14:B114)**.
In cell C7, enter the formula **=$F$7/$H$9**. This corresponds to R/a, the equilibrium value of the prey population (see Equation 3).

Cells B7 and C7 are the coordinates of the right-hand end of the prey ZNGI. Of course, this line extends infinitely to the right, but we cut it off even with the maximum actual value of the prey population so that we can graph our results.

In cell B8, enter the value 0. Copy the formula from cell C7 into cell C8.
Cells B8 and C8 are the coordinates of the point where the prey ZNGI intersects the predator (vertical) axis.

In cells B9 and C9, enter the value 0.
Cells B9 and C9 are the coordinates of the origin of the graph. This is a trick to get us from the prey ZNGI to the predator ZNGI without drawing extraneous lines on the graph.

In cell B10 enter the formula **=$H$7/($H$9*$H$8)**. This corresponds to q/af, the equilibrium value of the predator population (see Equation 4).
In cell C10, enter the value 0.
Cells B10 and C10 are the coordinates of the point where the predator ZNGI intersects the prey (horizontal) axis.

Copy the formula from cell B10 into cell B11.
In cell C11, enter the formula **=MAX(C14:C114)**.
Cells B11 and C11 are the coordinates of the upper end of the predator ZNGI. Like the prey ZNGI, this line is infinitely long, but we truncate it at the maximum predator population for convenience.

6. Enter a formula to calculate the size of the prey population at time 1.

In cell B15, enter the formula **=IF(B14+$F$7*B14-$H$9*C14*B14>0, B14+$F$7*B14-$H$9*C14*B14,0)**.

B14+$F$7*B14-$H$9*C14*B14 corresponds to Equation 1,

$$V_{t+1} = V_t + RV_t - aC_tV_t$$

However, if you simply use Equation 1, it is likely to produce negative population sizes, which make no sense biologically. We use the **IF()** function here to prevent this population from going negative. The formula says, "Calculate the prey population according to Equation 1, and if the result is greater than zero, use it. If the result is zero or less, use zero."

You can simplify the task of entering this formula if you type it in through the ">0," copy the part between the left parenthesis and the ">" sign, and paste it after the comma. Then type in the second comma, followed by a zero, and close the parentheses.

7. Enter a formula to calculate the size of the predator population at time 1.

In cell C15, enter the formula **=IF(C14+$H$8*$H$9*B14*C14$H$7*C14>0, C14+$H$8*$H$9*B14*C14-$H$7*C14,0)**.
C14+$H$8*$H$9*B14*C14-$H$7*C14 corresponds to Equation 2,

$$C_{t+1} + afV_tC_t - qC_t$$

Here again, we use the **IF()** function to prevent the population from going negative. You can use the same shortcut to enter this formula as in the previous step.

8. Copy the formulae from cells B16 and C16 down their columns.

Select cells B15 through C15. Copy.
Select cells B16 through C114. Paste.

9. Save your work.

B. Create graphs.

1. Graph prey and predator populations against time. Edit your graph for readability.

Select cells A14 through C114. Follow the usual procedure to make an XY graph. In the second Chart Wizard dialog box, click on the Series tab, and use the boxes to name Series 1 "Prey" and Series 2 "Predator."

Be aware that the predator population is plotted on a different scale (the right-hand y-axis) than the prey population (the left-hand y-axis). This is necessary because the two cover such different ranges.

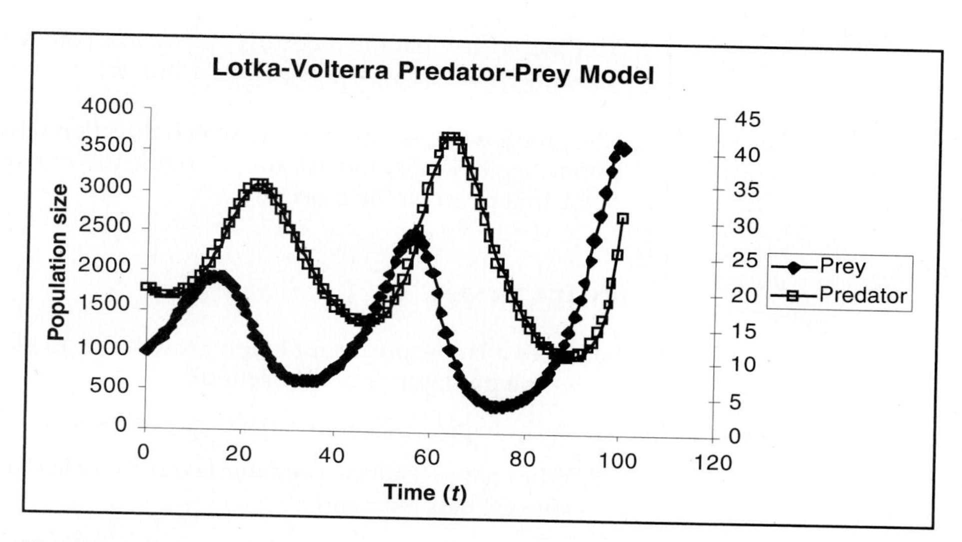

Figure 3

After you've finished the graph, double-click on a data point in the line for the predator population. This line will lie almost on top of the x-axis, so it may take several tries to select the data series rather than the axis. In the Format Data Series dialog box, click on the Axis tab, and select Secondary Axis. This will cause the predator population to be plotted on a separate y-axis, with a different scale from that of the prey population. Your graph should resemble Figure 3.

See Exercise 8, "Logistic Population Models," for details on creating a second y-axis.

2. Graph predator population (y-axis) against prey population (x-axis), as in the standard presentation of the Lotka-Volterra model. Edit your graph for readability.

Select cells B7 through C114 and make an XY graph.

In the third Chart Wizard dialog box, click the Legend tab and click in the Show Legend checkbox, to prevent the legend from being shown (the check mark in the box should disappear). Your graph should resemble Figure 4.

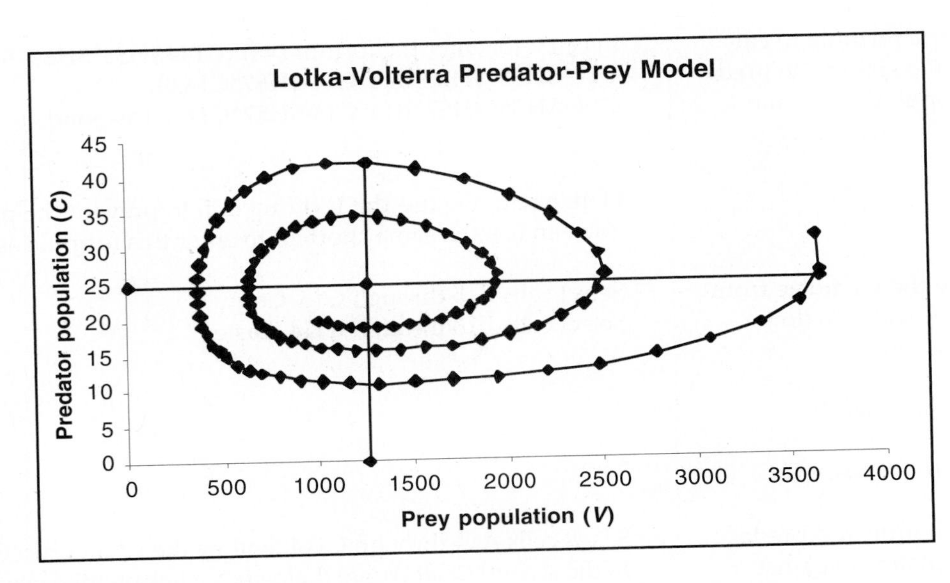

Figure 4

You should see that the trajectory spirals in a counterclockwise direction. Your graph will show the two ZNGIs, but unfortunately will not label their endpoints.

The graph will also not indicate which direction (clockwise or counterclockwise) the population trajectory moves. You can figure this out by locating the point (V_0, C_0), which is the first point on the trajectory.

QUESTIONS

1. Does a larger prey population growth rate (R) increase or decrease the stability of the predator-prey interaction?
2. What happens if the predators starve more quickly? Less quickly?
3. What happens if the predator is more efficient at converting prey into offspring? Less efficient?
4. What happens if the predator is better at finding prey? Worse?
5. Is the behavior of the model sensitive to starting populations? Begin with populations near the point where the isoclines cross, and move slowly farther out.
6. What is the ultimate outcome of the predator-prey interaction, regardless of parameter values? How does this compare to real predator and prey populations? What factors not included in the model may explain the differences between model predictions and reality?

Modifying the Model to Include Prey Refuges

In the model so far, predators are capable of hunting down every single prey individual. In reality, it is often the case that some prey individuals can escape predation by hiding in refuges, such as burrows, crevices in rocks or coral reefs, etc. Thus, there will always be at least a few prey individuals surviving. These survivors, of course, could potentially breed and replenish the prey population. Does the presence of prey refuges alter the outcome of the model?

INSTRUCTIONS	ANNOTATION
Part 2. Predator-Prey Model with Prey Refuges	
A. Set up the spreadsheet.	
1. Return the parameters to their original values (see Figure 2).	If you wish to retain your existing model, save it under a separate file name before making changes, or copy your spreadsheet to a new worksheet and make changes on the copy.
2. Modify your existing formula for the prey population at time 1 to include prey refuges.	Edit the formula in cell B15 by changing the zeros to tens. The new formula should read **=IF(B14+$F$7*B14-$H$9*B14*C14>10,B14+$F$7*B14-$H$9*B14*C14,10)**. This formula says to calculate the size of the prey population at time 1 based on its size at time 0 and losses to predation. If that size is greater than 10, use it; otherwise, make the prey population 10. The biological interpretation is that at least 10 prey individuals survive in refuges, regardless of the number or effectiveness of predators.
3. Copy the modified formula down its column.	Copy the formula in cell B15 into cells B16 through B114.
4. Try other values for the number of survivors.	Repeat steps 2 and 3, using some number other than 10.
B. Create graphs.	You do not need to make any new graphs or edit your existing ones. Your changes will be automatically reflected in your existing graphs.

QUESTIONS

7. Reinvestigate questions 1–6 on the preceding page, but based on your model with prey refuges.

Modifying the Model to Include a Prey Carrying Capacity

The classical continuous-time Lotka-Volterra predator-prey model predicts that prey and predator populations will cycle endlessly around their equilibrium values. Some real predator-prey systems, such as the snowshoe hare and Canada lynx, display cycles that resemble these, but others do not. Even in cases of cyclic population dynamics, ecologists seriously question whether the Lotka-Volterra model, with all its simplifying assumptions, accurately reflects reality. A recent model of the hare-lynx cycle (King and Schaffer 2001) includes 17 parameters and variables.

One obvious omission from the Lotka-Volterra model is any limitation on the prey population other than losses to predation. Surely, prey individuals require resources such as food and water, which could potentially limit the size of their population even in the absence of predators. Perhaps including a prey **carrying capacity** in the model would reduce its tendency to cycle, or in the case of the discrete-time model, its tendency toward increasing population fluctuations and eventual extinctions. In other words, if there were a cap on the size of the prey population, that number might also limit the predator population, which in turn might prevent the predators from hunting the prey to extinction and then starving.

We can modify our prey population equation, Equation 1, to include a carrying capacity in the same way we modified our geometric population equation in Exercise 5, "Logistic Population Models." If we let K_v represent the prey carrying capacity (in the absence of predators), we can write

$$V_{t+1} = V_t + RV_t\left(\frac{K_v - V_t}{K_v}\right) - aC_tV_t \qquad \text{Equation 5}$$

If the predator population (C_t) is zero, then losses to predation (aC_tV_t) will be zero, and the prey population will stabilize at K_v. If predators are present, losses to predation will reduce the prey population to some value less than K_v. We will leave the predator equation unchanged for now.

Equilibrium Solution. Because we have not changed the predator equation, its equilibrium solution remains unchanged. However, our change in the prey equation means we must solve the new equation for its equilibrium (ZNGI). We find this by setting $\Delta V_t = 0$.

$$\Delta V_t = V_{t+1} - V_t = RV_t\left(\frac{K_v - V_t}{K_v}\right) - aC_tV_t$$

$$0 = RV_t\left(\frac{K_v - V_t}{K_v}\right) - aC_tV_t$$

$$aC_tV_t = RV_t\left(\frac{K_v - V_t}{K_v}\right)$$

$$aC_t = R\left(\frac{K_v - V_t}{K_v}\right)$$

$$C_t = \frac{R}{a}\left(1 - \frac{V_t}{K_v}\right)$$

$$C_t = \frac{R}{a} - \frac{RV_t}{aK_v}$$

$$C_t = \frac{R}{a} - \frac{R}{aK_v}V_t$$

There's no easy way to express this equilibrium solution in words, but we can deduce some things about it. First, the equation is in the standard form of a straight line ($y = a + bx$), with a slope of $-R/(aK_v)$. Second, if we plug in $V_t = 0$, we find the y-intercept (C-intercept) to be R/a, just as in the classical Lotka-Volterra model. Third, if we plug in $C_t = 0$, we find the x-intercept (V-intercept) to be K_v (see below). This makes sense, because we would expect the prey population to go to K_v if there were no predators present.

$$0 = \frac{R}{a} - \frac{R}{aK_v}V_t$$

$$\frac{R}{aK_v}V_t = \frac{R}{a}$$

$$\frac{V_t}{K_v} = 1$$

$$V_t = K_v$$

INSTRUCTIONS	ANNOTATION
Part 3. Predator-Prey Model with Prey Carrying Capacity	
A. Set up the spreadsheet.	
1. Return the parameters to their original values.	To retain your existing model, save it under a separate file name before making changes, or copy your spreadsheet to a new worksheet and make changes on the copy.
2. Modify your existing spreadsheet headings to include a prey carrying capacity.	Edit the text in cell A2 to reflect the change to a logistically-growing prey population. In cell E8, enter the label "Kv". In cell F8, enter the value 2000.
3. Enter formulae and values into cells B7 through C11 to define the prey and predator ZNGIs.	Your graphs will look very odd while you are making these changes. Ignore them for now—the errors will disappear after you complete the changes to your spreadsheet. In cell B7, enter the formula **=$F$8**. In cell C7, enter the value 0. Cells B7 and C7 are the coordinates of the point where the prey ZNGI crosses the prey axis, (K_v,0). Leave cells B8 through C11 unchanged.
4. Modify the formula for the prey population at time 1 to include the prey carrying capacity.	In cell B15, enter the formula **=IF(B14+$F$7*B14*($F$8-B14)/$F$8-$H$9*B14*C14>0, B14+$F$7*B14*($F$8-B14)/$F$8-$H$9*B14*C14,0)**. **B14+$F$7*B14*($F$8-B14)/$F$8-$H$9*B14*C14** corresponds to the equation $V_{t+1} = V_t + RV_t\left(\frac{K_v - V_t}{K_v}\right) - aC_tV_t$ which is our logistic model of the prey population. Again, we use the **IF()** function to prevent the population from going negative. Note that we removed the refuges from the prey population by changing the >10 back to >0. We do this so we can see the effects of a prey carrying capacity without clouding the issue with refuges.
5. Copy the modified formula down its column.	Select cell B15. Copy. Select cells B16 through B115. Paste. Your spreadsheet should resemble Figure 5.

	A	B	C	D	E	F	G	H
1	*Predator-Prey Dynamics*							
2	Uses a logistically-growing prey population, with an additional term for losses to predators.							
3	Uses an exponentially-growing predator population with per capita pop growth rate determined							
4		by prey capture and conversion efficiency.						
5								
6		**Zero net growth isoclines**			**Prey parameters**		**Predator parameters**	
7		2000.000	0.000		*R*	0.250	Starvation rate (*q*)	0.100
8		0.000	25.000		K_v	2000.000	Conversion efficiency (*f*)	0.008
9		0.000	0.000				Attack rate (*a*)	0.010
10		1250.000	0.000					
11		1250.000	19.600					
12								
13	Time							
14	0	1000.000	20.000					
15	1	925.000	19.600					
16	2	867.997	19.090					

Figure 5

B. Create graphs.

You do not need to make any new graphs. Your existing graphs will automatically reflect the changes in your spreadsheet. Edit the graph titles to distinguish them from graphs of the classical Lotka-Volterra model. Your graphs should now resemble Figures 6 and 7.

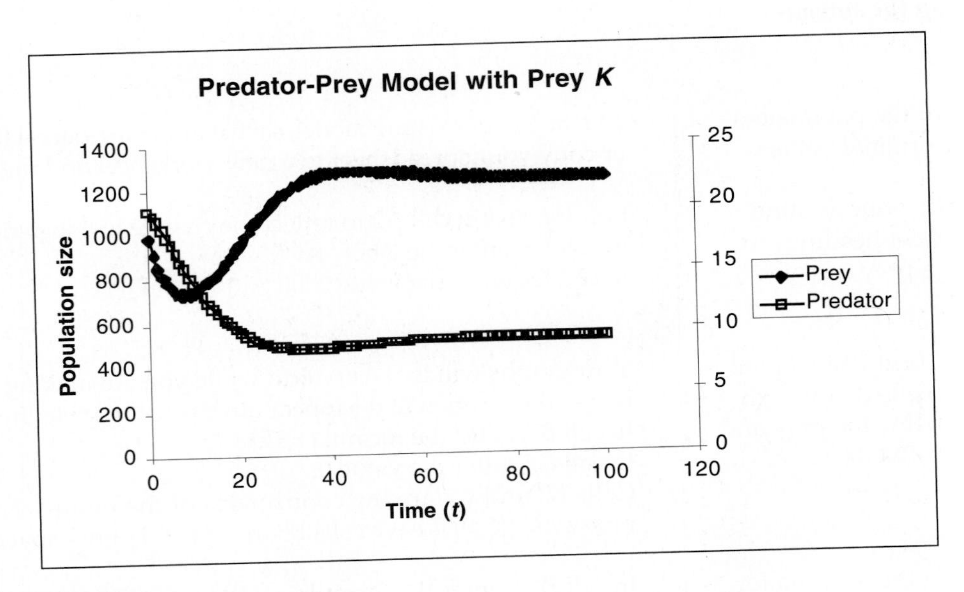

Figure 6

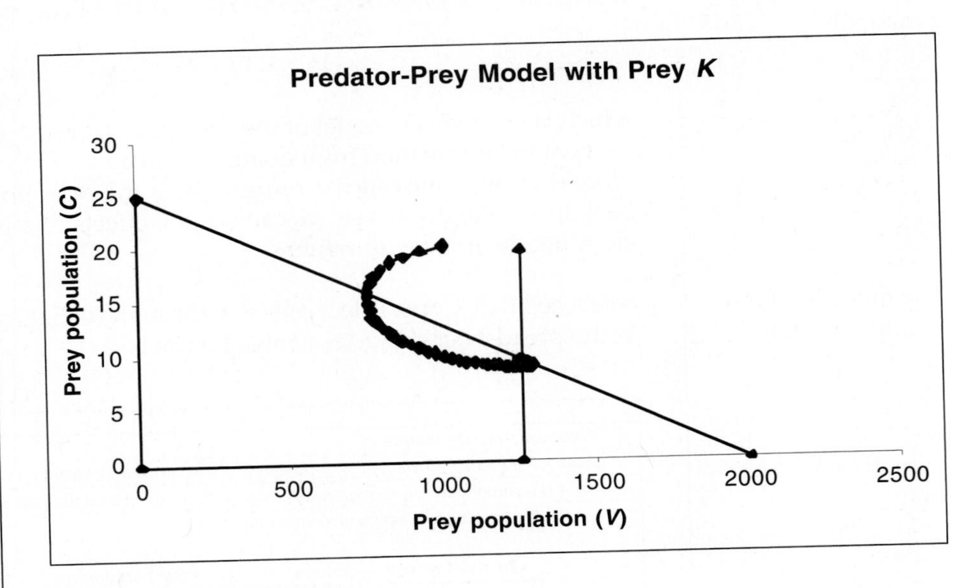

Figure 7

QUESTIONS

8. Reinvestigate questions 1–6 but based on your model with a carrying capacity for the prey population.

Modifying the Model to Include Carrying Capacities for Prey and Predator

It is quite conceivable that the predator population may have a carrying capacity imposed by environmental constraints other than prey availability. Factors imposing such a limitation might include mutual interference between predators (fighting over prey or hunting territories) or limited availability of other essential resources, such as water, burrow sites, or something else. If prey are superabundant (i.e., supply exceeds demand and no predators starve), then the predator population (C_t) will increase to its carrying capacity (K_c), but not beyond it.

We can include a predator carrying capacity in the same way we included a prey carrying capacity. We will modify the predator equation as follows:

$$C_{t+1} = C_t + afV_tC_t\left(\frac{K_C - C_t}{K_C}\right) - qC_t$$

Will the introduction of a predator carrying capacity change the behavior of the model? Try predicting the result before exploring it with the spreadsheet.

Equilibrium Solution. As before, we will have to re-derive our equilibrium solution for this modified equation. Letting $\Delta C_t = 0$, we get

$$0 = afV_tC_t\left(\frac{K_c - C_t}{K_c}\right) - qC_t$$

$$qC_t = afV_tC_t\left(\frac{K_c - C_t}{K_c}\right)$$

$$q = afV_t\left(\frac{K_c - C_t}{K_c}\right)$$

$$\frac{q}{af} = V_t\left(\frac{K_c - C_t}{K_c}\right)$$

$$\frac{qK_c}{af(K_c - C_t)} = V_t$$

In words, "Gadzooks!" But it turns out this produces a predator ZNGI that crosses the x-axis (V-axis) at the same point as before, $V = q/af$ (plug in $0 = C_t$ and solve). However, instead of a straight vertical line, it gives us a curve that leans over to the right, as you will see in the spreadsheet graph. The ZNGI equation makes no sense at $C_t = K_c$, because the denominator of the term on the left becomes undefined, and then negative.

INSTRUCTIONS	ANNOTATION
Part 4. Predator-Prey Model with Carrying Capacities for Prey and Predator	
A. Set up the spreadsheet.	If you wish to retain your existing model, save it under a separate file name before making changes, or copy your spreadsheet to a new worksheet and make changes on the copy.
1. Change your parameters to these values: $q = 0.25, f = 0.20, a = 0.005$	Enter the values given into cells H7, H8, and H9, respectively.
2. Modify your existing spreadsheet headings to include a predator carrying capacity.	Edit the text in cell A3 to reflect the change to a logistically growing predator population. In cell G10, enter the label "Kc". In cell H10, enter the value 100.
3. Change the initial population sizes to $V_0 = 100$, $C_0 = 10$.	Enter the given values into cells B14 and C14, respectively.
4. Enter formulae and values into cells B8 through C12 to define the prey and predator ZNGIs.	Your graphs will look very odd while you are making these changes. Ignore them for now—the errors will disappear after you have completed all the changes to your spreadsheet. Leave cells B8 through C10 unchanged. Delete the contents of cells B11 and C11.
5. Modify the formula for the predator population at time 1 to include the predator carrying capacity.	In cell C15, enter the formula **=IF(C14+$H$8*$H$9*B14*C14*($H$10-C14)/$H$10-$H$7*C14>0,C14+$H$8*$H$9*B14*C14*($H$10-C14)/$H$10-$H$7*C14,0)**. This corresponds to Equation 6: $$C_{t+1} = C_t + afV_tC_t\left(\frac{K_C - C_t}{K_C}\right) - qC_t$$ Again, we use the **IF()** function to prevent the population from going negative.
6. Copy the modified formula down its column.	Select cell C15. Copy. Select cells C16 through B114. Paste.
7. Set up a new data series in column D to graph the predator ZNGI.	We need to do this because this ZNGI is not a straight line, so we must calculate many points along it, and connect them with a line. We will use the formula we derived above to express the predator ZNGI as a function of prey population size: $$C_t = K_c - \frac{qK_c}{afV_t}$$ We must use a little spreadsheet trickery to make this come out right on the graph. Indeed, even with our trickery, the ZNGI may look a little strange with some parameter values.

In cell B13 enter the formula **=$H$7/($H$9*$H$8)**. This is equal to $q/(af)$. Leave cell C13 empty. In cell D13, enter the value 0.

In cell D14, enter the formula **=IF($H$10-($H$7*$H$10)/($H$9*$H$8*B14)>0,$H$10-($H$7*$H$10)/($H$9*$H$8*B14),0)**.
Use the same shortcut as before to enter this formula.

This formula requires a little explanation. It is the spreadsheet version of the equation for the predator ZNGI (derived above), rewritten as a function of V_t, so that we can plot it on the graph of predator population versus prey population. The derivation is:

$$\frac{qK_c}{af(K_c - C_t)} = V_t$$

$$qK_c = af(K_c - C_t)V_t$$

$$\frac{qK_c}{V_t} = afK_c - afC_t$$

$$afC_t = afK_c - \frac{qK_c}{V_t}$$

$$C_t = \frac{afK_c}{af} - \frac{qK_c}{afV_t}$$

$$C_t = K_c - \frac{qK_c}{afV_t}$$

Copy the formula from cell D14 into cells D15 through D114. Your spreadsheet should look like Figure 8.

	A	B	C	D	E	F	G	H
1	***Predator-Prey Dynamics***							
2	Uses an exponentially-growing prey population, with an additional term for losses to predators.							
3	Uses an exponentially-growing predator population with per capita pop growth rate determined							
4		by prey capture and conversion efficiency.						
5								
6		**Zero net growth isoclines**			**Prey parameters**		**Predator parameters**	
7		2000.000	0.000		R	0.250	Starvation rate (q)	0.250
8		0.000	50.000		K_v	2000.000	Conversion efficiency (f)	0.200
9		0.000	0.000				Attack rate (a)	0.005
10		250.000	0.000				K_c	100.000
11								
12								
13	Time	250.000		0.000				
14	0	100.000	10.000	0.000				
15	1	118.750	8.400	0.000				
16	2	141.687	7.214	0.000				

Figure 8

B. Create graphs.

1. Make a new graph of predator population versus prey population, including the new ZNGIs. Edit your graph for readability. It should resemble Figure 9.

It is possible to edit your existing graph, but that is difficult and prone to error, so it's easier just to start over.

Select cells B7 through D114 and make an XY graph.
Select the predator ZNGI by double-clicking on any data point along it. In the Format Data Series dialog box, click the Patterns tab and choose None for marker style. This will cause the predator ZNGI to be plotted as a line with no data markers, like the prey ZNGI.

2. Do not change your graph of population sizes versus time.

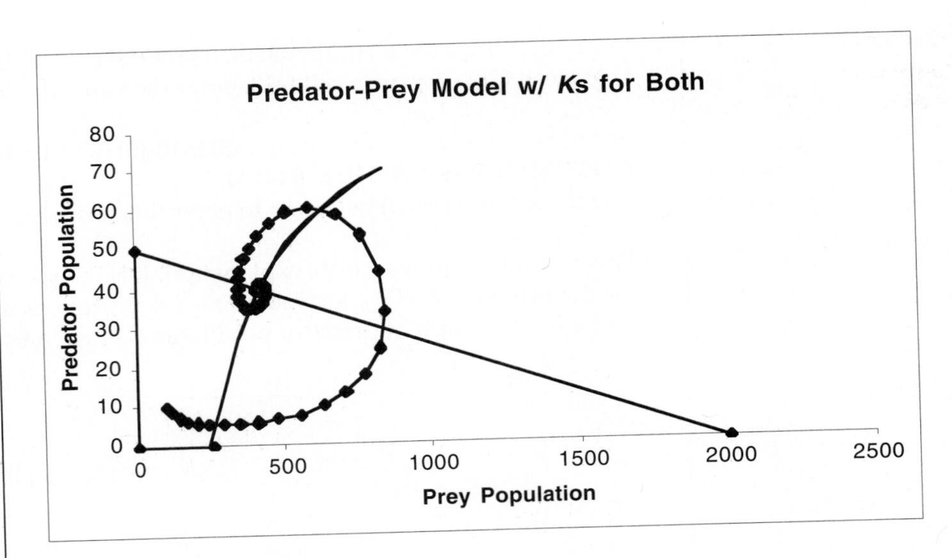

Figure 9

QUESTIONS

9. Reinvestigate questions 1–6 but based on your model with carrying capacities for both prey and predator populations.
10. Attempt to summarize the implications of all the models developed in this exercise.

LITERATURE CITED

King, A. A. and W. M. Schaffer. 2001. The geometry of a population cycle: A mechanistic model of snowshoe hare demography. *Ecology* 82: 814–830.

Rosenzweig, M. L. and R. H. MacArthur. 1963. Graphical representation and stability conditions of predator-prey interactions. *American Naturalist* 97: 209–223.

11 ISLAND BIOGEOGRAPHY

Objectives

- Explore the relationships of immigration and extinction rates and species richness to island area and distance from the mainland.
- Observe the accumulation of species on an island, and the approach of immigration and extinction rates and species richness values to equilibrium.
- Find equilibrium values of immigration and extinction rates and species richness, both graphically and algebraically.
- Understand species-area curves and the underlying mathematical relationships implied.
- Explore the interaction effects of area and distance.

INTRODUCTION

People have long known that larger islands, and islands closer to a mainland, support a greater number of species than smaller or more distant islands. Most ecology textbooks give examples of such species-area and species-distance relationships, not only for islands in the strict sense, but also for habitat islands such as mountaintops and lakes. Few books explicitly state the mathematical relationship between number of species and area or distance, but most show them as straight lines on log-log plots. This should indicate to you that the underlying relationships are power functions. (See Exercise 1, "Mathematical Functions and Graphs," for definitions and examples of power functions and other kinds of functions.) On linear axes, both relationships are curves, hence the term "species-area curve" and what could be called the "species-distance curve."

Having observed and quantified these relationships, ecologists proposed several hypotheses to explain them. One of the best-known hypotheses is the **equilibrium theory of island biogeography** developed by Robert MacArthur and Edward O. Wilson.

The MacArthur-Wilson Model of Island Biogeography

MacArthur and Wilson (1967) modeled **species richness** (the number of species present) on an island as the result of two processes: **immigration** and **extinction**. In their model, species immigrate to an island randomly from a mainland pool. The rate at which new species arrive at the island is determined by three factors:

- The distance of the island from the mainland
- The number of species remaining in the mainland pool that have not already established themselves on the island
- The probability that a given species will disperse from the mainland to the island

The rate at which species on the island go extinct is also determined by three different factors:

- The area of the island
- The number of species present on the island
- The probability that a given species on the island will go extinct

In the simplest version of the model, all species have equal probability of reaching the island and of going extinct once there. The model ignores interactions such as competition, predation, or mutualism between species on the island.

We will develop a spreadsheet model incorporating these ideas. Let us begin with immigration. It seems reasonable to suppose that the farther an island lies from the mainland, the lower the rate of immigration—in other words, immigration is inversely related to distance. Since immigrants are drawn from a finite pool, as more species establish themselves on the island, fewer species will remain in the pool that have not already established themselves on the island. Based on these considerations, we can write a simple equation for the rate of immigration to an island. Let

I = immigration rate (Note: This is overall immigration rate of species to the island, which is different from the probability that any one species will make that journey)
P = total number of species in the mainland pool
S = species richness of the island
D = distance of the island from the mainland
c = colonization probability, or the probability that a given species will make it to the island; here it is assumed to be equal for all species
f = a scaling factor for distance

Note that $(P - S)$ is the number of species in the mainland pool that have not already reached the island. Now we can write an equation for immigration:

$$I = \frac{c(P-S)}{fD} \qquad \text{Equation 1}$$

We must determine a values for c and f from actual data. Based on the work of MacArthur and Wilson, we can begin with reasonable values of $c = 0.10$ and $f = 0.01$. Note that Equation 1 is a power function, in which the variable D is raised to a constant power, –1.

Turning our attention to extinction, we can write a simple equation for that as well. Let

E = extinction rate
S = species richness of the island
A = area of the island
q = extinction probability for a given species (assumed to be equal for all species)
m = a power scaling factor for area

Now we can write an equation for extinction:

$$E = \frac{qS}{A^m} \qquad \text{Equation 2}$$

Values of q and m must be determined from actual data, and based on work by MacArthur and Wilson, we can begin with a reasonable values of $q = 0.20$ and $m = 0.25$.

Note that Equation 2 is also a power function, in which the variable A is raised to a constant power, m.

If you consider Equation 1, you can see that as species accumulate on an island (i.e., as S increases), the immigration rate, I, will decrease. Inspection of Equation 2 shows that as S increases, the extinction rate, E, will increase. At some value of S, immigration and extinction will become equal (i.e., $I = E$), and species richness will come to an equilibrium. This is an equilibrium because every new species immigrating to the island is balanced by one already-established species going extinct, and vice versa.

This is an important point of the model: *Equilibrium species richness is determined by a balance between immigration and extinction*. Note that this is a statement about the model, not about species richness on real islands, which is certainly affected by other factors in addition to immigration and extinction. However, like other simple models, this one has proven fruitful in stimulating thinking and research.

A second important point of the model is that the equilibrium in species richness is a **dynamic equilibrium**. At equilibrium, immigration and extinction rates are equal, but neither is zero. The rate of immigration or extinction at equilibrium species richness is called the **turnover rate**.

According to the model, then, the particular species inhabiting an island continue to change, or turn over, indefinitely—even after species richness has reached equilibrium. That is, species continue to go extinct and are replaced by an equal number of immigrating species. A biologist revisiting the same island at different times would, according to the model, find different sets of species present, but (at least roughly) the same total number of species.

This prediction of continuing turnover is an important feature of MacArthur and Wilson's model. This model is often used in conservation biology to predict the number of species that would be expected to persist or go extinct in nature reserves (which are often habitat islands). However, it is not useful in planning for protecting specific species, because of this prediction of continuing turnover.

PROCEDURES

This exercise is presented in four parts. In each part you will develop a spreadsheet model and make graphs. Between parts, we return to a little mathematical exposition to lay the groundwork for modeling.

First you will build a spreadsheet version of the MacArthur-Wilson model of island biogeography. Using Equations 1 and 2, you will graphically estimate the species richness of an island. In the second part, you explore how the island's area and distance from the mainland affect its species richness. In the third part, you will examine the time-course of species accumulation on an island. In the fourth part, we derive equilibrium solutions for species richness and turnover rate.

As always, save your work frequently to disk.

INSTRUCTIONS	ANNOTATION
A. The MacArthur-Wilson island biogeography model.	
1. Open a new spreadsheet and set up titles and column headings as shown in Figure 1.	Enter the text items and values shown for "Parameters" and "Scaling factors." These are all literals, so just select the appropriate cells and type them in.

	A	B	C	D	E	F	G
1	***Island Biogeography***						
2	Assumes all species have equal dispersal ability and risk of extinction.						
3	Assumes no interaction between species on an island.						
4							
5	**Parameters**					**Scaling factors**	
6	Species pool on mainland (*P*)		1000			For distance (*f*)	0.01
7	Area of island (*A*)		200			For area (*m*)	0.25
8	Distance from mainland (*D*)		300				
9	Colonization probability (*c*)		0.10				
10	Extinction probability (*q*)		0.20			**Equilibium values**	
11						Species richness	
12	**Immigration, extinction, and species richness**					Immigration rate	
13	Fraction of pool	Species richness	Immigration	Extinction		Extinction rate	
14	0.0						
15	0.1						
16	0.2						

Figure 1

2. Set up a series: 0.0, 0.1, 0.2, . . . , 0.9, 1.0 in cells A14–A24.

In cell A14 enter the value 0.
In cell A15 enter the formula **=A14+0.1**.
Copy the formula from cell A15 into cells A16–A24.
This series represents different fractions of the mainland pool present on the island, from 0% to 10% and so on to 100%.

3. In cell B14 enter a formula to calculate the actual number of species present on an island.

In cell B14 enter the formula **=A14*$C$6**. This formula is based on the fraction of the mainland pool in cell A14 and the total number of species in the mainland pool.

4. In cell C14 enter a formula to calculate the rate of immigration to an island already colonized by the number of species in cell B14.

In cell C14 enter the formula **=$C$9*($C$6-B14)/($G$6*$C$8)**. This corresponds to Equation 1:

$$I = \frac{c(P-S)}{fD}$$

5. In cell D14 enter a formula to calculate the rate of extinction on an island already colonized by the number of species in cell B14.

In cell D14 enter the formula **=$C$10*B14/$C$7^$G$7**. This corresponds to Equation 2:

$$E = \frac{qS}{A^m}$$

6. Copy the formulae in cells B14–D14 down their columns to row 24.

Select cells B14–D14. Copy.
Select cells B15–D24. Paste.
Save your work!

7. Graph immigration and extinction rates against species richness.

Select cells B14–D24 and make an XY graph. Edit your graph for readability. It should resemble the graph in Figure 2.

8. Try changing the parameter values in cells C6–C10, one at a time, and observe how equilibrium species richnesss changes.

You should see that smaller or more distant islands have fewer species than larger or closer ones. We will examine these relationships more rigorously in the next part of the exercise.

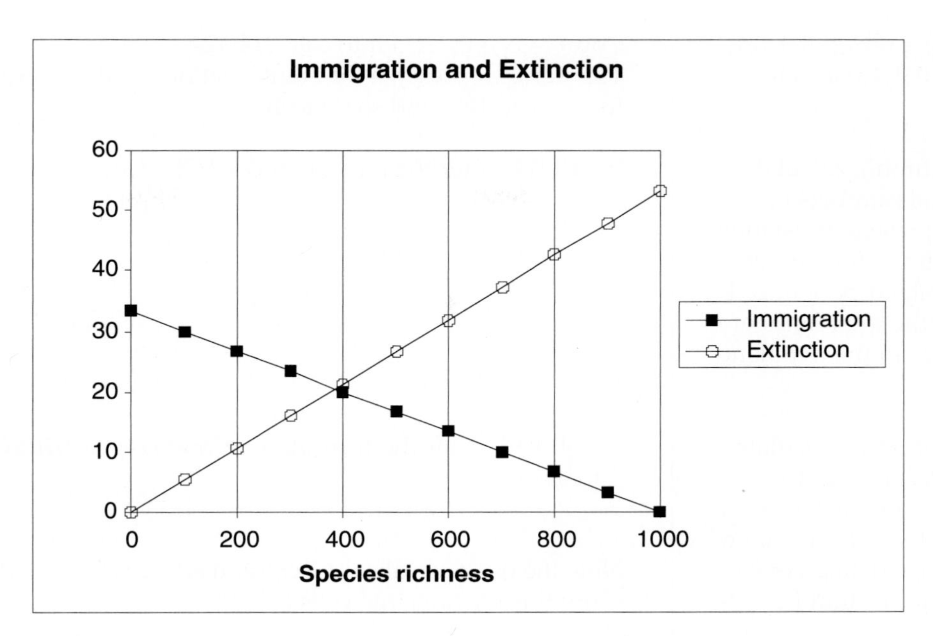

Figure 2

Effects of Island Area and Distance from the Mainland

In Step 8 of the preceding section of the exercise, you experimented with different parameter values to see the effects on species richness. In this section, we will examine the effects of an island's area and its distance from the mainland somewhat more rigorously.

To quantify these effects, let us compare three islands of the same area, but at three distances from the mainland: 0.5, 1.0, and 2.0 times some distance that you specify (in cell C8 of your spreadsheet). Looking at Equation 1, which models the immigration rate, you can see that it includes distance but not area. Accordingly, we will compute immigration rates on these three islands, and estimate the effects on species richness.

We will also compare three islands at the same distance from the mainland, but having three different areas: 0.1, 1.0, and 10.0 times the area that you specified in cell C7 of your spreadsheet. Looking at Equation 2, which models extinction rate, you can see that it includes area but not distance. Accordingly, we will compute extinction rates on these three islands and estimate the effects on species richness.

INSTRUCTIONS

ANNOTATION

B. The effects of distance and area on the MacArthur-Wilson model.

1. Add the column headings shown in Figure 3 to cells I12–P13 of the spreadsheet you set up in Part A (see Figure 1).

These are all literals, so just select the appropriate cells and type them in.

	I	J	K	L	M	N	O	P
12	Fraction	Species	Effect of distance on immigration			Effect of area on extinction		
13	of pool	richness	Imm near	Imm medium	Imm far	Ext small	Ext medium	Ext large
14	0.0							
15	0.1							

Figure 3

2. Set up a series: 0.0, 0.1, 0.2, … , 0.9, 1.0 in cells I14–I24.

Copy cells A14–A24 into cells I14–I24.
This series represents different fractions of the mainland pool present on the island, from 0% to 10% and so on to 100%.

3. In column J, calculate the actual numbers of species present on islands, based on the fraction of the mainland pool in cell I14 and the total number of species in the mainland pool.

In cell J14, enter the formula **= I14*$C$6**. Copy this formula into cells J15–J24.

4. In column L, calculate immigration rates to islands at the distance specified in cell C8, using the species richnesses calculated in column J.

In cell L14, enter the formula **=($C$6-$J14)*$C$9/($C$8*$G$6)**. This corresponds to Equation 1:

$$I = \frac{c(P - S)}{fD}$$

Note the use of an absolute column address for cell **$J14**.
Copy this formula into cells L14–L24.

5. In column K, calculate immigration rates to islands at half the distance specified in cell C8, using the species richnesses calculated in column J.

Copy the formula from cell L14 into cell K14, and edit it to multiply distance (cell C8) by 0.5.
The new formula should read **=($C$6-$J14)*$C$9/($C$8*0.5*$G$6)**.
Copy the formula from cell K14 into cells K15–K24.

6. In column M, calculate immigration rates to islands at 2.0 times the distance specified in cell C8, using the species richnesses calculated in column J.

Copy the formula from cell K14 into cell M14, and edit it to multiply distance (cell C8) by 2.0.
The new formula should read **=($C$6-$J14)*$C$9/($C$8*2.0*$G$6)**.
Copy the formula from cell M14 into cells M15–M24.

7. In column O, calculate extinction rates for islands of the area specified in cell C7, using the species richnesses calculated in column J.

In cell O14, enter the formula **=$J14*$C$10/$C$7^$G$7**. This corresponds to Equation 2:

$$E = \frac{qS}{A^m}$$

Again, note the use of an absolute column address for cell **$J14**.
Copy the formula from cell O14 into cells O15–O24.

8. In column N, calculate extinction rates for islands of 0.1 times the area specified in cell C7, with the species richnesses calculated in column J.

Copy the formula from cell O14 into cell N14, and edit it to multiply area by 0.1.
The new formula should read **=$J14*$C$10/($C$7*0.1)^$G$7**.
Copy the formula from cell N14 into cells N15–N24.

9. In column P, calculate extinction rates for islands of 10.0 times the area specified in cell C7, with the species richnesses calculated in column J.

Copy the formula from cell N14 into cell P14, and edit it to multiply area by 10.0.
The new formula should read **=$J14*$C$10/($C$7*10.0)^$G$7**.
Copy the formula from cell P14 into cells P15–P24.
Save your work!

10. Graph immigration rates for near, medium-distance, and far islands along with the extinction rate for a medium-sized island against species richness.

Select cells J13–M24.
Hold down the control key or while selecting cells O13–O24.
Make an XY graph. Edit your graph for readability. It should resemble the one in Figure 4.

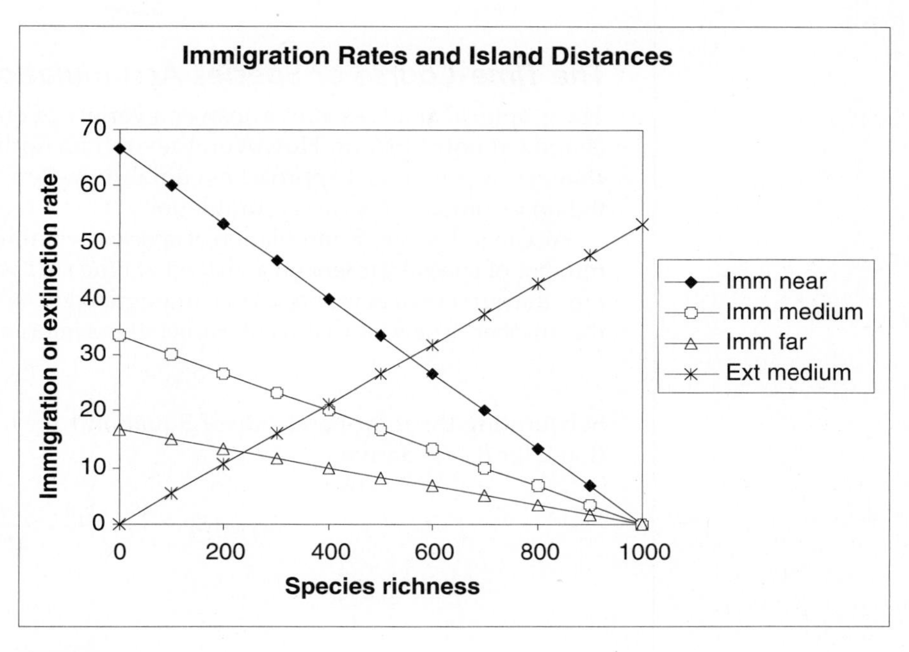

Figure 4

11. Graph extinction rates for small, medium-sized, and large islands, and the immigration rate for a medium-distance island, against species richness.

Select cells J13–J24.
Hold down the control key or and select cells L13–L24.
Hold down the control key or and select cells N13–P24.
Make an XY graph. Edit your graph for readability. It should resemble the one in Figure 5.

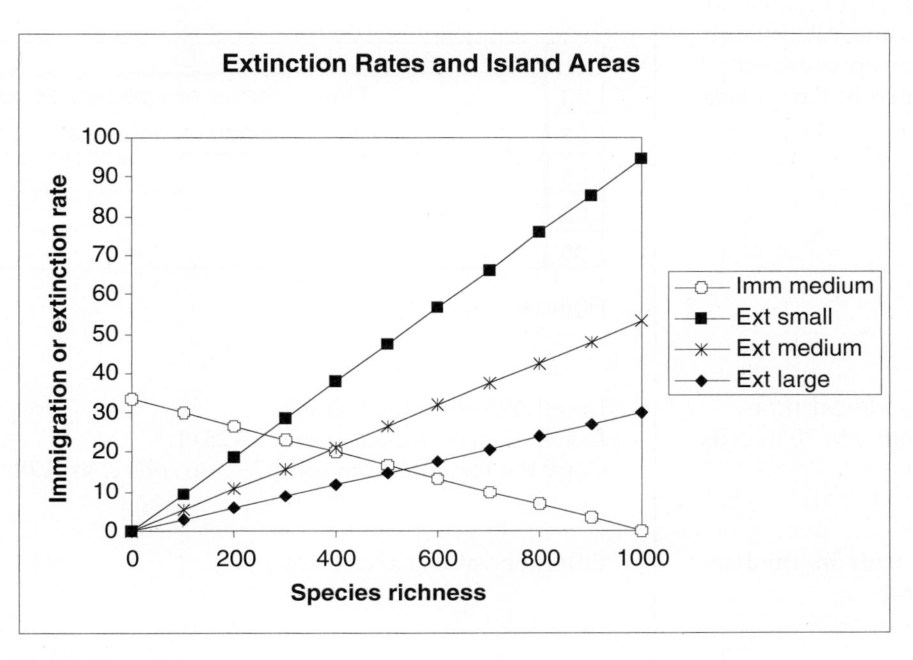

Figure 5

*12. As an OPTIONAL exercise, graph three immigration rates and three extinction rates on a single graph.

Select cells J13–P24, and make an XY graph. This will allow you to compare species richness and turnover rates on islands of three different sizes, at three different distances from the mainland. However, your graph might be rather cluttered and hard to read.

The Time-Course of Species Accumulation on an Island

The graphical analyses above answer a variety of questions about species richness on islands at equilibrium. However, they tell us nothing about how species richness changes over time as it approaches equilibrium. To find out about that, we must model the time-course of species accumulation.

We can follow the accumulation of species over time using a discrete-time model. The number of species present on an island at time $t + 1$ will be the number present at time t plus the number of new species that immigrated in the interval from time t to $t + 1$, minus the number of species that went extinct in the interval from t to $t + 1$. In symbols,

$$S_{t+1} = S_t + I_t - E_t$$

Substituting the right-hand side of Equation 1 for I_t and the right-hand side of Equation 2 for E_t, we derive

$$S_{t+1} = S_t + \frac{c(P - S_t)}{D} - \frac{qS_t}{A^m} \qquad \text{Equation 3}$$

INSTRUCTIONS / ANNOTATION

C. Model the time-course of species accumulation.

1. Add the column headings shown in Figure 6 to cells A26 and A27 though D27 of the spreadsheet you created in Part A (see Figure 1).

These are all literals, so just select the appropriate cells and type them in.

	A	B	C	D
26	**Time-course of species accumulation**			
27	Time	Species richness	Immigration	Extinction
28	0			
29	1			
30	2			

Figure 6

2. Set up a linear time series from 0 to 50 in cells A28–A78.

In cell A28 enter the value 0.
In cell A29 enter the formula **=A28+1**.
Copy the formula from cell A29 into cells A30–A78.

3. Begin with an uninhabited island.

Enter the value 0 in cell B28.

4. Enter a formula to calculate the number of species immigrating to the island in the interval from time 0 to time 1.

Copy the formula from cell C14 into cell C28. This corresponds to Equation 1.

5. Enter a formula to calculate the number of species going extinct on the island from time 0 to time 1.

Copy the formula from cell D14 into cell D28. This corresponds to Equation 2.

6. Enter a formula to calculate the number of species present on the island at time 1.

In cell B29 enter the formula **=B28+C28-D28**.
This corresponds to Equation 3:

$$S_{t+1} = S_t + \frac{c(P - S_t)}{D} - \frac{qS_t}{A^m}$$

The formula calculates the number of species on the island as the number already there plus the number immigrating to the island, minus the number going extinct, in the preceding time interval.

7. Copy the formulae in cells C28 and D28 into cells C29 and D29.

8. Copy the formulae in cells B29–D29 into cells B30–D78.

Save your work!

9. Graph species richness, immigration rate, and extinction rate against time.

Select cells A27–D78 and make an XY graph.
After you have made your graph, double-click on any data point in the species richness curve. In the Format Data Series dialog box, click on the Axis tab, and choose Secondary axis. Plot species richness on the secondary y-axis.

To label the second y-axis, open Chart|Chart Options|Titles.
Edit your graph for readability. Your graph should resemble the one in Figure 7.

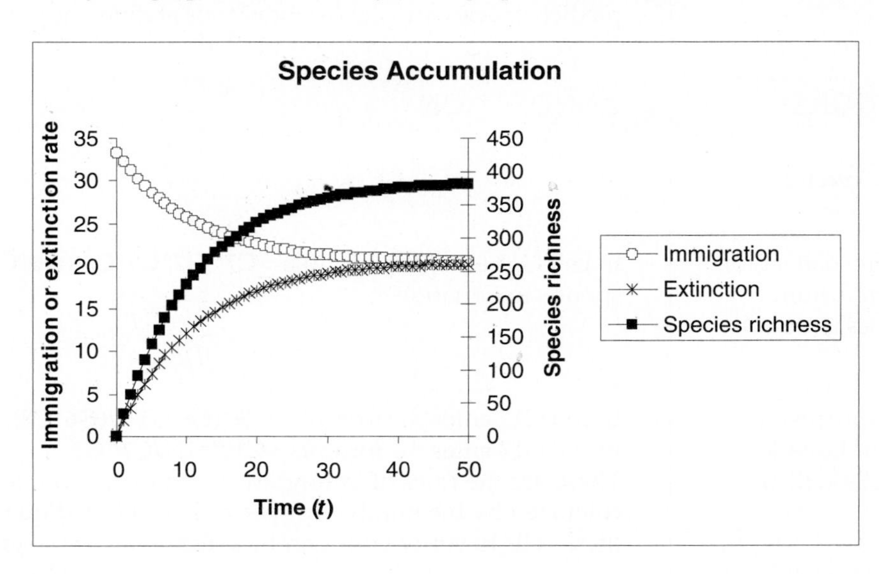

Figure 7

Equilibrium Solutions

So far, you have estimated equilibrium species richness using graphs. In the next section, we will calculate these quantities algebraically. We have two reasons for doing so.

First, calculations give us more precise results than estimating from a graph. Second, these calculations will allow us to close the loop, metaphorically, with the original motivation for MacArthur and Wilson's model. As we said at the beginning of this exercise, among the original observations from which this model sprang were the relationships of species richness to island area and distance from the mainland–the species-area curve. But nothing we have done so far explicitly shows a species-area curve. By finding equilibrium solutions, we can develop these curves, and briefly indicate how they have been used to test the model and to guide conservation decisions.

As we explained in the first section of this excercise, the MacArthur-Wilson model tells us that species accumulate by immigration and are removed by extinction, and that species richness reaches equilibrium when these two processes balance. Algebraically, we can find the equilibrium species richness of an island by solving for S_{eq} when $I = E$. So, let's do a little algrebra.

$$\text{Let } I = E$$

Substituting from Equations 1 and 2 above, we can derive the equation for S_{eq}:

$$\frac{c(P - S_{eq})}{fD} = \frac{qS_{eq}}{A^m}$$

$$A^m c(P - S_{eq}) = fDqS_{eq}$$

$$A^m cP - A^m cS_{eq} = fDqS_{eq}$$

$$A^m cP = fDqS_{eq} + A^m cS_{eq}$$

$$A^m cP = S_{eq}(fDq + A^m c)$$

$$\frac{A^m cP}{fDq + A^m c} = S_{eq} \qquad \text{Equation 4}$$

Equation 4 isn't very pretty, but you can use it in your spreadsheet model to see how equilibrium species richness relates to island area, to colonization and extinction probabilities, and to the richness of the mainland species pool. In particular, we will see how the model predicts species-area curves for islands at different distances from the mainland.

INSTRUCTIONS	ANNOTATION
D. Calculate species equilibrium.	
1. Enter a spreadsheet formula for equilibrium species richness into cell G11.	In cell G11 enter the formula **=C7^G7*C9*C6/(G6*C8*C10+C7^G7*C9)**. This corresponds to Equation 4: $\frac{A^m cP}{fDq + A^m c} = S_{eq}$
2. Enter the spreadsheet equivalents of Equations 1 and 2 into cells G12 and G13.	In cell G12 enter the formula **=C9*(C6-G11)/(G6*C8)**. In cell G13 enter the formula **=C10*G11/C7^G7**. These are the rates of immigration and extinction, respectively, on an island already colonized by the number of species in cell B14 (Equations 1 and 2). Use the values in these cells to verify your graphical estimates in the previous parts of this exercise.

3. Enter the row and column labels shown in Figure 8 into cells R11–X13.

We will use this part of the spreadsheet to calculate species area curves for islands at different distances from the mainland.

	R	S	T	U	V	W	X
11	**Species-Area Relationships at Different Distances**						
12		Equilibrium species richness			Turnover rates		
13	Area	Near	Medium	Far	Near	Medium	Far
14	10						
15	50						
16	100						

Figure 8

4. To represent a wide range of island areas, set up a series 10, 50, 100, 500, 1000 … , 500,000, 1,000,000 in cells R14–R24.

Enter the values10 and 50 into cells R14 and R15, respectively.
In cell R16, enter the formula **=R14*10**. Copy this formula into cells R17–R24.

5. In column T, calculate the equilibrium species richnesses of islands at the distance specified in cell C8, with the areas given in column R.

In cell T14 enter the formula **=$R14^$G$7*$C$9*$C$6/($G$6*$C$8*$C$10+$R14^$G$7*$C$9)**, which again corresponds to Equation 4. Copy this formula into cells T15–T24. Note that the address **$R14** has an absolute column reference but a relative row reference.

6. In column S, calculate the equilibrium species richnesses of islands at 0.1 times the distance specified in cell C8, with the areas given in column R.

Copy the formula from cell T14 into cell S14. Edit the formula to multiply distance by 0.1. The edited formula should read **=$R14^$G$7*$C$9*$C$6/($G$6*0.1*$C$8*$C$10+$R14^$G$7*$C$9)**.
Copy the edited formula from cell S14 into cells S15–S24.

7. In column U, calculate the equilibrium species richnesses of islands at 10 times the distance specified in cell C8, with the areas given in column R.

Copy the formula from cell S14 into cell U14. Edit the formula to multiply distance by 10. The edited formula should read **=$R14^$G$7*$C$9*$C$6/($G$6*10*$C$8*$C$10+$R14^$G$7*$C$9)**.
Copy the edited formula from cell U14 into cells U15–U24.

8. Graph equilibrium species richness against island area for islands at near, medium, and far distances from the mainland.

Select cells R13 though U24, and create an XY graph. Edit your graph for readability; It should resemble Figure 9. The three species-area curves will rise very quickly, almost following the vertical axis on the left, and then abruptly level out.

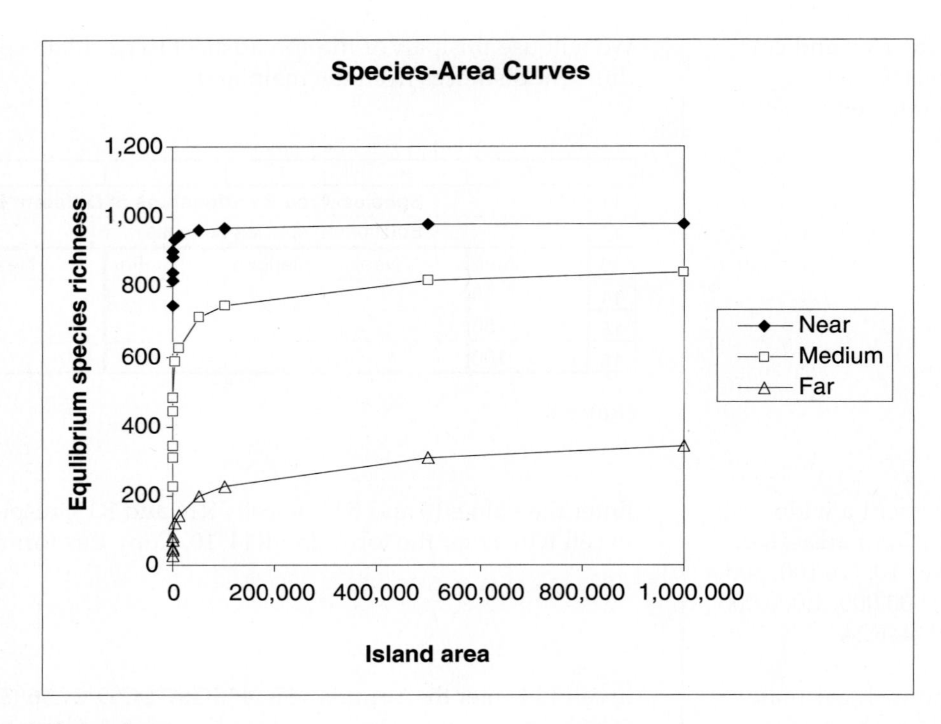

Figure 9

9. Change both vertical and horizontal axes to logarithmic scales.

As in Figure 10, the species-area curves should become almost straight lines on the log-log plot.

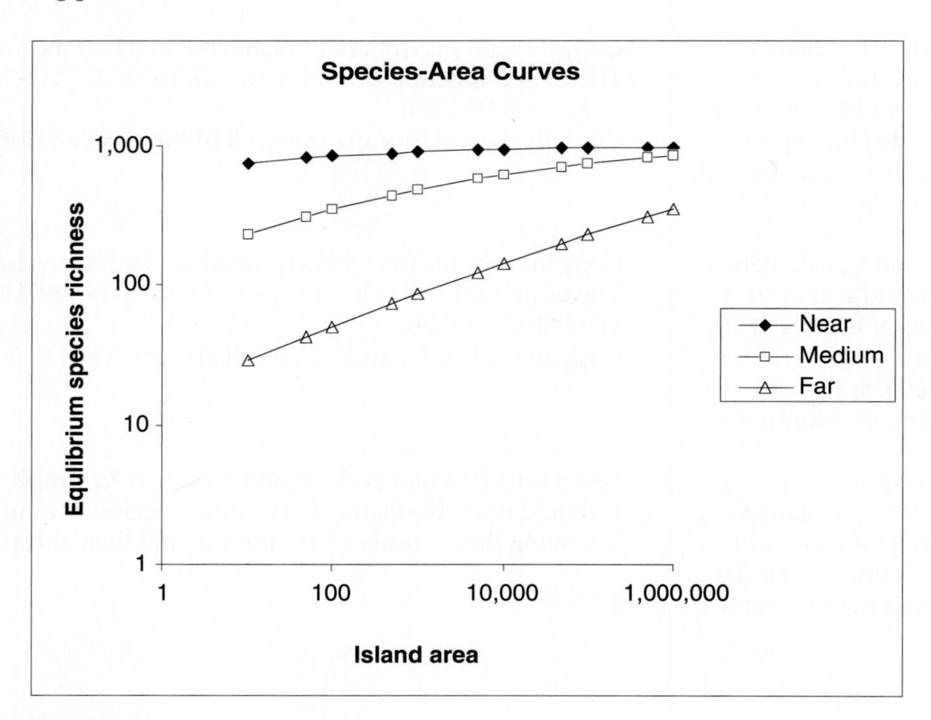

Figure 10

QUESTIONS

1. How can you estimate the equilibrium species richness of an island from Figure 2?
2. Is the equilibrium of species richness stable or unstable?
3. Is the equilibrium of species richness static or dynamic?
4. How does greater distance from the mainland affect species richness on an island?
5. How does greater distance from the mainland affect the turnover rate on an island?
6. How does larger area affect species richness on an island?
7. How does larger area affect the turnover rate on an island?
8. (OPTIONAL) How do area and distance from the mainland interact to determine species richness and turnover rate on an island?
9. How do species accumulate on an island over time? That is, does species richness increase linearly, exponentially, logarithmically, or otherwise?
10. What does Figure 7 tell us about the changing state of species equilibrium?
11. How is species richness related to island area?
12. How do the species-area curves differ for islands at different distances from the mainland?

LITERATURE CITED

MacArthur, R. H. and E. O. Wilson. 1967. *The Theory of Island Biogeography.* Princeton University Press, Princeton, NJ.

12 LIFE TABLES, SURVIVORSHIP CURVES, AND POPULATION GROWTH

Objectives

- Discover how patterns of survivorship relate to the classic three types of survivorship curves.
- Learn how patterns of survivorship relate to life expectancy.
- Explore how patterns of survivorship and fecundity affect rate of population growth.

Suggested Preliminary Exercise: Geometric and Exponential Population Models

INTRODUCTION

A **life table** is a record of survival and reproductive rates in a population, broken out by age, size, or developmental stage (e.g., egg, hatchling, juvenile, adult). Ecologists and demographers (scientists who study human population dynamics) have found life tables useful in understanding patterns and causes of mortality, predicting the future growth or decline of populations, and managing populations of endangered species.

Predicting the growth and decline of human populations is one very important application of life tables. As you might expect, whether the population of a country or region increases or decreases depends in part on how many children each person has and the age at which people die. But it may surprise you to learn that population growth or decline also depends on the age at which they have their children. A major part of this exercise will explore the effects of changing patterns of survival and reproduction on population dynamics.

Another use of life tables is in species conservation efforts, such as in the case of the loggerhead sea turtle of the southeastern United States (Crouse et al., 1987). We explore this case in greater depth in Exercise 14, "Stage-Structured Matrix Models," but generally speaking, the loggerhead population is declining and mortality among loggerhead eggs and hatchlings is very high. These facts led conservation biologists to advocate the protection of nesting beaches. When these measures proved ineffective in halting the population decline, compiling and analyzing a life table for loggerheads indicated that reducing mortality of *older* turtles would have a greater probability of reversing the population decline. Therefore, management efforts shifted to persuading fishermen to install turtle exclusion devices on their nets to prevent older turtles from drowning.

Life tables come in two varieties: cohort and static. A **cohort life table** follows the survival and reproduction of all members of a cohort from birth to death. A cohort is the set of all individuals born, hatched, or recruited into a population during a defined time interval. Cohorts are frequently defined on an annual basis (e.g., all individuals born in 1978), but other time intervals can be used as well.

A **static life table** records the number of living individuals of each age in a population and their reproductive output. The two varieties have distinct advantages and disadvantages, some of which we discuss below.

Life tables (whether cohort or static) that classify individuals by age are called **age-based life tables**. Such life tables treat age the same way we normally do: that is, individuals that have lived less than one full year are assigned age zero; those that have lived one year or more but less than two years are assigned age one; and so on. Life tables represent age by the letter x, and use x as a subscript to refer to survivorship, fecundity, and so on, for each age.

Size-based and **stage-based life tables** classify individuals by size or developmental stage, rather than by age. Size-based and stage-based tables are often more useful or more practical for studying organisms that are difficult to classify by age, or whose ecological roles depend more on size or stage than on age. Such analyses are more complex, however, and we will leave them for a later exercise.

Cohort Life Tables

To build a cohort life table for, let's say, humans born in the United States during the year 1900, we would record how many individuals were born during the year 1900, and how many survived to the *beginning* of 1901, 1902, etc., until there were no more survivors. This record is called the **survivorship schedule**. Unfortunately, different textbooks use different notations for the number of survivors in each age; some write this as S_x, some a_x, and some n_x. We will use S_x here.

We must also record the **fecundity schedule**—the number of offspring born to individuals of each age. The total number of offpsring is usually divided by the number of individuals in the age, giving the average number of offspring per individual, or per capita fecundity. Again, different texts use different notations for the fecundity schedule, including b_x (the symbol we will use) or m_x.[1]

Many life tables count only females and their female offspring; for animals with two sexes and equal numbers of males and females of each age, the resulting numbers are the same as if males and females were both counted. For most plants, hermaphroditic animals, and many other organisms, distinctions between the sexes are nonexistent or more complex, and life table calculations may have to be adjusted.

Static Life Tables

A static life table is similar to a cohort life table but introduces a few complications. For many organisms, especially mobile animals with long life spans, it can be difficult or impossible to follow all the members of a cohort throughout their lives. In such cases, population biologists often count how many individuals of each age are alive *at a given time*. That is, they count how many members of the population are currently in the 0–1-year-old class, the 1–2-year-old class, etc.

These counts can be used as if they were counts of survivors in a cohort, and all the calculations described below for a cohort life table can be performed using them. In doing this, however, the researcher must bear in mind that she or he is assuming that age-specific survivorship and fertility rates have remained constant since the oldest members of the population were born. This is usually not the case and can lead to some

[1] Some demographers use the term fecundity to be the physiological maximum number of *eggs* produced per female per year, and the term fertility to be the number of *offspring* produced per female per year. In this book, we will assume that the two are equivalent unless noted otherwise.

strange results, such as negative mortality rates. These are often resolved by averaging across several ages, or by making additional assumptions. We will avoid these complications by focusing this exercise on cohort life tables.

Quantities in a Life Table

Survivorship and fecundity schedules are the raw data of any life table. From them we can calculate a variety of other quantities, including age-specific rates of survival, mortality, fecundity, survivorship curves, life expectancy, generation time, net reproductive rate, and intrinsic rate of increase. Which of these quantities you calculate will depend on your goals in constructing the life table. Rather than presenting all the quantities that may appear in a life table, we will present two applications of life tables, using the quantities needed in each case. First you will build life tables that illustrate the three classic **survivorship curves**. These curves are a powerful visual tool for understanding the patterns of survivorship and mortality in populations.Then you will use a life table to predict the future growth or decline of a population. This kind of analysis is frequently used in studies of human populations, in management of fish and game, and in attempts to rescue endangered species.

Survivorship Curves

Ecology textbooks frequently present the three classic survivorship curves, called type I, type II, and type III (Figure 1). To understand survivorship curves you can use survivorship schedules (S_x) to calculate and graph **standardized survivorship** (l_x), **age-specific survivorship** (g_x), and **life expectancy** (e_x).

Standardized Survival Schedule (l_x). Because we want to compare cohorts of different initial sizes, we standardize all cohorts to their initial size at time zero, S_0. We do this by dividing each S_x by S_0. This proportion of original numbers surviving to the beginning of each interval is denoted l_x, and calculated as

$$l_x = \frac{S_x}{S_0} \qquad \text{Equation 1}$$

We can also think of l_x as the probability that an individual survives from birth to the beginning of age x. Because we begin with *all* the individuals born during the year (or other interval), l_x always begins at a value of one (i.e., S_0/S_0), and can only decrease with time. At the last age, k, S_k is zero.

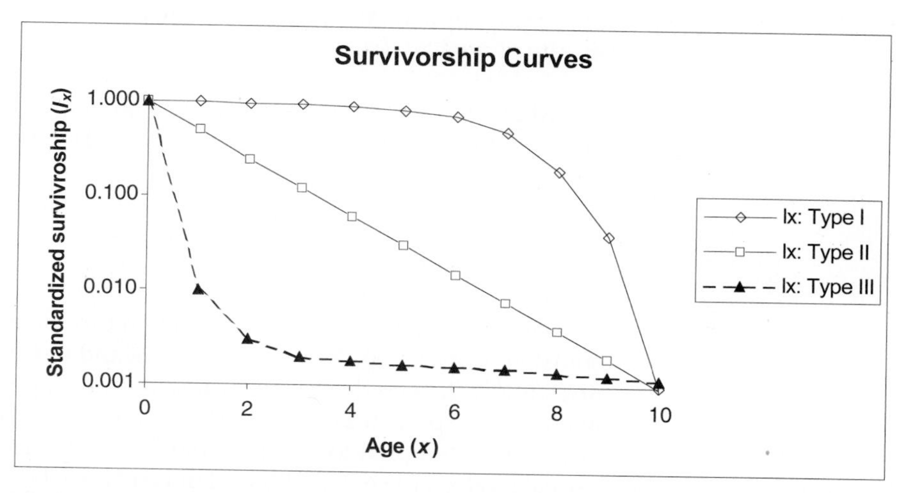

Figure 1 Hypothetical survivorship curves. Note that the y-axis has a logarithmic scale. Type 1 organisms have high surviorship throughout life until old age sets in, and then survivorship declines dramatically to 0. Humans are type 1 organisms. Type III organisms, in contrast, have very low survivorship early in life, and few individuals live to old age.

Age-Specific Survivorship (g_x). Standardized survivorship, l_x, gives us the probability of an individual surviving from birth to the beginning of age x. But what if we want to know the probability that an individual who has already survived to age x will survive to age $x + 1$? We calculate this age-specific survivorship as $g_x = l_{x+1}/l_x$, or equivalently,

$$g_x = \frac{S_{x+1}}{S_x} \qquad \text{Equation 2}$$

Life Expectancy (e_x). You may have heard another demographic statistic, life expectancy, mentioned in discussions of human populations. Life expectancy is how much longer an individual of a given age can be expected to live beyond its present age. Life expectancy is calculated in three steps. First, we compute the proportion of survivors at the mid-point of each time interval (L_x—note the capital L here) by averaging l_x and l_{x+1}; that is,

$$L_x = \frac{l_x + l_{x+1}}{2} \qquad \text{Equation 3}$$

Second, we sum all the L_x values from the age of interest (n) up to the oldest age, k:

$$T_x = \sum_{x=n}^{k} L_x \qquad \text{Equation 4}$$

Finally, we calculate life expectancy as

$$e_x = \frac{T_x}{l_x} \qquad \text{Equation 5}$$

(note the lowercase l_x).

Life expectancy is age-specific—it is the expected number of time-intervals remaining to members of a given age. The statistic most often quoted (usually without qualification) is the life expectancy at birth (e_0). As you will see, the implications of e_0 depend greatly on the survivorship schedule.

Population Growth or Decline

We frequently want to know whether a population can be expected to grow, shrink, or remain stable, given its current age-specific rates of survival and fecundity. We can determine this by computing the **net reproductive rate** (R_0). To predict long-term changes in population size, we must use this net reproductive rate to estimate the intrinsic rate of increase (r).

Net Reproductive Rate (R_0) We calculate net reproductive rate (R_0) by multiplying the standardized survivorship of each age (l_x) by its fecundity (b_x), and summing these products:

$$R_0 = \sum_{x=0}^{k} l_x b_x \qquad \text{Equation 6}$$

The net reproductive rate is the lifetime reproductive potential of the average female, adjusted for survival. Assuming survival and fertility schedules remain constant over time, if $R_0 > 1$, then the population will grow exponentially. If $R_0 < 1$, the population will shrink exponentially, and if $R_0 = 1$, the population size will not change over time. You may be tempted to conclude the $R_0 = r$, the intrinsic rate of increase of the exponential model. However, this is not quite correct, because r measures population change in absolute units of time (e.g., years) whereas R_0 measures population change in terms of generation time. To convert R_0 into r, we must first calculate generation time (G), and then adjust R_0.

Generation Time. Generation time is calculated as

$$G = \frac{\sum_{x=0}^{k} l_x b_x x}{\sum_{x=0}^{k} l_x b_x} \qquad \text{Equation 7}$$

For organisms that live only one year, the numerator and denominator will be equal, and generation time will equal one year. For all longer-lived organisms, generation time will be greater than one year, but exactly how much greater will depend on the survival and fertility schedules. A long-lived species that reproduces at an early age may have a shorter generation time than a shorter-lived one that delays reproduction.

Intrinsic Rate of Increase. We can use our knowledge of exponential population growth and our value of R_0 to estimate the intrinsic rate of increase (r) (Gotelli 2001). Recall from Exercise 7, "Geometric and Exponential Population Models," that the size of an exponentially growing population at some arbitrary time t is $N_t = N_0 e^{rt}$, where e is the base of the natural logarithms and r is the intrinsic rate of increase. If we consider the growth of such a population from time zero through one generation time, G, it is

$$N_G = N_0 e^{rG}$$

Dividing both sides by N_0 gives us

$$\frac{N_G}{N_0} = e^{rG}$$

We can think of N_G/N_0 as roughly equivalent to R_0; both are estimates of the rate of population growth over the period of one generation.

Substituting R_0 into the equation gives us

$$R_0 \approx e^{rG}$$

Taking the natural logarithm of both sides gives us

$$\ln(R_0) \approx rG$$

and dividing through by G gives us an estimate of r:

$$r \approx \frac{\ln(R_0)}{G} \qquad \text{Equation 8}$$

Euler's Correction to *r*. The value of r as estimated above is usually a good approximation (within 10%), and it will suffice for most purposes. Some applications, however, may require a more precise value. To improve this estimate, you must solve the Euler equation:

$$1 = \sum_{x=0}^{k} e^{-rx} l_x b_x \qquad \text{Equation 9}$$

The only way to solve this equation is by trial and error. We already know the values of $l_x b_x$, and e (it is the base of the natural logarithms, $e \approx 2.7183$), so we can plug in various guesses for r until Equation 9 comes up 1.0. That will tell us the corrected value of r. Fortunately, a spreadsheet is an ideal medium for such trial and error solution-hunting.

Finally, we can use our estimate of r (uncorrected or corrected) to predict the size of the population in the future. In this exercise, you will adjust survivorship and fecundity schedules and observe the effects on population growth or decline. This kind of analysis is done for human populations to predict the effects of changes in medical care and birth control programs. If we assume that all age groups are roughly equivalent in size, a similar analysis can be done for endangered species to determine what intervention may be most effective in promoting population growth. The same analysis can be applied to pest species to determine what intervention may be most effective in reducing population size.

PROCEDURES

Our purpose here is to show how survivorship curves are generated and what they mean. You will use survivorship schedules to calculate and graph l_x, g_x, and e_x, resulting in survivorship curves of type I, II, or III. In the final section of the exercise you will see how this information can be used to predict population rise and decline.

As always, save your work frequently to disk.

INSTRUCTIONS	ANNOTATION

A. Generate survivorship curves.

1. Open a new spreadsheet and set up titles and column headings as shown in Figure 2.

These are all literals, so just select the appropriate cells and type them in.

	A	B	C	D	E	F	G	H	I	J
1	*Life Tables and Survivorship Curves*									
2	**Survivorship curves**									
3	Age (x)	S_x: Type I	S_x: Type II	S_x: Type III	l_x: Type I	l_x: Type II	l_x: Type III	g_x: Type I	g_x: Type II	g_x: Type III
4	0	1000	2048	10000						
5	1	990	1024	100						
6	2	970	512	30						
7	3	940	256	20						
8	4	900	128	18						
9	5	850	64	17						
10	6	750	32	16						
11	7	500	16	15						
12	8	200	8	14						
13	9	40	4	13						
14	10	1	2	12						
15	11	0	0	0						

Figure 2

2. Set up a linear series from 0 to 11 in column A.

In cell A4 enter the value zero.
In cell A5 enter the formula **=A4+1**. Copy the formula in cell A5 into cells A6–A15.

3. Enter the values shown in Figure 2 for cells B4–D15.

These are the raw data of three survivorship schedules—one for each survivorship curve. Each number is the number of surviving individuals from a cohort at each age.

4. Enter formulae to calculate the standardized survivorship, l_x, for each survivorship schedule.

In cell E4 enter the formula **=B4/$B$4**. Copy this formula into cells E5–E15. This corresponds to Equation 1:

$$l_x = \frac{S_x}{S_0}$$

Note the use of a relative cell address in the numerator and an absolute cell address in the denominator. The formula in cell F4 should be **=C4/$C$4**, and the formula in cell G4 should be **=D4/$D$4**. Copy cells F4–G4 down to F15–G15.

5. Enter formulae to calculate age-specific survivorship, g_x, for each survivorship schedule.

In cell H4 enter the formula **=B5/B4**. Copy this formula into cells H5–H14. *Do not* copy it into cell H15, because the formula would attempt to divide by zero and thus generate an error. Copy cells H4–H14 into cells I4–J14.

This corresponds to Equation 2:

$$g_x = \frac{S_{x+1}}{S_x}$$

Note that all cell addresses are relative.

6. Enter titles and column headings in cells A17–J18 as shown in Figure 3.

These are all literals, so just select the appropriate cells and type them in.

	A	B	C	D	E	F	G	H	I	J
17	**Age-specific life expectancy**									
18	Age (x)	l_x: Type I	l_x: Type II	l_x: Type III	l_x: Type I	l_x: Type II	l_x: Type III	e_x: Type I	e_x: Type II	e_x: Type III
19	0	1.0000	1.0000	1.0000						
20	1	0.9900	0.5000	0.0100						
21	2	0.9700	0.2500	0.0030						
22	3	0.9400	0.1250	0.0020						
23	4	0.9000	0.0625	0.0018						
24	5	0.8500	0.0313	0.0017						
25	6	0.7500	0.0156	0.0016						
26	7	0.5000	0.0078	0.0015						
27	8	0.2000	0.0039	0.0014						
28	9	0.0400	0.0020	0.0013						
29	10	0.0010	0.0010	0.0012						
30	11	0.0000	0.0000	0.0000						

Figure 3

7. Copy the values of age from cells A4–A15 into cells A19–A30.

Select cells A4–A15. Copy.
Select cell A19. Paste.

8. Echo the values of l_x from cells E4–G15 in cells B19–D30.

In cell B19 enter the formula **=E4**. Copy this formula into cells C19 and D19. Copy cells B19–D19 into cells B20–D30.
Doing it this way, rather than copying and pasting the values, will automatically update this part of the spreadsheet if you change any of the S_x values in cells B4–D15.

9. Enter formulae to calculate the number of survivors at the midpoint of each age, L_x.

In cell E19 enter the formula **=(B19+B20)/2**. Copy this formula into cells F19 and G19. Copy cells E19–G19 into cells E30–G30.
This corresponds to Equation 3:

$$L_x = \frac{l_x + l_{x+1}}{2}$$

10. Enter formulae to calculate life expectancy, e_x, for each age.

In cell H19 enter the formula **=SUM(E19:E$30)/B19**. Copy the formula from cell H19 into cells I19 and J19.
Copy cells H19–J19 into cells H20–J29.

The portion **SUM(E19:E$30)** corresponds to Equation 4:

$$T_x = \sum_{x=n}^{k} L_x$$

The entire formula corresponds to Equation 5:

$$e_x = \frac{T_x}{l_x}$$

Do not copy the formula into row 30, because l_x there is zero, and so Equation 5 would be undefined.

11. Your spreadsheet is complete. Save your work.

12. Graph standardized survivorship, l_x, against age.

Select cells A3–A15. Select cells E3–G15 and create an XY graph. Edit your graph for readability. It should resemble Figure 4.

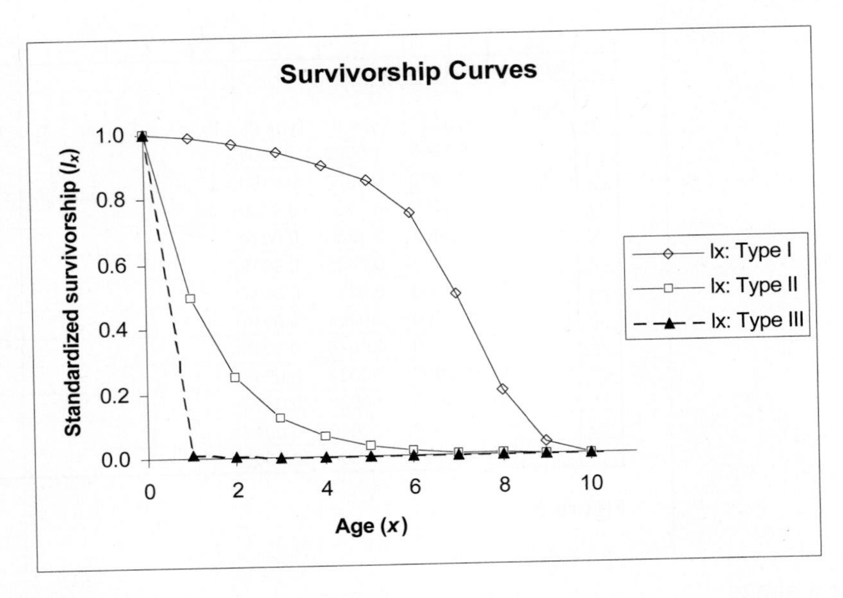

Figure 4

13. Change the y-axis to a logarithmic scale.

Double-click on the y-axis and choose the Number tab in the resulting dialog box. Set the number of decimal places to 3. Choose the Scale tab. Check the box for Logarithmic Scale. Set the Major unit to 10, and set Value (X) axis Crosses at to 0.0001. Your graph should resemble Figure 5.

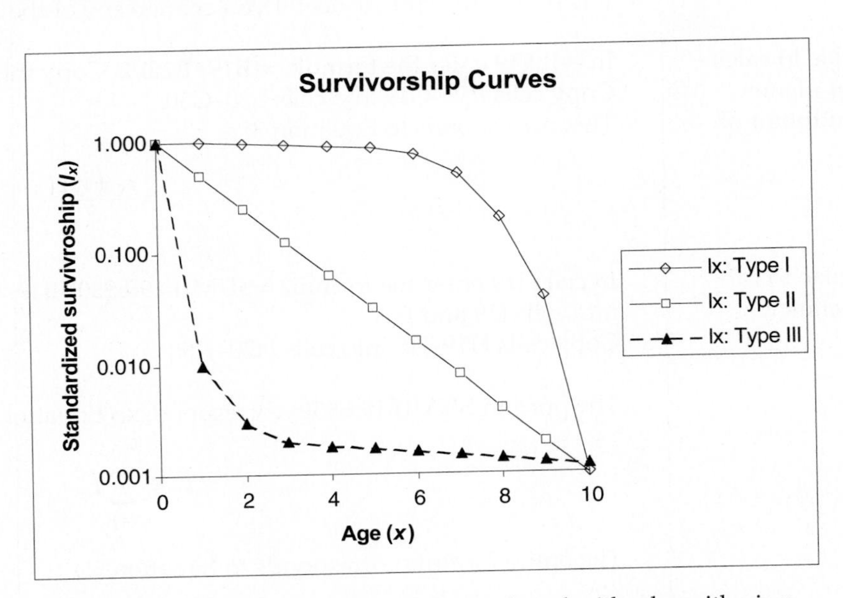

Figure 5 Survivorship curves are always plotted with a logarithmic y-axis. Can you see why?

14. Graph age-specific survival g_x, against age.

Select cells A3–A14 and cells H3–J14. Make an XY graph. Your graph should resemble Figure 6.

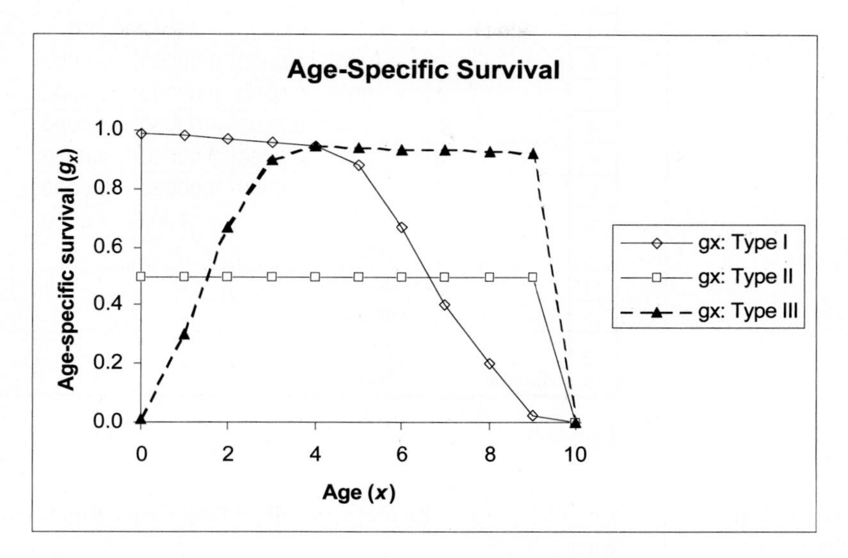

Figure 6

15. Graph life expectancy, e_x, against age.

Select cells A18–A29 and cells H18–J29. *Do not* include row 30 in either block. Make an XY graph. Your graph should resemble Figure 7.

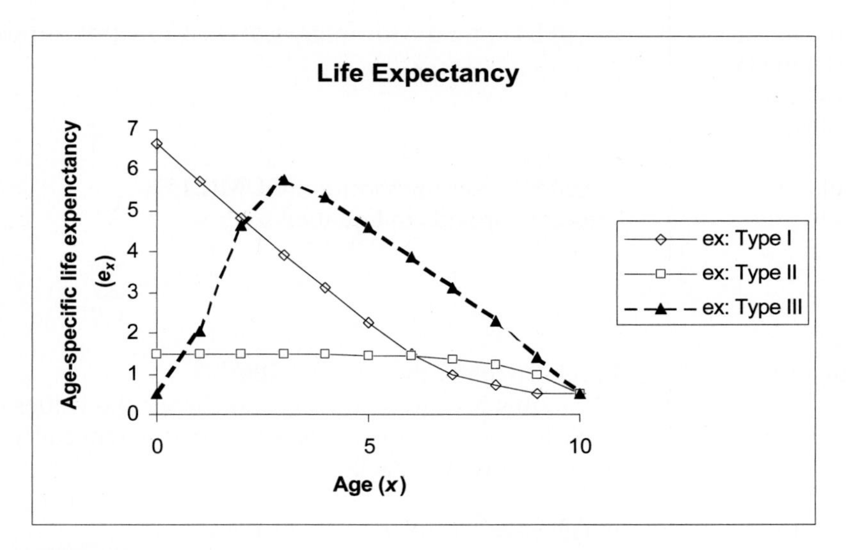

Figure 7

B. Population growth and decline.

1. Open a new spreadsheet and set up titles and column headings as shown in Figure 8. Set up a linear series of ages from 0 to 4 in column A. Enter the values shown for S_x.

We will use fewer ages here to simplify the manipulations that you will do later.

	A	B	C	D	E	F	G
1	**Cohort Life Table: Fertility, Survival, and Population Growth**						
2							
3	Age (x)	S_x	l_x	b_x	(l_x)(b_x)	(x)(l_x)(b_x)	(e^-rx)(l_x)(b_x)
4	0	1000	1.0000	0.00000	0.0000	0.0000	0.0000
5	1	900	0.9000	0.00000	0.0000	0.0000	0.0000
6	2	250	0.2500	4.00000	1.0000	2.0000	1.0000
7	3	10	0.0100	0.00000	0.0000	0.0000	0.0000
8	4	0	0.0000	0.00000	0.0000	0.0000	0.0000
9				Total	1.0000	2.0000	1.0000
10	R_0	1.00000					
11	G	2.00000					
12	r est.	0.00000					
13	r adj.	0.00000					
14	Should be 1	1.00000					

Figure 8

2. Enter a formula to calculate standardized survival, l_x.

In cell C4 enter the formula **=B4/$B$4**. Copy this formula into cells C5–C8. *Do not* copy into cell C9.
Again, this corresponds to Equation 1. Note the use of a relative cell address in the numerator, and an absolute cell address in the denominator.

3. Enter the values shown for age-specific fertility, b_x.

Enter the value 0.00 into cells D4, D5, D7, and D8, Enter the value 4.00 into cell D6.

4. Enter a formula to calculate the product of standardized survival times age-specific fertility, $l_x b_x$.

In cell E4 enter the formula **=C4*D4**. Copy this formula into cells E5–E8.

5. Enter a formula to calculate net reproductive rate, R_0.

In cell E9 enter the formula **=SUM(E4:E8)**.
This corresponds to Equation 6:

$$R_0 = \sum_{x=0}^{k} l_x b_x$$

6. Echo the value of R_0 in cell B10.

In cell B10 enter the formula **=E9**.
We do this because you will soon change the values of S_x and b_x, and this layout will make it easier to compare the effects of different survival and fertility schedules on population growth or decline.

7. Enter a formula to calculate the product $l_x b_x x$.

In cell F4 enter the formula **=E4*A4**.
This is an intermediate step in calculating generation time, G.
Copy the formula from cell F4 into cells F5–F8.

8. Enter a formula to calculate the sum of the products $l_x b_x x$.

In cell F9 enter the formula **=SUM(F4:F8)**.
This is another intermediate step in calculating generation time, G.

9. Enter a formula to calculate generation time, G.

In cell B11 enter the formula **=F9/E9**.
This corresponds to Equation 7:

$$G = \frac{\sum_{x=0}^{k} l_x b_x x}{\sum_{x=0}^{k} l_x b_x}$$

10. Enter a formula to estimate the intrinsic rate of increase, r.

In cell B12 enter the formula **=LN(B10)/B11**.
This corresponds to Equation 8:

$$r \approx \frac{\ln(R_0)}{G}$$

11. Your spreadsheet is complete. Save your work.

12. Create a survivorship curve from your S_x values.

Follow the procedures in Steps 12 and 13 of Section A. Your graph should resemble Figure 9.

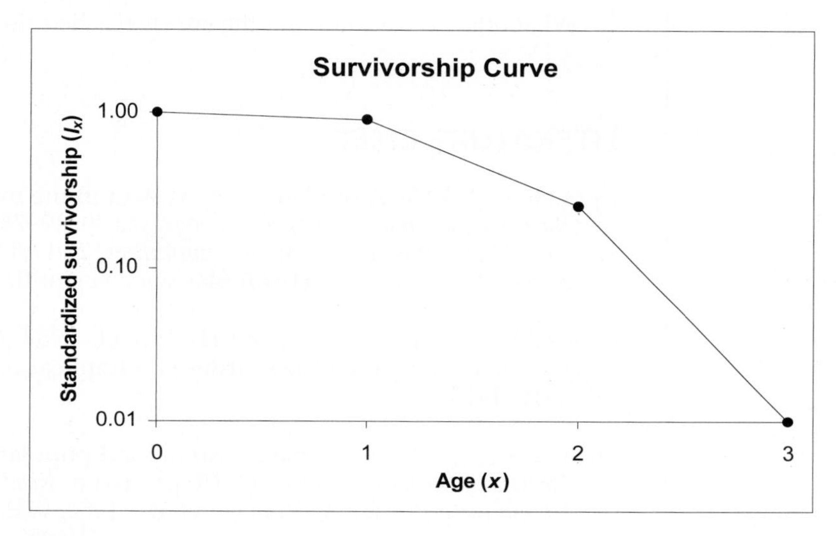

Figure 9

C. Euler's correction (Optional)

1. (*Optional) Enter a guess for the correct value of r into cell B13.

Start by entering the estimated value of r from cell B12. You will see how to use this guess below.

2. Enter a formula to calculate $e^{-rx}l_x b_x$.

In cell G4 enter the formula **=EXP(-$B$13*A4)*E4**.
This is an intermediate step in applying Euler's correction to the estimate of r calculated in Step 11 of Section B. Note that the formula uses your guess for the value of r.

3. Enter a formula to compute Euler's equation.

In cell G9 enter the formula **=SUM(G4:G8)**.
This corresponds to the right side of Equation 9:

$$\sum_{x=0}^{k} e^{-rx} l_x b_x$$

If your guess for r is correct, this formula will yield a value of 1.0.

4. In cell B14, echo the result of the formula in cell G9.

In cell B14 enter the formula **=G9**.
Again, this is simply a convenient layout for comparing the effects of changing S_x and b_x.

QUESTIONS

1. Why do we plot survivorship curves on a semi-log graph?
2. What do the shapes of the survivorship curves tell us about patterns of survival and mortality? Compare each curve to the corresponding graph of age-specific survivorship.
3. How can we interpret the graph of life expectancies?
4. Use the S_x values for real populations provided in the Appendix at the end of this exercise to compare survivorship curves between animal species. You may also wish to visit the U.S. Census Bureau's web site (http://www.census.gov/), from which you can download survivorship data for human populations in most of the countries of the world.
5. What effect does changing the fecundity schedule have on R_0, G, and r?
6. What effect does changing the survival schedule have on R_0, G, and r?

LITERATURE CITED

Connell, J. H. 1970. A predator-prey system in the marine intertidal region. I. *Balanus glandula*. *Ecological Monographs* 40: 49–78. (Reprinted in *Ecology: Individuals, Populations and Communities*, 2nd edition, M. Begon, J. L. Harper and C. R. Townsend.. (1990) Blackwell Scientific Publications, Oxford.)

Crouse, D. T., L. B. Crowder, and H. Caswell. 1987. A stage-based population model for loggerhead sea turtles and implications for conservation. *Ecology* 68: 1412–1423.

Deevey, E. S., Jr. 1947. Life tables for natural populations of animals. *The Quarterly Review of Biology* 22: 283–314. (Reprinted in *Readings in Population and Community Ecology*, W.E. Hazen (ed.). 1970, W.B. Saunders, Philadelphia.)

Gotelli, Nicholas J. 2001. *A Primer of Ecology*, 3rd Edition. Sinauer Associates, Sunderland, MA.

Appendix: SAMPLE SURVIVORSHIP SCHEDULES FROM NATURAL POPULATIONS OF ANIMALS

In all cases, assume S_x for the next age after the oldest in the table is 0.

Table A. Survivorship schedule for Dall Mountain Sheep (*Ovis dalli dalli*).

Age (years)	S_x	Age (years)	S_x
0	1000	7	640
1	801	8	571
2	789	9	439
3	776	10	252
4	764	11	96
5	734	12	6
6	688	13	3

Data from Deevey (1947). Numbers have been standardized to $S_0 = 1000$.

Table B. Survivorship schedule for the Song Thrush.

Age (years)	S_x	Age (years)	S_x
0	1000	5	30
1	444	6	17
2	259	7	6
3	123	8	3
4	51		

Data from Deevey (1947). Numbers have been standardized to $S_0 = 1000$.

Table C. Survivorship and fertility schedules for the barnacle *Balanus glandula*.

Age (years)	S_x	Age (years)	S_x
0	1000	7	640
1	801	8	571
2	789	9	439
3	776	10	252
4	764	11	96
5	734	12	6
6	688	13	3

Data are from Connell (1970). Values of S_4 and S_6 have been interpolated and rounded to the next integer.

13

AGE-STRUCTURED MATRIX MODELS

Objectives

- Set up a model of population growth with age structure.
- Determine the stable age distribution of the population.
- Estimate the finite rate of increase from Leslie matrix calculations.
- Construct and interpret the age distribution graphs.

Suggested Preliminary Exercises: Geometric and Exponential Population Models; Life Tables and Survivorship Curves

INTRODUCTION

You've probably seen the geometric growth formula many times by now (Exercise 7). It has the form

$$N_{t+1} = N_t + (b - d)N_t$$

where b is the per capita birth rate and d is the per capita death rate for a population that is growing in discrete time. The term $(b - d)$ is so important in population biology that it is given its own symbol, R. It is called the **intrinsic** (or **geometric**) **rate of natural increase**, and represents the per capita rate of change in the size of the population. Substituting R for $b - d$ gives

$$N_{t+1} = N_t + RN_t$$

We can factor N_t out of the terms on the right-hand side, to get

$$N_{t+1} = (1 + R)N_t$$

The quantity $(1 + R)$ is called the finite rate of increase, λ. Thus we can write

$$N_{t+1} = \lambda N_t \qquad \text{Equation 1}$$

where N is the number of individuals present in the population, and t is a time interval of interest. Equation 1 says that the size of a population at time $t + 1$ is equal to the size of the population at time t multiplied by a constant, λ. When $\lambda = 1$, the population will remain constant in size over time. When $\lambda < 1$, the population declines geometrically, and when $\lambda > 1$, the population increases geometrically.

Although geometric growth models have been used to describe population growth, like all models they come with a set of assumptions. What are the assumptions of the geometric growth model? The equations describe a population in which there is no genetic structure, no age structure, and no sex structure to the population (Gotelli 2001), and all individuals are reproductively active when the population census is taken. The model also assumes that resources are virtually unlimited and that growth is unaffected by the size of the population. Can you think of an organism whose life history meets these assumptions? Many natural populations violate at least one of these assumptions because the populations have **structure**: They are composed of individuals whose birth and death rates *differ* depending on age, sex, or genetic make-up. All else being equal, a population of 100 individuals that is composed of 35 prereproductive-age individuals, 10 reproductive-age individuals, and 55 postreproductive-age individuals will have a different growth rate than a population where all 100 individuals are of reproductive age. In this exercise, you will develop a **matrix model** to explore the growth of populations that have age structure. This approach will enable you to estimate λ in Equation 1 for structured populations.

Model Notation

Let us begin our exercise with some notation often used when modeling populations that are structured (Caswell 2001; Gotelli 2001). For modeling purposes, we divide individuals into groups by either their **age** or their **age class**. Although age is a continuous variable when individuals are born throughout the year, by convention individuals are grouped or categorized into discrete time intervals. That is, the age class of 3-year-olds consists of individuals that just had their third birthday, plus individuals that are 3.5 years old, 3.8 years old, and so on. In age-structured models, *all* individuals within a particular age group (e.g., 3-year-olds) are assumed to be equal with respect to their birth and death rates. The age of individuals is given by the letter x, followed by a number within parentheses. Thus, newborns are $x(0)$ and 3-year-olds are $x(3)$.

In contrast, the age class of an individual is given by the letter i, followed by a subscript number. A newborn enters the first age class upon birth (i_1), and enters the second age class upon its first birthday (i_2). Caswell (2001) illustrates the relationship between age and age class as:

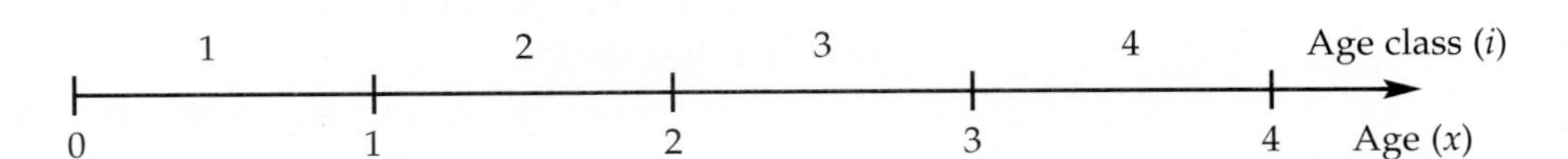

Thus, whether we are dealing with age classes or ages, individuals are grouped into discrete classes that are of equal duration for modeling purposes. In this exercise, we will model age classes rather than ages. A typical life cycle of a population with age-class structure is:

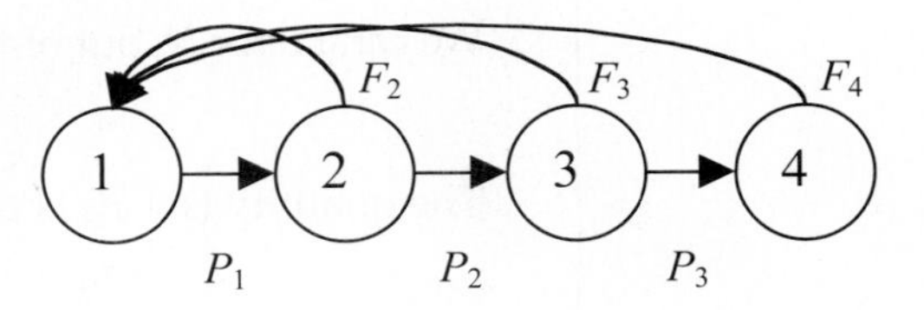

The age classes themselves are represented by circles. In this example, we are considering a population with just four age classes. The horizontal arrows between the circles represent survival probabilities, P_i—the probability that an individual in age class i will survive to age class $i + 1$. Note that the fourth age class has no arrow leading to a fifth age class, indicating that the probability of surviving to the fifth age class is 0. The curved arrows at the top of the diagram represent births. These arrows all lead to age class 1 because newborns, by definition, enter the first age class upon birth. Because "birth" arrows emerge from age classes 2, 3, and 4 in the above example, the diagram indicates that all three of these age classes are capable of reproduction. Note that individuals in age class 1 do not reproduce. If only individuals of age class 4 reproduced, our diagram would have to be modified:

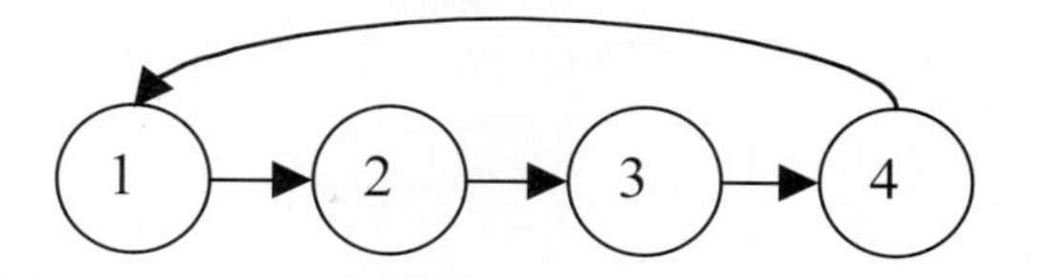

The Leslie Matrix

The major goal of the matrix model is to compute λ, the finite rate of increase in Equation 1, for a population with age structure. In our matrix model, we can compute the **time-specific growth rate** as λ_t. The value of λ_t can be computed as

$$\lambda_t = N_{t+1}/N_t \quad \text{or} \quad \lambda_t = \frac{N_{t+1}}{N_t} \qquad \text{Equation 2}$$

This time-specific growth rate is not necessarily the same λ in Equation 1. (We will discuss this important point later.) To determine N_t and N_{t+1}, we need to count individuals at some standardized time period over time. We will make two assumptions in our computations. First, we will assume that the time step between N_t and N_{t+1} is one year, and that age classes are defined by yearly intervals. This should be easy to grasp, since humans typically measure time in years and celebrate birthdays annually. (If we were interested in a different time step—say, six months—then our age classes would also have to be 6-month intervals.) Second, we will assume for this exercise that our population censuses are completed once a year, immediately after individuals breed (a **postbreeding census**). The number of individuals in the population in a census at time $t + 1$ will depend on how many individuals of each age class were in the population at time t, as well as the birth and survival probabilities for each age class.

Let us start by examining the **survival probability**, designated by the letter P. P is the probability that an individual in age class i will survive to age class $i + 1$. The small letter l gives the number of individuals in the population at a given time:

$$P_i = \frac{l(i)}{l(i-1)}$$

This equation is similar to the $g(x)$ calculations in the life table exercise. For example, let's assume the probability that individuals in age class 1 survive to age class 2 is $P_1 = 0.3$. This means 30% of the individuals in age class 1 will survive to be censused as age class 2 individuals. By definition, the remaining 70% of the individuals will die. If we consider survival alone, we can compute the number of individuals of age class 2 at time $t + 1$ as the number of individuals of age class 1 at time t multiplied by P_1. If we denote the number of individuals in class i at time t as $n_i(t)$, we can write the more general equation as

$$n_{i+1}(t + 1) = P_i n_i(t) \qquad \text{Equation 3}$$

This equation works for calculating the number of individuals at time $t + 1$ for each age class in the population except for the first, because individuals in the first age class arise only through birth. Accordingly, let's now consider birth rates.

There are many ways to describe the occurrence of births in a population. Here, we will assume a simple **birth-pulse model**, in which individuals give birth the moment they enter a new age class. When populations are structured, the birth rate is called the **fecundity**, or the average number of offspring born per unit time to an individual female of a particular age. If you have completed the exercise on life tables, you might recall

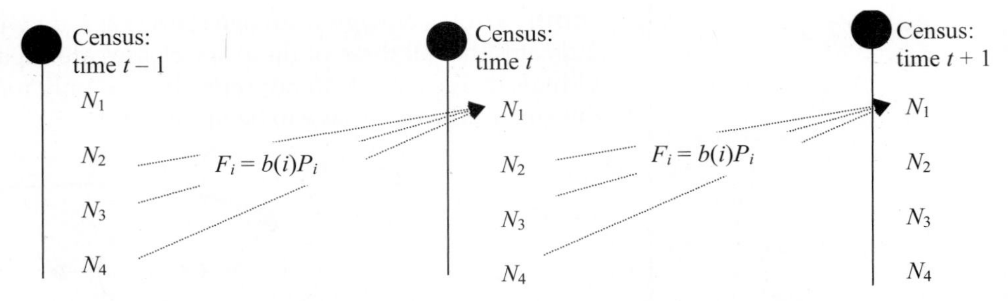

Figure 1 In this population, age classes 2, 3, and 4 can reproduce, as represented by the dashed arrows that lead to age class 1 in the next step. Births occur in a birth pulse (indicated by the filled circle and vertical line) and individuals are censused immediately after young are born. (After Akçakaya et al. 1997.)

that fecundity is labeled as $b(x)$, where b is for birth. Individuals that are of prereproductive or postreproductive age have fecundities of 0. Individuals of reproductive age typically have fecundities > 0.

Figure 1 is a hypothetical diagram of a population with four age classes that are censused at three time periods: time $t-1$, time t, and time $t+1$. In Figure 1, all individuals "graduate" to the next age class on their birthday, and since all individuals have roughly the same birthday, all individuals counted in the census are "fresh"; that is, the newborns were just born, individuals in age class 2 just entered age class 2, and so forth. With a postbreeding census, Figure 1 shows that the number of individuals in the first age class at time t depends on the number of breeding adults in the previous time step.

If we knew how many adults actually bred in the previous time step, we could compute fecundity, or the average number of offspring born per unit time per individual (Gotelli: 2001). However, the number of adults is not simply N_2 and N_3 and N_4 counted in the previous time step's census; these individuals must survive a long period of time (almost a full year until the birth pulse) before they have another opportunity to breed. Thus, we need to discount the fecundity, $b(i)$, by the probability that an adult will actually survive from the time of the census to the birth pulse (P_i), (Gotelli 2001). These adjusted estimates, which are used in matrix models, are called **fertilities** and are designated by the letter F.

$$F_i = b(i)P_i \qquad \text{Equation 4}$$

The adjustments are necessary to account for "lags" between the census time and the timing of births. Stating it another way, F_i indicates the number of young that are produced per female of age i in year t, given the appropriate adjustments. Be aware that various authors use the terms fertility and fecundity differently; we have followed the notation used by Caswell (2001) and Gotelli (2001). The total number of individuals counted in age class 1 in year $t+1$ is simply the fertility rate of each age class, multiplied by the number of individuals in that age class at time t. When these products are summed together, they yield the total number of individuals in age class 1 in year $t+1$. Generally speaking,

$$n_1(t+1) = \sum_{i=1}^{k} F_i n_i(t)$$

Once we know the fertility and survivorship coefficients for each age class, we can calculate the number of individuals in each age at time $t+1$, given the number of individuals in each class at time t:

$$n_1(t+1) = F_1 n_1(t) + F_2 n_2(t) + F_3 n_3(t) + F_4 n_4(t)$$
$$n_2(t+1) = P_1 n_1(t)$$
$$n_3(t+1) = P_2 n_2(t)$$
$$n_4(t+1) = P_3 n_3(t)$$

$$\mathbf{A} = \begin{bmatrix} F_1 & F_2 & F_3 & F_4 \\ P_1 & 0 & 0 & 0 \\ 0 & P_2 & 0 & 0 \\ 0 & 0 & P_3 & 0 \end{bmatrix}$$

Figure 2 The specific form of a Leslie matrix, based on a population with four age classes. The letters used to designate a mathematical matrix are conventionally uppercase, boldface, and not italic. The rows and columns of the matrix are enclosed in large brackets. See P. H. Leslie's original paper (Leslie 1945) for the classic discussion.

How can we incorporate the equations in Equation 4 into a model to compute the constant, λ, from Equation 1? Leslie (1945) developed a matrix method for predicting the size and structure of next year's population for populations with age structure. A **matrix** is a rectangular array of numbers; matrices are designated by uppercase, bold letters. **Leslie matrices**, named for the biologist P. H. Leslie, have the form shown in Figure 2.

Since our population has only four age classes, the Leslie matrix in Figure 2 is a four row by four column matrix. If our population had five age classes, the Leslie matrix would be a five row by five column matrix. The fertility rates of age classes 1 through 4 are given in the top row. Most matrix models consider only the female segment of the population, and define fertilities in terms of female offspring. The survival probabilities, P_i, are given in the subdiagonal; P_1 through P_3 are survival probabilities from one age class to the next. For example, P_1 is the probability of individuals surviving from age class 1 to age class 2. All other entries in the Leslie matrix are 0. The composition of our population can be expressed as a column **vector**, $\mathbf{n}(t)$, which is a matrix that consists of a single column. Our column vector will consist of the number of individuals in age classes 1, 2, 3, and 4:

$$\mathbf{n}(t) = \begin{bmatrix} w \\ x \\ y \\ z \end{bmatrix}$$

When the Leslie matrix, **A**, is multiplied by the population vector, $\mathbf{n}(t)$, the result is another population vector (which also consists of one column); this vector is called the **resultant vector** and provides information on how many individuals are in age classes 1, 2, 3, and 4 in year $t + 1$. The multiplication works as follows:

$$\begin{bmatrix} a & b & c & d \\ e & f & g & h \\ i & j & k & l \\ m & n & o & p \end{bmatrix} \times \begin{bmatrix} w \\ x \\ y \\ z \end{bmatrix} = \begin{bmatrix} aw + bx + cy + dz \\ ew + fx + gy + hz \\ iw + jx + ky + lz \\ mw + nx + oy + pz \end{bmatrix}$$

$$\mathbf{A} \times \mathbf{n} = \text{Resultant vector}$$

The first entry in the resultant vector is obtained by multiplying each element in the first row of the **A** matrix by the corresponding element in the **n** vector, and then summing the products together. In other words, the first entry in the resultant vector equals the total of several operations: multiply the first entry in the first row of the **A** matrix by the first entry in **n** vector, multiply the second entry in the first row of the **A** matrix by the second entry in the **n** vector, and so on until you reach the end of the first row of the **A** matrix, then add all the products. In the example above, a 4×4 matrix on the left is multiplied by a column vector (center). The resultant vector is the vector on the right-hand side of the equation.

Rearranging the matrices so that the resultant vector is on the left, we can compute the population size at time $t + 1$ by multiplying the Leslie matrix by the population vector at time t.

$$\begin{bmatrix} n_1(t+1) \\ n_2(t+1) \\ n_3(t+1) \\ n_4(t+1) \end{bmatrix} = \begin{bmatrix} F_1 & F_2 & F_3 & F_4 \\ P_1 & 0 & 0 & 0 \\ 0 & P_2 & 0 & 0 \\ 0 & 0 & P_3 & 0 \end{bmatrix} \times \begin{bmatrix} n_1(t) \\ n_2(t) \\ n_3(t) \\ n_4(t) \end{bmatrix} \quad \text{Equation 4}$$

For example, assume that you have been following a population that consists of 45 individuals in age class 1, 18 individuals in age class 2, 11 individuals in age class 3, and 4 individuals in age class 4. The initial vector of abundances is written

$$\begin{bmatrix} 45 \\ 18 \\ 11 \\ 4 \end{bmatrix}$$

Assume that the Leslie matrix for this population is

$$\begin{bmatrix} 0 & 1 & 1.5 & 1.2 \\ .8 & 0 & 0 & 0 \\ 0 & .5 & 0 & 0 \\ 0 & 0 & .25 & 0 \end{bmatrix}$$

Following Equation 4, the number of individuals of age classes 1, 2, 3, and 4 at time $t + 1$ would be computed as

$$\begin{bmatrix} n_1(t+1) \\ n_2(t+1) \\ n_3(t+1) \\ n_4(t+1) \end{bmatrix} = \begin{bmatrix} 0 & 1 & 1.5 & 1.2 \\ .8 & 0 & 0 & 0 \\ 0 & .5 & 0 & 0 \\ 0 & 0 & .25 & 0 \end{bmatrix} \times \begin{bmatrix} 45 \\ 18 \\ 11 \\ 4 \end{bmatrix} = \begin{bmatrix} 0\times45+1\times18+1.5\times11+1.2\times4 \\ 0.8\times45+0\times18+0\times11+0\times4 \\ 0\times45+.5\times18+0\times11+0\times4 \\ 0\times45+0\times18+.25\times11+0\times4 \end{bmatrix} = \begin{bmatrix} 39.3 \\ 36 \\ 9 \\ 2.75 \end{bmatrix}$$

The time-specific growth rate, λ_t, can be computed as the total population at time t + 1 divided by the total population at time t. For the above example,

$$\lambda_t = (39.3 + 36 + 9 + 2.75)/(45 + 18 + 11 + 4) = 87.05/78 = 1.116$$

As we mentioned earlier, λ_t is not necessarily equal to λ in Equation 1.The Leslie matrix not only allows you to calculate λ_t (by summing the total number of individuals in the population at time t + 1 and dividing this number by the total individuals in the population at time t), but also to evaluate how the composition of the population changes over time. If you multiply the Leslie matrix by the new vector of abundances, you will project population size for yet another year. Continued multiplication of a vector of abundance by the Leslie matrix eventually produces a population with a **stable age distribution**, where the proportion of individuals in each age class remains constant over time, and a stable (unchanging) time-specific growth rate, λ_t. When the λ_t's converge to a constant value, this constant is an estimate of λ in Equation 1. Note that this λ has no subscript associated with it. Technically, λ is called the **asymptotic growth rate** when the population converges to a stable age distribution. At this point, if the population is growing or declining, all age classes grow or decline at the same rate. In this exercise you'll set up a Leslie matrix model for a population with age structure. The goal is to project the population size and structure into the future, and examine properties of a stable age distribution. As always, save your work frequently to disk.

INSTRUCTIONS	ANNOTATION
A. Set up the spreadsheet.	
1. Set up new column headings as shown in Figure 3.	(Figure 3, below)
2. Enter values in the Leslie matrix in cells B5–E8 as shown.	Remember that the Leslie matrix has a specific form. Fertility rates are entered in the top row. Survival rates are entered on the subdiagonal, and all other values in the Leslie matrix are 0.
3. Enter values in the initial population vector in cells G5–G8 as shown.	The initial population vector, **n**, gives the number of individuals in the first, second, third, and fourth age classes. Thus our population will initially consist of 45 individuals in age class 1, 18 individuals in age class 2, 11 individuals in age class 3, and 4 individuals in age class 4.
4. Set up a linear series from 0 to 25 in cells A12–A37.	We will track the numbers of individuals in each age class over 25 years. Enter 0 in cell A12. Enter **=1+A12** in cell A13. Copy your formula down to cell A37.
5. Enter formulae in cells B12–E12 to link to values in the initial vector of abundances (G5–G8).	Enter the following formulae: • B12 **=G5** • C12 **=G6** • D12 **=G7** • E12 **=G8**
6. Sum the total number of individuals at time 0 in cell F12.	Enter the formula **=SUM(B12:E12)**. Your result should be 78.
7. Compute λ_t for time 0 in cell G12.	Enter the formula **=F13/F12**. Your result will not be interpretable until you compute the population size at time 1. You can generate the λ symbol by typing in the letter l, highlighting this letter in the formula bar, and then changing its font to the symbol font.
8. Save your work.	

	A	B	C	D	E	F	G
1	***Age-Structured Matrix Models***						
2							
3				**Age class**			**n**
4		**1**	**2**	**3**	**4**		
5		0	1	1.5	1.2		45
6	**A** =	0.8	0	0	0		18
7		0	0.5	0	0		11
8		0	0	0.25	0		4
9							
10							
11	**Time**	**1**	**2**	**3**	**4**	**Total pop**	λ_t

Figure 3

B. Project population size over time.

1. In cells B13–E13, enter formulae to calculate the number of individuals in each age class in year 1. In your formulae, use the initial vector of abundances listed in row 12 instead of column G.

Now we are ready to project the population sizes into the future. Remember, we want to multiply the Leslie matrix by our initial set of abundances to generate a resultant vector (which gives the abundances of the different age classes in the next time step). Recall how matrices are multiplied to generate the resultant vector:

$$\begin{bmatrix} a & b & c & d \\ e & f & g & h \\ i & j & k & l \\ m & n & o & p \end{bmatrix} \times \begin{bmatrix} w \\ x \\ y \\ z \end{bmatrix} = \begin{bmatrix} aw+bx+cy+dz \\ ew+fx+gy+hz \\ iw+jx+ky+lz \\ mw+nx+oy+pz \end{bmatrix}$$

See if you can follow how to calculate the resultant vector, and enter a formula for its calculation in the appropriate cell—it's pretty easy to get the hang of it. The cells in the Leslie matrix should be absolute references, while the cells in the vector of abundances should be relative references. We entered the following formulae:

- B13 **=$B$5*B12+$C$5*C12+$D$5*D12+$E$5*E12**
- C13 **=$B$6*B12+$C$6*C12+$D$6*D12+$E$6*E12**
- D13 **=$B$7*B12+$C$7*C12+$D$7*D12+$E$7*E12**
- E13 **=$B$8*B12+$C$8*C12+$D$8*D12+$E$8*E12**

2. Copy the formula in cell F12 into cell F13.

3. Copy the formula in cell G12 into cell G13. Your spreadsheet should now resemble Figure 4.

	A	B	C	D	E	F	G
11	Time	1	2	3	4	Total pop	λ_t
12	0	45	18	11	4	78	1.1153846
13	1	39	36	9	3	87	0

Figure 4

4. Select cells B13–G13, and copy their formulae into cells B14–G37. Save your work.

This will complete your population projection over 25 years.

C. Create graphs.

1. Graph the number of individuals in each age class over time, as well as the total number of individuals over time. Use the scatter graph option, and label your axes clearly. Your graph should resemble Figure 5.

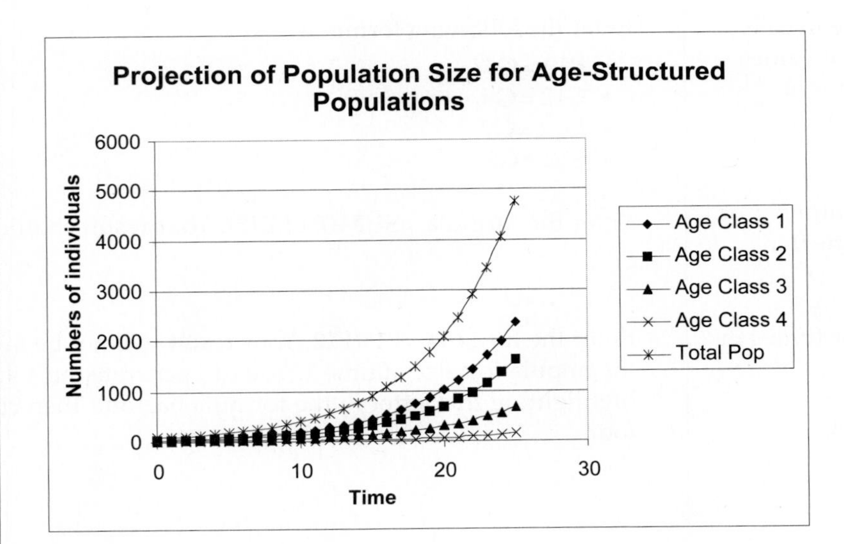

Figure 5

It's often useful to examine the logarithms of the number of individuals instead of the raw data. This takes the bending nature out of a geometrically growing or declining population (see Exercise 1). To adjust the scale of the y-axis, double click on the values in the y-axis. A dialog box (Figure 6) will appear:

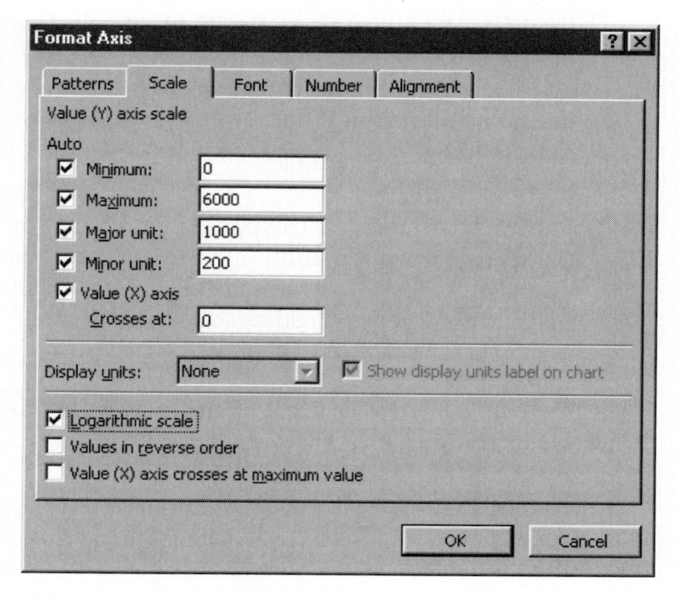

Figure 6

Toward the bottom of the screen is a box labeled Logarithmic scale. Click on that box, and then click the OK button, and your scale will be automatically adjusted. It's sometimes easier to interpret your population projections with a log scale.

2. Generate a new graph of the same data, but use a log scale for the y-axis.

3. Save your work.

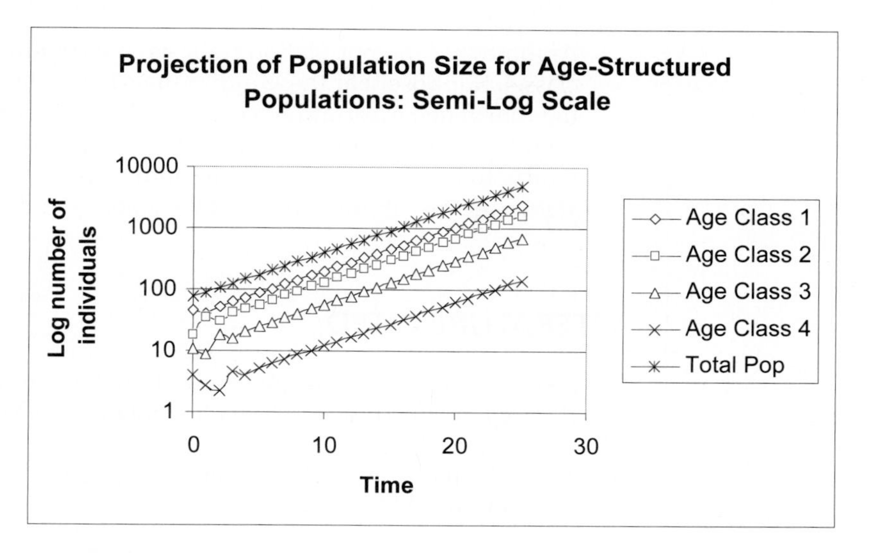

Figure 7

QUESTIONS

1. Examine your first graph (Figure 5). What is the nature of the population growth? Is the population increasing, stable, or declining? How does λ_t change with time?

2. Examine your semi-log graph (Figure 7) and your spreadsheet projections (column G). At what point in the 25-year projection does λ_t not change (or change very little) from year to year? When the λ_t's do not change over time, they are an estimate of λ, the asymptotic growth rate, or an estimate of λ in Equation 1 in the Introduction. What is λ for your population, and how does this affect population growth? If you change entries in your Leslie matrix, how does λ change?

3. Return your Leslie matrix parameters to their original values. What is the composition of the population (the proportion of individuals in age class 1, age class 2, age class 3, and age class 4) when the population has reached a stable distribution? Set up headings as shown:

	H	I	J	K
10	**Stable Age Distribution**			
11	**1**	**2**	**3**	**4**

In cell H12, enter a formula to calculate the proportion of the total population in year 25 that consists of individuals in age class 1. Enter formulae to compute the proportions of the remaining age classes in cells I12–K12. Cells H12–K12 should sum to 1 and give the stable age distribution.

4. How does the initial population vector affect λ_t, λ and the stable age distribution? How does it affect λ_t and the age distribution prior to stabilization? Change the initial vector of abundances so that the population consists of 75 individuals in age class 1, and 1 individual in each of the remaining age classes. Graph and interpret your results. Do your results have any management implications?

5. What are the assumptions of the age-structured matrix model you have built?

6. Assume that the population consists of individuals that can exist past age class 4. Suppose that these individuals have identical fertility functions (F) as the fourth age class and have a probability of surviving from year t to year $t + 1$ with a probability of 0.25. Draw the life cycle diagram, and adjust your Leslie matrix to incorporate these older individuals. How does this change affect the stable age distribution and λ at the stable age distribution?

LITERATURE CITED

Akçakaya, H. R., M. A. Burgman, and L. R. Ginzburg. 1997. *Applied Population Ecology*. Applied Biomathematics, Setauket, NY.

Caswell, H. 2001. *Matrix Population models*, Second Edition. Sinauer Associates, Inc. Sunderland, MA.

Gotelli, N. 2001. *A Primer of Ecology*, Third Edition. Sinauer Associates, Sunderland, MA.

Leslie, P. H. 1945. On the use of matrices in certain population mathematics. *Biometrika* 33: 183–212.

14 STAGE-STRUCTURED MATRIX MODELS

Objectives

- Set up a model of population growth with stage structure.
- Determine the stable stage distribution of the population.
- Estimate the finite rate of increase from Lefkovitch matrix calculations.
- Construct and interpret the stage distribution graphs.

Suggested Preliminary Exercise: Geometric and Exponential Population Models; Life Tables and Survivorship Curves

INTRODUCTION

Recall from Exercise 7 that the geometric model describes a population growing in discrete time. That is, the model treats time as if it moved in steps rather than flowing continuously. This kind of model is realistic for many populations that have seasonal, synchronous reproduction. For example, insectivorous songbirds in North America typically breed during the spring and summer months, when their major food sources peak in abundance. The geometric growth model has the form

$$N_{t+1} = N_t + (R)N_t$$

where R is the per capita change in population size, or intrinsic (or geometric) rate of natural increase. You might recall that R is equal to $b - d$ for discretely growing populations, or the difference in the per capita birth and death rates. We can factor N_t out of the terms on the right-hand side, to get

$$N_{t+1} = (1 + R)N_t$$

The quantity $(1 + R)$ is called the finite rate of increase, λ, and so we can write

$$N_{t+1} = \lambda N_t \qquad \text{Equation 1}$$

where N is the number of individuals present in the population, and t is a time interval of interest. Equation 1 says that the size of a population at time $t + 1$ equals the size of the population at time t multiplied by a constant, λ. When $\lambda = 1$, the population will remain constant in size over time. When $\lambda < 1$, the population declines geometrically, and when $\lambda > 1$, the population increases geometrically.

Equation 1 predicts change in numbers in a population over time, given the numbers of individuals currently in the population and λ. Simplistically speaking, the models assume that all individuals in the population make equal contributions to population change, regardless of their size, age, stage, genetic make-up, or sex. Many natural populations violate at least one of these assumptions because the populations are **structured**—they are composed of individuals whose birth and death rates *differ* depending on age, size, sex, stage, or genetic make-up. For example, small fish in a population differ in mortality rates from large fish, and larval insects differ in birth rates from adult insects.

Differences among individuals in a population are a cornerstone of ecology and evolutionary biology, and can greatly affect the population's finite rate of increase (λ). In this exercise, you will develop a **matrix model** to explore the growth of populations that have size or stage structure. This approach will enable you to estimate λ in Equation 1 for size- or stage-structured populations.

If you have completed the Life Tables exercise, you learned that *age structure* is often a critical variable in determining the size of a population over time. In fact, a primary goal of life table analysis and Leslie matrix modeling is to estimate the population's growth rate, λ, when the population has age structure. For many organisms, however, age is *not* an accurate predictor of birth or death rates. For example, a small sugar maple in a northeastern forest can be 50 years old and yet have low levels of reproduction. In this species, size is a better predictor of birth rate than age. In other species, birth and death rates are a function of the *stage in the life cycle* of an organism. For instance, death rates in some insect species may be higher in the larval stages than in the adult stage. Such organisms are best modeled with **size-** or **stage-structured** matrix models.

Model Notation

We begin our exercise with some notation often used when modeling structured populations (Caswell 2001; Gotelli 2001). For modeling purposes, the first decision is whether to develop a stage-structured or a size-structured model for the organism of interest. This in turn depends on whether size or life-history stage is a better state variable.

The second step is to assign individuals in the population to either stage or size **classes**. It is fairly straightforward to categorize individuals with stage structure, such as insects—simply place them in the appropriate stage, such as larva, pupa, or adult. Size, however, is a *continuous variable* because is not an either/or situation, but can take on a range of values. In our sugar maple example, size classes might consist of seedlings, small-sized individuals, medium-sized individuals, and large-sized individuals. The number of size classes you select for your model would depend on how "different" groups are in terms of reproduction and survival. If medium- and large-sized individuals have the same reproductive and survival rates, we might choose to lump them into a single class.

Note that the projection interval (the amount of time that elapses between time t and time $t + 1$) and the stage durations can be different. For instance, the larval stage may typically last 4 months, and the pupa stage might typically last only 2 months, and the projection interval may also be different. This is quite different from the age-based matrix model, in which the interval of the different classes, as well as the projection interval from time step to time step, must be equal.

A typical life cycle model for a species with stage or size structure is shown in Figure 1. The horizontal arrows between each stage (circles) represent **survival probabilities**, or the probability that an individual in stage/size class i will survive and move into stage/size class $i + 1$, designated by the letter P followed by two different subscripts.*

The curved arrows at the bottoms of the diagrams in Figure 1 represent the proba-

*Caswell 2001 calls these "graduation probabilities," designated by the letter G.

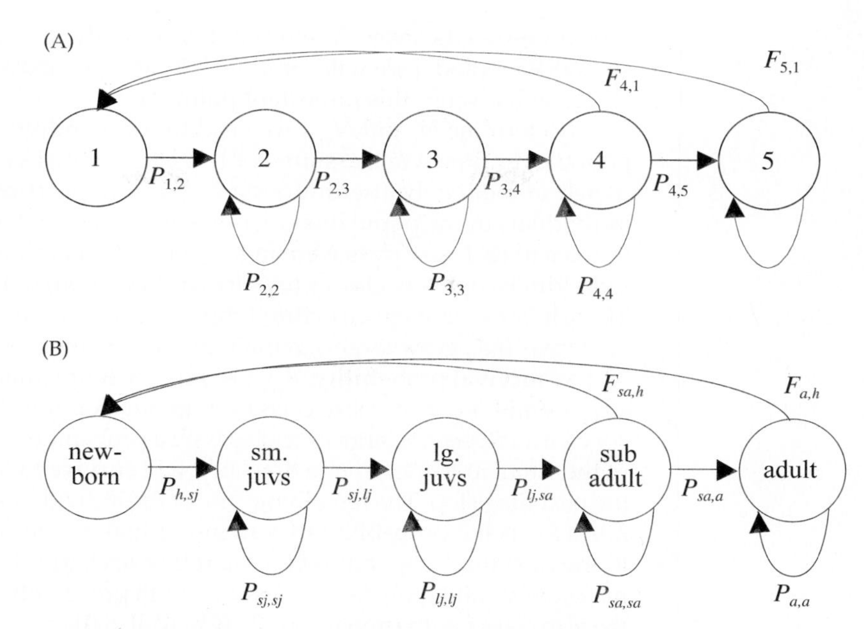

Figure 1 (A) A theoretical life history model for an organism with a stage- or size-structured life history. Classes are represented by circles. The arrows between the stages are called transitions, indicating the probability P of transitioning from one class to the next (horizontal arrows) or of remaining in the same class (lower curved arrows). The curved arrows at the top, labeled F, represent births. (B) Model of an organism with five specific size/stage structures (in this case, a combination of the two), as labeled. Two classes (subadults and adults) are capable of reproduction, so arrows associated with birth emerge from both classes returning to newborns (hatchlings). If only adult individuals reproduced, there would be a single arrow from adult to newborn.

bility of individuals surviving and remaining in their class from time t to time $t + 1$, designated by the letter P followed by two identical subscripts. For instance, the loop at the bottom of the small juvenile class represents the probability that a small juvenile in time t will be alive and counted as a small juvenile in time $t + 1$. The loop at the bottom of the adult class represents the probability that an adult counted in time t will be alive and counted as an adult in time $t + 1$. These self-loops are absent from age-based matrix models because individuals *must* move from one class to the next (you can't have two twentieth birthdays).

The curved arrows at the top of the diagrams represent births, designated by the letter F followed by two different subscripts. The arrows all lead to the first class because newborns, by definition, enter the first class upon birth.

Note that for both P and F, the subscripts have a definite pattern: the first subscript is the class *from* which individuals move, and the second subscripts indicate the class *to* which individuals move (Gotelli 2001).

Matrix Models

Now let's move on and discuss the computations of P, F, and λ for a population with stage structure. The major goal of the matrix model is to compute λ, the finite rate of increase, for a population with stage structure (Equation 1). In our matrix model, we can compute the **time-specific growth rate**, λ_t, by rearranging terms in Equation 1:

$$\lambda_t = \frac{N_{t+1}}{N_t} \qquad \text{Equation 2}$$

This time-specific growth rate is not necessarily the same λ in Equation 1, but in our spreadsheet model we will compute it in order to arrive at λ (no subscript) in Equation 1. (We will discuss this important point later).

To determine N_t and N_{t+1}, we need to count individuals at some standardized time period over time. We'll assume for this exercise that our population censuses are completed immediately after individuals breed (a **postbreeding census**). The number of individuals in the population in a census at time $t + 1$ will depend on how many individuals of each size class were in the population at time t, as well as the movements of individuals into new classes (by birth or transition) or out of the system (by mortality). Thus, in size- or stage-structured models, an individual in any class may move to the next class (i.e., grow larger), remain in their current class, or exit the system (i.e., die).

The **survival probability**, $P_{i,i+1}$, is the probability that an individual in size class i will survive and move into size class $i + 1$. In our example, let's assume that small juveniles survive and become large juveniles with a probability of $P_{sj,lj} = 0.3$. This means that 30% of the small juveniles in one time step will survive to be censused as large juveniles in the next time step. The remaining 70% of the individuals either die or remain small juveniles. $P_{i,i}$, is the probability that an individual in size class i will survive to be counted in the next time step, but will remain in size class i. Thus, an individual in size class i may survive and grow to size class $i + 1$ with probability $P_{i,i+1}$, or may survive and remain the size class i with probability $P_{i,i}$ (Caswell 2001).

In order to keep track of how many individuals are present in a given class at a given time, we must consider both kinds of survival probabilities to account for those individuals that graduated into the class, plus those individuals that remained in the class (i.e., did not graduate). For example, we can compute the number of individuals in the large juvenile size class at time $t + 1$ as the number of small juveniles at time t multiplied by $P_{sj,lj}$ (this gives the number of small juveniles in year t that graduated to become large juveniles in year $t + 1$), plus the number of large juveniles at time t multiplied by $P_{lj,lj}$ (this gives the number of large juveniles in year t that remained in the large juvenile class in year $t + 1$). More generally speaking, the number of individuals in class i in year $t +$ 1 will be

$$n_i(t + 1) = [P_{i,i}n_i(t)] + [P_{i-1,i},n_{i-1}(t)] \qquad \text{Equation 3}$$

Equation 3 works for calculating the number of individuals at time $t + 1$ for each size class in the population except for the first, because individuals in the first stage class at time $t + 1$ will include those individuals in class 1 that remain in class 1 in the next time step, plus any new individuals that arise through birth. Accordingly, let's now consider birth rates.

There are many ways to describe the births in a population. Here we will assume a simple **birth-pulse model**, in which individuals give birth as soon as they enter a new stage class. On this day, not only do births occur, but transitions from one size class to another also occur. When populations are structured, the birth rate is often called the **fecundity**, or the average number of offspring born per unit time to an individual female of a particular age (Gotelli 2001). If you have completed the exercise on life tables, you might recall that fecundity is labeled as $b(x)$, where b is for birth. Individuals that are of pre- or postreproductive age have fecundities of 0. Individuals of reproductive age typically have fecundities greater than 0.

To illustrate the concept of birth pulse and postbreeding census, consider a hypothetical diagram for a sea turtle (*Caretta caretta*) population with five stage classes that are censused at three time periods: $t - 1$, t, and $t + 1$ as shown in Figure 2. Since all individuals are born during the birth pulse, they have the same birthday. The birthday is also the "graduation day" for those individuals that move from one size class to the next. With a postbreeding census, the diagram shows that the number of individuals in the first size class (hatchlings, h) at time t depends on the number of subadults and adults in the previous time step, $t - 1$.

If we knew how many breeders were producing those hatchlings, we could compute fecundity as the number of offspring produced per individual per year (Gotelli 2001).

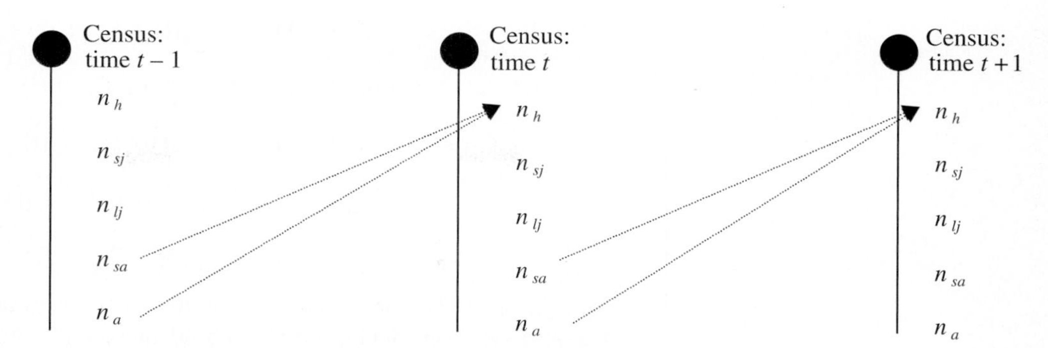

Figure 2 In this population, subadults (*sa*) and adults (*a*) can reproduce, represented by the dashed arrows that lead to the first class in the next time step. Births occur in a birth pulse (indicated by the filled circle and vertical line) and individuals are censused postbreeding (i.e., immediately after the young are born). (After Akçakaya et al. 1997.)

However, the number of breeders is not simply N_{sa} and N_a counted in time step $t - 1$; these individuals must survive a long period of time (almost a full year) until the birth pulse in time step $t - 1$ occurs. In other words, not all of the subadults and adults counted in year $t - 1$ will survive to the birth pulse and produce offspring that will be counted as hatchlings in year t. Thus, we need to discount the fecundity, $b(i)$, by the probability that an individual will actually survive from the time they were censused to the time they breed (Gotelli 2001). These adjusted estimates are used in matrix models and are commonly called **fertilities** (often defined as realized reproduction), designated by the letter F. (Be aware that various authors use the terms fertility and fecundity in different ways.) The adjustment is necessary to account for "lags" between the census and the timing of births.

These adjustments are a bit trickier for stage-based than for age-based models because both kinds of survival probabilities ($P_{i,i+1}$ and $P_{i,i}$) come into play. For example, suppose we want to compute the fertility rate of subadults, F_{sa} that were censused in year t. We need to ask, "How many offspring are produced, on average, per subadult censused in year t?" To answer this question, we need to know how many subadults were counted during the census for year t, how many of those individuals survived to the birth pulse in the same time step, and the total number of young produced by those individuals. Keeping in mind that the graduation day is the same day as the birth pulse, the young produced by the breeding individuals comes from two sources: (1) those subadults that survived to the birth pulse and reproduced at the rate of subadults ($b_i \times P_{i,i}$), and (2) those subadults that survived to the birth pulse and graduated to adulthood and reproduced as adults ($b_{i+1} \times P_{i,i+1}$). Accordingly, we can compute the fertility rate as

$$F_i = (b_i \times P_{i,i}) + (b_{i+i} \times P_{i,i+1}) \quad \text{Equation 4}$$

Thus, F_i indicates the number of young that are produced per female of stage i in year t, given the appropriate adjustments (Caswell 2001). The total number of individuals counted in stage 1 (newborns or hatchlings) in year $t + 1$ is simply the fertility rate of each age class, multiplied by the number of individuals in that size class at time t, plus any individuals that remained in the first size class from one time step to the next. Generally speaking,

$$n_1(t+1) = P_{1,1}n_1(t) + \sum_{i=1}^{k} F_i n_i(t) \quad \text{Equation 5}$$

Once we know the fertility and survivorship coefficients for each age class, we can calculate the number of individuals in each age at time $t + 1$, given the number of individuals in each class at time t (Gotelli 2001):

$$n_h(t+1) = P_{h,h}n_h(t) + F_{sj,h}n_{sj}(t) + F_{lj,h}n_{lj}(t) + F_{sa,h}n_{sa}(t) + F_{a,h}n_a(t)$$
$$n_{sj}(t+1) = P_{sj,sj}n_{sj}(t) + P_{h,sj}n_h(t)$$
$$n_{lj}(t+1) = P_{lj,lj}n_{lj}(t) + P_{sj,lj}n_{sj}(t) \quad \text{Equation 6}$$
$$n_{sa}(t+1) = P_{sa,sa}n_{sa}(t) + P_{lj,sa}n_{lj}(t)$$
$$n_a(t+1) = P_{a,a}n_a(t) + P_{sa,a}n_{sa}(t)$$

Equation 6 can be converted into a matrix form. A **matrix** is a rectangular array of numbers and symbols, designated by a bold-faced letter. Matrices that describe populations with stage or size structure are often called Lefkovitch matrices, after biologist L. P. Lefkovitch (1965).

$$\mathbf{L} = \begin{bmatrix} P_{h,h} & F_{sj} & F_{lj} & F_{sa} & F_a \\ P_{h,sj} & P_{sj,sj} & 0 & 0 & 0 \\ 0 & P_{sj,lj} & P_{lj,lj} & 0 & 0 \\ 0 & 0 & P_{lj,sa} & P_{sa,sa} & 0 \\ 0 & 0 & 0 & P_{sa,a} & P_{a,a} \end{bmatrix}$$

Since our population has only five classes, the matrix, denoted by the letter **L**, is a five-row × five-column matrix. The fertility rates are given in the top row. The survival probabilities, $P_{i,i+1}$, are given in the subdiagonal, which represent the survival from one class to the next. For example, $P_{sj,lj}$ is the probability of small juveniles will become large juveniles in year $t + 1$. The survival probabilities, $P_{i,i}$, are given in the diagonal, which represent the probability that an individual in a given class will survive, but will remain in the same class in year $t + 1$. The upper left entry in the **L** matrix gives the probability that a hatchling will remain a hatchling. If hatchlings could reproduce, we would add $F_{h,h}$ to $P_{h,h}$ for this matrix entry. Note that the $P_{i,i} + P_{i,i+1,}$ gives the total rate of survival for individuals in a particular stage.

Vectors and Matrix Multiplication

The composition of our population can be expressed as a column **vector**, $\mathbf{n}(t)$, which is a matrix that consists of a single column. Our column vector will consist of the number of individuals in the newborn, small juvenile, large juvenile, subadult, and adult classes. When the Leftkovitch matrix, **L**, is multiplied by the population vector, $\mathbf{n}(t)$, the result is another population vector (which also consists of 1 column); this vector is called the **resultant vector** and provides information on how many individuals are in each size class in year $t + 1$.

Multiplying each element in first row of the **L** matrix by the corresponding element in the **n** vector, and then repeating the process for the remaining elements in the first row and summing the products together generate the first entry in the resultant vector. In other words, the first entry in the first row of the **L** matrix is multiplied by the first entry in **n** vector, plus the second entry in the first row of the **L** matrix by the second entry in the **n** vector, and so on. In the example below, a 4 × 4 matrix on the left is multiplied by column vector (center). The resultant vector is on the right-hand side of the equation (note that summing the components would compress this vector to a single column).

$$\begin{bmatrix} a & b & c & d \\ e & f & g & h \\ i & j & k & l \\ m & n & o & p \end{bmatrix} \times \begin{bmatrix} w \\ x \\ y \\ z \end{bmatrix} = \begin{bmatrix} aw+bx+cy+dz \\ ew+fx+gy+hz \\ iw+jx+ky+lz \\ mw+nx+oy+pz \end{bmatrix} \quad \text{Equation 7}$$

Rearranging the matrices so that the resultant vector is on the left, we can compute the population size at time $t + 1$ by multiplying the Leftkovitch matrix by the population vector at time t:

$$\begin{bmatrix} n_h(t+1) \\ n_{sj}(t+1) \\ n_{lj}(t+1) \\ n_{sa}(t+1) \\ n_a(t+1) \end{bmatrix} = \begin{bmatrix} P_{h,h} & F_{sj} & F_{lj} & F_{sa} & F_a \\ P_{h,sj} & P_{sj,sj} & 0 & 0 & 0 \\ 0 & P_{sj,lj} & P_{lj,lj} & 0 & 0 \\ 0 & 0 & P_{lj,sa} & P_{sa,sa} & 0 \\ 0 & 0 & 0 & P_{sa,a} & P_{a,a} \end{bmatrix} \times \begin{bmatrix} n_h(t) \\ n_{sj}(t) \\ n_{lj}(t) \\ n_{sa}(t) \\ n_a(t) \end{bmatrix} \quad \text{Equation 8}$$

For example, assume that you have been following a population that consists of 45 newborns, 18 small juveniles, 56 large juveniles, 10 subadults, and 8 adults. The initial vector of abundances is written

$$\begin{bmatrix} 45 \\ 18 \\ 56 \\ 4 \\ 1 \end{bmatrix}$$

Assume that the Leftkovitch matrix for this population is

$$\mathbf{L} = \begin{bmatrix} 0 & 0 & 0 & 4.6 & 61.8 \\ .6 & .7 & 0 & 0 & 0 \\ 0 & .05 & .66 & 0 & 0 \\ 0 & 0 & .02 & .68 & 0 \\ 0 & 0 & 0 & .02 & .8 \end{bmatrix}$$

The number of newborns, small juveniles, large juveniles, subadults, and adults in year $t + 1$ (rounded) would be computed as

$$\begin{bmatrix} 0 & 0 & 0 & 4.6 & 61.8 \\ .6 & .7 & 0 & 0 & 0 \\ 0 & .05 & .66 & 0 & 0 \\ 0 & 0 & .02 & .68 & 0 \\ 0 & 0 & 0 & .02 & .8 \end{bmatrix} \times \begin{bmatrix} 45 \\ 18 \\ 56 \\ 4 \\ 1 \end{bmatrix} = \begin{bmatrix} 0 \times 45 + 0 \times 18 + 0 \times 56 + 4.6 \times 4 + 61.8 \times 1 \\ .6 \times 45 + .7 \times 18 + 0 \times 56 + 0 \times 4 + 0 \times 1 \\ 0 \times 45 + .05 \times 18 + .66 \times 56 + 0 \times 4 + 0 \times 1 \\ 0 \times 45 + 0 \times 18 + .02 \times 56 + .68 \times 4 + 0 \times 1 \\ 0 \times 45 + 0 \times 18 + 0 \times 56 + .02 \times 4 + .8 \times 1 \end{bmatrix} = \begin{bmatrix} 80 \\ 40 \\ 38 \\ 4 \\ 1 \end{bmatrix}$$

Upon inspection, you will see that the Lefkovitch matrix computes population numbers in year t + 1 in the manner of Equation 6. For this population, the time-specific growth rate is

$$\lambda_t = \frac{N_{t+1}}{N_t} = \frac{80 + 40 + 38 + 4 + 1}{45 + 18 + 56 + 4 + 1} = \frac{163}{124} = 1.31$$

The Lefkovitch matrix not only allows you to calculate λ_t (by summing the total number of individuals in the population at time $t + 1$ and dividing this number by the total individuals in the population at time t), but also lets you evaluate how the composition of the population changes from one time step to the next. If you continued projecting the population dynamics into the future, you would be able to ascertain how the population "behaves" if the present conditions (P's and F's) were to be maintained indefinitely (Caswell 2001). Continued multiplication of a vector of abundance by the Lefkovitch matrix eventually produces a population with a **stable size** or **stable stage**

distribution, where the proportion of individuals in each stage remains constant over time, and there is a stable (unchanging) finite rate of increase, λ_t. When the λ_t's converge to a constant value, this constant is an estimate of λ in Equation 1, and is called the **asymptotic growth rate.** At this point, if the population is growing or declining, all stage classes grow or decline at the same rate, even if the numbers of individuals in each class are different. You will see how this happens as you work through the exercise.

PROCEDURES

In this exercise, you will develop a stage-based model for sea turtles (*Caretta caretta*). In this population, the size stages are hatchlings (*h*), small juveniles (*sj*), large juveniles (*lj*), subadults (*sa*), adults (*a*). Turtles are counted every year in a postbreeding census, where the numbers of individuals in each stage class are tallied.

As always, save your work frequently to disk.

INSTRUCTIONS / ANNOTATION

A. Set up the model population.

1. Open a new spreadsheet and enter column headings as shown in Figure 3.

	A	B	C	D	E	F	G	H
1	*Stage-Structured Matrix Models*							**Initial**
2	*Five-stage matrix model of the loggerhead sea turtle*							**population**
3		*F*(*h*)	*F*(*sj*)	*F*(*lj*)	*F*(*sa*)	*F*(*a*)		**vector**
4		0	0	0	4.665	61.896		2000
5		0.675	0.703	0	0	0		500
6		0	0.047	0.657	0	0		300
7		0	0	0.019	0.682	0		300
8		0	0	0	0.061	0.8091		1

Figure 3

2. Enter the values shown in cells B4–F8. Write your interpretation of what each cell value means in the chart.

These are the matrix values derived by Crowder et al.1994. For example, the value in cell B5 is the probability that a hatchling in year *t* will become a small juvenile in year *t* + 1.

B4:
C4:
D4:
E4:
F4:
B5:
C5:
C6:
D6:
D7:
E7:
E8:
F8:

3. Enter the values shown in cells H4–H8.

These values make up the initial population vector, or how many individuals of each stage are currently in the population.

4. Set up new column headings as shown in Figure 4 and set up a linear series in column A that will track abundances of individuals for 100 years.

	A	B	C	D	E	F	G	H
10	Year	Hatchlings	S. juvs	L. juvs	Subadults	Adults	Total	λ_t
11	0							
12	1							

Figure 4

Enter 0 in cell A11.
Enter the formula **=A11+1** in cell A12. Copy this formula down to cell A111 to simulate 100 years of population growth.
(You can generate the λ symbol by typing in the letter l, then select this letter in the formula bar and change its font to Symbol.)

5. Link the initial vector abundances to the appropriate cells in B11–F11.

Enter the formula **=H4** in cell B11 to indicate that at year 0, the population consists of 2000 hatchlings. Enter a similar formula in cells C11–F11 to link the initial population vector with the proper stages.

- Cell C11 **=H5**
- Cell D11 **=H6**
- Cell E11 **=H7**
- Cell F11 **=H8**

6. In cell G11, use the **SUM** function to obtain the total population size for year 0.

Enter the formula **=SUM(B11:F11)** in cell G11.

7. In cell H11, enter a formula to compute λ_t.

Enter the formula = **G12/G11** in cell H11.
Remember that λ_t can be computed as N_{t+1}/N_t. Your result will not make sense until you compute the total population size for year 1.

8. Save your work.

B. Project the population sizes over time.

We will use matrix multiplication to project the population size and structure at year 1. Multiply your matrix of fecundities and survival values by your initial vector of abundances (given in year 0, row 11). The result will be the number of individuals in the next generation that are hatchlings, small juveniles, large juveniles, subadults, and adults. Recall how matrices are multiplied: The **L** matrix is located on the left, and is multiplied by the initial vector of abundances (**v**). The result is a new vector of abundances for the year $t + 1$. Refer to Equations 7 and 8. (Equation 7 is a 4×4 matrix; you will carry out the multiplication for a 5×5 matrix.)

1. Enter a formula in cell B12 to obtain the number of hatchlings in year 1.

Enter the formula **=$B$4*B11+$C$4*C11+$D$4*D11+$E$4*E11+$F$4*F11** in cell B12. Make sure you refer to the initial abundances listed in row 11 in your formula, rather than the initial abundances listed in column H.

2. Enter formulae in cells C12–F12 to obtain the number of small juveniles, large juveniles, subadults, and adults in year 1.

We used the following formulae:

- Cell C12 **=$B$5*B11+$C$5*C11+$D$5*D11+$E$5*E11+$F$5*F11**
- Cell D12 **=$B$6*B11+$C$6*C11+$D$6*D11+$E$6*E11+$F$6*F11**
- Cell E12 **=$B$7*B11+$C$7*C11+$D$7*D11+$E$7*E11+$F$7*F11**
- Cell F12 **=$B$8*B11+$C$8*C11+$D$8*D11+$E$8*E11+$F$8*F11**

3. In cell G12, use the **SUM** function to sum the individuals in the different stages.

Enter the formula **=SUM(B12:F12)** in cell G12.

4. In cell H12, calculate the time-specific growth rate, λ_t.

Copy cell H11 into H12. When $\lambda_t = 1$, the population remained constant in size. When $\lambda_t < 1$, the population declined, and when $\lambda_t > 1$, the population increased in numbers.

5. Select cells B12–H12, and copy the formulae down to cells B111–H111.

This will complete a 100-year simulation of stage-structured population growth. Click on a few random cells and make sure you can interpret the formulae and how they work.

6. Save your work.

C. Create graphs.

1. Graph your population abundances for all stages over time.

Use the line graph option and label your axes fully. Your graph should resemble Figure 5.

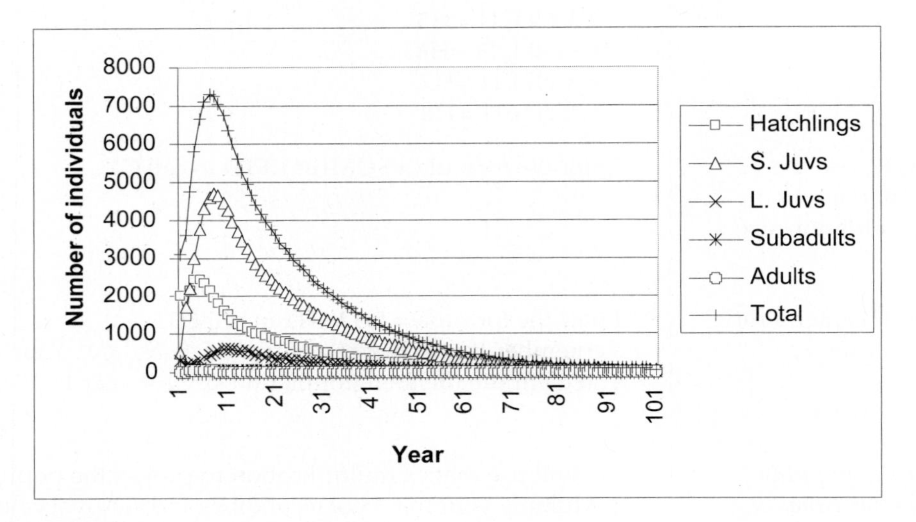

Figure 5

2. Copy the graph in Figure 5. Change the y-axis to a log scale.

To adjust the scale of the y-axis, double click on the values in the y-axis. You'll see the dialog box in Figure 6 on the facing page. Click on the Logarithmic scale box in the lower part of the dialog box. Your scale will be automatically adjusted.

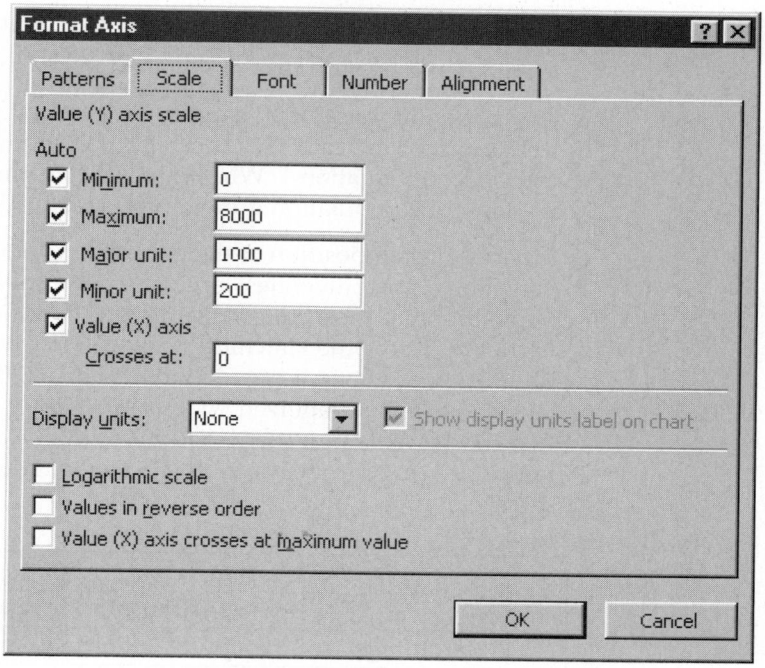

Format Axis

Patterns | Scale | Font | Number | Alignment

Value (Y) axis scale

Auto

- [x] Minimum: 0
- [x] Maximum: 8000
- [x] Major unit: 1000
- [x] Minor unit: 200
- [x] Value (X) axis Crosses at: 0

Display units: None ▾ [x] Show display units label on chart

- [ ] Logarithmic scale
- [ ] Values in reverse order
- [ ] Value (X) axis crosses at maximum value

OK | Cancel

Figure 6

Your graph should resemble Figure 7. It is sometimes easier to interpret your population projections with a log scale.

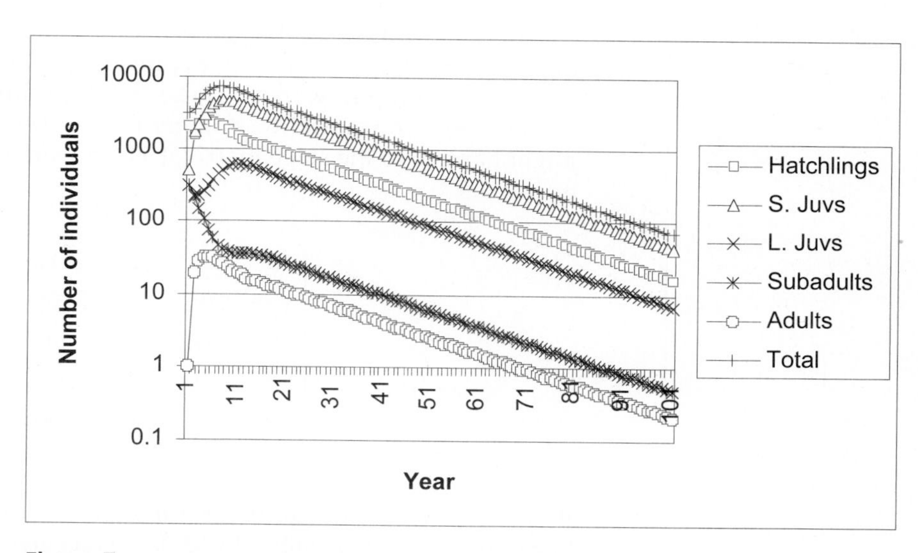

Figure 7

3. Save your work.

QUESTIONS

1. What are the assumptions of the model you have built?
2. At what point in the 100-year simulation does λ_t not change (or change very little) from year to year? This constant is an estimate of the asymptotic growth rate, λ, from Equation 1. What value is λ? Given this value of λ, how would you describe population growth of the sea turtle population?
3. What is the composition of the population (proportion of individuals that are hatchlings, small juveniles, large juveniles, subadults, and adults) when the population has reached a stable distribution? Set up the headings shown below. In the cell below the Hatchlings cell (cell I11) enter a formula to calculate the proportion of the total population in year 100 that consists of hatchlings (assuming λ_t has stabilized by year 100). Enter formulae to compute the proportions of the remaining stage classes in cells below the other stage-class headings. The five proportions calculated should sum to 1, and give the stable stage distribution.

	I	J	K	L	M
9	Proportion of individuals in class				
10	**Hatchlings**	**Small juvs**	**Large juvs**	**Subadults**	**Adults**

4. How does the initial population vector affect λ and the stable age distribution? How does it affect λ_t and the stage distribution prior to stabilization? Change the initial vector of abundances so that the population consists of 75 hatchlings, and 1 individual in each of the remaining stage classes. Graph and interpret your results. Do your results have any management implications?
5. One of the threats to the loggerhead sea turtle is accidental capture and drowning in shrimp trawls. One way to prevent this occurrence is to install escape hatches in shrimp trawl nets. These "turtle exclusion devices" (TEDS) can drastically reduce the mortality of larger turtles. The following matrix shows what might happen to the stage matrix if TEDS were widely installed in existing trawl nets:

$$\begin{bmatrix} 0 & 0 & 0 & 5.448 & 69.39 \\ .675 & .703 & 0 & 0 & 0 \\ 0 & .047 & .767 & 0 & 0 \\ 0 & 0 & .022 & .765 & 0 \\ 0 & 0 & 0 & 0.068 & .876 \end{bmatrix}$$

If the initial abundance is 100,000 turtles, distributed among stages at 30,000 hatchlings, 50,000 small juveniles, 18,000 large juveniles, and 2,000 subadults, and 1 adult, how does the use of TEDS influence population dynamics? Provide a graph and discuss your answer in terms of population size, structure, and growth. Discuss how the use of TEDS affects the *F* and *P* parameters in the Lefkovitch matrix.

6. Another important source of mortality for most marine turtles occurs in the very beginning of their lives, between the time the eggs are laid in a nest in the beach, and the time they hatch and are able to reach a safe distance into the sea. Most turtle conservation efforts in the past have concentrated on enhancing egg survival by protecting nests on beaches or removing eggs to protected hatch-

eries. If TEDS are not used, how much must fertilities increase in order to produce the population dynamics that would have been achieved with TEDS?

*7. (Advanced) Add stochasticity to the model by letting the P_i's, and F_i's vary stochastically with each time step.

LITERATURE CITED

Akçakaya, H. R., M. A. Burgman, and L. R. Ginzburg. 1997. *Applied Population Ecology*. Applied Biomathematics, Setauket, NY.

Caswell, H. 2001. *Matrix Population Models*, 2nd Edition. Sinauer Associates, Sunderland, MA.

Crowder, L. B., D. T. Crouse, S. S. Heppell, and T. H. Martin. 1994. Predicting the impact of turtle excluder devices on loggerhead turtle populations. *Ecological Applications* 4: 437–445.

Gotelli, N. 2001. *A Primer of Ecology*, 3rd Edition. Sinauer Associates, Sunderland, MA.

Lefkovitch, L. P. 1965. The study of population growth in organisms grouped by stages. *Biometrika* 35: 183–212.

15 REPRODUCTIVE VALUE: MATRIX APPROACH

Objectives

- Develop a Leslie matrix population growth model.
- Calculate reproductive values from the matrix model with the "inoculate" method.
- Calculate reproductive values from the matrix model with the "transpose vector" method.
- Evaluate how life history strategy affects reproductive values.

Suggested Preliminary Exercise: Age-Structured Matrix Models

INTRODUCTION

A basic premise in ecology and evolution is that not all individuals are created equal. In ecology, some individuals in a population are more "valuable" than others in terms of the number of offspring they are expected to produce over their remaining lifespan. Take, for example, a hypothetical population that consists of newborns, reproductively active 1-year-olds, reproductively active 2-year-olds, and postreproductive 3-year-olds. Which individuals are likely to produce the greatest number of offspring in the future?

If our population consisted solely of postreproductive individuals, it would go extinct because they are too old to reproduce. Clearly, this age class is not the most valuable in terms of future offspring production. Newborns may be valuable to the population in terms of future offspring because, although they cannot reproduce right now, they have their entire reproductive life ahead of them. However, they must survive to a reproductive age, and their value right now may be low if their chances of making it to a reproductive age in the future are slim. The 1-year-olds are valuable because they have already "made it" to the age of reproduction and are producing young. They may even be more valuable than the 2-year-olds because 2-year-olds are in their final year of breeding. But they may be less valuable than 2-year-olds if they have a slim chance of surviving to a second year *and/or* if they produce fewer offspring than the 2-year-olds.

Biologists are often interested in knowing the value of the different individuals from a practical standpoint because this information can suggest which individuals should be harvested, killed, transplanted, and so forth from a conservation or wildlife management perspective. For example, assuming the numbers of individuals in each age or stage class were equal, if you were trying to eliminate or control a pest species, you would attempt to kill individuals with the highest value because those individuals affect future population size more than any other age group. Conversely, if you were trying to save a threatened species by introducing it into a new area, you would want to "inoculate" the area with individuals of the highest value because those individuals will allow more rapid establishment of a population than other individuals.

An individual's potential for contributing offspring to future generations is called its **reproductive value**. R. A. Fisher introduced the concept of reproductive value in 1930, and defined it as the number of future offspring expected to be produced by an individual of age x over its remaining life span, adjusted by the growth rate of the population. Why the adjustment? To Fisher, the expected number of future offspring wasn't quite the same thing as the "value" of those offspring. Fisher treated offspring like money. If the economy is growing, a dollar received today is worth more than a dollar received next week, because that same dollar will be "diluted" by all the extra money around next week, and even more so in the following year. Similarly, if the population size is changing, the value of future individuals depends on whether the population is increasing, decreasing, or remaining constant over time. The value of each offspring produced by individuals in the future is diluted when the population is increasing (i.e., when the finite rate of increase, λ, is greater than 1), and the value of each offspring is increased when the population is decreasing ($\lambda < 1$). When the population remains constant over time ($\lambda = 1$), no adjustments are needed. (Refer to the next exercise, "Reproductive Value: Life Table Approach," for more details.) To make these adjustments, we divide the expected number of future offspring by the amount the population will have grown or declined when those offspring are produced. The discrete-time version of Fisher's formula to compute v_i, the reproductive value of an individual of age i, is

$$v_i = \sum_{j=i}^{s} \left(\prod_{h=i}^{j-1} P_h \right) F_j \lambda^{i-j-1} \qquad \text{Equation 1}$$

This equation is not so daunting as it might at first appear. Recall that F_j is the fertility of an individual in age class j, and P_h is the probability that an individual in age class h will survive to age class $h + 1$. The Σ symbol indicates that we are summing values starting with the current age class of our individual (i) and going up to the oldest age class (s). Thus, if we are calculating the reproductive value of an individual in age class 2 ($i = 2$), and this species has four age classes ($s = 4$), there will be only three values of j to consider in the summation ($j = 2$, $j = 3$, and $j = 4$). Using these values for i, s, and j, we can expand Equation 1 as follows:

$$\begin{aligned} v_2 &= \sum_{j=2}^{4} \left(\prod_{h=2}^{j-1} P_h \right) F_j \lambda^{2-j-1} \\ &= \left(\prod_{h=2}^{2-1} P_h \right) F_2 \lambda^{2-2-1} + \left(\prod_{h=2}^{3-1} P_h \right) F_3 \lambda^{2-3-1} + \left(\prod_{h=2}^{4-1} P_h \right) F_4 \lambda^{2-4-1} \\ &= \left(\prod_{h=2}^{1} P_h \right) F_2 \lambda^{-1} + \left(\prod_{h=2}^{2} P_h \right) F_3 \lambda^{-2} + \left(\prod_{h=2}^{3} P_h \right) F_4 \lambda^{-3} \end{aligned}$$

The Π symbol is a shorthand for repeated multiplication in the same way that the Σ symbol is a shorthand for repeated addition. For example,

$$\prod_{h=2}^{3} P_h = P_2 P_3$$

Note that in the first product of our expanded expression for v_2 (when $j = 2$), h goes from 2 to 1—a step backwards. In this case, we just consider the product to be equal to 1. We can now complete our expansion of Equation 1 for v_2:

$$v_2 = F_2\lambda^{-1} + P_2F_3\lambda^{-2} + P_2P_3F_4\lambda^{-3}$$

Translating this equation into English, the reproductive value of an individual in age class 2 is its fertility at age class 2 adjusted for one year's population change ($F_2\lambda^{-1}$) plus its fertility at age class 3 adjusted for the probability that it will survive age class 2 and for two years' population change ($P_2F_3\lambda^{-2}$) plus its fertility at age class 4 adjusted for the probability that it will survive age classes 2 *and* 3 and for three years' population change ($P_2P_3F_4\lambda^{-3}$).

As Caswell (2001) states, "The amount of future reproduction, the probability of surviving to realize it, and the time required for the offspring to be produced all enter into the reproductive value of a given age or stage class. Typical reproductive values are low at birth, increase to a peak near the age of first reproduction, and then decline." Individuals that are postreproductive have a reproductive value of 0 since their contribution to future population growth is 0. Newborns also might have low reproductive value because they may have several years of living (and hence mortality risk) before they can start producing offspring. In this exercise, you will calculate reproductive value with matrix calculations. We will begin with a brief review of the major Leslie matrix calculations, and then discuss the reproductive value computations.

Leslie Matrix Calculations

You might recall that an age-based (Leslie) matrix has the form

$$A = \begin{bmatrix} F_1 & F_2 & F_3 & F_4 \\ P_1 & 0 & 0 & 0 \\ 0 & P_2 & 0 & 0 \\ 0 & 0 & P_3 & 0 \end{bmatrix} \qquad \text{Equation 2}$$

The matrix shown is a 4×4 square, which indicates that there are four age classes under consideration. The fertility rates of age classes 1 through 4 are given in the top row. The survival probabilities, P, are given in the subdiagonal; P_1 through P_3 are survival probabilities from one age class to the next. For example, P_1 is the probability of individuals surviving from age class 1 to age class 2. All other entries in the Leslie matrix are 0.

The composition of our population can be expressed as a **column vector**, $\mathbf{n}(t)$, which is a matrix that consists of a single column. Our column vector will consist of the number of individuals in age classes 1, 2, 3, and 4. When the Leslie matrix, $\mathbf{A}$, is multiplied by the population vector, $\mathbf{n}(t)$, the result is another population vector (which also consists of one column); this vector is called the **resultant vector** and provides information on how many individuals are in age classes 1, 2, 3, and 4 in year $t + 1$. The new resultant vector is then multiplied by the Leslie matrix to generate the vector of abundances in the next time step. When this process is repeated over time, eventually the population reaches a stable age distribution, in which the proportion of individuals in each age class remains constant over time.

$$\begin{bmatrix} N_{1(t+1)} \\ N_{2(t+1)} \\ N_{3(t+1)} \\ N_{4(t+1)} \end{bmatrix} = \begin{bmatrix} F_1 & F_2 & F_3 & F_4 \\ P_1 & 0 & 0 & 0 \\ 0 & P_2 & 0 & 0 \\ 0 & 0 & P_3 & 0 \end{bmatrix} \times \begin{bmatrix} N_{1(t)} \\ N_{2(t)} \\ N_{3(t)} \\ N_{4(t)} \end{bmatrix} \qquad \text{Equation 3}$$

There are two ways to examine reproductive value with matrices. One way is what we call the **inoculate method** (Case 2000). In this method, assume that a number of individuals can be introduced into a completely empty habitat. Should you introduce (inoculate) the habitat with individuals from age class 1, 2, 3, or 4? This approach answers the question *"Which age class will produce the largest population size after the population has reached a stable distribution?"* The answer is the age class with the highest reproductive value. For example, suppose a population has a Leslie matrix with fertilities and survival probabilities as shown in Figure 1. If the habitat was inoculated with 200 individuals from age class 1, the vector of abundances would be 200 individuals from age class 1 and 0 individuals for age classes 2, 3, and 4.

	B	C	D	E	F	G
3		**Leslie matrix**				
4	*1*	*2*	*3*	*4*		***Initial vector***
5	0.0	30.0	100.0	0.0	N_1=	200
6	0.2	0.0	0.0	0.0	N_2=	0
7	0.0	0.2	0.0	0.0	N_3=	0
8	0.0	0.0	0.5	0.0	N_4=	0

Figure 1

We then determine the long-term (asymptotic) λ by running the matrix model until the population has reached a stable distribution. We could repeat the process with a different inoculate, say 200 individuals in age class 2, and 0 individuals in age classes 1, 3, and 4. Although the asymptotic λ will be the same, we can compare the overall size of the population to determine the reproductive value of each age class. The age class "seed" with the highest reproductive value will generate the largest population size. We used this method to generate a hypothetical example in Figure 2, where numbers of individuals were tracked over 10 years for different kinds of inoculates. Age class 1 has the highest reproductive value, followed closely by age class 2.

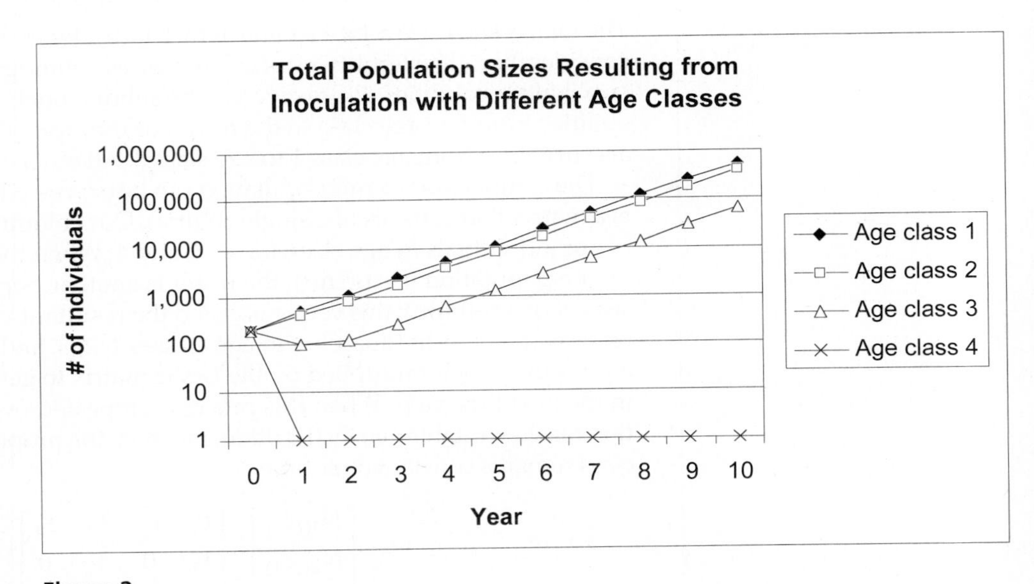

Figure 2

The inoculate method demonstrates clearly the concept of reproductive value, but it is not usually used to *calculate* reproductive value. A faster way to calculate reproductive value involves transposing the Leslie matrix vector and then calculating the proportion of the population that consists of age classes 1, 2, 3, and 4 when the population has reached a stable distribution. This method generates reproductive values very quickly.

Think back once again to your Leslie matrix exercise and how you computed the stable age distribution. You ran your model until λ_t stabilized over time, and computed the proportion of the population that consisted of each age class. These proportions can be written as a vector, **w**. This vector is called a **right eigenvector** of the matrix **A**. For example, the **w** vector for a population that consists of four age classes might be

$$\mathbf{w} = \begin{bmatrix} 0.70 \\ 0.20 \\ 0.05 \\ 0.05 \end{bmatrix}$$

which indicates that when the population growth rate (λ_t) has stabilized, 70% of the total population consists of individuals from age class 1, 20% of the total population consists of individuals from age class 2, 5% of the total population consists of individuals from age class 3, and 5% consists of individuals from age class 4. Thus, the right eigenvector (**w**) of the matrix **A** reveals the stable-age distribution of the population.

In contrast to the right eigenvector, the **left eigenvector** (**v**) of the matrix **A** reveals the reproductive value for each class in the matrix model (Caswell 2001). The simplest way to compute **v** for the **A** matrix is to transpose the **A** matrix (we call the transposed matrix $\mathbf{A}^T$), run the model until the population reaches a stable distribution, and then record the proportions of individuals that make up each class as you did with your original Leslie matrix model. Transposing a matrix simply means switching the columns and rows around—make the rows columns and the columns rows, as in Figure 3.

Original matrix				Transposed matrix		
A	B	C		A	D	G
D	E	F		B	E	H
G	H	I		C	F	I

Figure 3

When λ_t has stabilized for the transposed matrix, $\mathbf{A}^T$, the right eigenvector of $\mathbf{A}^T$ gives the reproductive values for each class. This same vector is called the **left eigenvector** for the original matrix, **A**. (Yes, it is confusing!) A left eigenvector, **v**, for a hypothetical population with four age classes is written as a row vector:

$$\mathbf{v} = \begin{bmatrix} 0.01 & 0.04 & 0.25 & 0.70 \end{bmatrix}$$

Note that the values sum to 1. This vector gives, in order, the reproductive values of age classes 1, 2, 3, and 4. In this hypothetical population, individuals in age class 4 have the greatest reproductive value, followed by individuals in age class 3. The first two age classes have very small reproductive values. Frequently, the reproductive value is **standardized** so that the first stage or age class has a reproductive value of 1. We can standardize the **v** vector above by dividing each entry by the reproductive value of the first age class. Our standardized vector would look like this:

$$\mathbf{v} = \begin{bmatrix} \frac{0.01}{0.01} & \frac{0.04}{0.01} & \frac{0.25}{0.01} & \frac{0.70}{0.01} \end{bmatrix} = \begin{bmatrix} 1 & 4 & 25 & 70 \end{bmatrix}$$

In this example, an individual in age class 4 is 70 times more "valuable" to the population in terms of (adjusted) future offspring production than an individual in age class 1. Let's now go back and consider how Fisher's computation of reproductive value (Equation 1) was derived. Recall that Equation 1 computes v_i, the reproductive value of an individual currently in age class i:

$$v_i = \sum_{j=i}^{s} \left(\prod_{h=i}^{j-1} P_h \right) F_j \lambda^{i-j-1}$$

Since our computations for reproductive value assume that λ_t has stabilized, multiplying **v**, the vector with reproductive values of each age class, by the original Leslie matrix, **A**,

$$(v_1 \quad v_2 \quad v_3 \quad v_4) \times \begin{pmatrix} F_1 & F_2 & F_3 & F_4 \\ P_1 & 0 & 0 & 0 \\ 0 & P_2 & 0 & 0 \\ 0 & 0 & P_3 & 0 \end{pmatrix} \qquad \text{Expression 1a}$$

is the same thing as multiplying **v** by λ:

$$\lambda(v_1 \quad v_2 \quad v_3 \quad v_4) \qquad \text{Expression 1b}$$

To multiply a matrix or vector by a single value, simply multiply each element of the matrix or vector by that value. Thus, Expression 1b is equal to the vector (λv1 λv2 λv3 λv4). Let's assume that reproductive values are standardized such that the reproductive value of age class 1 is 1 (v1 = 1). Since Expression 1a is equal to Expression 1b, we can write

$$\lambda v_1 = v_1 F_1 + v_2 P_1 + v_3 0 + v_4 0 = F_1 + v_2 P_1 \qquad \text{Expression 1c}$$

$$\lambda v_2 = v_1 F_2 + v_2 0 + v_3 P_2 + v_4 0 = F_2 + v_3 P_2 \qquad \text{Expression 1d}$$

$$\lambda v_3 = v_1 F_3 + v_2 0 + v_3 0 + v_4 P_3 = F_3 + v_4 P_3 \qquad \text{Expression 1e}$$

$$\lambda v_4 = v_1 F_4 + v_2 0 + v_3 0 + v_4 0 = F_4 \qquad \text{Expression 1f}$$

Now let's solve for v_1 in terms of only F's, P's and λ to see how these four equations are equivalent to Equation 1. Starting with Expression 1f (and recalling that $1/\lambda = \lambda^{-1}$), we can compute v_4 as

$$v_4 = F_4 \lambda^{-1} \qquad \text{Expression 1g}$$

Now let's plug Expression 1g back into Expression 1e:

$$v_3 = \frac{F_3 + F_4 \lambda^{-1} P_3}{\lambda} = F_3 \lambda^{-1} + P_3 F_4 \lambda^{-2} \qquad \text{Expression 1h}$$

Now let's plug Expression 1h back into Expression 1d:

$$v_2 = \frac{F_2 + v_3 P_2}{\lambda} = \frac{F_2 + (F_3 \lambda^{-1} + P_3 F_4 \lambda^{-2}) P_2}{\lambda} = F_2 \lambda^{-1} + P_2 F_3 \lambda^{-2} + P_2 P_3 F_4 \lambda^{-3} \qquad \text{Expression 1i}$$

Note that Expression 1i is the expansion of Equation 1 that we worked out earlier for $i = 2$ and $s = 4$. Finally, substituting Expression 1i into Expression 1c gives:

$$v_1 = 1 = \frac{F_1 + v_2 P_1}{\lambda} = \frac{F_1 + (F_2 \lambda^{-1} + P_2 F_3 \lambda^{-2} + P_2 P_3 F_4 \lambda^{-3}) P_1}{\lambda} = F_1 \lambda^{-1} + P_1 F_2 \lambda^{-2} + P_1 P_2 F_3 \lambda^{-3} + P_1 P_2 P_3 F_4 \lambda^{-4}$$

which is the expansion of Equation 1 when $i = 1$ and $s = 4$:

$$v_1 = \sum_{j=1}^{s}\left(\prod_{h=1}^{j-1} P_h\right) F_j \lambda^{-j}$$

PROCEDURES

In this exercise, you'll learn how to calculate the reproductive value of different individuals in a population. You will then be able to alter the Leslie matrix to reflect different life history schedules, and determine how such changes affect the reproductive value of different age classes. As always, save your work frequently to disk.

INSTRUCTIONS	ANNOTATION

A. Set up a Leslie matrix.

1. Open a new spreadsheet and set up headings as shown in Figure 4.

	A	B	C	D	E	F	G
1	***Reproductive Value Model: Matrix Approach***						
2							
3				**Leslie matrix**			
4		***1***	***2***	***3***	***4***		***Initial vector***
5		1.6	1.5	0.25	0		200
6		0.8	0	0	0		0
7		0	0.5	0	0		0
8		0	0	0.25	0		0
9							
10				**Age class**			
11	**Time**	**1**	**2**	**3**	**4**	**Total pop**	**λ**

Figure 4

2. Complete the entries in the Leslie matrix in cells B5–E8.

Describe each cell's entry in the space below:

D5 ______________________________

E5 ______________________________

B6 ______________________________

C7 ______________________________

D8 ______________________________

3. Enter the vector of abundances shown in cells G5–G8.

4. Set up a linear series from 0 to 50 in cells A12–A62.

Enter the value 0 in cell A12.
In cell A13, enter **=A12+1**.
Copy your formula down to cell A62.
This will track the growth of our age-structured population for 50 years.

5. Enter formulae in cells B12–E12 that link abundance at time 0 to the initial vector of abundances.

Enter the formulae

- B12 =**G5**
- C12 =**G6**
- D12 =**G7**
- E12 =**G8**

6. Calculate the total population size in Year 0 in cell F12. Copy your formula down one row.

Enter the formula =**SUM(B12:E12)**.

7. Enter formulae in cells B13–E13 to project population growth for Year 1.

Enter the following formulae:

- B13 =**$B$5*B12+$C$5*C12+$D$5*D12+$E$5*E12**
- C13 =**$B$6*B12+$C$6*C12+$D$6*D12+$E$6*E12**
- D13 =**$B$7*B12+$C$7*C12+$D$7*D12+$E$7*E12**
- E13 =**$B$8*B12+$C$8*C12+$D$8*D12+$E$8*E12**

8. Calculate lambda, λ, as N_{t+1}/N_t in cell G12. Copy this formula down to cell G13.

Enter the formula =**F13/F12**.

9. Select cells B13:G13, and copy their formulae down to row 62.

This completes your 50-year projection. Your spreadsheet should look like the one in Figure 5.

	A	B	C	D	E	F	G
1	*Reproductive Value Model: Matrix Approach*						
2							
3			**Leslie matrix**				
4		*1*	*2*	*3*	*4*		*Initial vector*
5		1.6	1.5	0.25	0		200
6		0.8	0	0	0		0
7		0	0.5	0	0		0
8		0	0	0.25	0		0
9							
10			**Age class**				
11	**Time**	**1**	**2**	**3**	**4**	**Total pop**	λ
12	0	200	0	0	0	200	2.4
13	1	320	160	0	0	480	2.266666667
14	2	752	256	80	0	1088	2.166176471

Figure 5

10. Graph population growth over time (graph the first 10 years).

Use a semi-log graph and the line graph option, and label your axes fully. The resulting graph should resemble Figure 6. (You may wish to *not* use a semi-log graph because the spreadsheeet will generate a message, "zero values cannot be plotted correctly on log charts." This message will appear frequently if you choose to use a semi-log graph where some entries are 0.)

11. Save your work.

B. Calculate the reproductive value: Inoculate method.

Now we are ready to compute the reproductive values using a matrix approach. There are two ways to generate reproductive values from matrices, the "inoculation" approach and the "transpose vector" approach. We'll start with the inoculation approach. To get an idea of what reproductive value means, we'll inoculate our population with 200 individuals from age class 1 (the other age classes will have 0 individuals), and then record

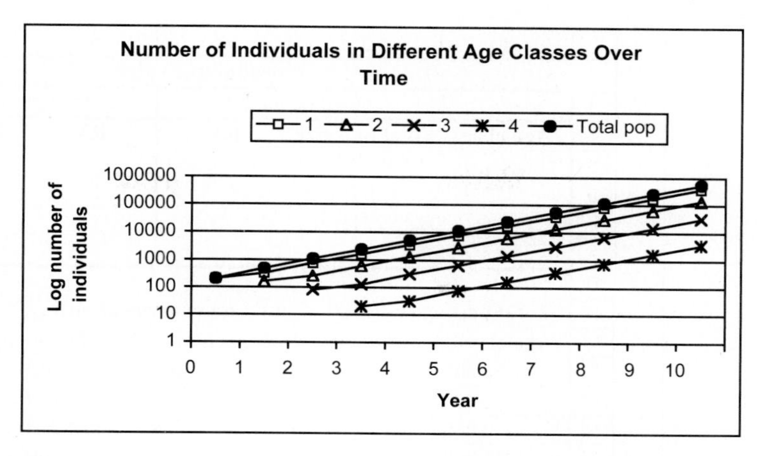

Figure 6

the total population size over 50 years of time in cells I12–I62 (Figure 7). We'll also record final population size at year 50 in cell J5. We'll repeat the process for inoculate of the remaining age classes. For example, for age class 2, our inoculate will consist of 200 individuals of age class 2 (the other age classes will have 0 individuals). We'll record the total population size over 50 years of growth in cells J12–J62. We'll record the final population size at Year 50 in cell J6. The process will be repeated for age classes 3 and 4.

1. Set up new headings as shown in Figure 7.

	I	J	K	L	M
2	**Reproductive value**				
3		Inoculate method		Transpose method	
4	**Age class**	**Final pop size**	**RV**	**RV**	**Standardized**
5	1				
6	2				
7	3				
8	4				
9					
10	**Total pop when initial population consists of only:**				
11	**Age class 1**	**Age class 2**	**Age class 3**	**Age class 4**	

Figure 7

2. Set cell G5 to 200, and the other vector elements in cells G6–G8 to 0.

First we'll inoculate our population with 200 individuals from age class 1. Your projections should be automatically updated. If not, make sure that your Calculation setting is set to automatic (Tools | Options | Calculation).

3. Copy cells F12 to F62 into cells I12 and down.

Use the Paste Special option and paste the values.
By copying the total population size with an inoculate of age class 0, we can determine how "fast" the population grows relative to other kinds of inoculates.

4. Select cell F62; copy and paste its value into cell J5.

Cell F62 gives the total population size at Year 50 when our inoculate consists of 200 individuals from age class 1.

5. Repeat steps 2-4 for the remaining age class inoculates and enter results into appropriate cells.

Your finished spreadsheet should look like Figure 8.

	I	J	K	L	M
2	**Reproductive value**				
3		Inoculate method		Transpose method	
4	**Age class**	**Final pop size**	**RV**	**RV**	**Standardized**
5	1	1.64938E+19			
6	2	1.18204E+19			
7	3	1.89731E+18			
8	4	0			
9					
10	**Total pop when initial population consists of only:**				
11	**Age class 1**	**Age class 2**	**Age class 3**	**Age class 4**	
12	200	200	200	200	
13	480	400	100	0	
14	1088	770	120	0	
15	2356.8	1692	272	0	
16	5124.48	3671.2	589.2	0	

Figure 8

6. Graph population growth from year 0 to year 10 for each of the inoculates.

Use the line graph option and label your axes. Your graph should resemble Figure 9.

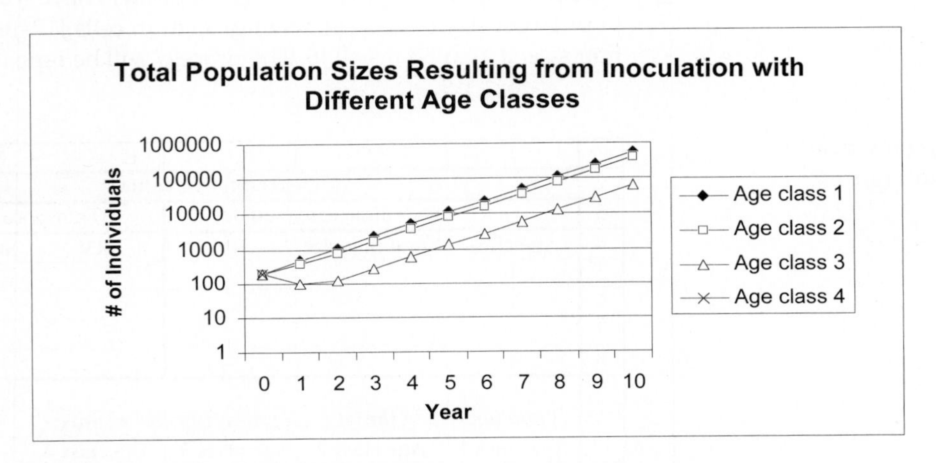

Figure 9

Interpret your graph. Which age class inoculate generated the largest population size after 10 years?

7. Enter the formula **=J5/$J$5** in cell K5; copy your formula down to cell K8. Interpret your results.

Now we can compute reproductive values. As mentioned in the Introduction, reproductive values can be scaled so that the reproductive value of the first age class is 1. The formula **=J5/$J$5** does this scaling. We set cell K5 = 1, then the reproductive values indicate the value of each age class compared to age class 1 (Figure 10).

	I	J	K
3		Inoculate method	
4	**Age class**	**Final pop size**	**RV**
5	1	1.64938E+19	1
6	2	1.18204E+19	0.71665214
7	3	1.89731E+18	0.11503129
8	4	0	0

8. Save your work.

Figure 10

C. Calculate the reproductive value: Transpose vector method.

The second method for computing reproductive values using a matrix approach is the transpose vector approach. It is perhaps quicker than the first approach, and is the method commonly used to compute reproductive values with matrices (Caswell 2001).

The first step is to transpose your Leslie matrix by inverting the rows and columns. For example, if your Leslie matrix has the form

$$\begin{bmatrix} 1.6 & 1.5 & 0.25 & 0 \\ 0.8 & 0 & 0 & 0 \\ 0 & 0.5 & 0 & 0 \\ 0 & 0 & 0.25 & 0 \end{bmatrix}$$

then the transposed matrix is

$$\begin{bmatrix} 1.6 & 0.8 & 0 & 0 \\ 1.5 & 0 & 0.5 & 0 \\ 0.25 & 0 & 0 & 0.25 \\ 0 & 0 & 0 & 0 \end{bmatrix}$$

1. Modify the spreadsheet from section A.

Select cells A3–G12 and open Edit | Copy. Select cell N3 and paste the cells. Modify the heading in row 3 to read "Transposed Leslie Matrix."

2. Set up a linear series from 0 to 50 in cells N12–N62.

3. Select cells O5–R8, and use the **TRANSPOSE** function to transpose the original Leslie matrix (cells B5–E8).

The **TRANSPOSE** formula is an array formula because it is entered into a block of cells rather than a single cell. You may want to review the mechanics of working with an array formula, described on pages 10–11.

Select cells O5–R8 with your mouse. Use the f_x key to select the **TRANSPOSE** function. The dialog box will ask you to define an array that you wish to transpose. Use your mouse to highlight cells B5–E8, or enter this by hand. Instead of clicking OK, press <Control><Shift><Enter> (or <Enter>) and the function will return your transposed matrix.

Once you've obtained your results, examine the formulae in cells O5–R8. This formula should read **{=TRANSPOSE(B5:E8)}**. (Remember that the { } symbols indicate the formula is part of an array. If for some reason you get "stuck" in an array formula, press the <Escape> key and start over.) Your spreadsheet should now look like Figure 11.

	N	O	P	Q	R	S	T
3		**Transposed Leslie matrix**					
4		*1*	*2*	*3*	*4*		***Initial vector***
5		1.6	0.8	0	0		200
6		1.5	0	0.5	0		0
7		0.25	0	0	0.25		0
8		0	0	0	0		0
9							
10			**Age class**				
11	**Time**	**1**	**2**	**3**	**4**	**Total pop**	λ
12	0	200	0	0	0	200	

Figure 11

4. Compute λ in cell T12.

In cell T12 enter the equation **=S13/S12** to compute λ.

5. Enter formulae to project the population over time in cells O13–R13, as you did in Part A.

Enter the formulae

- O13 **=$O$5*O12+$P$5*P12+$Q$5*Q12+$R$5*R12**
- P13 **=$O$6*O12+$P$6*P12+$Q$6*Q12+$R$6*R12**
- Q13 **=$O$7*O12+$P$7*P12+$Q$7*Q12+$R$7*R12**
- R13 **=$O$8*O12+$P$8*P12+$Q$8*Q12+$R$8*R12**

6. Copy cells O13–T13 down to row 62 to complete the projection.

At this point, your population projection should show the same λ values as before. If λ is *not* the same value, you made a mistake somewhere.

7. Calculate the proportion of individuals in age class 1 after 50 years of population growth in cell L5.

Enter the formula **=O62/$S$62** in cell L5. The result should be the unstandardized reproductive value for individuals in age class 1.

8. Compute the reproductive values for the other age classes in cells L6–L8.

Enter the formulae

- L6 **=P62/$S$62**
- L7 **=Q62/$S$62**
- L8 **=R62/$S$62**

The results are your reproductive values.

9. Compute the standardized reproductive values in cells M5–M8.

Once again we need to standardize so that the reproductive value for the first age class is equal to 1. By dividing each value by the value in the first age class, you will set age class 1 to a value of 1, so that the reproductive values of the remaining age classes indicate the reproductive value of a particular class compared to age class 1. We used the following formula:

- M5 **=L5/$L$5**
- M6 **=L6/$L$5**
- M7 **=L7/$L$5**
- M8 **=L8/$L$5**

10. Save your work.

Your results should match the values obtained with the inoculate method, and your spreadsheet should resemble Figure 12.

	I	J	K	L	M
2	Reproductive value				
3	Inoculate method			Transpose method	
4	**Age class**	**Final pop size**	**RV**	**RV**	**Standardized**
5	1	1.64938E+19	1	0.54594587	1
6	2	1.18204E+19	0.71665214	0.39125327	0.716652136
7	3	1.89731E+18	0.11503129	0.06280086	0.11503129
8	4	0	0	0	0

Figure 12

D. Create graphs.

1. Graph the reproductive values for the various age classes.

Use the column graph option and label your axes fully. Your graph should resemble Figure 13.

Figure 13

QUESTIONS

1. Interpret the graph from the inoculate method. In what way does the graph show the reproductive values for the various age classes?
2. Interpret the reproductive values from your models from the standpoint of conservation of a game species whose populations are harvested and maintained at a high level, versus a pest species whose populations you would like to reduce or eliminate, versus a threatened species that is being reintroduced to an area. For each situation, which actions would you recommend based on your knowledge of reproductive values (e.g., which age class should be harvested; which age class should be reintroduced?) Does it matter how abundant each age class is when the population stabilizes?
3. Change the Leslie matrix to reflect a population with a Type I survival curve. Compare the reproductive value of the different age classes with a Type I survival schedule versus the original schedule (which was a Type II curve). Use the transpose method to assess reproductive value because your results will automatically be calculated.
4. Change the Leslie matrix to reflect a population with a Type III survival curve. Compare the reproductive value of the different age classes with a Type I and Type II schedule. Use the transpose method to assess reproductive value because your results will automatically be calculated.
5. Find the life history schedule of an organism of interest to you and enter Leslie matrix parameters to the best of your knowledge. How do small changes in different matrix elements affect reproductive value? How might the environment in which your organism resides help shape its life history?

LITERATURE CITED

Case, T. 2000. *An Illustrated Guide to Theoretical Ecology*. Oxford University Press, New York.

Caswell, H. 2001. *Matrix Population Models*, 2nd Edition. Sinauer Associates, Sunderland, MA.

Gotelli, N. 2001. *A Primer of Ecology*, 3rd Edition. Sinauer Associates, Sunderland, MA.

16 REPRODUCTIVE VALUE: LIFE TABLE APPROACH

Objectives

- Perform standard life table calculations on a hypothetical data set.
- Compute the stable age distribution and reproductive values for individuals of age x from life table data.
- Evaluate how life history strategy affects reproductive values.

Suggested Preliminary Exercise: Life Tables and Survivorship Curves; Reproductive Value: Matrix Approach

INTRODUCTION

As we discussed in the previous exercise, the idea that different individuals have different "value" in terms of their contribution to future generations is called their *reproductive value* (Fisher 1930). As Caswell (2001) states, "The amount of future reproduction, the probability of surviving to realize it, and the time required for the offspring to be produced all enter into the **reproductive value** of an age-class."

The reproductive value of an individual of age x is designated at V_x, and is the number of offspring that an individual is expected to produce over its remaining life span (after adjusting for the growth rate of the population). Biologists are often interested in knowing the "value" of the different individuals from a practical standpoint because knowing something about the reproductive value can suggest which individuals should be harvested, killed, transplanted, etc. from a conservation or management perspective

The reproductive value of different ages is strongly tied to an organism's life history. Typically, reproductive value is low at birth, increases to a peak near the age of first reproduction, and then declines (Caswell 2001). In this exercise, you will calculate reproductive value of individuals of various ages from life table calculations. We will start with a brief review of the major calculations in the life table, and then move on to the calculations and explanations of reproductive value. We will then modify the life history schedule of organisms to compare how reproductive value changes under different life history scenarios.

Life Table Calculations

A typical life table is shown in Figure 1. If we were to build a cohort life table for a population born during the year 1900, we would record how many individuals were born during the year 1900, and how many survived to the *beginning* of 1901, 1902, etc., until there were no more survivors. This record is called the **sur-**

vivorship schedule, or S_x. We would also record the **fecundity schedule**: the number of offspring born to members of each age class. The total number of offspring is usually divided by the number of individuals in the age class, giving the average number of offspring per individual, which is represented by b_x. The survivorship and fecundity schedules are the raw data of a life table. From these data, age-specific rates of survival, life expectancy, generation time, and net reproductive rate can be calculated.

You might recall that l_x is the proportion of original numbers surviving to the beginning of each interval, and is calculated as

$$l_x = \frac{S_x}{S_0} \qquad \text{Equation 1}$$

We can also think of l_x as the probability that an individual survives from birth to the beginning of age-class x. Column E in Figure 1 is simply l_x multiplied by b_x, and Column F is simply l_x times b_x times x (the age class). The sum of Column E generates R_0, the net reproductive rate, which can be written mathematically as

$$R_0 = \sum_{x=0}^{k} l_x b_x \qquad \text{Equation 2}$$

The net reproductive rate is the lifetime reproductive potential of the average female, adjusted for mortality. Assuming mortality and fertility schedules remain constant over time, if $R_0 > 1$, then the population will grow exponentially. If $R_0 < 1$, the population will shrink exponentially, and if $R_0 = 1$, the population size will not change over time. You might recall from the life table exercise that R_0 measures population change in terms of generation time. To convert R_0 into an intrinsic rate of increase (r) or finite rate of increase (λ), we must first calculate generation time, and then adjust R_0 accordingly.

	A	B	C	D	E	F	G
1	Cohort Life Table: Fecundity Schedule and Population Growth						
2							
3	Age class (x)	S_x	b_x	l_x	$(l_x)(b_x)$	$(l_x)(b_x)(x)$	(e^-rx)$(l_x)(b_x)$
4	0	3751	0.00	1.0000	0.0000	0.0000	0.0000
5	1	357	10.51	0.0952	1.0003	1.0003	1.0002
6	2	159	0.00	0.0424	0.0000	0.0000	0.0000
7	3	59	0.00	0.0157	0.0000	0.0000	0.0000
8	4	57	0.00	0.0152	0.0000	0.0000	0.0000
9	5	53	0.00	0.0141	0.0000	0.0000	0.0000
10	6	29	0.00	0.0077	0.0000	0.0000	0.0000
11	7	19	0.00	0.0051	0.0000	0.0000	0.0000
12	8	17	0.00	0.0045	0.0000	0.0000	0.0000
13	9	13	0.00	0.0035	0.0000	0.0000	0.0000
14	10	7	0.00	0.0019	0.0000	0.0000	0.0000
15	11	0		0.0000	0.0000	0.0000	0.0000
16				Total	1.0003	1.0003	1.0002
17	R_0	1.0003					
18	G	1.0000					
19	r est.	0.0003					
20	r Euler	0.0001					
21	Should be 1	1.0002					

Figure 1 A cohort of 3751 individuals tracked over time. The number alive at the beginning of each year is given in Column B, and the average number of offspring per female is given in Column C. Columns D through G are calculated from information in columns A through C.

Generation time is calculated as the sum of Column F divided by the sum of Column E, or

$$G = \frac{\sum_{x=0}^{k} l_x b_x x}{\sum_{x=0}^{k} l_x b_x} \qquad \text{Equation 3}$$

With G and R_0 calculated, we can estimate r, the intrinsic rate of increase, as

$$\frac{\ln R_0}{G} \approx r \qquad \text{Equation 4}$$

We need to know r in order to calculate the reproductive value of each age class. However, Equation 4 provides only an estimate of r. To obtain a more precise estimate of r, we need to solve for r in the following equation:

$$1 = \sum_{x=0}^{k} e^{-rx} l_x b_x \qquad \text{Equation 5}$$

This is called the **Euler equation**, named after the Swiss mathematician Leonhard Euler (Gotelli 2001). In the life table exercise, you solved the Euler equation by plugging numbers in until the equation was solved. In this exercise, you will use the Solver option in Excel to solve the Euler equation. You might remember that when $r = 0$, the population remains constant in numbers over time; when $r < 0$, the population declines exponentially, and when $r > 0$ the population increases exponentially. When a population has a **stable age structure**, it means that all age classes increase or decrease at a constant rate of r, even if the numbers of individuals in each age class differ.

With an estimate of r for our population, we are ready to calculate the **reproductive value** for individuals of age x (Fisher 1930), which can be calculated from a life table as

$$v_x = \frac{e^{rx}}{l_x} \sum_{y=x+1}^{\Omega} e^{-ry} l_y b_y \qquad \text{Equation 6}$$

where $y = x + 1$ is the first age class subsequent to age class x, and Ω is the final age class into the future. Equation 6 can be written out in full as

$$\begin{aligned} v_x = \frac{e^{rx}}{l_x} [& e^{-r(x+1)} l_{x+1} b_{x+1} \\ & + e^{-r(x+2)} l_{x+2} b_{x+2} \\ & + e^{-r(x+3)} l_{x+3} b_{x+3} \\ & + \ldots] \end{aligned}$$

This equation assumes that the next reproductive bout for individuals of age x will occur at age $x + 1$, i.e., individuals of age x have already reproduced as x year olds. In order for us to understand how Equation 6 was derived, its useful to recall that the reproductive value of an individual of age x is the expected number of offspring that this individual will produce over the rest of its life, adjusted by population growth. Let's start by computing the expected number of offspring that an individual of age x will produce over the rest of its life. If we let any age class beyond age x be denoted with the letter y, the total number of future offspring can be calculated as:

$$\sum_{y=x+1}^{\Omega} \frac{l_y b_y}{l_x} \qquad \text{Expression 6.1}$$

This term can be written out in full as

$$\frac{l_{x+1} b_{x+1}}{l_x} + \frac{l_{x+2} b_{x+2}}{l_x} + \frac{l_{x+3} b_{x+3}}{l_x} + \ldots$$

Thus, for each age class following age class x, compute the probability that an individual of age class x will survive to a given future class as $\Pr = l_y/l_x$ and multiply by the corresponding birth rate, b_y. It should be fairly straightforward why l_y/l_x and b_x must be computed to calculate the expected number of offspring that an individual of age x will produce in the future: in order to produce future offspring in year $x + 3$ (for example), you must survive from age x to age $x + 3$ to realize the reproduction.

But this expected number of future offspring isn't quite the same thing as the "value" of those offspring. Ronald A. Fisher (1930) got the idea of treating offspring like money. If the economy is growing, a dollar received today is worth more to me than a dollar received next week, because that same dollar will be "diluted" by all the extra money around next week, and even more so in the following year. Similarly, if the population size is different when the future offspring are produced, their values depend on whether the population is increasing or decreasing: the value of each offspring produced by individuals in the future is "diluted" when the population is increasing, and the value of each offspring is "concentrated" when the population is decreasing.

So Fisher discounted the value of the offspring produced at later ages by the amount by which the population will have grown by the time they are produced. Since the population is growing at the rate r, by the time our x-year-old individual reaches age y, the population will have grown by a factor

$$e^{r(y-x)}$$

Thus, to compute the value of future offspring, we need to "adjust" the number of future offspring by dividing by the factor by which the population will have grown. We can compute the "adjusted" number of expected future offspring for an individual of age x as

$$\frac{\sum_{y=x+1}^{\Omega} \frac{l_y b_y}{l_x}}{e^{r(y-x)}} \qquad \text{Expression 6.2}$$

Expression 6.2 can be written out in full as

$$\frac{\frac{l_{x+1}b_{x+1}}{l_x}}{e^{r(x+1-x)}} + \frac{\frac{l_{x+2}b_{x+2}}{l_x}}{e^{r(x+2-x)}} + \frac{\frac{l_{x+3}b_{x+3}}{l_x}}{e^{r(x+3-x)}} + \dots$$

If we graph $e^{r(y-x)}$ for various levels of r, we can visualize the denominator of Expression 6.2 and see how the adjustment works. This is shown in Figure 2.

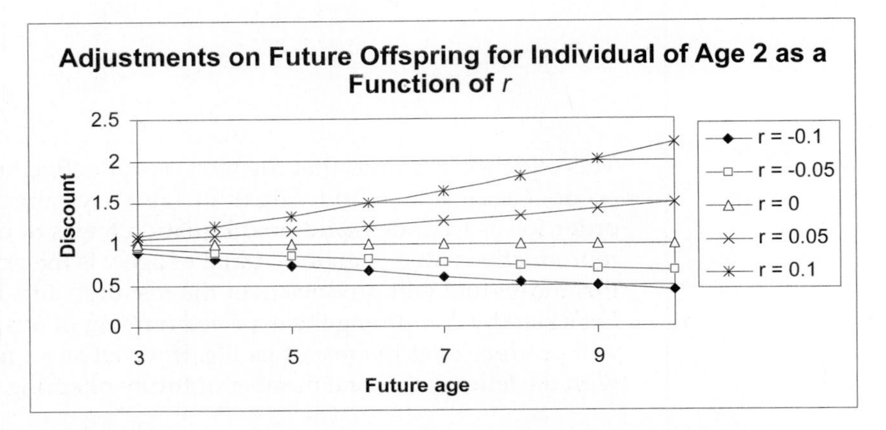

Figure 2 For an individual of age 2, the graph shows how offspring produced in the future are adjusted under various levels of r. When $r > 1$, the adjustment, $e^{r(y-x)}$, is positive and increases as ever more distant age classes are considered. This makes the denominator of Expression 6.2 large, which *decreases* the value of future offspring. When $r < 0$, the population is decreasing and the adjustment is below 1. This makes the denominator of Expression 6.2 small, which *increases* the value of future offspring. When $r = 0$, no adjustment is made.

We have now arrived at the number of future offspring expected to be produced by an individual of age x, adjusted by population growth (i.e., the reproductive value for an individual of age x). From here, we can arrive back at Fisher's computation of reproductive value (Equation 6) with a few simple mathematical steps. It might be helpful to recall certain mathematical principles before we proceed:

- If n is a positive integer, then a^{-n} is $1/a^n$.
- For any number a, and any integers m and n, $a^m \times a^n = a^{m+n}$.
- Any term expressed as $\dfrac{a/b}{c/d}$ can be written as $\dfrac{ad}{bc}$

Now let's proceed with Expression 6.2 and work our way towards Fisher's formula for computing reproductive value (Equation 6). With the mathematical principles in mind, we can rewrite Expression 6.2 as

$$\sum_{y=x+1}^{\Omega} \frac{l_y b_y}{l_x e^{ry} e^{-rx}} \qquad \text{Expression 6.3}$$

which can be written out in full as

$$\frac{l_{x+1} b_{x+1}}{l_x e^{r(x+1)} e^{-rx}} + \frac{l_{x+2} b_{x+2}}{l_x e^{r(x+2)} e^{-rx}} + \frac{l_{x+3} b_{x+3}}{l_x e^{r(x+3)} e^{-rx}} + \ldots$$

We can then pull two common terms out of the denominator, l_x and e^{-rx} and re-write Expression 6.3 as

$$\frac{1}{l_x e^{-rx}} \sum_{y=x+1}^{\Omega} \frac{l_y b_y}{e^{ry}} \qquad \text{Expression 6.4}$$

which is the same thing as:

$$\frac{e^{rx}}{l_x} \sum_{y=x+1}^{\Omega} \frac{l_y b_y}{e^{ry}} \qquad \text{Expression 6.5}$$

Expression 6.5 can be written out in full as:

$$\frac{e^{rx}}{l_x}\left[\frac{l_{x+1} b_{x+1}}{e^{r(x+1)}} + \frac{l_{x+2} b_{x+2}}{e^{r(x+2)}} + \frac{l_{x+3} b_{x+3}}{e^{r(x+3)}} + \ldots\right]$$

Finally, we can move the term e^{ry} from the denominator to the numerator (in Expression 6.5) and arrive at Fisher's equation (Equation 6):

$$v_x = \frac{e^{rx}}{l_x} \sum_{y=x+1}^{\Omega} e^{-ry} l_y b_y \qquad \text{Equation 6}$$

Hopefully, Equation 6 will now make some sense to you. Equation 6 is specifically for populations in which there is a birth pulse and in which individuals are censused immediately after the breeding season (individuals of age x have already given birth). If individuals of age x have not yet given birth, the summation would begin with $y = x$ in Equation 6, rather than $y = x + 1$. In this case, reproductive value can be partitioned into current (imminent) reproduction, as well as future reproduction (Williams 1966). Although the equation might look a bit cumbersome, we'll walk you step by step through the calculations so that you can see exactly how the values are computed. Reproductive value can also be computed with a matrix approach (see the previous exercise). The critical pieces of information from a life table are r, the intrinsic rate of growth, l_x, or the survivorship schedule, and b_x, the fecundity schedule. If we know these values for each age (with ages denoted by x), we can identify the reproductive value for each age.

In addition to reproductive value, we will also calculate the stable age distribution of the population. The stable age distribution gives the proportion of the population that consists of 0, 1, 2, 3, and 4 year olds, given that the population has reached an equilibrium growth rate. In other words, no matter what r is for the population, each age group will increase or decrease by a constant amount. For example, if the stable population is made up of 55% 0-year olds, 22% 1-year olds, 33% 2-year olds, and 0% 3-year olds, the stable age distribution is 0.55, 0.22, 0.33, and 0, respectively. These proportions are calculated from the following equation (Mertz 1970):

$$c_x = \frac{e^{-rx} l_x}{\sum_{x=0}^{k} e^{-rx} l_x} \quad \text{Equation 7}$$

where c_x is the proportion of the population that consists of individuals of age class x when the population has stabilized.

PROCEDURES

In this exercise, you'll learn how to calculate reproductive value for individuals in a population, as well as the stable age distribution. In setting up this model, we have followed the life table computations Gotelli (2001) used to compute reproductive value. As a result, some steps in the computation have not been explained in the introductory material here, but the final results do indeed reflect the reproductive values from Equation 6.

After the model is completed, you will be able to change the life history schedule of the population to evaluate how life history schedules affect reproductive value. As always, save your work frequently to disk.

INSTRUCTIONS / ANNOTATION

A. Set up the life table spreadsheet.

1. Open a new spreadsheet and set up column headings as shown in Figure 3.

	A	B	C	D	E	F	G
1	*Reproductive Value: Life Table Approach*						
2				Cohort life table			
3	x (age)	S_x	b_x	l_x	g_x	l_x*b_x	l_x*b_x*x
4	0	500	0				
5	1	400	2				
6	2	200	3				
7	3	50	1				
8	4	0	0				
9					R_0 =		
10							
11		Outputs					
12		R_0 =					
13		G =					
14		r (estimate) =					
15		Euler equation =					
16		Euler r (adj) =					
17		λ =					

Figure 3

2. Enter the values shown in cells B4–C8 as shown.

We will start with 500 newborns (cell B4) and follow their numbers over time. S_x gives the number of individuals that are counted at the beginning of each age class. The fecundity schedule, b_x, gives the average number of female offspring per female per year for each age class.

3. Enter formulae in cells D4–G4 to compute the standard life table data, and copy your formulae down to row 8.

Refer back to the "Life Tables and Survivorship Curves" exercise if you cannot remember the formulae. We used the following formulae:

- D4 =**B4/$B$4**
- E4 =**D5/D4**
- F4 =**D4*C4**
- G4 =**F4*A4**

4. Graph the survivorship curve.

Select cells A3–B8. Use the scattergraph option and label your axes fully. Your graph should resemble Figure 4.

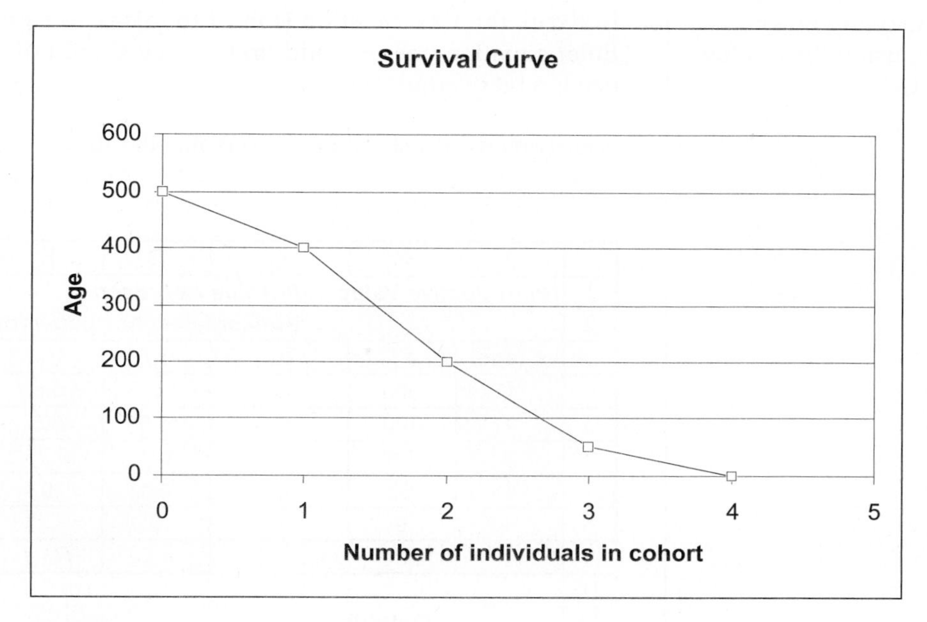

5. Save your work.

Figure 4

B. Compute life table outputs.

1. Enter a formula in cell F9 and C12 to compute R_0.

Enter the formula =**SUM(F4:F8)** in cells F9 and C12.
R_0 is the **net reproductive rate**. It reveals the mean number of offspring produced per female over her lifetime (Gotelli 2001). R_0 can be calculated by multiplying $l_x \times b_x$ for each age class, and then summing up the values over age classes; this corresponds to **Equation 2**.

2. In cell C13, enter a formula to compute G.

Enter the formula =**SUM(G4:G8)/C12** in cell C13.
G is the generation time. It reveals the average age of the parents of all the offspring produced by a single cohort (Caughley 1977). G can be calculated by multiplying $l_x \times b_x \times x$ for each age class, and then summing up the values over age classes. This sum is then divided by (or adjusted for) R_0. This correspondes to **Equation 3**.

3. In cell C14, enter a formula to estimate r.

Enter the formula =**LN(C12)/C13** in cell C14.
This corresponds to **Equation 4**.

4. **Manually** enter the estimated value of r in cell C16.

Enter 0.72 in cell C16.

You might remember that r can be more precisely estimated by using the Euler equation (Gotelli 2001). The exact solution for r can be found by solving for r in the Euler equation:

$$1 = \sum_{x=0}^{k} e^{-rx} l_x b_x \qquad \text{Equation 5}$$

You should have reached $r = 0.72$ as an estimate. Knowing that r is approximately 0.72, you can plug various values of r (a bit higher or lower) until the equation is solved (as you did in the "Life Tables" exercise), or you can use the Solver spreadsheet tool to solve the problem for you. For now, you've entered 0.72 into cell C16. The Solver will change this value to the precise estimate in the next couple of steps.

5. Enter a formula in cell C15 to calculate the right-hand side of the Euler equation, using the r value in cell C16.

Enter the formula **=SUM(EXP(-C16*A4)*F4,EXP(-C16*A5)*F5,EXP(-C16*A6)*F6,EXP(-C16*A7)*F7,EXP(-C16*A8)*F8)** in cell C15.

In Excel, the **EXP** function is used to raise e to a given power. You'll see that your Euler equation does not add up to 1 as it should (it adds up to 1.07), which means r needs a bit of adjusting.

Your spreadsheet should now look like Figure 5.

	A	B	C	D	E	F	G
1	***Reproductive Value: Life Table Approach***						
2			**Horizontal (cohort) life table**				
3	***x* (age)**	S_x	b_x	l_x	g_x	$l_x{*}b_x$	$l_x{*}b_x{*}x$
4	0	500	0	1	0.8	0	0
5	1	400	2	0.8	0.5	1.6	1.6
6	2	200	3	0.4	0.25	1.2	2.4
7	3	50	1	0.1	0	0.1	0.3
8	4	0	0	0		0	0
9					R_0 =	2.9	4.3
10							
11		**Outputs**					
12		R_0 =	2.9				
13		G =	1.4827586				
14		*r* (estimate) =	0.7180607				
15		Euler equation =	1.0746494				
16		Euler *r* (adj) =	0.72				
17		λ =					

Figure 5

C. Use the Solver function to adjust the value of r.

1. Access Solver.

Go to Tools | Solver and select Solver. If Solver does not appear in the menu, go to Tools | Add-ins and select the Solver add-in. (Your computing administrator may need to help you with the installation.) The dialog box in Figure 6 will appear.

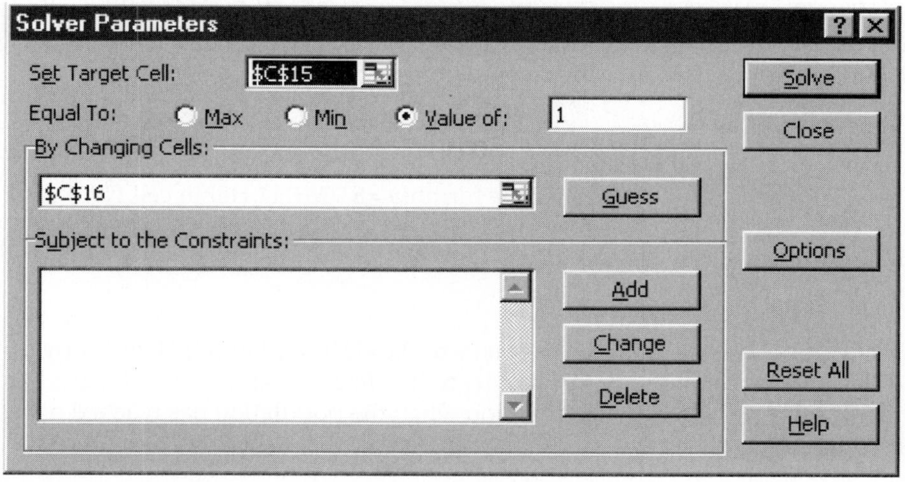

Figure 6

2. Use the Solver function to set cell C15 (the Euler equation) to 1 by changing cell C16.

Enter **$C$15** in the Set Target Cell box
Set the target cell equal to a Value of 1.
Enter **$C$16** in the By Changing Cells box.

3. Press Solve to return the precise estimate of r in cell C16.

You should get a value of $r = 0.776$, and you'll see that cell C15 is very close to 1.

4. Calculate λ, the finite rate of increase, in cell C17.

Enter the formula **=EXP(C16)** in cell C17.
Lambda is the finite rate of increase. It can be calculated from r as $\lambda = e^r$.

5. Save your work.

D. Calculate the stable age distribution.

The stable age distribution gives the proportion of the population that consists of 0, 1, 2, 3, and 4 year olds, given that the population has reached an equilibrium growth rate. For example, if the stable population is made up of 50% 0-year olds, 22% 1-year olds, 33% 2-year olds, and 0% 3-year olds, the stable age distribution is 0.50, 0.22, 0.33, and 0, respectively. These proportions are calculated from **Equation 7**, the Mertz equation:

$$c_x = \frac{e^{-rx} l_x}{\sum_{x=0}^{k} e^{-rx} l_x} \qquad \text{Equation 7}$$

1. Set up new spreadsheet headings as shown in Figure 7.

	H	I	J	K	L	M
2	**Stable age distribution**		**Reproductive value distribution**			
3	$l_x e^{-rx}$	c_x	e^{rx}/l_x	$e^{-rx} l_x b_x$	$\Sigma e^{-ry} l_y b_y$	v_x
4						
5						
6						
7						
8						
9						

Figure 7

2. In cells H4–H8, enter a formula in cell H4 to calculate the numerator of the Mertz equation for each age class.

Enter the formula =**D4*EXP(-$C$16*A4)** in cell H4 to calculate the numerator of the Mertz equation for age class 0. Copy this formula down to cell H8 to obtain this value for the remaining age classes.

3. In cell H9, sum cells H4–H8 to obtain the denominator of the Mertz equation.

Enter the formula **=SUM(H4:H8)** in cell H9.

4. Calculate c_x for age class 0 in cell I4. Copy this formula down for the remaining ages.

Enter the formula **=H4/$H$9** in cell I4 and copy down the column.
The results of this formula give, for each age class, the proportionate makeup of the population when the population has reached a stable distribution.

5. Sum the c_x values in cell I9.

Enter the formula **=SUM(I4:I8)** in cell I9.
This is to double-check your results. The values should sum to 1.

6. Save your work.

Your spreadsheet should now resemble Figure 8.

	A	B	C	D	E	F	G	H	I
2	Horizontal (cohort) life table							Stable age distribution	
3	x (age)	s_x	b_x	l_x	g_x	l_x*b_x	l_x*b_x*x	$l_x e^{-rx}$	c_x
4	0	500	0	1	0.8	0	0	1.000	0.684
5	1	400	2	0.8	0.5	1.6	1.6	0.368	0.252
6	2	200	3	0.4	0.25	1.2	2.4	0.085	0.058
7	3	50	1	0.1	0	0.1	0.3	0.010	0.007
8	4	0	0	0		0	0	0.000	0.000
9					R_0 =	2.9	4.3	1.463	1.000

Figure 8

7. Graph the stable age distribution for the population.

Use the column graph option, and label your axes fully. Your graph should resemble Figure 9.

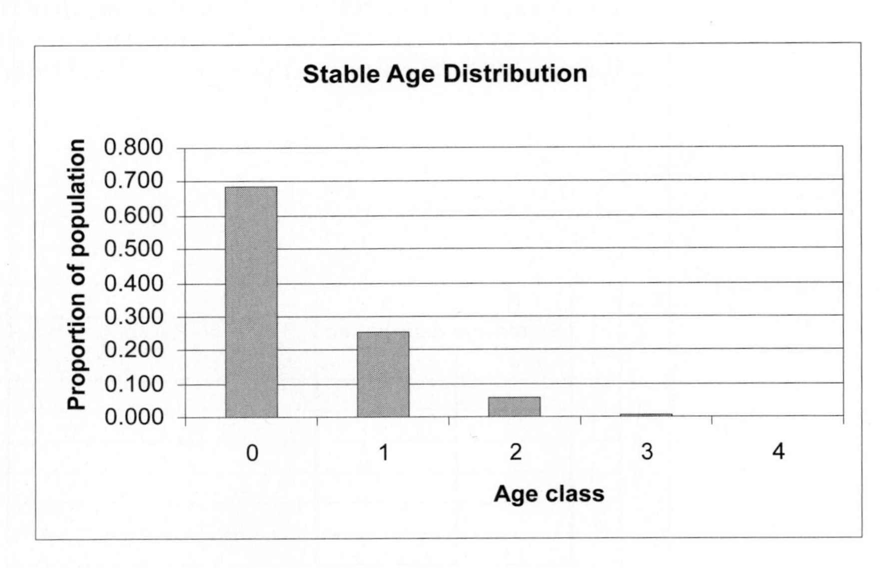

Figure 9

8. Save your work. Review your results and computations and make sure you understand the spreadsheet thus far.

E. Calculate the reproductive value distribution.

Remember that reproductive value can be computed by Fisher's equation (**Equation 6**):

$$v_x = \frac{e^{rx}}{l_x} \sum_{y=x+1}^{\Omega} e^{-ry} l_y b_y \qquad \text{Equation 6}$$

1. In cells J4–J8, enter a formula to compute the left-hand side of Fisher's equation (e^{rx}/l_x).

Enter the formula **=EXP($C$16*A4)/D4** in cell J4. Copy your formula down to cell J8.

2. In cells K4–K8 enter an equation to calculate $e^{-rx}l_x b_x$.

Enter the formula **=EXP(-$C$16*A4)*F4** in cell K4. Copy your formula down to cell K8. This calculation is an intermediate step that will be helpful for future calculations

3. In cells L4–L7, enter a formula in cell L4 to calculate the right-hand side of the reproductive value equation,

$$\sum_{x}^{\Omega} e^{-ry} l_y b_y$$

Enter the formula **=SUM(K4:$K$7)** in cell L4. Copy this formula down to cell L7. Now that we have $e^{-rx}l_x b_x$ for each age class, we are able to sum these values over age classes into the future. Note that we include the individual of age x as well as individuals of any age class in the future (denoted by the letter y) in the computations. We have added this step to facilitate the computations in the next step.

4. In cells M4–M7, enter a formula to calculate v_x.

Enter the formula **=J4*L5** in cell M4. Copy your formula down to cell M7.
Finally, we can compute v_x, the reproductive value, for each age class. The formula **=J4*L5** offsets the formula by one row, so that the reproductive value is computed as e^{rx}/l_x times the sum of $e^{-ry}l_y b_y$ for any age classes into the future. The result gives the expected number of offspring to be produced by an individual of age x over its remaining life span, adjusted by the population growth rate, r.

5. Save your work.

Your spreadsheet should now resemble Figure 10.

	H	I	J	K	L	M
2	**Stable age distribution**		**Reproductive value distribution**			
3	$l_x e^{-rx}$	c_x	e^{rx}/l_x	$e^{-rx} l_x b_x$	$\Sigma e^{-ry} l_y b$	v_x
4	1.000	0.684	1.000	0.000	1.000	1.000
5	0.368	0.252	2.717	0.736	1.000	0.717
6	0.085	0.058	11.808	0.254	0.264	0.115
7	0.010	0.007	102.653	0.010	0.010	0.000
8	0.000	0.000				
9	1.463	1.000		1.000		

Figure 10

QUESTIONS

1. Interpret your model results fully. Which age class has the highest reproductive value, which age class has the lowest reproductive value? Interpret your results in terms of r, and the birth and survivorship data from the life table.
2. Interpret the reproductive values from your models from the standpoint of conservation of a game species whose populations are harvested and maintained at a high level, versus a pest species whose populations you would like to control, versus a threatened species that is being reintroduced to an area. Based on your knowledge of reproductive value, does your decision also depend on the proportion of the population that occurs in the various age classes? Why or why not?
3. The model currently computes reproductive value for a population that is increasing. Adjust the birth rates values in cells C5–C7 in various ways to generate different values of r, (keep the S_x column the same). For each of your model runs, interpret how the birth schedule, and r, affect reproductive values. For each run, remember to use the Solver again to generate a correct r.
4. Change the life history parameters in the life table (cells B4–C8) to generate a different life history schedule (a Type III survival curve). Set up new life table entries as follows:

	A	B	C
3	*x* (age)	S_x	b_x
4	0	500	0
5	1	200	0
6	2	100	0
7	3	50	8
8	4	0	0

   This life history schedule represents a Type III survival curve in which reproduction occurs once and then organisms die (semelparous or annual). For such a life history, which individuals have the highest reproductive value? You will need to use the Solver again to obtain a correct r in cell C16 so that your reproductive value calculations are correct.

5. Compare this life history with a species with a Type I survival curve, in which reproduction is delayed but occurs over different age classes. Set up new life table entries as follows:

	A	B	C
3	*x* (age)	S_x	b_x
4	0	500	0
5	1	499	0
6	2	400	2
7	3	300	2.1
8	4	0	0

   For such a life history, which individuals have the highest reproductive value? You will need to use the Solver again to obtain a correct r in cell C16 so that your reproductive value calculations are correct.

LITERATURE CITED

Begon, M., J. L. Harper, and C. R. Townsend. 1986. *Ecology*. Blackwell Scientific, Oxford.

Caswell, H. 2001. *Matrix Population Models*, 2nd Ed. Sinauer Associates, Sunderland, MA.

Caughley, G. 1977. *Analysis of Vertebrate Populations*. Wiley, New York.

Fisher, R. A. 1930. *The Genetical Theory of Natural Selection*. Clarendon Press, Oxford.

Gotelli, N. 2001. *A Primer of Ecology*, 3rd Ed. Sinauer Associates, Sunderland, MA.

Mertz, D. B. 1970. Notes on methods used in life-history studies. *In* R. M. May (ed.), *Theoretical Ecology: Principles and Applications*, pp. 4–25. W. B. Saunders, Philadelphia.

17 DEMOGRAPHIC STOCHASTICITY

Objectives

- Evaluate effects of stochastic processes in small versus large populations.
- Develop a macro to simulate several trials.
- Compute standard statistics, such as means, variances, coefficients of variation.

Suggested Preliminary Exercise: Geometric and Exponential Population Models; Statistical Distributions

INTRODUCTION

In a seminal book in conservation biology, Mark Shaffer (1987) wrote, "Given an expanding human population with rising economic expectations, competition for the use of the world's remaining resources will be intense. Conservationists will often face the problem of determining just how little habitat a species can have and yet survive. At the same time, biologists are increasingly coming to recognize that extinction may often be the result of chance events and that the likelihood of extinction may increase dramatically as population size diminishes."

Just how does chance play a role in the ability of a species to persist or go extinct, and how can we characterize the "risk of extinction" due to chance? This very question was asked by D. Saltz (1996), who was interested in determining how many Persian fallow deer (*Dama dama mesopotamica*), a critically endangered species, should be introduced into an area in western Asia as part of a species reintroduction program.

Stochasticity means random variation. In population biology, stochasticity refers to the random changes that influence the growth rate of a population (Akçakaya et al. 1997). Such variation is pervasive in the ecology of natural populations and operates at many levels. If you have completed the exercise on genetic drift, you know that chance plays a role in changing the *allele frequencies* in a population. Unpredictable changes in weather, food supply, and populations of competitors, predators, and parasites act on the *population as a whole* and may contribute to chance extinction. A third kind of chance event operates on *individuals*. This uncertainty is called "demographic stochasticity," and in this exercise you will learn how demographic stochasticity can cause unpredictable population fluctuations and can lead to extinction.

Demographic stochasticity is the variation in average survivorship and reproduction that occurs because a population is made up of an integer number of individuals. For example, we might determine that a *population* has a birth rate b of 0.4 individuals per individual per year and a survival rate s of 0.6 individuals per individual per year. This indicates that, on average, individuals in the population produce 0.4 offspring per year and 0.6 individuals survive to the next year. But of course, an *individual* cannot partially die and there is no such thing as 0.4 of an offspring. The *population* has a growth rate, but *individuals* either live or die, and they reproduce an integer number of offspring. This interplay between the *finite* characteristics that describe individuals and the global characteristics that describe the collection of individuals in the population is the realm of demographic stochasticity.

Let's begin our explorations with a very brief review of modeling births and deaths in a population, and then discuss how demographic stochasticity can affect the population's growth over time. We will let

N_t represent the size of the population at some arbitrary time t
N_{t+1} represent population size one time-unit later
B_t represent the total number of births in the interval from time t to time $t + 1$
D_t represent the total number of deaths in the same time interval

We are assuming here, as we did in Exercise 7 on population growth, that the population is "closed" to immigration and emigration; thus we can write

$$N_{t+1} = N_t + B_t - D_t \qquad \text{Equation 1}$$

If we assume that B (total births) and D (total deaths) are governed by the per capita birth and death rates, we can substitute bN_t for B_t and dN_t for D_t, and rewrite our equation as

$$N_{t+1} = N_t + bN_t - dN_t \qquad \text{Equation 2}$$

Thus, if we know what the per capita birth and death rates (b and d) are at time step t, we can compute the total number of births and deaths (B and D) in Equation 1, and calculate the population size in the next time step, $t + 1$.

How does demographic stochasticity affect B and D, even if b, d, and N_t are known? Consider a population of 10 individuals, with $b = 0.4$ and $d = 0.4$ as described previously. The survival rate, s, equals $1 - d$, so $s = 0.6$. If there were no demographic stochasticity in this population, the total number of births would be

$$B = bN = 0.4 \times 10 = 4$$

and the total number of deaths would be

$$D = dN = 0.4 \times 10 = 4$$

The total number of survivors would be:

$$s = (1 - d)N = 0.6 \times 10 = 6$$

However, if we follow the fates of *individuals* in the population and determine whether each individual lives or produces offspring, we may not end up with B and D as computed because partial death and reproduction is generally not possible. We can evaluate this problem by modeling the fates of individuals, utilizing the per capita birth rate, b, and the per capita survival rate, s, in a process that determines whether an individual will live or die, and whether it will reproduce or not. This is often done with a random-number generator, where a random number between 0 and 1 is drawn from a uniform distribution. To determine whether an individual dies or survives, we can compare the random number to s and let all individuals with a random number less than s survive. To determine whether an individual reproduces offspring, we can compare a different random number to b and let all individuals with a random number less than b reproduce. This is quite easy to do on a spreadsheet such as the one shown in Figure 1.

	A	B	C	D	E
1	***Demographic Stochasticity***				
2		**Survival rate = *s* =**	0.6	**Death rate = *d* =**	0.4
3		**Birth rate = *b* =**	0.4		
4					
5			**POPULATION 1**		
6	Individual	Random #	Survive?	Random #	Reproduce?
7	1	0.38	1	0.11	1
8	2	0.91	0	0.86	0
9	3	0.16	1	0.56	0
10	4	0.78	0	0.78	0
11	5	0.98	0	0.62	0
12	6	0.59	1	0.44	0
13	7	0.23	1	0.89	0
14	8	0.61	0	0.28	1
15	9	0.48	1	0.44	0
16	10	0.61	0	0.94	0
17			5		2

Figure 1 In this population, $s = 0.6$ and we expect the total number of survivors to be 6, but we see that only 5 individuals actually survived. And although $b = 0.4$ and we expect B to be 4, only 2 individuals produced offspring. This variation or departure from the population birth and death rates is demographic stochasticity.

With $b = 0.4$ and $d = 0.4$, the population of 10 individuals in Figure 1 should theoretically remain at 10 individuals, since $r = 0$. (Remember that $r = b - d$.) However, in this instance, the population declined from time step 1 (10 individuals) to time step 2 (7 individuals). Occasionally, by chance, the total number of births will be 0 and the total number of survivors will be 0, in which case the population has gone extinct due to demographic stochasticity.

We can characterize the nature of demographic stochasticity under various population sizes, birth rates, and death (or survival) rates by simulating the fates of individuals as we have just done in Figure 1, and then recording the outcome (such as 5 total survivors and 2 total births). For instance, suppose that the probability of survival is 0.6, and we repeat the experiment in Figure 1 100 different times, recording only the total number of survivors for each trial. We will do this for two populations, the first of which consists of 10 individuals and the second of which consists of 25 individuals. Figure 2 shows the results of one such experiment. For population 1, 27 of the trials resulted in 6 survivors (the expected result), but the remaining trials deviated from this result. For population 2, 17 trials resulted in 15 survivors (60% of 25 individuals), but the remaining trials deviated from this expected result.

Which population shows a greater scatter, or more variation, in the trial results? If you have completed Exercise 3, "Statistical Distributions," you know that the **standard deviation** (S) is commonly used to measure the amount of variation from the mean in a data set. The standard deviation is calculated as

$$S = \sqrt{\frac{\sum (x - \bar{x})^2}{N - 1}} \qquad \text{Equation 3}$$

where $(x - \bar{x})^2$ represents the square of the difference between each data point (x) and the mean ($\bar{x}$), and N is the total number of data points. In Figure 2, the standard deviation for population 1 turns out to be about 1.6, and the standard deviation for popu-

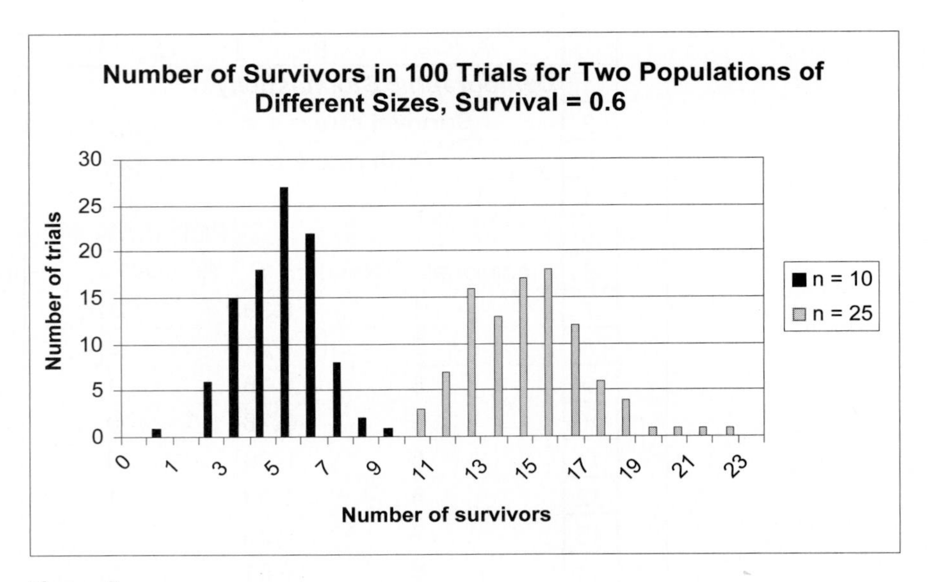

Figure 2

lation 2 is about 2.3, so by this measure, population 2 shows more variation than population 1. But let's think about this for a moment. Note that for each trial, population 1 had only 11 possible outcomes (0–10 survivors), almost all of which occurred, but population 2 had 26 possible outcomes (0–25 survivors), only half of which occurred. In fact, a general property of data sets is that the mean and standard deviation tend to change together—the lower the mean, the lower the standard deviation, and the higher the mean, the higher the standard deviation. Population 2 has a higher mean than population 1, so the difference in their standard deviations might not be as significant as it at first appears. To compare populations whose means are quite different, we "adjust" the standard deviations by dividing each one by its corresponding mean to get the **coefficient of variation (CV)**:

$$CV = \frac{S}{\bar{x}} \qquad \text{Equation 4}$$

The coefficient of variation is the ratio of the standard deviation to the mean, and it provides a relative measure of data dispersion compared to the mean. The CV has no units. It may be reported as a simple decimal value or it may be reported as a percentage by multiplying by 100. In the example presented in Figure 2, the CV for population 1 is about 0.27, and the CV for population 2 is about 0.15, so by this measure (which takes into account that we *expect* less data scatter when the mean is small than when it is large), population 1 showed *more* variation than population 2.

Demographic stochasticity has important biological implications. Shaffer (1987) has demonstrated that the chance of extinction through demographic stochasticity increases dramatically as population size diminishes. Mating systems (Legendre et al. 1999) and age structure (Saltz 1996) have also been shown to be affected by demographic stochasticity.

PROCEDURES

In this exercise, you will set up a spreadsheet model to investigate the effects of demographic stochasticity on two populations. Population 1 is a small population (10 individuals), while population 2 is large (100 individuals). The values of *b* and *d* remain fixed throughout the exercise. After the exercise is completed, Questions 1 and 2 will ask you to change the values of *b* and *d* to explore how their relative differences, and absolute values, affect demographic stochasticity.

As always, save your work frequently to disk.

INSTRUCTIONS	ANNOTATION

A. Calculate birth and survival rates for population 1.

1. Open a new spreadsheet and set up titles and column headings as shown in Figure 3.

	A	B	C	D	E
1	***Demographic Stochasticity***				
2		**Survival rate = *s* =**	0.6	**Death rate = *d* =**	0.4
3		**Birth rate = *b* =**	0.2		
4					
5			**POPULATION 1**		
6	Individual	Random #	Survive?	Random #	Reproduce?
7	1				
8	2				
9	3				
10	4				
11	5				
12	6				
13	7				
14	8				
15	9				
16	10				
17					

Figure 3

2. In cells C2 and C3, enter the values shown for *s* and *b*.

Enter 0.6 in cell C2.
Enter 0.2 in cell C3.
Here the survival rate, *s*, is 0.6 individuals/individual/year) and the per capita birth rate, *b*, 0.2 individuals/individual/year). Remember that the death rate, *d*, is 1 – *s*, so *d* = 0.4. These values will remain fixed for the purposes of this exercise. You will vary them to answer the questions at the end of the exercise.

3. In cells A7–A16, set up a linear series from 1 to 10.

Enter 1 in cell A7.
Enter the formula **=A7+1** in cell A8. Copy this formula down to cell A16.
These numbers designate the 10 individuals that make up population 1.

4. In cells B7–B16, use the **RAND** function to assign a random number between 0 and 1 to each individual in population 1.

Enter **=RAND()** in cells B7–B16
This formula generates a random number between 0 and 1. Note that the spreadsheet generates new random numbers each time the calculate shortcut key, **F9**, is pressed.

5. In cells C7–C16 enter an **IF** formula to determine whether each individual survives (1) or dies (0).

Enter the formula **=IF(B7<$C$2,1,0)** in cell C7. Copy this formula down to cell C16.
Whether an individual survives or dies is based on the population survival rate in cell C2 and the random number associated with each individual in cells B7–B16. In cell C7, if the random number in cell B7 is less than the survival rate in cell C2, the individual receives a score of 1 (survives); otherwise it receives a score of 0 (dies). Copy this formula down for the remaining nine individuals in population 1.

6. Enter a random number in cells D7–D16.

Enter **=RAND()** in cells D7–D16.

7. In cells E7–E16, enter an **IF** formula to determine whether each individual reproduces (1) or not (0), based on the birth rate given in cell C3.

Enter the formula **=IF(D7<$C$3,1,0)** in cell E7. Copy this formula down to cell E16. In this exercise, you will assume that individuals that reproduce have just one offspring. Whether an individual reproduces is based on the birth rate given in cell C3 and the random numbers in column D; the formula is analogous to the one in Step 5.

8. In cells C17 and E17, use the **SUM** function to tabulate the total number of survivors and births, respectively.

Enter the formula **=SUM(C7:C16)** in cell C17.
Enter the formula **=SUM(E7:E16)** in cell E17.
You can also use the "Autosum" button on your toolbar, which looks like a sigma (Σ). Based on the survival and birth rates entered in cells C2 and C3, how many total survivors and total births do you expect for population 1?

9. Press the F9 key to generate a new set of random numbers, and hence a new total number of survivors and total number of births in population 1.

How did your total survivors and total births change with the new set of random numbers? The difference between your results and the population's birth and survival rates is an example of demographic stochasticity. Although the rates are "fixed" in cells C2 and C3, the numbers of survivors and births vary due to chance and because individuals cannot reproduce 0.2 individuals, nor can they partially die. What is the likelihood of obtaining the same results again? Characterize the nature of demographic stochasticity based on your two "trials."

10. Save your work.

B. Write a macro to simulate 150 trials.

By conducting a great number of trials, you can determine how likely a certain outcome is by calculating the means and variances of the survivors and births produced in population 1 and characterize the nature of demographic stochasticity more effectively.

1. Set up new column headings as shown in Figure 4, but extend the trial numbers to 150 (cell A170).

Enter the number 1 in cell A21.
In cell A22, enter **=1+A21**. Copy this formula down to cell A170.

	A	B	C
18		POPULATION 1	
19	Trial	Total	Total
20	number	survivors	births
21	1		
22	2		
23	3		

Figure 4

2. Repeat your "experiment" 150 times.

You can either push F9, the calculate key, 150 more times and manually enter how many individuals survived and reproduced in each trial (keeping track of your results in the appropriate cell labeled "trial"), or you can write a macro to do this for you.

From the menu, select Tools | Options | Calculations. Select Manual Calculation. (In Macintosh programs, the sequence is Tools and then Preferences.) From this point on you will need to press F9 when you want the spreadsheet to recalculate numbers generated by your macro. Then open the Macro function to the Record mode and assign a shortcut key (see Exercise 2 for details). Enter the following steps in your macro:

- Press F9 to obtain a new set of random numbers, and hence a new set of total survivors and total births.
- Select cell C17, then open Edit | Copy.
- Select cell B20, the column labeled "Total Survivors."

- Open Edit | Find. The dialog box in Figure 5 will appear. Leave the Find What box empty, searching by columns and formulas, and then select Find Next and Close.

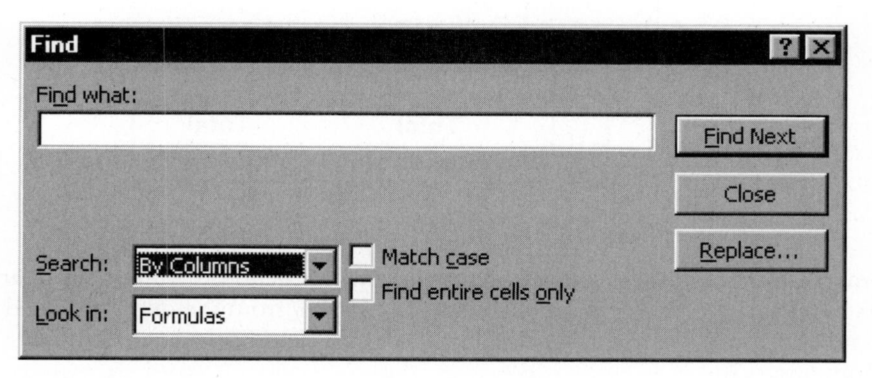

Figure 5

- Open Edit | Paste Special | Paste Values. Click OK.
- Select cell E17.
- Open Edit | Copy.
- Select cell C20, the column labeled "Total Births."
- Open Edit | Find. Leave the Find What box empty, searching by columns and formulas. Select Find Next and Close.
- Open Edit | Paste Special | Paste Values. Click OK.

The macro is finished; stop recording (Tools | Macro | Stop Recording). Now when press your shortcut key 150 times; each trial will run automatically.

3. Switch back to Automatic Calculation. Save your work

From the menu, select Tools | Options | Calculations. Select Automatic Calculation.

C. Calculate birth and survival rates for population 2.

Population 2 is larger, consisting of 100 individuals. In this section, we will repeat the steps you've just completed for the larger population.

1. Enter the column headings shown in Figure 6.

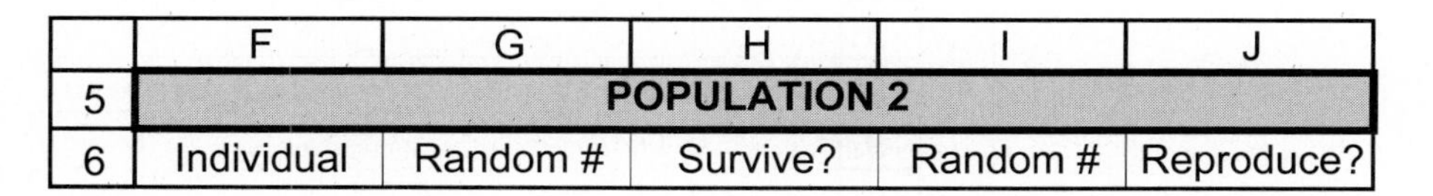

	F	G	H	I	J
5	POPULATION 2				
6	Individual	Random #	Survive?	Random #	Reproduce?

Figure 6

2. Set up a linear series from 1 to 100 in cells F7–F106.

Enter 1 in cell F7.
Enter **=1+F7** in cell F8. Copy your formula down to cell F106.

3. Repeat the steps in Part A to fill in survival and birth outcomes for population 2, and sum the total survivors and total births.

You should generate numbers and outcomes in cells G7–J106.

4. Enter column headings as shown in Figure 7.

Count the total number of survivors and total number of births for population 2, and record the results of each "simulation" as we did for population 1.

	D	E
18	POPULATION 2	
19	Total	Total
20	survivors	births

Figure 7

5. Record a macro to track the total survivors and total births and run 150 trials with population 2.

Follow the instructions in Section B. Make sure your macro for population 2 has a *different name and shortcut key* from the ones you used in population 1. Press your new macro 150 times to run 150 trials.

6. Switch back to Automatic Calculation. Save your work.

From the menu, select Tools | Options | Calculations. Select Automatic Calculation.

D. Construct a frequency histogram of results.

1. Set up column headings as shown in Figure 8.

These are the headings for a frequency histogram for population 1, which consists of 10 individuals. For any trial, the number of survivors could be between 0 and 10, and the total number of births could be between 0 and 10.

	A	B	C	D
175	# of survivors and breeders in 150 trials for population 1			
176	# Survivors	Frequency	# Births	Frequency
177	0		0	
178	1		1	
179	2		2	
180	3		3	
181	4		4	
182	5		5	
183	6		6	
184	7		7	
185	8		8	
186	9		9	
187	10		10	

Figure 8

2. In cells B177–B187, enter a **COUNTIF** formula to count the number of trials in which there were 0 survivors, 1 survivor, etc.

Enter the formula **=COUNTIF($B$21:$B$170,A177)** in cell B177. Copy the formula down to cell B187.

This formula examines the range of numbers in cells **$B$21:$B$170** and counts the number of times 0 (listed in cell A177) appears. Fill this formula down to obtain frequency counts for the number of trials in which 1, 2, 3 … 10 survivors were recorded. Double-check your results upon completion; the sum of the numbers generated by the cells with the **COUNTIF** formula should be 150 because there were 150 trials.

3. In cells D177–D187, enter a **COUNTIF** formula to count the number individuals that reproduced a single offspring in the various trials.

Enter the formula **=COUNTIF($C$21:$C$170,C177)** in cell D177. Copy the formula down to cell D187.

4. Construct a frequency histogram of the number of survivors and the number of breeders for population 1.

Your histogram should resemble Figure 9.

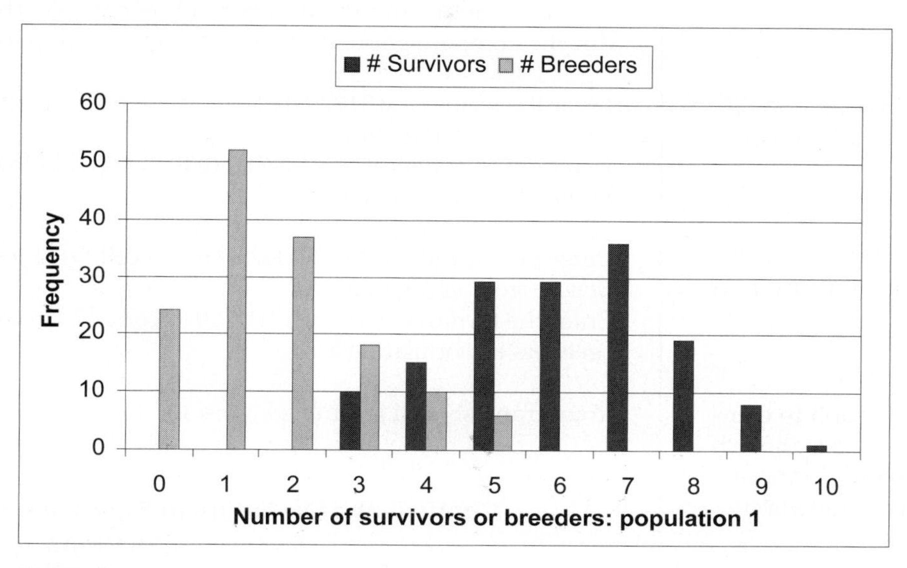

Figure 9

5. Construct a frequency histogram for population 2.

Adapt the preceding steps (1–4) to the values for population 2. Remember that population 2 consists of 100 individuals, so the total number of survivors or births in any trial can range between 0 and 100. Interpret your results.

E. Compute means and standard deviations.

Review Exercise 3, "Statistical Distributions," if you are unsure about the use of means and standard deviations.

1. In cells B171 and B172, enter **AVERAGE** and **STDEV** formulae, respectively, to calculate the mean and standard deviation of the number of survivors in population 1.

Enter the formula **=AVERAGE(B21:B170)** in cell B171.
Enter the formula **=STDEV(B21:B170)** in cell B172.

2. In cells C171–E172, enter **AVERAGE** and **STDEV** formulae to calculate the mean and standard deviation of the number of breeders in population 1, and the number of breeders in population 1 and the number of survivors and breeders in population 2.

Select cells B171–B172 and copy them over to cells E171–E172. The resulting formulae should be:

- C171 **=AVERAGE(C21:C170)**
- D171 **=AVERAGE(D21:D170)**
- E171 **=AVERAGE(E21:E170)**
- C172 **=STDEV(C21:C170)**
- D172 **=STDEV(D21:D170)**
- E172 **=STDEV(E21:E170)**

3. Examine the histograms you made for each population in Section D and answer the questions at right.

Which population appears to exhibit greater stochasticity (i.e., greater variation in the number of births and survivors)? Pay attention to the standard deviations, which measure dispersion of variation in results. Now reflect on the mean values you computed in the previous two steps. Is it useful to compare the variation in two populations that have such different mean values? Why or why not?

F. Calculate and graph coefficients of variation.

The **coefficient of variation**, or **CV**, is calculated as standard deviation divided by the mean, which is then multiplied by 100. We perform this caculation for both the number of survivors and the number of breeders. Analysis of the CV will allow you to directly compare populations 1 and 2 by adjusting for their means.

1. In cells B173 and C173, compute the CVs for population 1.

Enter the formula **=(B172/B171)*100** in cell B173 to compute the CV for the number of survivors in population 1.
Enter the formula **=(C172/C171)*100** in cell C173 to compute the CV for the number of breeders in population 1.

2. In cells D173 and E173, compute the CVs for population 2.

Enter the formula **=(D172/D171)*100** in cell D173 to compute the CV for the number of survivors in population 2.
Enter the formula **=(E172/E171)*100** in cell E173 to compute the CV for the number of breeders in population 2.

3. Create a graph to compare the CVs, a standardized measure of variation, for the two populations.

Your graph should resemble Figure 10.

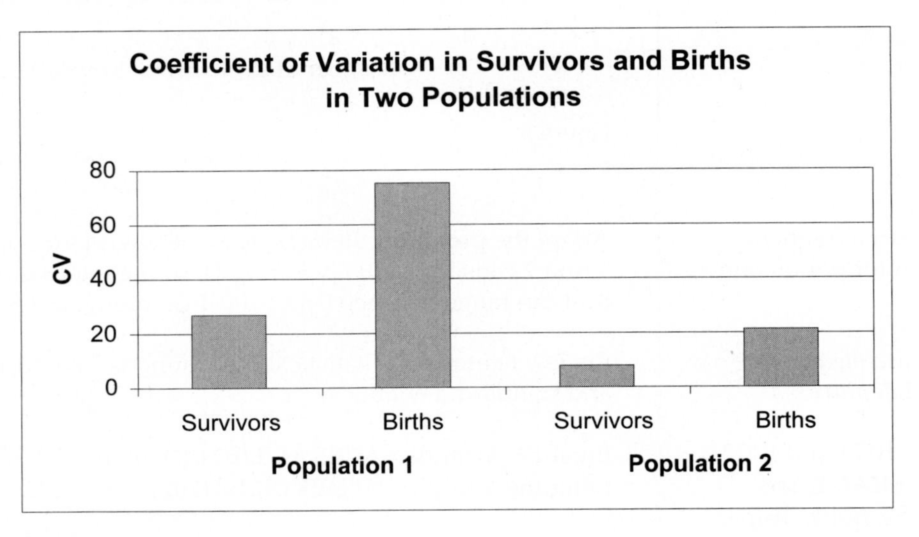

Figure 10

You should see that the smaller population, population 1, has a much higher CV in both number of survivors and number of births than population 2. This reflects a greater amount of unpredictable variation (demographic stochasticity) in small populations.

QUESTIONS

1. Focus on population 1 (the small population), cells C17 and E17 (the total survivors and total births). Press F9 20 times and record the number of times the population goes extinct (number of survivors *and* number of births = 0) Then compute the extinction risk, *P*(extinction), as the number of times the population went extinct divided by 20 trials. (Your result is likely to be 0). Enter differing values of *s* and *b* in cells C2 and C3 (except 0). What levels of *s* and *b* are likely to produce higher extinction rates due to demographic stochasticity?

2. Compare the stochasticity in the larger population, population 2, under different survival and birth rates, while keeping r constant (remember that $r = b - d$). In the first scenario, let the population have a high birth rate ($b = 0.9$) and low survival rate ($s = 0.1$). In the second scenario, let the population have a low birth rate ($b = 0.1$) and a high survival ($s = 0.9$). Note that in both cases, $r = 0$. Set up spreadsheet headings as shown, and modify the survival and birth rates in cells C2 and C3. For each scenario, develop a macro in which the number of survivors and births are recorded in 100 trials. For each trial, compute ΔN as the change in population size (number of births minus number of deaths). (You can generate the delta symbol, Δ, by typing in a capital D and then changing the font to "Symbol.") Then compare the coefficient of variation in ΔN over 100 trials when the population size is 100. How do the *absolute* birth and death rates affect stochasticity when population size is relatively large?

	L	M	N	O	P	Q	R
2		High birth and low survival (b = 0.9, s = 0.1)			Low birth and high survival (b = 0.1, s = 0 .9)		
3	Trial	# survivors	# births	ΔN	# survivors	# births	ΔN

3. Variation is pervasive in nature. For example, birth rates and death rates are rarely constant over time. How do you think demographic stochasticity differs from a more commonly noted type of variation, environmental stochasticity? With environmental stochasticity, b, s, and d *vary* with some randomness as opposed to remaining fixed (cells C2, E2, C3). Can you think of ways in which you might add an element of environmental stochasticity to your model?

*4. (Advanced) In your model, you've discovered that demographic stochasticity is different between populations consisting of 10 and 100 individuals. As population size increases, in what fashion do the effects of demographic stochasticity decrease? (For example, does it decrease linearly as population size increases, or is there some threshold at which increasing population size has little effect on stochastic processes?) Develop your model more fully to answer this question (you may want to copy your entire model onto a new sheet for this question, so that you do not alter your original model).

*5. (Advanced) Examine the Visual Basic for Applications code that was used to write your macro. See if you can follow through the code and match the action of your keystrokes outlined in step 2 of Section B to the code. It should look something like this:

MACROS

```
Sub trial()
'
' trial Macro
' Macro recorded 8/31/99 by Authorized User
'
' Keyboard Shortcut: Ctrl+t
'
 Application.Goto Reference:="R21C2:R21C3"
 Selection.Insert Shift:=xlDown
 Application.Goto Reference:="R17C3:R17C4"
 Calculate
 Selection.Copy
 Application.Goto Reference:="R21C2:R21C3"
 Selection.PasteSpecial Paste:=xlValues, Operation:=xlNone,
  SkipBlanks:= _
 False, Transpose:=False
End Sub
```

*6. (Advanced) The binomial distribution could have been used to estimate the various probabilities that x number of survivors and x number of breeders would have occurred in the 150 trials (see Exercise 3, "Statistical Distributions"). Use the **BINOMDIST** function to obtain survivorship probabilities for Population 1, and compare your trial results with those predicted by the binomial distribution. Does the binomial distribution also reflect greater "stochasticity" when sample sizes are small?

LITERATURE CITED

Akçakaya, H. R., M. A. Burgman, and L. R. Ginzburg. 1997. *Applied Population Biology*. Applied Biomathematics, Setauket, New York.

Legendre, S., J. Clobert, A. P. Moller, and G. Sorci. 1999. Demographic stochasticity and social mating system in the process of extinction of small populations: The case of passerines introduced to New Zealand. *American Naturalist* 153(5): 449–463.

Saltz, D. 1996. Minimizing extinction probability due to demographic stochasticity in a reintroduced herd of Persian fallow deer *Dama dama mesopotamica*. *Biological Conservation* 75: 27–33.

Shaffer, M. 1987. Minimum viable populations: Coping with uncertainty. In M. E. Soulé (ed.), *Viable Populations for Conservation*, pp. 69–86. Cambridge University Press, Cambridge.

18

KEY FACTOR ANALYSIS

In collaboration with David Bonter

Objectives

- Simulate a population that has nonoverlapping generations.
- Use the beta distribution.
- Calculate the stage-specific mortality, K_x, for each stage in the life cycles.
- Conduct a key stage analysis of the various stages in the life cycle.

Suggested Preliminary Exercise: Life Tables, Survivorship Curves, and Population Growth

INTRODUCTION

Let's assume you've been tracking the population dynamics of an annual plant through its life cycle. You tediously count the number of seeds the plant sets, then count the number of seedlings to estimate the germination rate, then count the number of vegetative rosettes, the number of flowering adults, then the number of fruiting adults. Thus you have tracked the fates of individuals in one stage and counted how many individuals survived to the next stage. If this was an endangered plant, you might want to know the stage of the life cycle in which the highest mortality occurs. For example, you might find out that the total mortality across the life cycle is strongly influenced by the failure of seeds to germinate, or by the failure of flowering plants to produce fruit. With such information, you can potentially target your efforts to reducing mortality at that particular stage.

The attempt to identify factors responsible for population change and to assess the magnitude of their effects is called **key factor analysis**. This analysis was developed by Morris (1959) to study spruce budworm outbreaks in forests in eastern Canada. Key factor analysis is specifically for organisms with discrete (nonoverlapping) generations, in which a single age class is present at any given time. The analysis, for example, could be applied to an insect population that moves from egg to larval to pupae to adult stages. The method also assumes that a series of different mortality factors operate on the population sequentially. For example, if two parasites and one disease kill larval insects, key factor analysis assumes that parasite A acts first to kill a sample, then parasite B kills a portion, then disease C acts to kill some of the remaining individuals (Krebs 1999: 511).

Modeling Key Factors

To set up a spreadsheet model of key factor analysis, we will let

- N_x denote the number of individuals alive at any given stage.
- N_{x+1} denote the number of individuals at the next stage.
- b_x denote the per capita birth rate of reproducing adults.
- k denote the stage-specific mortality, or "killing power."
- K denote the total generational mortality, or the sum of all the k's.

The main idea behind key factor analysis is that by comparing N_x in one stage to N_x in the previous stage, we can identify which stage has the largest mortality. We can also add up the k's to calculate K, the total mortality across generations. The k factors indicate the importance of a particular stage to the total generational mortality, and the k factor that most strongly affects generational mortality, K, is called the **key factor**.

The steps in a key factor analysis (Varley and Gradwell 1960) include:

1. Computing the observed fecundity, which is the per capita birth rate times the number of females in the population
2. Computing the population size for each stage, or N_x in a life table
3. Computing the absolute losses of individuals from one stage to the next. For stage x, the losses are computed as
$$N_x - N_{x+1}$$
4. Converting the absolute losses of individuals from one stage to the next into proportional losses. This is accomplished by taking the log of N_x. Age- or stage-specific mortality, then, is calculated as
$$\log(N_x) - \log(N_{x+1})$$
5. Defining age-specific mortality, k_x, as
$$k_x = \log(N_x) - \log(N_{x+1}) \qquad \text{Equation 1}$$
6. Computing total generational loss, K, as
$$K = k_0 + k_1 + k_2 + k_3 + \cdots + k_x. \qquad \text{Equation 2}$$

This analysis is done over several generations, where each generation consists of a complete life cycle and where the life cycles from one generation to the next do not overlap with each other. Each generation that is studied is a "replicate" of the key factor analysis, and these replicates are important because they let you know if a certain factor is normally a key factor, or if it is a key factor in some conditions or years but not in others.

As an example, suppose that k's and K were computed for an insect population for 10 generations. Figure 1 shows that K, the total generational mortality, fluctuates from generation to generation. The stage-specific mortalities (small ks) are also plotted for each year and reveal the losses that occur within a stage for a single generation. The little k that most closely mimics K over time is the key factor. In this case, graphed in Figure 1, the pupal stage is the key stage. Note that a pattern could not be detected if only a single generation were studied. Which k factor is most closely tied to K can be hard to discern, especially if the k factors have similar values. In this case, the key factor can be identified by plotting the k factor against K for every single k; Figure 2 does this for egg-stage mortality vs. total mortality.

Problems with Key Factor Analysis

You probably know by now that populations change over time through birth, death, immigration, and emigration. In fact, the equation for population growth given in Exercise 7,

$$N_{t+1} = N_t + B + I - D - E$$

is the basis for many exercises in this book. But because key factor analysis focuses on losses to a population, only death and emigration are properly represented by k fac-

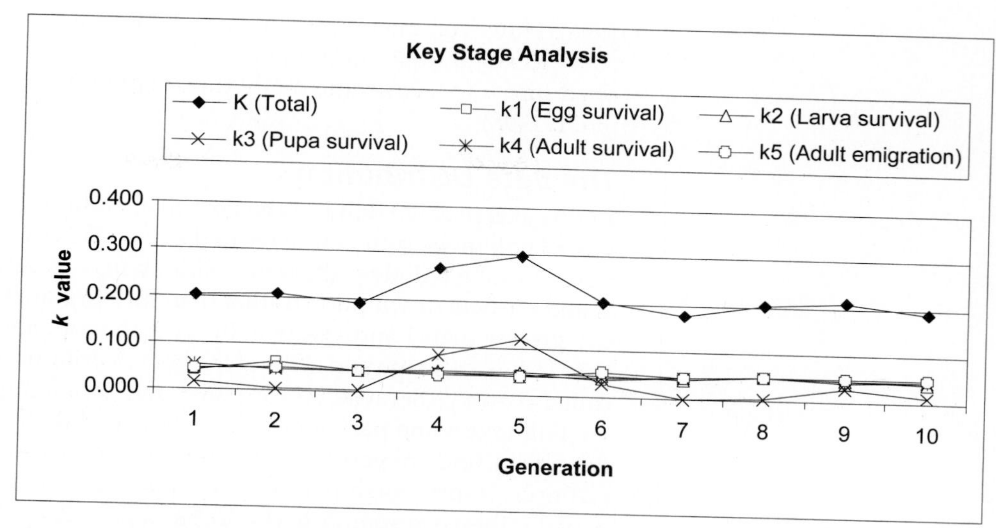

Figure 1 The total generational mortality, K, for each year is the sum of all the stage-specific mortalities for that year. The stage-specific mortalities (k_1–k_5) are also plotted for each year, and reveal the losses that occur within a stage for a given year. The k that most closely mimics K over time is the key factor. In this case, the pupae stage is the key stage. Note that a pattern could not be detected if only a single generation was studied.

tors. The analysis also does not specifically identify factors that are responsible for population change, only the stages that are correlated with total generational mortality. The analysis gives no indication of what might be *causing* such mortality, only the stage in which it occurs. For this reason, the analysis may be more properly named key *stage* analysis. Additionally, the assumptions of key factor analysis are often violated, and many ecologists have criticized the use of traditional key factor analysis (e.g., Royama

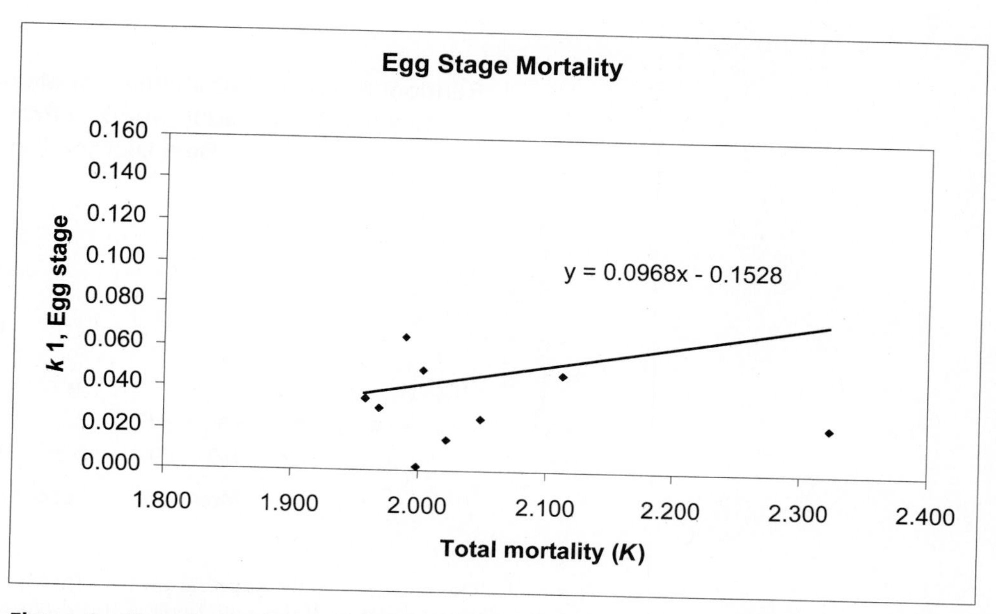

Figure 2 The relationship between k for the egg stage and total generational mortality, K, for 10 years. The slope of the regression equation is +0.0968. Similar graphs can be constructed for the other stages, and the slopes can then be compared. Were we to construct similar graphs for K and each of the other four k's in Figure 1, the k factor that generates the highest slope with K would be the key factor.

1996). However, the traditional analysis is often used as the first step in the analysis of census data from natural populations, and several new methods have been developed that improve on the method presented here (e.g., Brown et al. 1993; Sibly and Smith 1998).

The Beta Distribution

In this exercise, we will use the beta distribution to assign probabilities that an individual will move from one stage to the next. This distribution is not used in other exercises, and we will describe it only briefly here. All probabilities range between 0 and 1, and the beta distribution (rather than the normal distribution, which can take on values greater than 1 and less than 0) is much more appropriate for modeling probabilities. The exact shape and scale of the beta distribution is controlled by two parameters, called α and β. Because you are (by now) very familiar with the normal distribution, we will take some parameters from a normal distribution that you are familiar with (μ and σ^2), and convert them into parameters from the beta distribution, α and β. For example, if survivorship is known to have a mean, $\bar{x}$, of 0.6, and a standard deviation, S, of 0.1, this corresponds to $\alpha = 13.8$ and $\beta = 9.2$. A beta distribution with these parameters will show that most probabilities are 0.6, but there is substantial variation from sample to sample. The values of α and β can be calculated as follows, where the sample mean and standard deviation, $\bar{x}$ and S^2, estimate μ and σ^2:

$$\alpha = \frac{\beta * \mu}{1-\mu} \qquad \beta = \mu - 1 + \frac{\mu * (1-\mu)^2}{\sigma^2}$$

In this way, we can include variation in survival probabilities with an appropriate distribution (the beta distribution). However, you can intuitively visualize the probabilities based on your experience from working with normal distributions. (Thanks to Jeff Buzas at the University of Vermont, who provided these conversions). Figure 3 shows how the conversion works for a mean survival = 0.6 and a standard deviation = 0.1. These parameters translate into a beta distribution whose $\alpha = 13.8$ and $\beta = 9.2$. If we changed α and β in Figure 3, the distribution would take on a new shape.

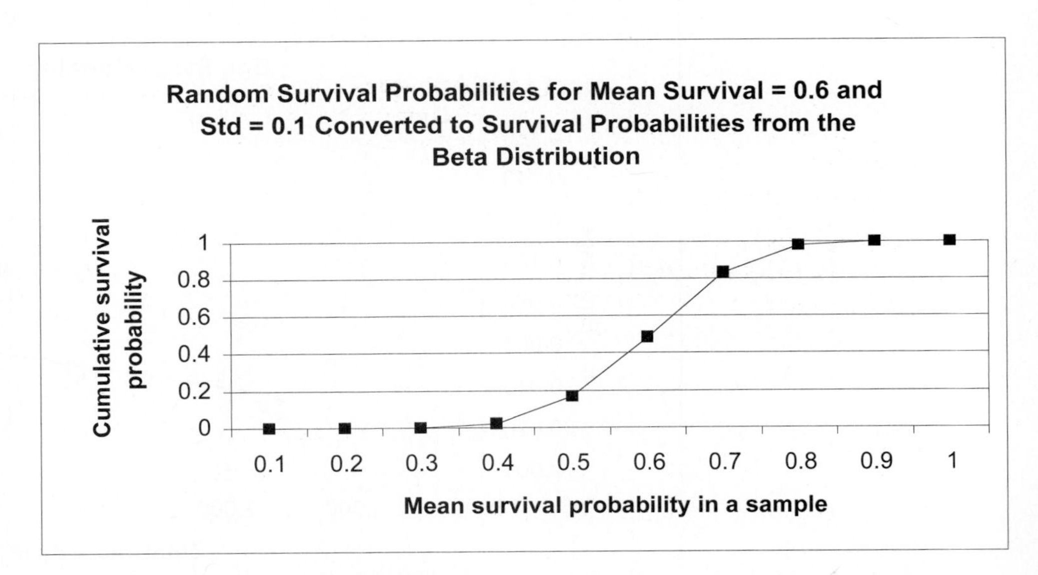

Figure 3 The x-axis gives the survival probability for a single sample. The cumulative probabilities are given on the y-axis. Look at the x-axis and note where the cumulative survival probabilities change very little. You should see that they change very little for probabilities $P < 0.4$ and $P > 0.8$. In between these values, the cumulative survival probabilities increase dramatically, suggesting that most of the data points in this distribution fall between 0.5 and 0.7. At $P = 0.6$, the cumulative probability equals 0.5, suggesting that half of the observations in the data set fall above 0.6 and half fall below 0.6, as expected.

PROCEDURES

In this exercise, you'll model a hypothetical insect population that moves through several stages in its life cycle. We'll assume the population is large and that we can track the total number of individuals alive at each stage. We'll assign probabilities that individuals move from one stage to the next, and then calculate the *k* factors and identify the key mortality factor. You'll assign probabilities that individuals move from one stage to the next with the beta distribution, and then calculate the *k* factors and identify the key mortality factor.

As always, save your work frequently to disk.

INSTRUCTIONS / ANNOTATION

A. Set up the model population.

1. Open a new spreadsheet and set up headings as shown in Figure 4.

	A	B	C	D	E	F	G	H
1	*Key Factor Analysis*							
2		Population variables						
3		Eggs	Egg	Larva	Pupa	Adult	Adult	Number of
4		laid	survival	survival	survival	survival	fidelity	females
5	Mean =	100	0.90	0.40	0.60	0.80	0.95	1000.00
6	Standard deviation =	30.00	0.10	0.05	0.10	0.10	0.01	
7	Beta distribution α =							
8	Beta distribution β =							

Figure 4

2. Enter parameter estimates (means and standard deviations) in cells B5–G6.

Enter the values shown in Figure 4 in cells B5–G6.

We'll consider an insect with nonoverlapping generations and whose life cycle consists of a series of mortality factors that operate in a linear sequence with no interaction. Eggs are laid by adults and hatch with some probability, then move to the larvae stage, and pupate to become adults. The probability of moving from one stage to the next is defined by a probability between 0 and 1. Some adults are capable of moving away from the study area population (emigration). The probability of remaining in the population and not emigrating is given in the column labeled "Adult fidelity."

3. Draw random values from a beta distribution for survival probability at each stage.

We'll add an element of stochasticity to our model by establishing means and variances for each parameter, and then "drawing" a random number from these distributions. In previous exercises, you may have used the **NORMINV** function to draw a random probability from a normal distribution with a given mean and standard deviation. The spreadsheet then converts this probability into a data point. This function won't work for survival probabilities, though, because our survival probabilities can only take on values between 0 and 1. For survival probabilities, the distribution we must draw at random from a beta distribution. The parameters in the beta distribution are α and β (made by typing "a" or "b" in on your keypad and then changing the font to the Symbol font).

4. In cells C8–G8 enter a formula for the β parameter of a beta distribution. In cells C7–G7, enter a formula that will calculate the α parameter.

Although α and β are not the same thing as means and standard deviations, we can enter these formula based on the conversion equations

$$\alpha = \frac{\beta * \mu}{1-\mu} \qquad \beta = \mu - 1 + \frac{\mu * (1-\mu)^2}{\sigma^2}$$

In cell C8, enter the formula **=C5-1+((C5*(1-C5)^2)/C6^2)** . Copy the formula across to cell G8.

In cell C7 enter the formula **=(C8*C5)/(1-C5)**. Copy the formula across to cell G7.

Now α and β are mathematical functions of the means and standard deviations specified in rows 5 and 6. As a result, we can draw random probabilities between 0 and 1 that have the means and standard deviations we specify.

5. Your spreadsheet should now resemble Figure 5. Save your work!

	A	B	C	D	E	F	G	H
1	*Key Factor Analysis*							
2		Population variables						
3		Eggs	Egg	Larva	Pupa	Adult	Adult	Number of
4		laid	survival	survival	survival	survival	fidelity	females
5	Mean =	100	0.90	0.40	0.60	0.80	0.95	1000.00
6	Standard deviation =	30.00	0.10	0.05	0.10	0.10	0.01	
7	Beta distribution =		7.20	38.00	13.80	12.00	450.30	
8	Beta distribution =		0.80	57.00	9.20	3.00	23.70	

Figure 5

B. Determine model inputs for Years 1-10.

1. Set up new headings as shown in Figure 6, but extend years to year 10 in cell K10.

	A	B	C	D	E
10	**Model Inputs**	**Year 1**	**Year 2**	**Year 3**	**Year 4**
11	Eggs laid				
12	Egg survival				
13	Larva survival				
14	Pupa survival				
15	Adult survival				
16	Adult fidelity				

Figure 6

2. In cells B11–K11, enter a formula to give the mean number of eggs laid in year 1.

In cell B11 enter the formula **=NORMINV(RAND(),$B$5,$B$6)**. Copy this formula over to cell K11.
Here we do use the **NORMINV** function. The **NORMINV** function returns the *inverse* of the normal cumulative distribution, given a mean and standard deviation. It has the syntax **NORMINV(probability,mean,standard_dev)**. The B11 formula draws a random probability from a distribution whose mean is given in cell B5 and whose standard deviation is given in cell B6. The spreadsheet then converts this probability into an actual data point from the distribution, which is the number of eggs laid in year 1. Note that when you press F9, the calculate key, the spreadsheet will generate a new random number, which means that a new random number is drawn from the distribution and hence a new average fecundity computed.

3. In cells B12–K12, enter a formula to give the probability that eggs will hatch.

In cell B12 enter the formula **=BETAINV(RAND(),$C$7,$C$8)** . Copy the formula over to cell K12.
The B12 formula gives the probability that eggs will hatch. Since this is a probability whose values must fall between 0 and 1, we use the beta distribution (instead of the normal distribution). The **BETAINV** formula functions like the **NORMINV** formula, except that the distribution is a beta distribution instead of a normal distribution. The formula in cell B12 tells the spreadsheet to draw a random cumulative probability from the beta distribution whose parameters are α (cell C7) and β (cell C8). (Remember, you entered formulae to compute α and β based on the means and standard deviations entered in rows 5 and 6.) The spreadsheet converts the cumulative probability into a data point, which is the probability that eggs will hatch in year 1. Press F9 to generate a new estimate.

4. Enter formulae in cells B13–B16 to determine random probabilities, drawn from the beta distribution. Copy your formulae across to column K.

We used the following formulae:

- Cell B13 **=BETAINV(RAND(),$D$7,$D$8)**
- Cell B14 **=BETAINV(RAND(),$E$7,$E$8)**
- Cell B15 **=BETAINV(RAND(),$F$7,$F$8)**
- Cell B16 **=BETAINV(RAND(),$G$7,$G$8)**

5. Double-check results.

6. Save your work.

Your spreadsheet should now resemble Figure 7, although your numbers will probably be different due to the random sampling from the normal and beta distributions.

	A	B	C	D	E
10	**Model Inputs**	**Year 1**	**Year 2**	**Year 3**	**Year 4**
11	Eggs laid	96.9	151.5	72.1	119.5
12	Egg survival	0.9	1.0	0.9	0.9
13	Larva survival	0.4	0.3	0.4	0.4
14	Pupa survival	0.6	0.7	0.7	0.6
15	Adult survival	0.9	0.8	0.7	0.6
16	Adult fidelity	1.0	0.9	0.9	0.9

Figure 7

C. Calculate model outputs and project growth for 10 years.

1. Set up new spreadsheet headings as shown in Figure 8, but extend your years to year 10.

	A	B	C	D	E
18	**Model outputs (*Nx*)**	**Year 1**	**Year 2**	**Year 3**	**Year 4**
19	Eggs				
20	Eggs hatched				
21	Larvae				
22	Pupae				
23	Adults				
24	Nonemigrants				

Figure 8

2. In cell B19, enter a formula to calculate the actual number of eggs laid in year 1.

In cell B19 enter the formula **=$H$5*B11**.
The actual number of eggs laid is the average fecundity × the number of females.

3. In cell B20, enter a formula to calculate the number of eggs hatched in year 1.

In cell B20 enter the formula **=B12*B19**.

The number of eggs hatched is a function of hatching probability calculated in cell B12.

4. In cells B21–B24, enter formulae to compute numbers of individuals in various stages. Copy each formula across to column K to complete 10-year simulation.

We used the following formulae:

- Cell B21 **=B20*B13**
- Cell B22 **=B21*B14**
- Cell B23 **=B22*B15**
- Cell B24 **=B23*B16**

D. Set up the k factor analysis.

1. Set up new headings as shown in Figure 9, but extend your years to year 10.

Now we can estimate the the stage-specific mortalities—the *k* factors ("little *k*'s")—for each stage in the life cycle.

	A	B	C	D	E
26	**Key factor analysis**	**Year 1**	**Year 2**	**Year 3**	**Year 4**
27	K (Total)				
28	k1 (Egg survival)				
29	k2 (Larva survival)				
30	k3 (Pupa survival)				
31	k4 (Adult survival)				
32	k5 (Adult emigration)				

Figure 9

2. In cell B28, enter a formula to calculate the mortality due to number of eggs that failed to hatch.

In cell B28 enter the formula **=LOG(B19)-LOG(B20).**

As shown in Equation 1, age-specific mortality is calculated by subtracting each log of the population size from the previous one:

$$k_x = \log(N_x) - \log(N_{x+1})$$

Thus, the formula in cell B28 gives the *k* value or the mortality due to the number of eggs that failed to hatch.

3. In cells B29–B32, enter formulae to compute *k* for the remaining stages.

We used the following formulae:

- Cell B29 **=LOG(B20)-LOG(B21)**
- Cell B30 **=LOG(B21)-LOG(B22)**
- Cell B31 **=LOG(B22)-LOG(B23)**
- Cell B32 **=LOG(B23)-LOG(B24)**

Cell B32 does not give a mortality value *per se*, because it reflects the loss of individuals due to emigration rather than death. However, emigration has the same effect on the population as mortality in that emigrants will not contribute to the next generation.

4. In cell B27, sum the *k* values for year 1 to generate the *K* value.

In cell B27 enter the formula **=SUM(B28:B32)**.

5. Compute the *k* and *K* values for years 1–10. Save your work.

Copy the formula in cell B27 across columns to column K.

E. Create graphs.

1. Graph *K* and the *k*'s as a function of time. *Which* k *factor appears to "track"* K *the most?*

Use the line graph option and label your axes fully. Your graph should resemble Figure 10.

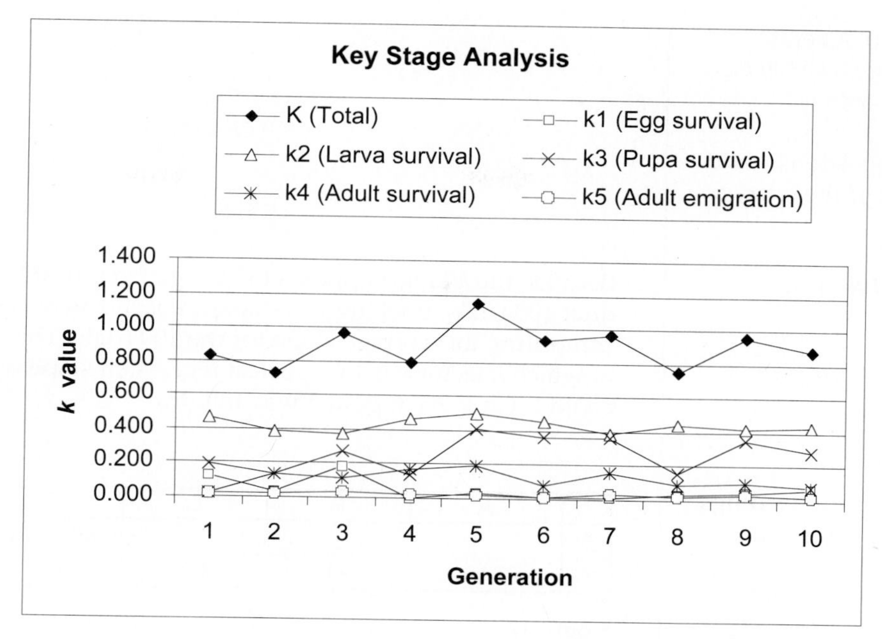

Figure 10

2. Press F9, the calculate key, to simulate new conditions over time. *Does your key factor appear to change?*

3. For each *k*, construct a scatter graph that plots *k* against *K*. Add trendlines (slope) for each graph.

Add trendlines by selecting the chart; then go to the Chart menu, select Add Trendline, and add a Linear trendline. Then click on Options | Display equation on chart. Your graph should look something like Figure 11.

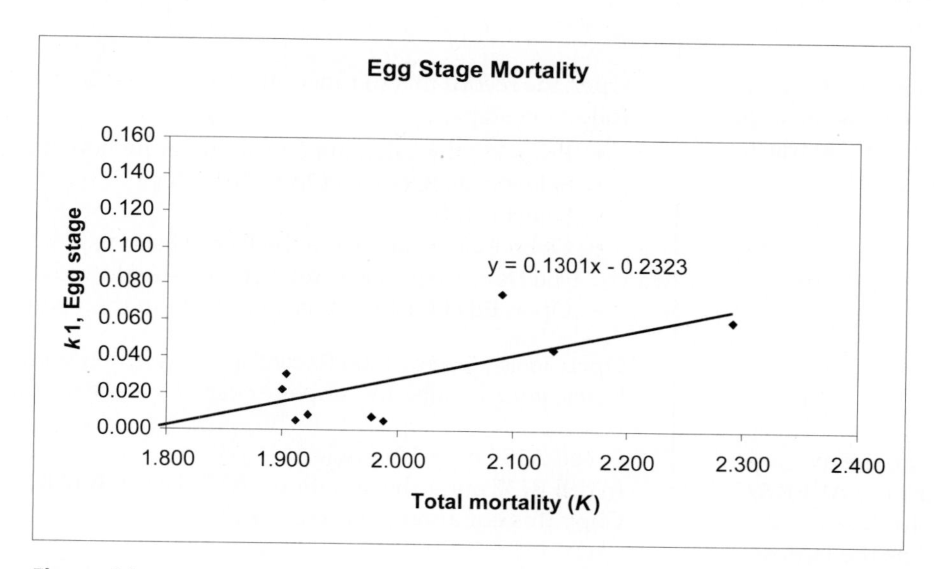

Figure 11

Compare the slopes of *K* versus *k* for each stage. The *k* value that has the greatest slope is the key factor.

4. Press F9 to generate new data. *Does the key factor appear to change?*

5. Answer questions 1 and 2 at the end of the exercise.

F. Conduct 100 trials.

Because the k factor appears to change from trial to trial, it would be useful to conduct 100 trials, tracking the slopes of each k versus K regression equation, and then computing the average slope for the 100 trials. This will give you a better indication of which k factor has the greatest regression slope with K. There are many ways you could do this. A suggested way follows.

1. Set up column headings as shown in Figure 12.

	A	B	C	D	E	F
34				Regression slopes		
35	Trial	k1	k2	k3	k4	k5

Figure 12

2. In cell B36, use the **SLOPE** function to compute the slope of the regression between k_1 and K.

In cell B36 enter the equation **=SLOPE(B28:K28,B27:K27)**. Your answer should match the slope displayed on your graph that is analogous to Figure 11.

3. In cells C36–F36, compute the slopes of other k regressions.

- Cell C36 **=SLOPE(B29:K29,B27:K27**
- Cell D36 **=SLOPE(B30:K30,B27:K27**
- Cell E36 **=SLOPE(B31:K31,B27:K27**
- Cell F36 **=SLOPE(B32:K32,B27:K27**

4. Set up a linear series from 1 to 100 in cells A37–A136.

5. Write a macro to track k versus K regression slopes and run it for 100 trials. Save your work!

Open the record macro function (see Exercise 2). Assign a shortcut key, then record the following steps:

- Press F9, the calculate key, to generate new data, and hence new slopes.
- Select cells B36–F36. Open Edit | Copy.
- Select cell B35.
- Open Edit | Find. Leave the Find What box blank and search by columns. Select Find Next, then Close. Your cursor should move down to cell B37.
- Open Edit | Paste Special and paste in the values.

Open Tools | Macro | Stop Recording. Now when your press your shortcut key 99 more times, your results (the slopes of each k versus K) will be recorded for each trial.

6. Compute the average slope with the **AVERAGE** function to determine which k has the largest slope with K. This is the key stage.

In cell A137 type "Average."
In cell B137 enter the equation **=AVERAGE(B37:B136)**.
Copy this equation over to cell F137.

QUESTIONS

1. Fully interpret the *k* factors in your figures. Which factor appears to by the key factor in your model?
2. Press F9 to generate new sets of data, and inspect your plot of *k*'s and *K* over generations. Does your key factor change with new simulations?
3. Compute the average of the regression slope estimates from your 100 trials. Which *k* factor has the highest regression coefficient when regressed against *K*?
4. Based on the original population variables, and assuming our hypothetical insect population is endangered, did the key factor analysis assist you in developing management recommendations? If so, how?
5. Change the parameter values in cells B5–H6 so that the standard deviation of all parameters is 0.001 (little variation over generations). Clear your macro results (cells B37–F135) and run your macro again. When the parameters do not vary from generation to generation, which stage is the key factor?
6. Change the parameter values in cells B5–H6 so that all survival probabilities equal 0.7. Increase one of the standard deviations (e.g., cell D6) to 0.1. Clear your macro results (cells B37–F135) and run your macro again. When the parameters are equal but one stage is variable, which stage is the key factor?

	A	B	C	D	E	F	G	H
2		Population variables						
3		Eggs	Egg	Larva	Pupa	Adult	Adult	Number of
4		laid	survival	survival	survival	survival	fidelity	females
5	Mean =	100	0.70	0.70	0.70	0.70	0.70	1000.00
6	Standard deviation =	30.00	0.01	0.10	0.01	0.01	0.01	

LITERATURE CITED

Brown, D., N. D. E. Alexander, R. W. Marrs and S. Albon. 1993. Structured accounting of the variance of demographic change. *Journal of Animal Ecology* 62: 490–502.

Krebs, C. J. 1999. *Ecological Methodology*. Addison-Wesley, New York.

Morris, R. F. 1959. Single factor analysis in population dynamics. *Ecology* 40: 580–588.

Royama, T. 1996. A fundamental problem in key factor analysis. *Ecology* 77: 87–93.

Sibly, R. M. and R. H. Smith. 1998. Identifying key factors using λ contribution analysis. *Journal of Animal Ecology* 67: 17–24.

Varly, G. C., and G. R. Gradwall. 1960. Key factors in population studies. *Journal of Animal Ecology* 29: 399–401.

19 SENSITIVITY AND ELASTICITY ANALYSES

Objectives

- Using the stage-based matrix model for a sea turtle population, conduct a sensitivity analysis of model parameters to determine the absolute contribution of each demographic parameter to population growth rate.
- Conduct an elasticity analysis on model parameters to determine the relative contribution of each demographic parameter to population growth rate.
- Interpret the meaning of the sensitivity and elasticity analyses from a conservation and management perspective.

Prerequisite Exercise: Stage-Structured Matrix Models
Suggested Preliminary Exercises: Reproductive Value Exercises

INTRODUCTION

Let's imagine that you are a biologist working for an international conservation organization, and that your task is to suggest the best ways to manage the population of an endangered marine reptile, the sea turtle *Caretta caretta*. You have already constructed a stage-based matrix model (Exercise 14) for the population, and you want to manage it so that population growth, λ, increases. You know that the sea turtle has a complex life cycle, and that individuals can be classified into 1 of 5 classes: hatchlings (h), small juveniles (sj), large juveniles (lj), subadults (sa), and adults (a). Individuals in each class have a specific probability of surviving; they can *either*: (1) survive and remain in the same class, denoted by the letter P followed by two identical subscripts (i.e., the probability that a small juvenile remains a small juvenile in the next year is $P_{sj,sj}$); (2) survive and move into the next group, denoted by the letter P followed by two different subscripts (the probability that a small juvenile will become a large juvenile in the next year is $P_{sj,lj}$); or (3) die, thus exiting the population. Only subadults and adults can breed, and the letter F_i denotes their fertilities. In this population, turtles are counted every year with a postbreeding census. The matrix for this population (Crowder et al. 1994) has the following form:

$$\mathbf{L} = \begin{bmatrix} P_{h,h} & F_{sj} & F_{lj} & F_{sa} & F_a \\ P_{h,sj} & P_{sj,sj} & 0 & 0 & 0 \\ 0 & P_{sj,lj} & P_{lj,lj} & 0 & 0 \\ 0 & 0 & P_{lj,sa} & P_{sa,sa} & 0 \\ 0 & 0 & 0 & P_{sa,a} & P_{a,a} \end{bmatrix}$$

Given the above **L** matrix, the population reaches a stable stage distribution with all stage classes declining by 5% per year, or $\lambda = 0.95$. Your task is to suggest the best ways to manage the turtle population to increase the long-term asymptotic λ, and hence increase the population size. But λ can be increased in a variety of ways! Should you focus your efforts on increasing adult fertility? Should you focus your efforts on increasing the probability that hatchlings in year t will become small juveniles in year $t + 1$? Or should you focus on increasing survivorship of adults? Finances and resources are limited, so it is not likely that you can do all these things at once.

In this exercise, you will extend the stage-based model you developed for *Caretta caretta* to conduct a **sensitivity** and/or **elasticity** analysis of each model parameter. These analyses will tell you how λ, population size, and the stable distribution might change as we alter the values of F_i and P_i in the **L** matrix.

Sensitivity Analyses

Sensitivity analysis reveals how very small changes in each F_i and P_i will affect λ when the other elements in the **L** matrix are held constant. These analyses are important from several perspectives. From a conservation and management perspective, sensitivity analysis can help you identify the life-history stage that will contribute the most to population growth of a species. From an evolutionary perspective, such an analysis can help identify the life-history attribute that contributes most to an organism's fitness.

Conducting sensitivity analysis requires some basic knowledge of matrix algebra. While we will not delve into matrix formulations in detail here (see Caswell 2001), we will very briefly overview the concepts associated with sensitivity analysis. In the stage-based matrix models you developed earlier, you projected population size from time t to time $t + 1$ by multiplying the **L** matrix by a vector of abundance, **n**, at time t. (Remember that uppercase boldface letters indicate a matrix, and lowercase boldface letters indicate a vector.) The result was a vector of abundances, **n**, at time $t + 1$:

$$\mathbf{n}(t+1) = \mathbf{L} \times \mathbf{n}(t) \qquad \text{Equation 1}$$

After attaining the new vector of abundances, you repeated the process for the next time step and attained yet another vector of abundances. When the process was repeated over many time steps, eventually the system reached a stable stage distribution, where λ_t remained constant from one time step to the next. This stabilized λ_t is called the long-term or **asymptotic population growth rate**, λ. In the sea turtle exercise, the population stabilized within 100 years. If $\lambda > 1$, the numbers of individuals in the population increase geometrically; if $\lambda < 1$, the numbers of individuals in the population decline geometrically; and when $\lambda = 1$, the numbers of individuals in the population remain constant in numbers over time. Since $\lambda = 0.95$ for the sea turtle population, number of individuals in the population decreases geometrically at 5% per time step. Graphically, the point in time in which the population reaches a stable stage distribution is the point where the population growth lines for *each* class become parallel (Figure 1). When λ_t has stabilized, the population can be described in terms of the *proportion* of each stage in the total population. When the population stabilizes, these proportions remain constant regardless of the value of λ.

Thus, given a matrix, **L**, you can determine the stable stage distribution of individuals among the different classes, and the value of λ at this point. The value of λ when the population has stabilized is called an **eigenvalue** of the matrix. An **eigenvalue** is a

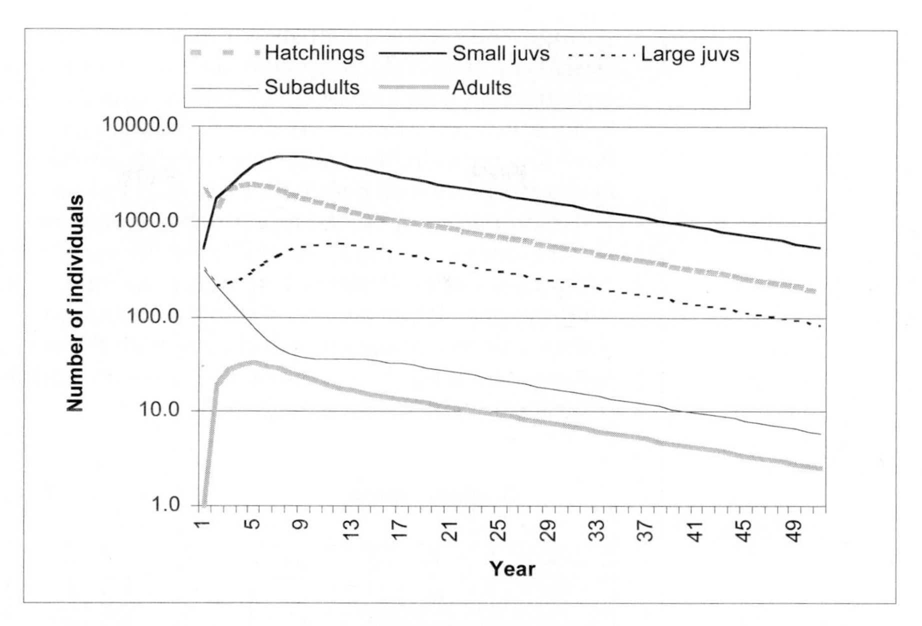

Figure 1 The stage distribution of a population becomes stable when changes in numbers over time for each growth stage are parallel, regardless of the value of λ. At this point the proportion of each stage in the population remains the same into the future.

number (numbers in matrix algebra are called scalars) that, when multiplied by a vector of abundances, yields the same result as the **L** matrix multiplied by the same vector of abundances. For example, if λ is 1.15, the numbers of individuals in each class will increase by 15% from time step t to time step $t + 1$. If λ instead is 0.97, the numbers of individuals in each class will decrease by 3% from time step t to time step $t + 1$.

In order to conduct a sensitivity analysis on the parameters in the **L** matrix, we need to determine the stable-stage distribution of the population. For sea turtles, this was 23.9% hatchlings, 64.8% small juveniles, 10.3% large juveniles, 0.7% subadults, and 0.3% adults. We can convert these percentages into the proportions 0.239, 0.648, 0.103, 0.007, and 0.003. This vector of proportions is called a **right eigenvector** of the **L** matrix. The right eigenvector is represented by the symbol **w**. The **w** vector for the sea turtle population can be written as a column vector, where the first entry gives the proportion of the stabilized population that consists of hatchlings, and the last entry gives the proportion of the stabilized population that consists of adults:

$$\mathbf{w} = \begin{bmatrix} .239 \\ .648 \\ .103 \\ .007 \\ .003 \end{bmatrix}$$

Note that the values sum to 1.

The final piece of information needed for compute sensitivities for the values of F_i and P_i in the **L** matrix is the **left eigenvector**, represented by the symbol **v**. The left eigenvector of the **L** matrix reveals the reproductive value for each class in the model. If you have completed the exercises on reproductive value, you know that reproductive value computes the "worth" of individuals of different classes (age, stage, or size) in terms of future offspring it is destined to contribute to the next generation, adjusted for the growth rate of the population (Fisher 1930). As Caswell (2001) states, "The amount of future reproduction, the probability of surviving to realize it, and the time required for the offspring to be produced all enter into the reproductive value of a given age or stage class. Typical reproductive values are low at birth, increase to a peak near the age of first reproduction, and then decline." Individuals that are postreproductive have a value of 0, since their contribution to future population growth is 0. Sea turtle newborns also may have low reproductive value because they probably have several years of living (and hence mortality risk) before they can start producing offspring

We need to compute the reproductive values for each class in order to conduct a sensitivity analysis of the F_i's and P_i's for the sea turtle population. The simplest way to compute **v** for the **L** matrix is to transpose the **L** matrix, called **L′**, then run the model until the population reaches a stable distribution, and then record the proportions of individuals that make up each class as with the **w** vector. Transposing a matrix simply means switching the columns and rows around: Make the rows columns and the columns rows, as shown in Figure 2.

Original matrix				Transposed matrix		
A	B	C		A	D	G
D	E	F		B	E	H
G	H	I		C	F	I

Figure 2

When λ is computed for the transposed matrix **L′**, the right eigenvector of **L′** gives the reproductive values for each class. This same vector is called the left eigenvector for the original matrix, **L**. (Yes, it is confusing!) The **v** vector for the sea turtle population is written as a row vector:

$$\mathbf{v} = [.002 \quad .003 \quad .013 \quad .207 \quad .776]$$

This vector gives, in order, the reproductive values of hatchlings, small juveniles, large juveniles, subadults, and adults. In this population, adults have the greatest reproductive value, followed by subadults. Large juveniles, small juveniles, and hatchlings have very small reproductive values. Oftentimes the reproductive value is standardized so that the first stage or age class has a reproductive value of 1. We can standardize the **v** vector above by dividing each entry by 0.002 (the reproductive value of hatchlings) to generate standardized reproductive values. Our standardized vector would look like this:

$$\mathbf{v} = \left[\frac{.002}{.002} \quad \frac{.003}{.002} \quad \frac{.013}{.002} \quad \frac{.207}{.002} \quad \frac{.776}{.002}\right] = [1 \quad 1.4 \quad 7.5 \quad 115.6 \quad 434.4]$$

Thus, an adult individual is 434.4 times more "valuable" to the population in terms of future, adjusted offspring production than a single hatchling.

Computing Sensitivities

Now we are ready to explore how the sensitivities of each P_i and F_i in the **L** matrix are computed. Remember that sensitivity analyses reveal how very small changes in each F_i and P_i will affect λ when the other elements in the **L** matrix are held constant. The steps for conducting a sensitivity analysis include: (1) running the projection model until the population reaches a stable distribution, (2) calculating the stable stage structure of the population, which is given by the vector **w**, and (3) calculating the reproductive values for the different size classes, which is given by the vector **v**. The sensitivity, s_{ij}, of an element in the **L** matrix, a_{ij}, is given by

$$s_{ij} = \frac{v_i w_j}{<\mathbf{w},\mathbf{v}>} \qquad \text{Equation 2}$$

where v_i is the ith element of the reproductive value vector, w_j is the jth element of the stable stage vector, and $<\mathbf{w},\mathbf{v}>$ is the product of the **w** and **v** vectors, which is a single number (a **scalar**). Thus, the sensitivity of λ to changes in a_{ij} is proportional to the product of the ith element of the reproductive value vector and the jth element of the stable stage vector (Caswell 2001). You'll see how these calculations are made as you work through the exercise. We can also write Equation 2 as a partial derivative, because all but one of the variables of which λ is a function are being held constant:

$$s_{ij} = \frac{\partial \lambda}{\partial a_{ij}} = \frac{v_i w_j}{<\mathbf{w},\mathbf{v}>} \quad \text{Equation 3}$$

How are the s_{ij}'s to be interpreted? A sensitivity analysis, for example, on the $P_{a,a}$ and F_{sa} might yield values of 0.1499 and 0.2287, respectively. These values answer the question, "If we change P_{aa} by a small amount in the **L** matrix and hold the remaining matrix entries constant, what is the corresponding change in λ?" The sensitivity of the P_{aa} matrix entry means, for example, that a small unit change in P_{aa} results in a change in λ by a factor of 0.1499. In other words, *sensitivity is represented as a slope.*

The most sensitive matrix elements produce the largest slopes, or the largest changes in the asymptotic growth rate λ. In our example above, where sensitivities were 0.1499 for the P_{aa} entry and 0.2287 for the F_{sa} entry, small changes in adult survival will not have as large an effect as changes in subadult fertility in terms of increasing growth, so you would recommend management efforts that aim to increase subadult fertility values.

Elasticity Analysis

One challenge in interpreting sensitivities is that demographic variables are measured in different units. Survival rates are probabilities and they can only take values between 0 and 1. Fertility, on the other hand, has no such restrictions. Therefore, the sensitivity of λ to changes in survival rates may be difficult to compare with the sensitivities of fertility rates. This is where elasticities come into play. Elasticity analysis estimates the effect of a *proportional* change in the vital rates on population growth. The elasticity of a matrix element, e_{ij}, is the product of the sensitivity of a matrix element (s_{ij}) and the matrix element itself (a_{ij}), divided by λ. In essence, elasticities are proportional sensitivities, scaled so that they are dimensionless:

$$e_{ij} = \frac{a_{ij} s_{ij}}{\lambda} \quad \text{Equation 4}$$

Thus, you can directly compare elasticities among all life history variables. An elasticity analysis, for example, on the parameters hatchling survival and adult fecundity might yield values of 0.047 and 0.538, respectively. This means that a 1% increase in hatchling survival will cause 0.047 % increase in λ, while a 1% increase in adult fecundity will cause a 0.538% increase in λ. In this situation, you would recommend management efforts that aim to increase adult fecundity values.

PROCEDURES

The goal of this exercise is to introduce you to matrix methods of computing sensitivities and elasticities for the vital population parameters, P and F, for a population with stage structure. As always, save your work frequently to disk.

INSTRUCTIONS	ANNOTATION

A. Set up the spreadsheet.

1. Open the stage-based matrix model you created in Exercise 14 and save it under a new name. Retitle cell A1 to "Sensitivity and Elasticity Analysis."

Your spreadsheet headings should resemble Figure 3.

	A	B	C	D	E	F	G	H
1	*Sensitivity and Elasticity Analysis*							
2	Loggerhead Sea Turtle Population							Initial population
3		*F*(*h*)	*F*(*sj*)	*F*(*lj*)	*F*(*sa*)	*F*(*a*)		vector
4	Hatchlings:	0	0	0	4.665	61.896		2000
5	Small juveniles:	0.675	0.703	0	0	0		500
6	Large juveniles:	0	0.047	0.657	0	0		300
7	Subadults:	0	0	0.019	0.682	0		300
8	Adults:	0	0	0	0.061	0.8091		1
9								
10	Year	Hatchlings	Small juvs	Large juvs	Subadults	Adults	Total	λ_t

Figure 3

2. Enter the values shown in cells B4–F8. (You may have changed these values in your previous exercise).

B. Calculate w, the stable-stage vector.

The stable stage distribution vector, **w**, is simply the proportion of individuals in the population that is made up of the different stage classes.

1. Set up new column headings as shown in Figure 4.

	X
3	**Stable stage distribution**
4	**vector, *w***
5	
6	
7	
8	
9	

Figure 4

The first entry, cell X5, is the proportion of the population that is made up of hatchlings (given that the population has reached a stable distribution). The second entry, cell X6, is the proportion of the population that is made up of small juveniles. Cells X7 and X8 will contain the proportions of large juveniles and subadults, and the last entry, cell X9, will contain the proportions of adults.

2. In cell X5, calculate the proportion of total population in year 100 that consists of hatchlings.

Enter the formula **=B111/$G$111** in cell X5.
In Exercise 14, you calculated the number of individuals in each class when the population has stabilized (remains constant over time). You might recall that the population stabilized at $\lambda = 0.95$, and that the stable population consists of 16.22 hatchlings, 44.05 small juveniles, 7.03 large juveniles, 0.50 subadults, and 0.21 adults. To calculate the **w** vector, we need to present these numbers in terms of proportions of the total population size. Rather than entering these values by hand, the above formula references the *proportion* of hatchlings listed in the last year of the projection.

3. In cells X6-X9, compute the proportions in the remaining classes.

We entered the formulae

- X6 =**C111/$G$111**
- X7 =**D111/$G$111**
- X8 =**E111/$G$111**
- X9 =**F111/$G$111**

These equations assume the population has stabilized by year 100.

4. Save your work. Your spreadsheet should now resemble Figure 5.

	X
3	**Stable stage distribution**
4	**vector, *w***
5	0.239
6	0.648
7	0.103
8	0.007
9	0.003

Figure 5

C. Calculate v, the reproductive value vector.

The **v** vector gives the reproductive values for members in different stages of the population. The easiest way to do this is to transpose your original population matrix, then run the same type of analysis you ran to determine the **w** vector. *Transposing* a matrix simply means you interchange the rows and columns.

1. Set up new column headings as shown in Figure 6. Enter only the headings for now.

	I	J	K	L	M	N	O	P
1								
2			**Reproductive value: transposed matrix**					
3								
4		*F*(*h*)						
5		*F*(*sj*)						
6		*F*(*lj*)						
7		*F*(*sa*)						
8		*F*(*a*)						
9								
10	**Year**	**Hatchlings**	**Small juvs**	**Large juvs**	**Subadults**	**Adults**	**Total**	λ_t

Figure 6

2. Use the **TRANSPOSE** function to transpose the original matrix, given in cells B4–F8, into cells K4–O8. Your spreadsheet should resemble Figure 7.

The **TRANSPOSE** function in Excel is an array function. The mechanics of entering an array formula are a bit different than the typical (single cell) formula entry. Instead of selecting a single cell to enter a formula, you need to select a series of cells, then enter a formula, then press <Control>+<Shift>+<Enter> (Windows machines) to enter the formula for all of the cells you have selected. This function works best when you use the f_x key and follow the cues for entering a formula.

Select cells K4–O8 with your mouse, then use your f_x key to select Transpose. A dialog box will appear asking you to define an array that you wish to transpose. Use your mouse to highlight cells B4–F8, or enter this by hand. Instead of clicking OK, press <Control>+<Shift>+<Enter>, and the spreadsheet will return your transposed matrix. After you've obtained your results, examine the formulae in cells K4–O8. Your formula should look like this: **{=TRANSPOSE(B4:F8)}**. The { } symbols indicate that the formula is part of an array. If for some reason you get "stuck" in an array formula, press the Escape key and start over.

	J	K	L	M	N	O
2		Reproductive value: transposed matrix				
3						
4	*F*(*h*)	0	0.675	0	0	0
5	*F*(*sj*)	0	0.703	0.047	0	0
6	*F*(*lj*)	0	0	0.657	0.019	0
7	*F*(*sa*)	4.665	0	0	0.682	0.061
8	*F*(*a*)	61.896	0	0	0	0.8091

Figure 7

3. Set up a linear series from 0 to 100 in cells I11–I111.

Enter 0 in cell I11.
Enter **=1+I11** in cell I12. Copy this formula down to cell I111.

4. Link the starting number of individuals of each class in year 0 to the original vector of abundances in cells H4–H8.

You'll need to stick with the same initial population vector of abundances you used earlier in the exercise. We used the following formulae:

- J11 **=H4**
- K11 **=H5**
- L11 **=H6**
- M11 **=H7**
- N11 **=H8**

5. In cell O11, compute the total number of individuals in year 0.

Enter the formula **=SUM(J11:N11)** in cell O11.

6. In cell P11, enter a formula to compute λ_t for year 0.

Enter the formula **=O12/O11** in cell P11.

7. Project the population over time as you did in your turtle matrix model, using the values from the transposed matrix for your calculations.

We used the following formulae:

- J12 **=$K$4*J11+$L$4*K11+$M$4*L11+$N$4*M11+$O$4*N11**
- K12 **=$K$5*J11+$L$5*K11+$M$5*L11+$N$5*M11+$O$5*N11**
- L12 **=$K$6*J11+$L$6*K11+$M$6*L11+$N$6*M11+$O$6*N11**
- M12 **=$K$7*J11+$L$7*K11+$M$7*L11+$N$7*M11+$O$7*N11**
- N12 **=$K$8*J11+$L$8*K11+$M$8*L11+$N$8*M11+$O$8*N11**
- O12 **=SUM(J12:N12)**
- P12 **=O13/O12**

8. Compute λ_t for Year 1. Copy cells J12–P12 down to row 111 to complete the projection.

You should see that λ_t stabilizes at the same value it did for your original projections.

9. Set up new column headings as shown in Figure 8.

	Q	R	S	T	U	V	W
3				Small	Large		
4			Hatchlings	juveniles	juveniles	Subadults	Adults
5		*v* = reproductive value vector =					
6		Standardized reproductive value =					

Figure 8

10. In cell S5 enter a formula to compute the reproductive value of the hatchling stage.

Enter the formula **`=J111/$O$111`** in cell S5.
As you did in computing the **w** vector, enter formula in these cells to reference the proportions listed in the last year of the projection. Thus, cell S5 gives the proportion of "hatchlings" in Year 100.

11. In cells T5–W5, enter formulae to compute the reproductive value of the remaining stages.

We entered the following formulae:

- T5 **`=K111/$O$111`**
- U5 **`=L111/$O$111`**
- V5 **`=M111/$O$111`**
- W5 **`=N111/$O$111`**

12. Double-check your work.

Cells S5–W5 should sum to 1.

13. In cells S6–W6, calculate the *standardized reproductive value* for each stage class.

Enter the formula **`=S5/$S$5`** in cell S6. Copy this formula across to cell W6. Reproductive values are often standardized such that the reproductive value of the first class (hatchlings) is 1. To standardize the reproductive values, divide each value by the value obtained for hatchlings. Your spreadsheet should now resemble Figure 9.

	Q	R	S	T	U	V	W
3				Small	Large		
4			Hatchlings	juveniles	juveniles	Subadults	Adults
5		*v* = reproductive value vector =	0.0018	0.0025	0.0133	0.2065	0.7759
6		Standardized reproductive value =	1.0	1.4	7.5	115.6	434.4

Figure 9

14. Save your work.

D. Calculate sensitivities of matrix parameters.

Now that you have calculated the **w** and **v** vectors, you are ready to perform a sensitivity analysis.

1. Set up new column headings as shown in Figure 10. Enter only the headings (literals) for now.

	R	S	T	U	V	W
7						
8	*X* = <*w*,*v*> =					
9						
10		Sensitivity matrix				
11			*F*(*sj*)	*F*(*lj*)	*F*(*sa*)	*F*(*a*)
12	Hatchlings					
13	Small juveniles					
14	Large juveniles					
15	Subadults					
16	Adults					
17						
18						
19		Elasticity matrix				
20		*F*(*h*)	*F*(*sj*)	*F*(*lj*)	*F*(*sa*)	*F*(*a*)
21	Hatchlings					
22	Small juveniles					
23	Large juveniles					
24	Subadults					
25	Adults					
26						

Figure 10

2. In cell S8, use the **MMULT** (matrix multiplication) function to multiply the **v** vector by the **w** vector.

Enter the formula **=MMULT(S5:W5,X5:X9)** in cell S8.
The **MMULT** function returns the matrix product of two arrays. The result is an array with the same number of rows as array 1 and the same number of columns as array 2. In our case, it ends up being a single digit (since our **v** vector consists of one row and our **w** vector consists of one column). This value is the denominator **<w,v>** of the formula for calculating sensitivity values (Equation 3). This single-digit result is called a scalar; for purposes of the spreadsheet, we will call this value X.

Now you are ready to calculate the numerator of the sensitivities, and compute the sensitivity values for each entry in your matrix. Note that sensitivities are computed for all matrix entries, even those that are 0 in the original **L** matrix. For example, you will compute the sensitivity of subadult fertility ($F_{sa,h}$) even though subadults cannot reproduce. This sensitivity value will allow you to answer, "If I could make subadults reproduce, it would increase λ at this rate. You may wish to shade the **L** matrix entries that have original cell entries that are equal to 0 a different color (as shown in Step 1).

3. In cell S12–W12, enter formulae to compute the sensitivity of fertility rates for each stage over time.

Sensitivity of a population growth rate to changes in the a_{ij} element is simply the *i*th entry of **v** times the *j*th entry of **w**, divided by X. For example, to calculate the sensitivity of fertility rate of subadults (row 1, column 4), we would multiply the first element in the **v** vector by the fourth element in the **w** vector, and then divide that number by X. The formula in cell V12 would be **=(X8*S5)/S8**. Enter formula in the remainder of the sensitivity matrix. Below are the formulae we used (note that we used absolute references for some cell addresses).

- S12 **=($X$5*S5)/$S$8**
- T12 **=($X$6*S5)/$S$8**
- U12 **=($X$7*S5)/$S$8**
- V12 **=($X$8*S5)/$S$8**
- W12 **=($X$9*S5)/$S$8**

4. Copy cells S12–W12 down to cells S16–W16. Save your work.

Adjust your formulae in the formula bar to reference the appropriate cells in the v and w vectors. For example, in row 13, replace the reference to cell 56 with T5. In row 14, replace the reference to cell S7 with V5, etc. This completes the sensitivity analysis.

E. Calculate elasticities of matrix parameters.

1. In cell S21–W21, enter formulae to calculate the elasticity values for fertility at each stage for year 0.

Enter the formula **=(B4*S12)/$H$110** in cell S21. Copy this formula across to cell W21. The elasticity of a_{ij} is the sensitivity of a_{ij} times the value of a_{ij} in the original matrix, divided by λ when λ_t has stabilized. For example, the elasticity calculation of fecundities of the subadults would be **=(E4*V12)/$H$110**. If the original matrix element was a 0 (such as the fecundities of the hatchling stage), the elasticity should be 0.

2. Copy the formulae over the remaining years of the analysis.

Copy the formulae in cells S21–W21 down to cells S25–W25. This will complete the elasticity analysis. The sum of the elasticities should add to be 1, since each elasticity value measures the proportional contribution of each element to λ (yours might be off by a bit due to rounding error).

3. Save your work.

F. Create graphs.

1. Graph the elasticity values for fertility of the various stage classes.

Use a column graph and label your axes fully. Your graph should resemble Figure 11.

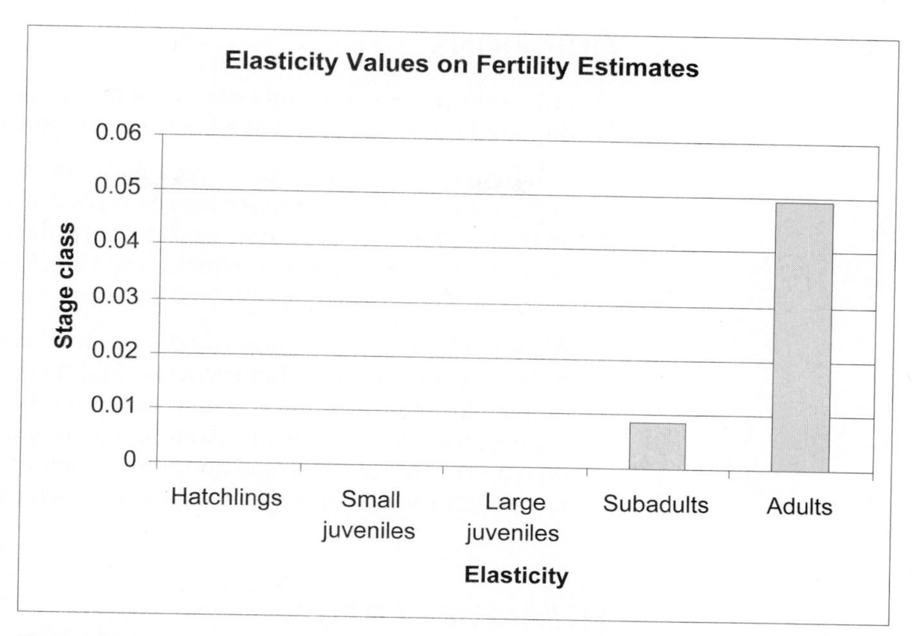

Figure 11

2. Graph the elasticity values for the survival values, $P_{i,i}$ and $P_{i,i+1}$ for each stage class.

You will have to manually select bars within the graph and color-code them to reflect within-stage survival ($P_{i,i}$) or survival to the next stage ($P_{i,i+1}$). Your graph should resemble Figure 12.

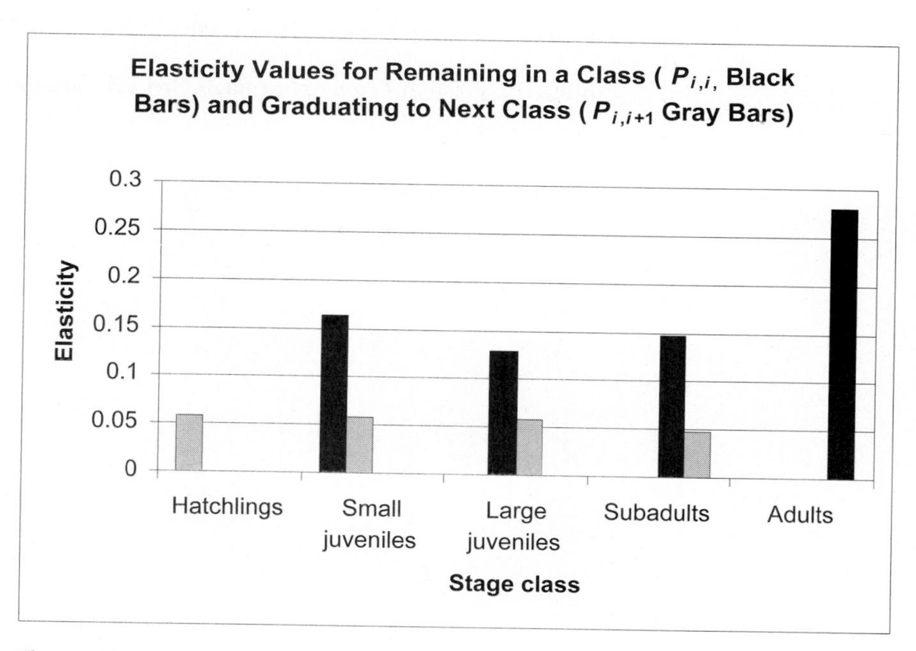

Figure 12

3. Save your work.

QUESTIONS

1. Fully interpret the meaning of your sensitivity analysis. What management recommendations can you make for sea turtle conservation given your analysis?
2. Fully interpret the meaning of your elasticity analysis. What management recommendations can you make for sea turtle conservation given your elasticity analysis? Would your recommendations be different if you simply examined the sensitivies, and ignored elasticities? Which do you think is more appropriate for guiding management decisions?
3. As with all models in ecology and evolution, elasticity and sensitivity analyses have their assumptions (and weaknesses). Let's say you make some recommendations for sea turtle conservation based on the matrix parameters provided in the exercise. What kinds of assumptions are implicit in the model parameters? (What do you need to know about how the data were collected and the environmental and biological conditions in which the data were collected?)

LITERATURE CITED

Caswell, H. 2001. *Matrix Population Models*, 2nd Ed. Sinauer Associates, Sunderland, MA.

Crowder, L. B., D. T. Crouse, S. S. Heppell and T. H. Martin. 1994. Predicting the impact of turtle excluder devices on loggerhead sea turtle populations. *Ecological Applications* 4: 437–445.

Fisher, R. A. 1930. *The Genetical Theory of Natural Selection*. Clarendon Press, Oxford.

Gotelli, N. 2001. *A Primer of Ecology*, 3rd Ed. Sinauer Associates, Sunderland, MA.

20 METAPOPULATION DYNAMICS

Objectives

- Determine how extinction and colonization parameters influence metapopulation dynamics.
- Determine how the number of patches in a system affects the probability of local extinction and probability of regional extinction.
- Compare "propagule rain" versus "internal colonization" metapopulation dynamics.
- Evaluate how the "rescue effect" affects metapopulation dynamics.

INTRODUCTION

Can you think of any species where the entire population is situated within one patch, where all individuals potentially interact with each other? You will probably be hard pressed to come up with more than a few examples. Most species have distributions that are discontinuous at some spatial scale. In some species, subdivided populations may be linked to each other when individuals disperse from one location to another. For example, butterflies may progress from egg to larvae to pupa to adult on one patch, then disperse to other patches in search of mates, linking the population on one patch to a population on another. This "population of populations" is often called a **metapopulation**, and in this exercise we will explore the dynamics of such interacting systems.

Metapopulation theory was first formalized by Richard Levins in 1969 (Levins 1969, 1970). In Levins' model, a metapopulation exists in a network of habitat patches, some occupied and some unoccupied by subpopulations of individuals. The dynamics of metapopulations can be explored by examining patch occupancy patterns over time. In the left-hand side of Figure 1, the 100 squares represent 100 patches in a metapopulation at time t. The right-hand side of the figure shows the pattern of patch occupancy at time $t + 1$.

In the traditional metapopulation model (Levins 1970), each subpopulation has a finite lifetime and each subpopulation has the same probability of extinction. Additionally, all unoccupied patches have the same probability of being colonized. At equilibrium, the proportion of patches that are occupied remains constant, although the pattern of occupancy continually shifts as some subpopulations suffer extinction

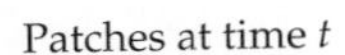

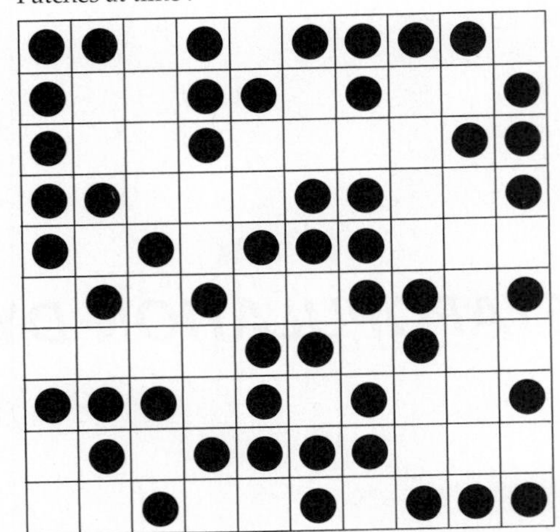

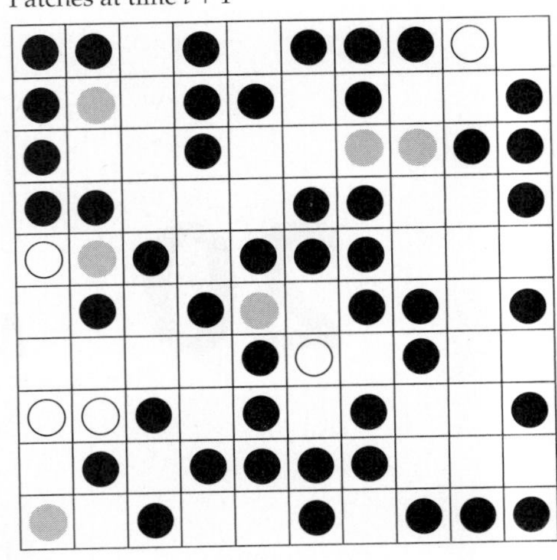

Figure 1 At time t, occupied habitat patches are represented with filled circles; empty squares represent currently unoccupied patches. At time t + 1, some of the patches that were occupied in time t are vacant (open circles), some patches that were vacant at time t are now occupied (gray circles), and some patches maintain their "occupancy status" from time t to time t + 1 (filled circles).

followed by recolonization. This is sometimes referred to as the "winking" nature of metapopulations, as newly colonized patches "wink in" and extirpated patches "wink out." Thus, the classic metapopulation model (sensu Levins 1970) is a "presence-absence" model that examines whether a population is present or absent on a given patch over time, how presence and absence changes over time, and how the entire metapopulation system can persist. In other words, metapopulation models explain and predict the distribution of occupied and unoccupied habitat patches, factors that affect dispersal between patches, and the persistence of the greater metapopulation (Hanski and Gilpin 1997).

Metapopulation Dynamics: Colonization and Extinction

Let's begin our exploration of metapopulation dynamics by defining extinction and colonization mathematically. Patches that are currently occupied in the system have a probability of going extinct, p_e, and a probability of persistence, $1 - p_e$. Patches that are currently empty in the system have a probability of being recolonized, p_i, and a probability of remaining vacant, $1 - p_i$. Since both p_e and p_i are probabilities, their values range between 0 and 1.

Metapopulation dynamics focus on the occupancy patterns of patches over time. We can think about the fate of a given patch over the course of time, and additionally we can consider the fate of the entire metapopulation over the course of time. For a given patch, the probability that a patch will persist for n years in a row is simply the probability of persistence, raised to the number of years in consideration (Gotelli 2001).

$$P_n = (1 - p_e)^n \qquad \text{Equation 1}$$

For example, if a patch has a probability of persistence = 0.8, and we are interested in computing the probability of that patch remaining occupied for 3 consecutive years, $P_3 = 0.8^3 = 0.512$. In other words, if we had 100 occupied patches in a metapopulation, approximately 51.2% of the patches would persist over a 3-year period; 48.8% would likely go extinct within that time period.

If we want to consider the fate of the *entire* metapopulation over time, we need to know the extinction probabilities of each patch, and the number of patches in the system. Given this information, we could compute the probability that all patches would go extinct simultaneously, leading to extinction of the entire metapopulation. Assuming that all patches have the same probability of extinction, the probability that the entire metapopulation will go extinct is simply the p_e raised to the number of patches in the system. Thus, when $p_e = 0.5$ and there are 6 patches in the system, the probability that all 6 patches will go extinct simultaneously is $0.5^6 = 0.0156$. Thus there is about a 1.5% chance that the system will go extinct. Similarly, we can compute the probability of metapopulation persistence as the probability of persistence raised to the power of the number of patches in the system.

$$P_x = 1 - (p_e)^x \qquad \text{Equation 2}$$

Now that we know a little bit about extinction and colonization of patches, let's focus on the dynamics of a metapopulation system, or how patch occupancy patterns change over time. The basic metapopulation model has the form

$$\frac{df}{dt} = I - E \qquad \text{Equation 3}$$

where f is the fraction of patches occupied in the system. For example, if our system contained 25 patches, and 5 of them are occupied, $f = 5/25 = 0.2$. By definition, 20/25 patches are vacant. Equation 3 simply states that the (instantaneous) *change* in the fraction of patches that are occupied depends on the rates of immigration (I) to empty sites and the rates of extinction (E) of occupied sites (Gotelli 2001). If you have completed the exercise on exponential growth, this equation has a form that might be familiar to you, but instead of births and deaths (B and D in the exponential growth model), we are now concerned with I and E. Two critical pieces of information determine I, the rate at which empty patches are recolonized: the number of patches that are currently empty and available for recolonization, and p_i, the probability that an empty patch will actually be recolonized. If f is the fraction of patches that are occupied, then $1 - f$ is the fraction of patches that are currently empty, and we can compute I as

$$I = p_i(1 - f)$$

Now let's focus on E, the rate at which currently occupied patches go extinct. E depends on the number of patches that are currently occupied and available for extinction, as well as p_e, the probability that an occupied patch will go extinct. If f is the fraction of patches that are currently occupied, we can compute E as

$$E = p_e f$$

Substituting the above two values for I and E into Equation 3, we now have a general model of metapopulation dynamics:

$$\frac{df}{dt} = p_i(1-f) - p_e f \qquad \text{Equation 4}$$

This model is called a **propagule rain model** or an **island-mainland model**, because the colonization rate does not depend on patch occupancy patterns—it is assumed that colonists are available to populate an empty patch and that these colonists can originate from either currently occupied patches or from patches outside the metapopulation system. At equilibrium, the fraction of patches remains constant over time, although patches continually "wink in" and "wink out" of existence. How do we solve for this equilibrium?

To solve for the equilibrium fraction of patches, set the left-hand side of Equation 4 to 0 (which indicates that the system is not changing, and the fraction of patches is therefore constant) and solve for f:

$$0 = p_i - p_i f - p_e f$$

$$f = \frac{p_i}{p_i + p_e} \qquad \text{Equation 5}$$

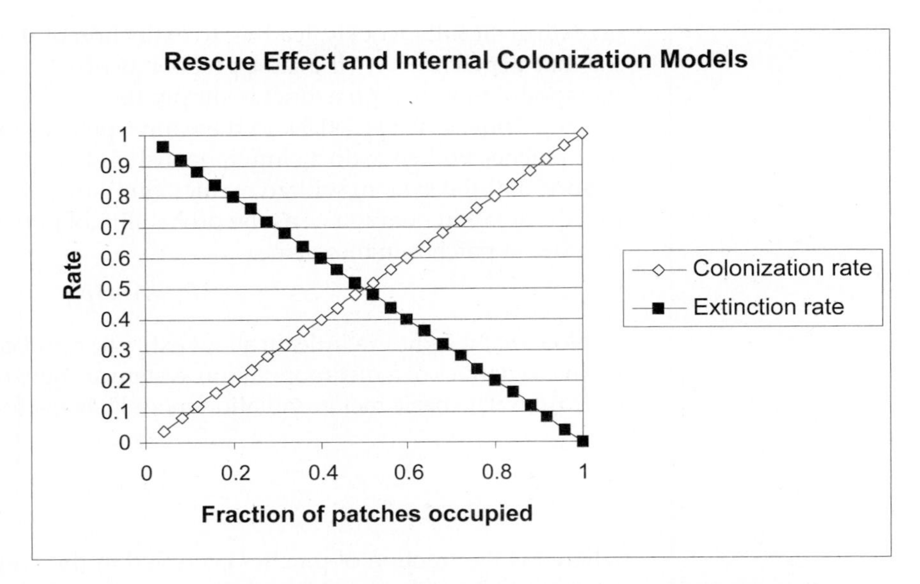

Figure 2 The colonization rate rises as a greater fraction of habitat patches are occupied (the internal coloniation model), whereas extinction rates are higher when fewer habitat patches are occupied (the rescue effect model).

As with all models, the metapopulation model has several assumptions, the most important being that all patches are created equal: p_e and p_i are constant over time and apply to patches regardless of their population size, habitat quality, or other factors. Additionally, this basic model assumes that the explicit location of any patch in relation to other patches is not an important factor in p_e or p_i (Gotelli 2001).

Clearly, some of these assumptions are violated in natural populations, where p_e and p_i are not independent of f, the fraction of patches in the metapopulation that are occupied. For example, colonization of an empty patch may be more likely when f is high than when f is low. When f is high, potentially more colonists are available to recolonize a vacant site. When f is low, colonists arise from only a few patches and may not be able to colonize empty patches efficiently. This kind of metapopulation model is often called an **internal colonization model** because colonization rates depend on current status (f) of the metapopulation system.

Similarly, extinction of a patch may depend on the fraction of patches occupied in the metapopulation system. When f is high, there are many potential colonists available to keep a patch from going extinct; when f is low, there are fewer potential colonists, and risk of extinction increases. This kind of metapopulation model is often called a **rescue effect model** because extinction rates depend on the current status (f) of the metapopulation system. Graphically, the "adjusted" colonization and extinction rates may be proportionally related to the fraction of patches occupied (Figure 2), although the exact relationship between rates and fraction of patches can take a variety of forms.

PROCEDURES

The metapopulation concept has become an important paradigm in conservation biology in recent years, and it is worth exploring some of its assumptions and predictions. In this exercise, you will develop a spreadsheet model of metapopulation dynamics. We will expand the model and explore the internal colonization and rescue effect models in the Questions section. As always, save your work frequently to disk.

INSTRUCTIONS	ANNOTATION

A. Set up the model.

1. Open a new spreadsheet file and fill in column and row headings as shown in Figure 3.

	A	B	C	D	E
1	***Introduction to Metapopulation Dynamics***				
2					
3	**Model parameters:**				
4	**x = number of patches in system**				25
5	**n = number of years under consideration**				10
6	**p_e = probability of local extinction**				0.3
7	**$1 - p_e$ = probability of local persistence**				
8	**p_n = probability of continued local persistence**				
9	**p_i = probability of local colonization**				0.9
10	**P_x = probability of regional extinction**				
11	**$1 - P_x$ = probability of regional persistence**				
12	**f = equilibrium number of patches occupied**				

Figure 3

2. Set up a scenario in which there are 25 habitat patches.

Enter the value 25 in cell E4. (The term *metapopulation* implies that there must be at least 2 habitat patches in the system. To begin, we will consider a system in which there are 25 patches.)

3. Consider what will happen to our metapopulation in the next 10 years.

Enter the value 10 in cell E5.

4. In cell E6, set p_e equal to 0.3.

Enter 0.3 in cell E6. Remember that p_e is the probability of local extinction—that is, the probability that any currently occupied patch in the system will go extinct. The value $p_e = 0.3$ means that any occupied patch has a 30% probability of going extinct. (This cell has been shaded to indicate that its value can be manipulated in the spreadsheet.)

5. In cell E7, enter a formula to caculcate the probability that any given occupied patch will persist.

The probability that any occupied patch will persist (i.e., *not* go extinct) is 1 – E6. Thus you can enter the formula **=1-E6** in cell E7.

6. In cell E8, enter a formula to calculate the probability that a patch will be occupied for 10 straight years.

This is simply E7 raised to the tenth power. For a population to persist 10 years in a row, we multiply the probability of persistence by itself for the number of years we are interested in projecting to the future. Recall that you entered the the value 10 in cell E5; thus the formula in cell E8 can be **=E7^E5**, where the ^ symbol indicates the power to which the value in cell E7 is raised.

7. In cell E9, set p_i equal to 0.9.

Enter 0.9 in cell E9. This is the colonization parameter, p_i—the probability that an unoccupied site will become colonized through immigration to that site. (This cell has been shaded to indicate that its value can be manipulated in the spreadsheet.)

8. In cells E10 and E11, enter formulae to calculate the probability of regional extinction and the probability of regional persistence, respectively.

Since you know the probability that each patch will go extinct, and you know how many patches there are in the system, the probability that all of the patches will simultaneously go extinct is simply the probability of local extinction raised to the number of patches in the system. Enter **=E6^E4** in cell E10.

The probably of persistence is simply 1 – E10; thus enter **=1-E10** in cell E11.

9. In cell E12, enter a formula to calculate *f*, the equibrium fraction of patches occupied.

Enter the formula **=E9/(E9+E6)**. This corresponds to Equation 5, $f = p_i/(p_i + p_e)$. Review your work to this point and interpret your results before proceeding.

10. Save your work.

B. Simulate the metapopulation dynamics from Year 0 to Year 1.

Now we are ready to simulate how metapopulations work. You should make sure that your calculation key is set to "Automatic" at this time. Go to Tools | Options | Calculation and select the Automatic button.

1. Set up new column headings as shown in Figure 4.

We'll start with a hypothetical system that consists of 25 patches, where each cell in A14–E18 represents a patch. The first block of cells in the figure below indicates the pattern of patch occupancy in Year 0. The second block of cells (A22–E26) indicates the patch occupancy pattern in Year 1.

	A	B	C	D	E
13	**Initial patch occupancy, year 0**				
14	0	1	1	1	1
15	1	0	0	1	1
16	0	0	1	1	1
17	0	1	0	0	0
18	0	1	1	0	0
19				*f* =	**0.52**
20					
21	**Landscape occupancy, year 1**				
22					
23					
24					
25					
26					
27				*f* =	

Figure 4

2. Enter 0s and 1s as shown in cells A14–E18.

Cells A14–E18 will represent the initial patch occupancy of the 25 patches in the metapopulation system (Year 0). Cell A14 is the upper-left patch in the system; cell C16 is the middle patch in the system, and so on. We let 0 indicate that the patch is currently unoccupied and 1 indicate that the patch is occupied.

3. Format cells A14–E18 so that occupied patches are a different color than the unoccupied patches.

To format the cells, select cells A14–E18 with your mouse, then select Format | Conditional Formatting. The dialog box similar to Figure 5 will appear. Follow the prompts to format your cells. For Condition 1, set the cell value to equal to 1, then click on the Format button, select the Patterns tab, and format the pattern of the cell to be shaded one color. Click OK. Then select the Add >> button to add a new Condition and format cells that are equal to 0 as a different color. When you are finished, click OK and continue to the next step.

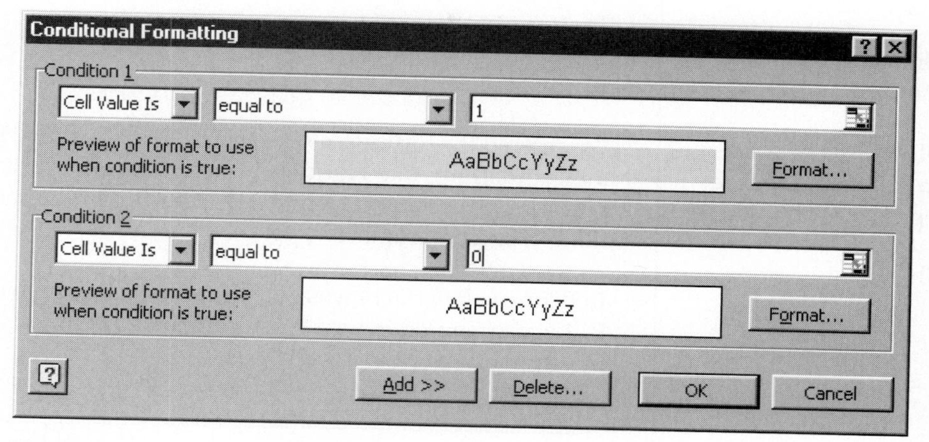

Figure 5

4. In cell E19, enter a formula to calculate the fraction of patches that are occupied, *f*.

We used the formula **=ROUND(SUM(A14:E18)/25,2)**. This formula nests two functions, **SUM** and **ROUND.** Remember that the formula within parentheses will be computed first. Thus the spreadsheet first sums the number of patches occupied and divides this number by the total number of patches in the system (25). The result is then rounded to 2 decimal places with the **ROUND** function.

5. In cell A22, enter a formula to simulate the fate of the upper-left patch (cell A14) in year 1, given its current status and extinction and colonization probabilities. Copy this formula across the 25 patch landscapes (cells A22–E26).

The upper-left patch (A14) in our initial (Year 0) landscape is currently unoccupied. Thus we need a formula that tells the spreadsheet to evaluate whether cell A14 is 0 (unoccupied) or 1 (occupied). If it's 0, then let the patch be colonized according to the colonization probability in cell **$E$9**. If it's 1, then let it go extinct according to the extinction probability in cell **$E$6**. We entered the following formula in cell A22:

=IF(A14=0,IF(RAND()<$E$9,1,0)**,IF(RAND()<$E$6,0,1))**

There are three **IF** formulae here, nested within each other; boldface type has been applied in a way that separates the three formulae. Let's walk through them carefully. Remember that the **IF** formula returns one value if a condition you specify is TRUE, and another value if the condition you specify is FALSE.

The overall structure of the formula in cell A20 tells the spreadshet to examine cell A14. If A14 is 0, then carry out the second IF statement (in light type); otherwise, carry out the third **IF** statement. Since cell A14 is 0 (unoccupied in year 0), the spreadsheet will carry out the second **IF** statement.

The second IF statement, IF(RAND()<$E$9,1,0), tells the program to draw a random number between 0 and 1 (the RAND() portion of the formula). If this random number is less than the colonization rate given in cell $E$9, then let the patch be colonized (i.e., assign it the value 1); otherwise, keep it uncolonized by assigning it the value 0.

If cell A14 had been occupied (=1), the spreadsheet would have computed the third **IF** statement, **IF(RAND()<$E$6,0,1)**. This portion of the formula tells the spreadsheet to draw a random number between 0 and 1. If this random number is less than the extinction rate given in cell **$E$6**, then let the patch go extinct (assign it the value of 0); otherwise, let it persist by assigning the cell the value 1.

Copy this formula across the landscape to see how patch occupancy changed from year 0 to year 1.

6. Conditionally format cells A22–E26 to add shading.

See Step 3 and Figure 5.

7. In cell E27, enter a formula to calculate the fraction of patches occupied in Year 1.

We entered the formula **=ROUND(SUM(A22:E26)/25,2)**. Your spreadsheet should now look something like Figure 6, although your landscape occupancy pattern for year 1 will likely differ from ours due the nature of the random number function in determining patch occupancy.

In Figure 6, Patch A14 was empty in year 0, but was colonized in year 1 (cell A22). Patch B14 was occupied in year 0 and remained occupied in year 1. Patch C14 was occupied in year 0 but went extinct in year 1.

	A	B	C	D	E
13	**Initial patch occupancy, year 0**				
14	0	1	1	1	1
15	1	0	0	1	1
16	0	0	1	1	1
17	0	1	0	0	0
18	0	1	1	0	0
19				*f*=	**0.52**
20					
21	**Landscape occupancy, year 1**				
22	1	1	0	1	1
23	1	0	1	1	1
24	0	0	0	1	1
25	0	1	0	1	1
26	0	0	0	1	0
27				*f*=	**0.56**

Figure 6

8. Press F9, the Calculate key, several times to simulate changes in patch occupancy from Year 0 to Year 1.

Each time you press F9 the spreadsheet generates a new set of random numbers, which in turn affects whether patches become colonized or go extinct. When you press F9, you should see under various scenarios how the fraction of patches in the landscape changes from year 0 to year 1. You should also see the "winking" nature of metapopulations: Patches "wink in" when they become colonized and "wink out" as they go extinct. Given a configuration of occupied patches in year 1, our next step is to determine what the occupancy pattern will be in year 2 and into the future. We will do this in the next step.

9. Save your work.

C. Simulate metapopulation dynamics over time.

Now we'll track "winking" over time, and determine the fraction of patches that remain occupied over time. When the *fraction occupied* no longer changes across generations, but the *pattern of occupancy* continually shifts, the metapopulation has reached an equilibrium state.

1. Set up new column headings as shown in Figure 7.

	A	B	C	D	E
29	**Landscape occupancy, year *t***				
30					
31					
32					
33					
34					
35				*f* =	
36					
37	**Landscape occupancy, year *t* + 1**				
38					
39					
40					
41					
42					
43				*f* =	

Figure 7

2. Copy cells A22–E26, and then go to Edit | Paste Special | Paste Values into cells A30–E34. Do not copy and paste the formulae.

We will now let the pattern of patch occupancy in year 1 be labeled *year t*. We want to predict what will happen in year *t* + 1—that is to say, in year 2. To continue simulating the metapopulation dynamics over time, the occupancy pattern in year 2 will then be pasted into year *t*, and year 3 will be year *t* + 1. After year 3 is calculated, year 3 will become year *t*, and year 4 will become year *t* + 1 (and so on). You can ignore the cells labeled "Landscape Occupancy, year 0" (cells A14–E18) and "year 1" (cells A22–E26) from this point forward.

3. In cell E35, enter a formula to calculate the fraction of patches that are occupied in year *t*.

We entered the formula **=ROUND(SUM(A30:E34)/25,2).**

4. In cell A38, enter a formula to determine the fate of the upper-left patch in the system (cell A30) for year *t* + 1 (refer to the formula entered in cell A22). Copy this formula across the landscape.

To predict the pattern of occupancy for year *t* + 1, we need to write a formula based on the occupancy patterns in year *t*. We used the formula **=IF(A30=0,IF(RAND()<$E$9,1,0),IF(RAND()<$E$6,0,1))**.

5. Calculate the fraction of patches that are occupied in cell E43.

Enter the formula **=ROUND(SUM(A38:E42)/25,2)**.

6. Set up new column headings as shown in Figure 8.

	G	H
2		**Fraction**
3	**Year**	**occupied**
4	0	
5	1	
6	2	
7	3	
8	4	
9	5	
10	6	
11	7	
12	8	
13	9	
14	10	

Figure 8

7. Enter **=E35** in cell H4.

This designates the occupancy rate in the initial landscape.

8. Write a macro to simulate patch occupancy over 10 years.

Under Tools | Options | Calculation, set your caculation key to Manual. Then record a macro to track *f* across years (see Excercise 2, "Spreadsheet Functions and Macros"). Once your macro is in the "Record" mode, do the following:

- Press F9, the calculate key, to determine the pattern of occupancy for Year $t + 1$ (cells A38–A42).
- Select cell E43, the new proportion of the landscape occupied, and select Edit | Copy.
- Select cell H4, then go to Edit | Find. Leave Find What *completely blank*, searching by columns, and select Find Next and then Close (Figure 9).

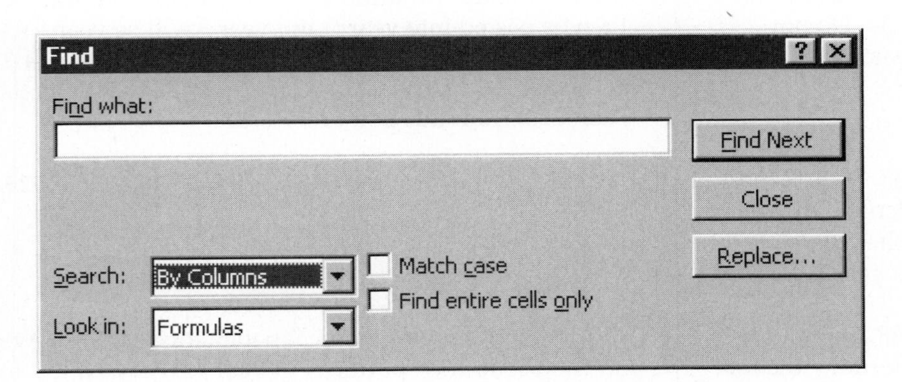

Figure 9

- Select Edit | Paste Special, and paste in the values, which are the proportion of the landscape that is occupied for that year.
- Use your mouse to highlight cells A38–E42 and select Edit | Copy.
- Now select cell A30, then select Edit | Paste Special and paste in the values. This is your new metapopulation configuration for the following year.
- Select Tools | Macro | Stop Recording.

Now when you press the shortcut key you assigned, the macro automatically determines the proportion of patches that are occupied and enters this value into the appropriate generation. Run your macro until you have tracked your metapopulation over 10 years.

9. Save your work.

D. Create graphs.

1. Graph the fraction of patches occupied over time. Use the line graph option and label your axes fully. Save your work.

Your graph should resemble Figure 10, although the exact fraction of patches will vary due to the random number function used to determine the fate of a given patch.

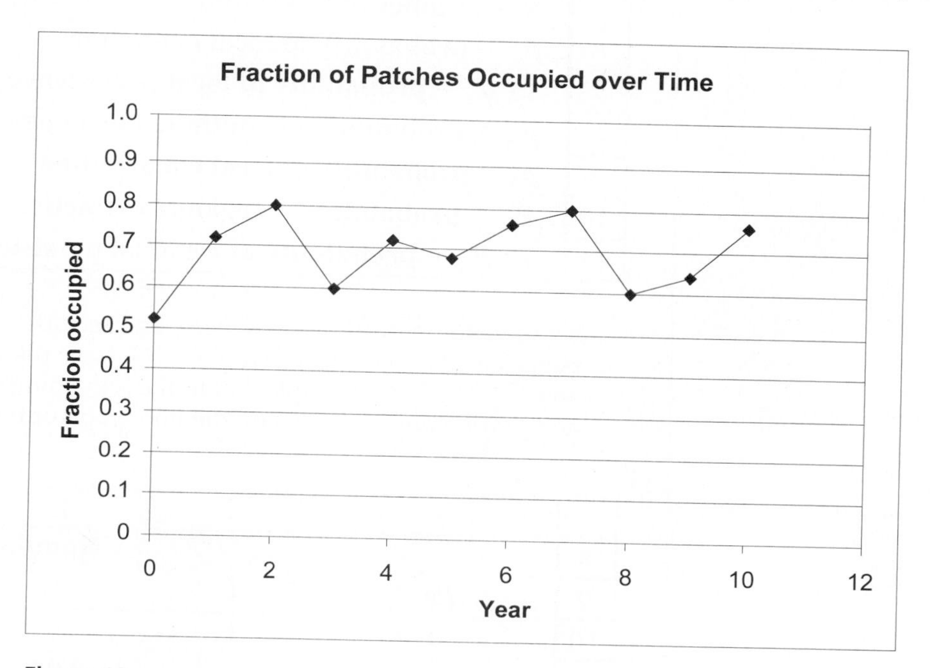

Figure 10

E. Explore the model.

1. Explore your model by changing the probability of extinction and the probability of colonization.

Choose any (reasonable) values you'd like. Run your macro again for 10 years to simulate the new conditions. Remember that as long as the calculation is set to manual, you will always have to press the F9 key to complete any calculations.

In your explorations, don't forget that you'll have to "reset" the cells labeled "Landscape occupancy, year *t*" (cells A30–E34) to reflect the initial conditions you desire. You will also want to clear the simulation results in cells H5–H14 before you run your new macro.

QUESTIONS

1. Compute *f*, the equilibrium number of patches occupied in the metapopulation system. (Refer to Equation 5.) Examine the graph of the metapopulation simulation. Has the population reached an equilibrium value, where the number of patches stays constant over time although the occupancy of each patch changes over time? Why or why not? Extend years in column G to 100. Run your macro until 100 simulations are completed. Is the system in equilibrium by year 100? Why or why not?

2. How does number of patches in a metapopulation system affect the probability of regional persistence (P_x) under a fixed level of local colonization but various scenarios of local extinction? Enter model parameters as shown. To address this question, change cells E4 and E6 according to the table below (cells J8–N15), then record the value in cell E11 in the appropriate cell.

	A	B	C	D	E
3	**Model parameters:**				
4	**x = number of patches in system**				1
5	**n = number of years under consideration**				1
6	**p_e = probability of local extinction**				0.2
7	**$1 - p_e$ = probability of local persistence**				0.8
8	**p_n = probability of continued local persistence**				0.8
9	**p_i = probability of local colonization**				0.5
10	**P_x = probability of regional extinction**				0.2
11	**$1 - P_x$ = probability of regional persistence**				0.8

Set up column headings as shown, and record $1 - P_x$ (the probability of regional persistence) in the appropriate cell. We have filled in the $1 - P_x$ values for $P_e = 0$ and $P_e = 0.2$ as an example. Fill in the remaining cells. Then select cells K10–N15 and graph your results using the line graph option. Interpret your graph.

	J	K	L	M	N
8		**Number of patches**			
9	***Pe***	**1**	**2**	**4**	**8**
10	0	1	1	1	1
11	0.2	0.8	0.96	0.998	0.999
12	0.4				
13	0.6				
14	0.8				
15	1				

3. How does f, the equilibrium fraction of patches occupied, change as function of p_e and p_i? Set up spreadsheet columns as shown:

	J	K	L	M	N	O	P
18		***P_i***					
19	***P_e***	$p_i = 0$	$p_i = 0.2$	$p_i = 0.4$	$p_i = 0.6$	$p_i = 0.8$	$p_i = 1$
20	0	0	1	1	1	1	1
21	0.2						
22	0.4						
23	0.6						
24	0.8						
25	1						

For each combination of p_i and p_e, enter f in the appropriate cell. For example, in cell L21 enter f (computed in cell E12) when $p_i = 0.2$ and $p_e = 0.2$. Graph and interpret your results. Use the line graph option and select the data series in columns option.

4. Set cell E6 to 1, and cell E9 to 0.9. This will make the probability of extinction, p_e, equal to 1, and the probability of colonization, p_i, equal to 0.9. Clear your old macro results and run a new simulation. Why has the population persisted, considering that all patches are doomed to extinction?

5. Set cell E6 to 0.3, and enter 1s and 0s in cells A30–E34 such that $f = 0.6$. Assume that p_i is now a function of the number of patches occupied (instead of the propagule rain model in question 4). As more patches are occupied, the colonization rate increases because a greater number of colonists will likely locate an empty patch. Write an equation in cell E9 to modify the model into an internal colonization model and re-run your simulation. How do your results differ from those of question 4?

6. Return cell E9 to 0.6 (propagule rain model), and enter 1s and 0s in cells A30–E34 such that $f = 0.6$. Assume now that p_e is now a function of the number of patches occupied. As more patches are occupied, the extinction rate decreases because more colonists are available to "rescue" the patch from extinction. The fewer patches that are occupied, the more likely a patch will go extinct because colonists are less available to "rescue" a patch from extinction. This metapopulation model is called the *rescue effect model* (Gotelli 2001), where the extinction rate *depends* on how many patches are currently occupied. Write an equation in cell E6 to modify your model into a rescue effect model, and re-run your simulation. How do your results compare to questions 4 (propagule rain model) and 5 (internal colonization model)?

7. *Advanced. How does number of patches in the system affect the "stochastic" behavior of a metapopulation? Set up a new system in which the number of patches is 10,000 (100 × 100 cells), and compare the two models.

LITERATURE CITED

Gotelli, N. 2001. *A Primer of Ecology*, 3rd Edition. Sinauer Associates, Sunderland, MA.

Hanski, I. A., and M. E. Gilpin. 1997. *Metapopulation Biology: Ecology, Genetics, and Evolution*. Academic Press, San Diego.

Levins, R. 1969. Some demographics and genetic consequences of environmental heterogeneity for biological control. *Bulletin of the Entomological Society of America* 15: 237–240.

Levins, R. 1970. Extinction. *In* M. Gerstenhaber (ed.), *Some Mathematical Questions in Biology: Lecture Notes on Mathematics in the Life Sciences*, pp. 75–107. The American Mathematical Society, Providence, RI.

21 SOURCE-SINK DYNAMICS

Objectives

- Set up a population model of two subpopulations that interact through dispersal.
- Determine how birth, death, and dispersal between source and sink habitat affect population persistence.
- Determine how the initial distribution of individuals among source and sink habitat affects population dynamics.
- Examine the conditions in which a source-sink system is in equilibrium.

Prerequisite Exercise: Geometric and Exponential Population Models

INTRODUCTION

If you could spend your life anywhere in the world, where would it be? A Hawaiian island? The Peruvian Andes? The French Riviera? Midtown Manhattan? A village in Bosnia? The rain forest of Madagascar? The Gobi Desert? New Zealand's South Island? In thinking about your choice, it becomes obvious that all habitat patches are not created equal.

For any given species, some habitats are superior to others for individual survival and reproduction. The fact that patch quality is heterogeneous (mixed) and that individuals of a population occupy different kinds of patches is an important consideration in predicting the population dynamics of a species. **Source-sink theory** addresses the issue of such heterogeneity. **Sources** are areas or locations where local reproductive success is greater than local mortality (Pulliam 1988). Alas, not all patches are optimal, and some individuals of a population may be forced to occupy poorer quality patches that lead to low birth rates and high death rates. These areas or locations are called **sinks**, because the populations occupying them will spiral "down the drain" to extinction unless they receive immigrants from other locations—usually a source.

Why would individuals disperse from a high-quality source habitat to a low-quality sink habitat? Because resources are limited, not all individuals can obtain breeding sites in the source. Individuals unable to find a breeding site in the source emigrate to the sink because, from a fitness perspective, even a poor-quality breeding site may be better than none at all (Pulliam 1988).

If we want to project the size of a population in which some individuals reside in source habitats and others reside in sink habitats, we need to consider the population dynamics of each source and sink **subpopulation**, and then consider how the distribution of individuals in sources and sinks influences the dynamics of the greater source-sink system. How can such a population be modeled? If you have completed Exercise 7, "Geometric and Exponential Population Models," you may recall that the most basic way to describe population growth is through the equation

$$N_{t+1} = N_t + B_t - D_t + I_t - E_t \qquad \text{Equation 1}$$

where

N_t represents the size or density of the population at some arbitrary time t
N_{t+1} represents the population size one arbitrary time unit later
B_t represents the total number of births in the interval from time t to time t+1
D_t represents the total number of deaths in the same time interval
I_t represents the total number of immigrants in the same time interval
E_t represents the total number of emigrants in the same time interval

Birth, death, immigration, and emigration are the four "biggies" in population dynamics. In concert, they determine whether a population will grow or decline over time, and are often called the **BIDE factors**. If you have completed the exercise "Geometric and Exponential Population Models", you modeled a population in which dispersal was neglible, and hence I and E were set to 0. However, in source-sink dynamics, the movement of individuals from one population to another must be considered, and changes in numbers over time must therefore include the movements of individuals into the population (immigration, I) and the movement of individuals out of the population (emigration, E). To make population projections of a source-sink system, we need to know the numbers of individuals in each habitat type, as well as the BIDE factors for each habitat type. Thus, two equations are needed: one for the source population, and one for the sink. We will consider these equations for a population that grows in discrete time, rather than for a continuously growing population.

To begin, let's think about a single habitat, say, the source. What controls the total number of births (B), immigrants (I), deaths (D), and emigrants (E) in the source habitat? If we switch from *total* numbers to *per capita* rates, we can do some fruitful modeling. A **per capita rate** is a per individual rate; the per capita birth rate is the number of births per individual in the population per unit time, and the per capita death rate is the number of deaths per individual in the population per unit time. Similarly, per capita immigration and emigration rates are the number of immigrants and emigrants per individual per unit time.

Per capita birth rate and immigration are easy to understand; they are the number of new individuals per individual that enter the population through birth or immigration. Per capita death and emigration rates may seem strange at first because they reflect the number of deaths or emigration events per individual per unit time, and usually these things happen to individuals only once. But you can think of these rates as each individual's risk of dying in a given unit of time, or the chance of exiting the population through dispersal in a given unit of time.

Keeping in mind that per capita rates are per individual rates, we can translate raw numbers (B_t, I_t, D_t, and E_t) into per capita rates, which we will represent with lowercase letters (b_t, i_t, d_t, and e_t) to distinguish them from raw numbers. All we have to do is divide the raw numbers by N_t, the population size at time t:

$$b_t = \frac{B_t}{N_t} \text{ and } B_t = b_t N_t$$

$$i_t = \frac{I_t}{N_t} \text{ and } I_t = i_t N_t$$

$$d_t = \frac{D_t}{N_t} \text{ and } D_t = d_t N_t$$

$$e_t = \frac{E_t}{N_t} \text{ and } E_t = e_t N_t$$

Because we assume constant per capita rates, we can make one further, minor modification to our equation by leaving off the time subscripts on b, i, d, and e. Thus,

$$N_{t+1} = N_t + bN_t + iN_t - dN_t - eN_t \qquad \text{Equation 2}$$

We can further simplify this model by factoring N_t out of the birth, immigration, death, and emigration terms:

$$N_{t+1} = N_t + (b + i - d - e)N_t \qquad \text{Equation 3}$$

The term $(b + i - d - e)$ is so important in population biology that it is given its own symbol, R and is called the **geometric rate of natural increase**. Thus*

$$R = b + i - d - e$$

Substituting R into Equation 3 gives us

$$N_{t+1} = N_t + RN_t \qquad \text{Equation 4}$$

We can calculate the *change* in population size, ΔN_t, by subtracting N_t from both sides of this equation:

$$N_{t+1} - N_t = RN_t$$

Because $\Delta N_t = N_{t+1} - N_t$, or the difference in population size over time, we can substitute and write

$$\Delta N_t = RN_t \qquad \text{Equation 5}$$

In words, the rate of change in population size is proportional to the population size, and the constant of proportionality is R. We can convert this to per capita rate of change in population size if we divide both sides by N_t:

$$\Delta N_t / N_t = R \qquad \text{Equation 6}$$

In words, the parameter R represents the per capita rate of change in the size of the population. If you'd like to determine how R will affect population size from one time step to the next, you can start with Equation 4, and then factor N_t out of the terms on the right side to get

$$N_{t+1} = (1 + R)N_t \qquad \text{Equation 7}$$

The quantity $(1 + R)$ is often given its own symbol, λ, or the **finite rate of increase**, and so we can write

$$N_{t+1} = \lambda N_t \qquad \text{Equation 8}$$

When $\lambda = 1$, the population size remains constant (unchanged) over time; when $\lambda > 1$, the population increases geometrically; and when $\lambda < 1$, the population declines geometrically.

Now let's return to the topic of sources and sinks. Without dispersal, a source can be defined as a subpopulation where $\lambda > 1$. This occurs only when $b > d$. A sink can be defined as a subpopulation where $\lambda < 1$, which occurs when $d > b$. With dispersal (immigration and emigration), a source or sink subpopulation is in dynamic equilibrium (not changing) when $B + I - D - E = 0$. Thus, because births are greater than deaths in a source population, to maintain an equilibrium number of individuals, the source must export individuals to other locations ($b > d$ and $e > i$). In contrast, for a sink to be in equilibrium, it must import individuals because deaths outnumber births ($d > b$ and $i > e$).

---

*In Exercise 7, R was defined simply as $b - d$ because in that exercise i and e were assumed to be 0.

How is the equilibrium size of the greater population (source and sink) determined? If there are many habitats, the population reaches equilibrium when the total surplus in all the source habitats equals the total deficit in all the sink habitats. Some basic take-home points from Pulliam's (1988) source-sink model are:

- At equilibrium, the *number of individuals* in the overall, greater population is not changing.
- Each source and sink subpopulation can be characterized by its "strength," depending on its intrinsic rate of growth and the number of individuals present. Within-subpopulation dynamics (b, i, d, e) are important in determining the overall equilibrium population size, since the numbers of individuals on each patch and their growth rates are implicit in the model.
- The source-sink status of a subpopulation may have little to do with the size (number of individuals) within the subpopulation. Sinks can support a vast number of individuals and sources can be numerically very small. However, sources must have enough individuals with a high enough per capita production to support sink populations.

PROCEDURES

In this exercise you will develop a simple source-sink model in which dispersal occurs from the source to the sink when the source reaches its carrying capacity. We will consider only the female portion of the population and assume that there are plenty of males available for reproductive purposes. Once the model is constructed, you will be able to explore how the different BIDE parameters, population sizes, and carrying capacities influence the source-sink system. As always, save your work frequently to disk.

INSTRUCTIONS / ANNOTATION

A. Set up the basic spreadsheet.

1. Open a new spreadsheet and set up headings as shown in Figure 1.

	A	B	C	D	E	F
1	***Source - Sink Model***					
2						
3				**Constants**		
4		**Source**			**Sink**	
5		N_0 =	10		N_0 =	100
6		K =	25		b =	0.4
7		b =	0.5		d =	0.5
8		d =	0.2			
9		i =	0.1			

Figure 1

2. Enter the starting number of individuals, N_0; carrying capacity, K; and BIDE rates for the source and sink habitat as shown in Figure 1.

In our source-sink model, we will assume that the source has a carrying capacity (see Exercise 8, "Logistic Population Models") because not all individuals can occupy prime habitat. In the source, the birth, death, and immigration rates are constants that can be modified. The emigration rate, e, is not a constant but is calculated as the per capita number of individuals that leave the source after the carrying capacity has been reached. We will assume that the sink has no carrying capacity and that "poor quality" habitat is plentiful. The immigration rate into the sink, i, is not a constant but is calculated as the per capita number of individuals that disperse from the source to the sink habitat.

B. Project population size in a source over time.

1. Set up new spreadsheet headings as shown in Figure 2.

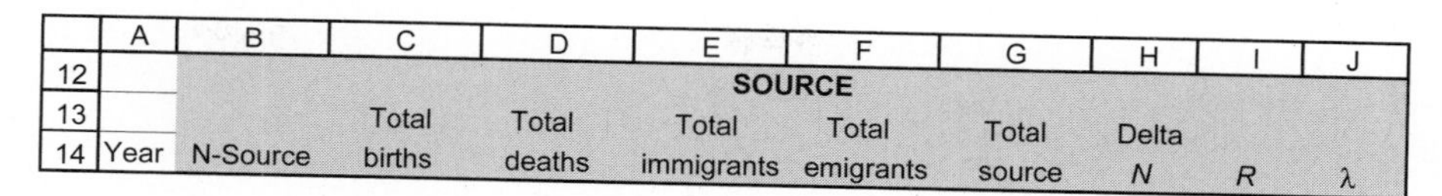

	A	B	C	D	E	F	G	H	I	J
12					SOURCE					
13			Total	Total	Total	Total	Total	Delta		
14	Year	N-Source	births	deaths	immigrants	emigrants	source	*N*	*R*	λ

Figure 2

2. Set up a linear series from 0 to 20 in cells A15–A35.

Enter 0 in cell A15.
Enter **=1+A15** in cell A16. Copy this formula down to cell A35.

3. In cell B15, link the starting number of individuals in the source population to cell C5.

Enter **=C5** in cell B15.

4. In cells C15–E15, enter formulae to compute *B*, *D*, and *I* (the total numbers of births, deaths, and immigrants) in the source population.

Remember that the numbers of births, deaths, and immigrants in the source depends on the per capita rates given in cells C7–C9, as well as the number of individuals currently in the population, N_t. Enter the following formulae:

- C15 **=B15*$C$7**
- D15 **=B15*$C$8**
- E15 **=B15*$C$9**

5. In cell F15, use an **IF** function to compute the total number of emigrants from the source as the number of individuals in excess of the source's carrying capacity.

Enter the formula **=IF(B15+C15-D15+E15>$C$6,B15+C15-D15+E15-$C$6,0)**. This formula is long, but is really a simple **IF** formula with three parts, each part separated by a comma.
The first part is the criterion; our criterion is **B15+C15-D15+E15>$C$6**, which tells the spreadsheet to evaluate whether $N_t + B - D + I$ is greater than the source's carrying capacity (which is given in cell **$C$6**). If this criterion is **TRUE**, the program carries out the second part of the formula. If this criterion is **FALSE**, it carries out the third part of the formula.
Thus, if the number of individuals in the source is below carrying capacity (i.e., the criterion is **FALSE**), the number of emigrants from the source will be 0, and the spreadsheet will return the number 0 in cell F15. If the number of individuals in the source is above *K* (the criterion is **TRUE**), the number of emigrants from the source is computed as **B15+C15-D15+E15-$C$6.**

6. In cell G15, compute the total number of individuals in the source as $N_0 + B + I - D - E$.

We entered the formula **=B15+C15-D15+E15-F15**.

7. In cell H15, enter a formula to compute ΔN.

Enter the formula **=C15-D15+E15-F15**. The formula **=G15-B15** gives the same result. (Remember that you can generate the delta symbol, Δ, by typing in a capital D, selecting it, and changing its font to Symbol.)

8. In cell I15, compute *R* as $\Delta N/N$ to generate the *per capita* rate of population change.

Enter the formula **=H15/B15**.

9. In cell J15, compute λ as $R + 1$.

Enter the formula **=1+I15**. Note that λ can also be computed as N_{t+1}/N_t. You can generate the λ symbol by typing in the letter l, then selecting this letter on the formula bar, and changing its font to the symbol font. Interpret your results before proceeding.

10. In cell B16, enter a formula to compute *N* in year 1.

Enter the formula **=B15+C15+E15-D15-F15**. You could also simply enter **=G15**.

11. Select cell B16 and copy its formula down to row 35. Select cells C15–J15 and copy their formulae down to year 20, row 35.

12. Save your work. The first portion of your spreadsheet should now look like Figure 3.

	A	B	C	D	E	F	G	H	I	J
12					SOURCE					
13			Total	Total	Total	Total	Total	Delta		
14	Year	N-Source	births	deaths	immigrants	emigrants	source	*N*	*R*	λ
15	0	10.0	5.0	2.0	1.0	0.0	14.0	4.0	0.40	1.40
16	1	14.0	7.0	2.8	1.4	0.0	19.6	5.6	0.40	1.40
17	2	19.6	9.8	3.9	2.0	2.4	25.0	5.4	0.28	1.28
18	3	25.0	12.5	5.0	2.5	10.0	25.0	0.0	0.00	1.00
19	4	25.0	12.5	5.0	2.5	10.0	25.0	0.0	0.00	1.00
20	5	25.0	12.5	5.0	2.5	10.0	25.0	0.0	0.00	1.00

Figure 3

C. Project population size in the sink over time.

1. Set up new spreadsheet headings as shown in Figure 4.

	K	L	M	N	O	P	Q	R
12				SINK				
13		Total	Total	Total	Total	Delta		
14	N-Sink	births	deaths	immigrants	sink	N	R	λ

Figure 4

2. In cell K15, link the starting number of individuals in the source population to cell F5.

Enter the formula **=$F$5** in cell K15.

3. Enter formulae in cells L15–M15 to compute the total births and deaths in the sink.

Enter the following formulae:

- L15 **=K15*$F$6**
- M15 **=K15*$F$7**

4. In cell N15, enter a formula to link emigants from the source to immigrants into the sink.

Enter the formula **=F15**.

5. In cells O15–R15, enter formulae to compute the total population size of the sink; ΔN; *R*; and λ.

We entered the following formulae:

- O15 **=K15+L15-M15+N15**
- P15 **=O15-K15** or **L15-M15+N15**
- Q15 **=P15/K15**
- R15 **=K16/K15** or **=Q15+1**

6. Enter an **IF** formula in cell K16 to compute the population size in year 1.

We entered the formula **=IF(K15+L15-M15+N15<0,0,K15+L15-M15+N15)**. This **IF** formula is used to keep the population from falling below 0 and generating negative population sizes. The formula simply says that if the total population in the sink is less than 0, return the number 0; otherwise, return the total population size of the sink.

7. Select cell K16 and cells L15–R15, and copy their formulae down to year 20 (row 35).

8. Your sink projections should now look something like those in Figure 5. Save your work.

	K	L	M	N	O	P	Q	R
12			SINK					
13		Total	Total	Total	Total	Delta		
14	N-Sink	births	deaths	immigrants	sink	*N*	*R*	λ
15	100	40	50.0	0.0	90.0	-10.0	-0.10	0.90
16	90.0	36	45.0	0.0	81.0	-9.0	-0.10	0.90
17	81.0	32.4	40.5	2.4	75.3	-5.7	-0.07	0.93
18	75.3	30.136	37.7	10.0	77.8	2.5	0.03	1.03
19	77.8	31.1224	38.9	10.0	80.0	2.2	0.03	1.03
20	80.0	32.01016	40.0	10.0	82.0	2.0	0.02	1.02

Figure 5

D. Project and graph population sizes for the source-sink system.

1. Set up new headings as shown in Figure 6

	S	T
12	SYSTEM	
13		
14	N-Total	λ-Total

Figure 6

2. In cell S15, compute the total population size as the sum of the source individuals and sink individuals.

Enter the formula **=G15+O15.**

3. In cell T15, enter a formula to compute λ for the entire source-sink system.

Enter the formula **=S16/S15.**

4. Copy cells S15–T15 down to year 20 (row 35).

5. Graph the numbers of individuals in the source, sink, and total population over time. Your graph should resemble Figure 7.

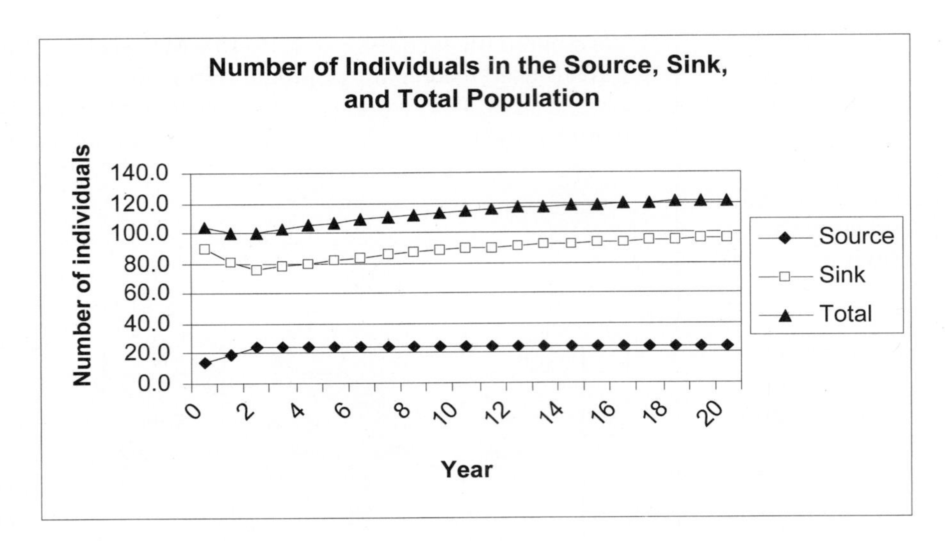

Figure 7

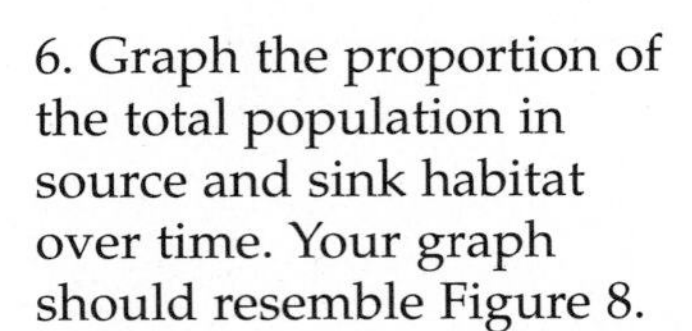
6. Graph the proportion of the total population in source and sink habitat over time. Your graph should resemble Figure 8.

Select cells G15–G35, then press the Control key or the key and select cells O15–O35. Use the 100% Stacked Column option, and label your axes fully.

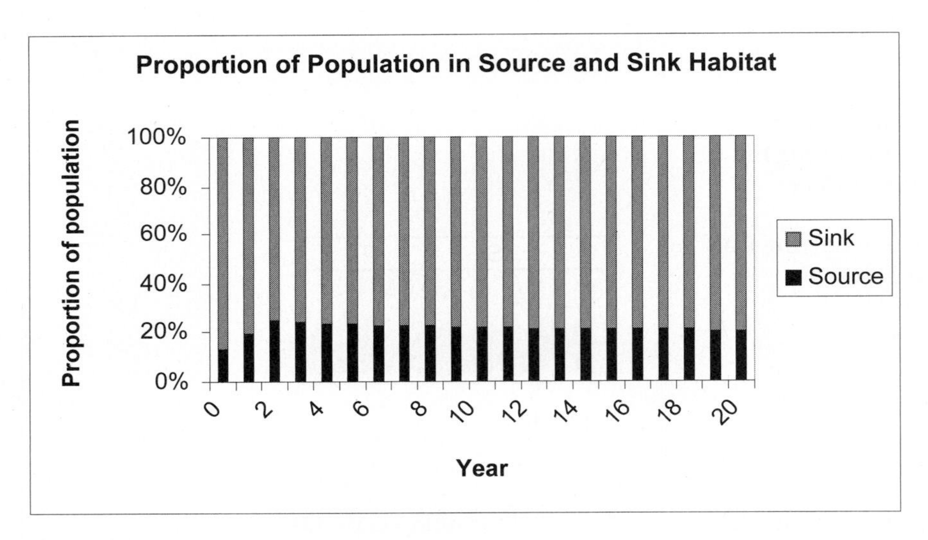

Figure 8

QUESTIONS

1. Keeping the parameters as you set them at the beginning of the exercise, and examining the graphs created in the last step, answer the following questions:

   - At what year does the source reach an equilibrium state?
   - At what year does the sink reach an equilibrium state?
   - How does the proportion of the total population in source and sink habitats change over time? Why do the proportions change?

2. Extend your population projections to 100 years. Copy the formula in row 35 down to row 115. Update your graphs to include the 100-year projection, and answer the questions in question 1 again.

3. The definition of a sink is that it is incapable of sustaining itself over time without the influx of individuals from source habitats. What happens if the source population is extirpated? Set cell C5 to 0, and interpret your model results.

4. With your model programmed, you can change various parameters and watch how the parameters effect the population over time.

- What happens to the greater population if you increase the starting number of individuals in the source? Increase the value of cell C5 from 10 to 100 in increments of 10. Interpret λ for the source, sink, and greater population.
- What happens to the greater population if you increase the starting number of individuals in the sink? Increase the value of cell F5 from 100 to 1000 in increments of 100. Interpret λ for the source, sink, and greater population over time. What is the equilibrium population size?
- What if you increase survival rate (i.e., lower the death rate, cell C8) in the source habitat?
- How does carrying capacity, K, affect overall population growth? What happens when you increase or decrease this factor in the source?

5. Field biologists seldom have the opportunity to estimate the birth and survival rates for many organisms. Instead of basing habitat quality on these parameters, quality is often associated with density (number of individuals per unit area). Modify your model to show that density may be a misleading indicator of habitat quality.

6. In Pulliam's 1988 model, b (cell C7) changes and is a function of number of breeding sites/total breeders. Thus, if the number of total breeders is large, b, the per capita birth rate, is low. And if the number of total breeders is less than total sites available in source, b is maximum. How can this be incorporated into your model, and how does this change affect your model results?

LITERATURE CITED

Pulliam, H. R. 1988. Sources, sinks, and population regulation. *American Naturalist* 132: 652–661.

22 NICHE BREADTH AND RESOURCE PARTIONING

Objectives

- Compute niche breadth for two organisms coexisting in a community.
- Compute niche overlap for the two coexisting organisms.
- Use the Solver function to evaluate how breadth and overlap between the two species can be maximized and minimized.

Suggested Preliminary Exercise: Interspecific Competition

INTRODUCTION

A **community** is an assemblage of different species that coexist in time and space (Gotelli 2001). Community dynamics can occur at any spatial scale. That is, we can study the interaction of different species within a community on our front lawn, inside the gut of a deer, in a pond after a rainstorm, or in a temperate rain forest.

Given that resources are not infinite within any ecosystem, a fundamental question in ecology is, How many species can occur together within a given community? The **competitive exclusion principle** states that if two species compete for critical resources in an environment, one of two outcomes results. Either both species coexist, or one species outcompetes the other and drives the other species to extinction (at least in that community). Coexistence can occur only if the species niches are different enough to limit competition between them. Thus, ecologists interested in community dynamics often ask, How do the different species partition the resources in this community? To answer this question, we need to know how organisms utilize their environment. One way to do this is to measure the **niche parameters** for one species and then compare it to the niche parameters of another.

As a hypothetical example, consider two species that occur together in a community. Both species consume a food resource that varies in size, such as seeds. Suppose both species 1 and species 2 consume a wide variety of seed sizes, but eat similar kinds of seed sizes. A graph of their resource consumption might look like Figure 1. Since both species eat a variety of seed sizes, *intraspecific* competition may not be that significant because individuals may not have to compete directly with members of their own species for a certain size of seed. However, the overlap in curves between species 1 and 2 suggests that *interspecific* competition may

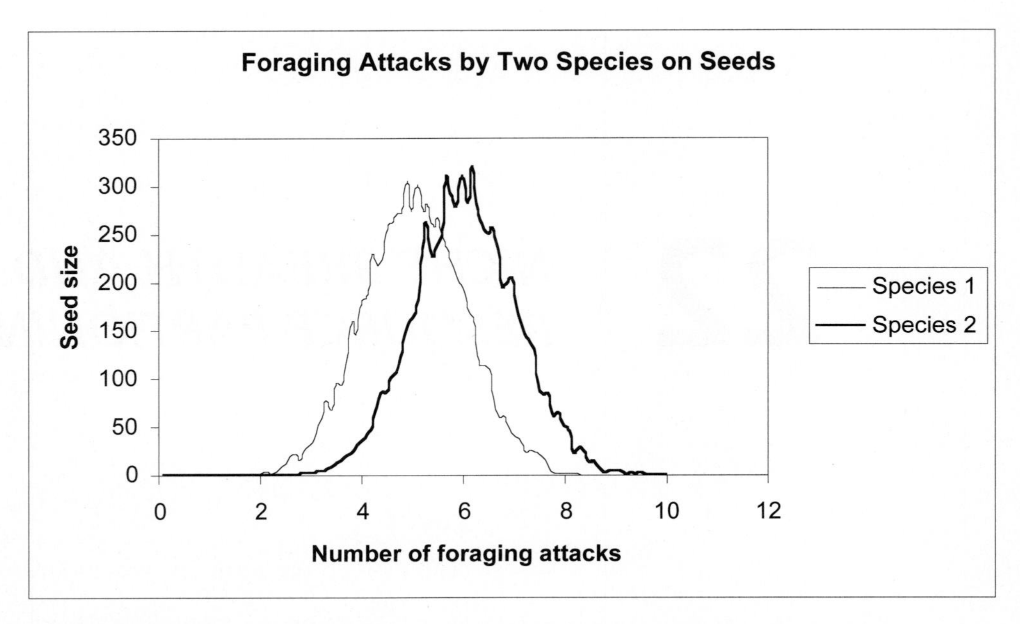

Figure 1 The two species whose foraging habits are charted here are competing for the same food resource and would not be able to coexist comfortably.

be significant. The competitive exclusion principle suggests that such competition may lead to the local extinction of one species.

Alternatively, assume that species 1 and species 2 consume different seed sizes, with a graph of seed consumption as shown in Figure 2. In this situation, *intraspecific* competition may be significant because each species specializes on only a small range of seed sizes in their diets. However, since the species do not overlap in their consumption of seeds of a certain size, *interspecific* competition is likely to be low, and the two species may coexist.

How can we determine quantitatively the degree to which two species can compete for a similar resource? We will consider two measures: niche breadth and niche over-

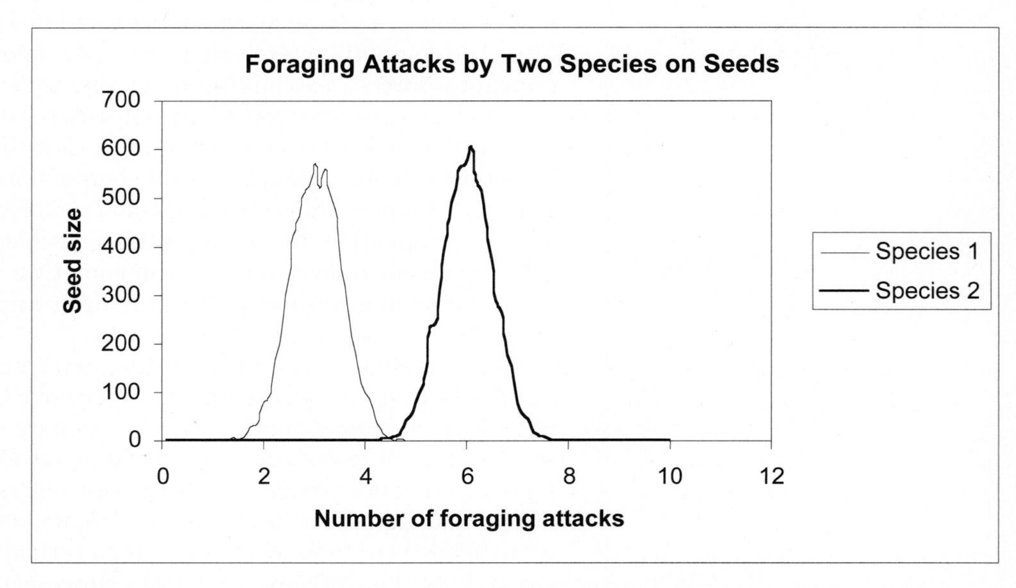

Figure 2 Individuals of the two species in this graph would face greater competition from members of their own species than from members of the second species.

lap. **Niche breadth** is a parameter that attempts to measure how specialized or unspecialized a species is within a given environment. A specialist that feeds on only one or two food sources will have a much smaller niche breadth than a generalist that feeds on many kinds of food items. Niche breadth is measured by observing how individuals in the community make use of the same set of resources. Food, for example, is a resource that can be measured by identifying the kind of food taken or the size of food taken. Habitat is also a resource whose use can be measured for niche analysis.

There are many ways to quantify niche breadth (Krebs 1999). One common measure is the Levins measure (1968), which measures how uniformly resources are being utilized by each species. The equation is

$$B = \frac{1}{\sum p_i^2} \qquad \text{Equation 1}$$

where B is the Levins measure of niche breadth and p_i is the proportion of individuals found using resource i. To derive measures of niche breadth for a species, an ecologist typically counts the number of resource items used by a set of individuals of that species.

Suppose we observed two species of lizards and quantified the food intake of 1000 individuals in both species. One species, the whiptail lizard (*Cnemidophorus tigris*), has a diet that consists of 20% grasshoppers, 30% termites, 20% insect larvae, 20% beetles, 5% vertebrates, and 5% roaches (data drastically modified from Pianka 1986). The second species, the side-blotched lizard (*Uta stansburiana*) has a diet that consists of 10% ants, 20% grasshoppers, 25% beetles, 15% termites, 10% insect larvae, 10% arthropods, and 10% spiders. The niche breadth for the whiptail lizard would be

$$B = \frac{1}{.20^2 + .30^2 + .20^2 + .20^2 + .05^2 + .05^2} = 4.65$$

and the niche breadth for the side-blotched lizard would be

$$B = \frac{1}{.10^2 + .20^2 + .25^2 + .15^2 + .10^2 + .10^2 + .10^2} = 6.06$$

Often, these measures are standardized on a scale of 0 to 1 by using the formula

$$B_A = \frac{B-1}{n-1} \qquad \text{Equation 2}$$

where B_A is the standardized niche breadth, and n is the total number of food items for the species of interest (in the whiptail example, six food types were observed in total, so $n = 6$).

In contrast to niche breadth, the parameter **niche overlap** measures the degree to which two different species overlap in their use of a particular resource. Estimating niche overlap is a way to answer the question, How do the different species partition the resources in the community? It might be obvious that some species do not overlap at all in their use of resources. For example, a hummingbird and an owl are very unlikely to compete for the same food resources, so measures of niche overlap seem trivial when it comes to food. However, estimating niche overlap and resource partitioning is often of interest when a number of species use resources in similar ways. Such a group of species is called a **guild**. Seed-eating finches on the Galápagos Islands are an example of a guild.

If species overlap in niches to a great extent, they may influence each other's population growth through interspecific competition. As with niche breadth, niche overlap can be measured in a variety of ways (Krebs 1999). One measure, developed by MacArthur and Levins (1967), is calculated as

$$M_{jk} = \frac{\sum p_{ij} p_{ik}}{\sum p_{ij}^2} \qquad \text{Equation 3}$$

where M_{jk} is the MacArthur and Levins niche overlap measure *of species k on species j* (keep track of the notation used), p_{ij} is the proportion that resource i is of the total resource that species j utilizes, and p_{ik} is the proportion that resource i is of the total resources that species k utilizes. Both summations are over the index i. Note that when we calculate niche overlap this way, the effect of species j on species k can be different from the effect of species k on species j. This formula was originally developed to estimate α and β coefficients in the Lotka-Volterra interspecific competition model (Exercise 10). However, most ecologists now agree that overlap measures are not appropriate for competition coefficients (Krebs 1999). A similar, but symmetrical, measure of overlap was developed by Pianka (1986), and is calculated as

$$O_{jk} = \frac{\sum p_{ij}p_{ik}}{\sqrt{\sum p_{ij}^2 p_{ik}^2}} \qquad \text{Equation 4}$$

where O_{jk} is Pianka's measure of overlap between species j and species k, p_{ij} is the proportion that resource i is of the total resources used by species j, and p_{ik} is the proportion that resource i is of the total resources used by species k. This measure ranges from 0 (no resources used in common) to 1 (complete overlap).

In our lizard example, we can plug in the numbers and calculate M to determine the extent to which whiptail lizards are overlapped by side-blotched lizards, and the extent to which side-blotched lizards are overlapped by whiptail lizards. We can compute Pianka's measure of overlap, O, as well. The results give us some indication of how food resources are partitioned between the two species in the community. Keep in mind that these measures suggest a potential for competition between species, which in turn may affect the diversity of species present at a site, but they do not provide direct evidence that the presence of one species can influence the population dynamics of the second.

PROCEDURES

In this exercise, you'll set up a spreadsheet to calculate both niche breadth and niche overlap of two hypothetical species. A primary goal is to be able to determine, in Questions 3 and 4, how diets must change in order to either maximize or minimize niche breadth and niche overlap.

As always, save your work frequently to disk.

INSTRUCTIONS	ANNOTATION
A. Set up the model community.	
1. Open a new spreadsheet and set up column headings as shown in Figure 3	
2. For each species, fill in the numbers of foraging attacks shown in Figure 1 for resources 1 through 10.	We'll focus on two species and assume that we can record how many times we observe foraging attacks on 10 major food resources, listed in cells A6–A15. Glancing at the raw data, which species do you think has a broader niche breadth?

	A	B	C	D	E	F	G	H	I	J
1	*Niche Breadth and Resource Partitioning*									
2										
3										
4		Species 1				Species 2				
5	Resource	# users	$\#^2$	p	p^2	# users	$\#^2$	p	p^2	$p1*p2$
6	1	7				0				
7	2	1				0				
8	3	286				38				
9	4	71				24				
10	5	0				30				
11	6	0				140				
12	7	0				5				
13	8	0				0				
14	9	0				0				
15	10	0				0				
16	Y =	365				237				
17										
18	n =					n =				
19	B =					B =				
20	B_A =					B_A =				
21	M_{12}=					M_{21}=				
22	O =					O =				

Figure 3

3. Use the **SUM** function in cell B16 and F16 to count the total number of foraging attacks observed for species 1 and 2, respectively.

Enter the formula **=SUM(B6:B15)** in cell B16.
Enter the formula **=SUM(F6:F15)** in cell F16.

4. In cells C6–C15, enter a formula that squares the number of foraging attacks on prey item 1 for species 1.

Enter the formula **=(B6^2)** in cell C6. Copy this formula down to cell C15. The columns labeled $\#^2$, p, and p^2 are simply steps that you need to compute in order to estimate niche breadth and overlap at a later point in time. The ^ symbol indicates that the value in cell **B6** is to be raised to a power (in this case, the power of **2**).

5. In cells D6–D15, calculate the proportion (p) of the total number of attacks for each resource type.

Enter the formula **=B6/$B$16** in cell D6. Copy this formula down to cell D15.

6. In cells E6–E15, square the values in column D.

Enter the formula **=D6^2** in cell E6. Copy this formula down to cell E15.

7. Sum your column values in cells C16–E16.

Enter the following formulae:

- Cell C16 **=SUM(C6:C15)**
- Cell D16 **=SUM(D6:D15)**. This result should be 1.
- Cell E16 **=SUM(E6:E15)**

8. Calculate $\#^2$, p, and p^2 for species 2.

Enter the following formulae:

- Cell G6 **=(F6^2)**
- Cell H6 **=F6/$F$16**
- Cell I6 **=H6^2**.

Copy these formulae down to row 15.

9. In cell J6, multiply p (species 1) by p (species 2), and copy your formula down to cell J15.

Enter the formula **=D6*H6** in cell J6.

10. Sum cells J6–J15 in cell J16.

Enter the formula **=SUM(J6:J15)** in cell J16.

11. Save your work.

Your spreadsheet should now resemble Figure 4.

	A	B	C	D	E	F	G	H	I	J
4			Species 1				Species 2			
5	Resource	# users	#²	*p*	*p*²	# users	#²	*p*	*p*²	*p*1**p*2
6	1	7	49	0.019178	0.000368	0	0	0	0	0
7	2	1	1	0.00274	7.51E-06	0	0	0	0	0
8	3	286	81796	0.783562	0.613969	38	1444	0.160338	0.025708	0.125634
9	4	71	5041	0.194521	0.037838	24	576	0.101266	0.010255	0.019698
10	5	0	0	0	0	30	900	0.126582	0.016023	0
11	6	0	0	0	0	140	19600	0.590717	0.348947	0
12	7	0	0	0	0	5	25	0.021097	0.000445	0
13	8	0	0	0	0	0	0	0	0	0
14	9	0	0	0	0	0	0	0	0	0
15	10	0	0	0	0	0	0	0	0	0
16	Y =	365	86887	1	0.652182	237	22545	1	0.401378	0.145333

Figure 5

B. Calculate niche statistics.

With the basic calculations in place, you are now ready to calculate n, B, B_A, M_{12}, M_{21}, and O. Take a moment to review the equations presented in the Introduction to this exercise.

1. In cells B18 and G18, enter formulae to calculate n for each species.

Enter the formula **=COUNTIF(B6:B15,">0")** in cell B18.
Enter the formula **=COUNTIF(F6:F15,">0")** in cell G18.
These formulae count the number of entries in cells B6–B15 and F6–F15 that are greater than 0, hence providing information on n.

2. In cells B19 and G19, enter formulae to calculate B for each species.

Enter the formula **=1/E16** in cell B19.
Enter the formula **=1/I16** in cell G19.

3. In cells B20 and G20, enter formulae to calculate B_A for each species.

Enter the formula **=(B19-1)/(B18-1)** in cell B20.
Enter the formula **=(G19-1)/(G18-1)** in cell G20.

4. In cells B21 and E21, enter formulae to calculate M_{12} and M_{21}, respectively.

Enter the formula **=J16/E16** in cell B21.
Enter the formula **=J16/I16** in cell G21.

5. In cells B22 and G22, enter a formula to calculate O.

Enter the formula **=J16/SQRT(E16*I16)** in cells B22 and G22.

6. Save your work.

C. Create graphs.

1. Graph the overlap statistics for the two species.

2. Save your work.

Use the column graph option and label your axes fully. Your graph should resemble Figure 5.

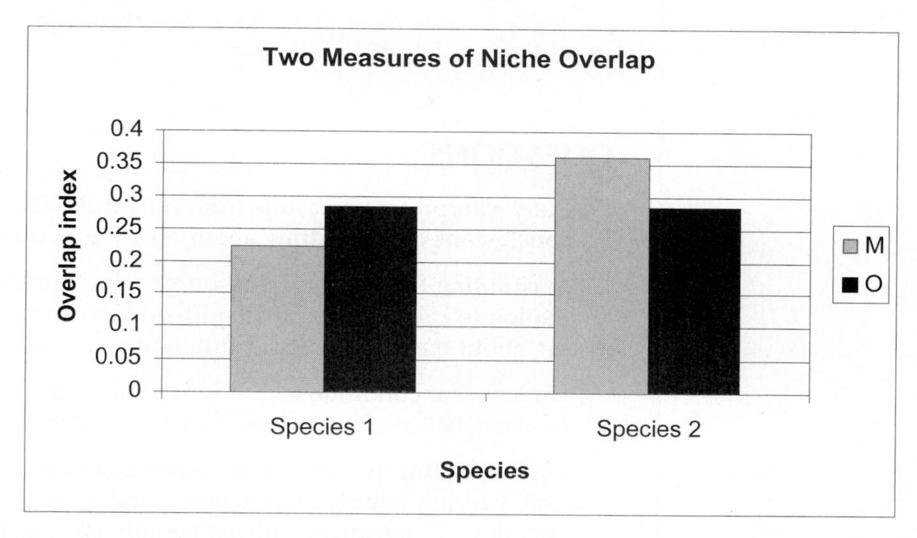

Figure 5

Using Solver

Questions 3 and 4 ask you to use the function **SOLVER** to mathematically *optimize* specific niche parameters. To access Solver, go to Tools | Solver and select Solver. (If Solver does not appear in the menu, go to Tools | Add-ins and select the Solver add-in. Your computing administrator may need to help you with the installation.) The dialog box in Figure 6 (see Question 3) will appear. In general, Solver works through the following steps:

- In the Set Target Cell box, enter a cell reference or name for the target cell. The target cell must contain a formula.
- To have the value of the target cell be as large as possible, click Max. To have the value of the target cell be as small as possible, click Min. To have the target cell be a certain value, click Value of, then type the value in the box.
- In the By Changing Cells box, enter a name or reference for each adjustable cell, separating nonadjacent references with commas. The adjustable cells must be related directly or indirectly to the target cell.
- In the Subject to the Constraints box, enter any **constraints** you want to apply. For instance, we will constrain the number of foraging attacks and the total observations. Table 1 lists of the operators that can be used in writing constraints.
- When you click Solve, Solver will run through several different scenarios with varying combinations of parameters, evaluating each combination given the constraints you identify. When the Solver finds a solution, a dialog box will

TABLE 1. Operators Used as Constraints in Solver

Operator	Meaning
<=	Less than or equal to
>=	Greater than or equal to
=	Equal to
int	Integer (applies only to adjustable cells)
bin	Binary (applies only to adjustable cells)

appear, asking you whether you want to keep the solution or to restore the original values.

- To keep the solution values on the worksheet, click Keep Solver Solution in the Solver Results dialog box. To restore the original data, click Restore Original Values.

QUESTIONS

1. Fully interpret your results from cells B18–B22 and cells G18–G22. What general conclusions can you draw about how the two species coexist in the community?
2. In counting the number of resource "hits" for each species and using this information to calculate niche breadth and overlap, what assumptions are you making about the availability of different resources in the environment?
3. Under what conditions would niche breadth, *B*, for species 1 be maximized? Under what conditions would niche breadth for species 1 be minimized?

   To answer this question, you could plug in some numbers for resource utilization, varying your scenarios from conditions in which a single item is utilized versus all 10 resources utilized equally versus all 10 resources utilized unequally. However, you can readily use the Solver function described at the end of the exercise to answer this question.

- To use the Solver, select cell B19, then go to Tools | Solver and the dialog box shown below.

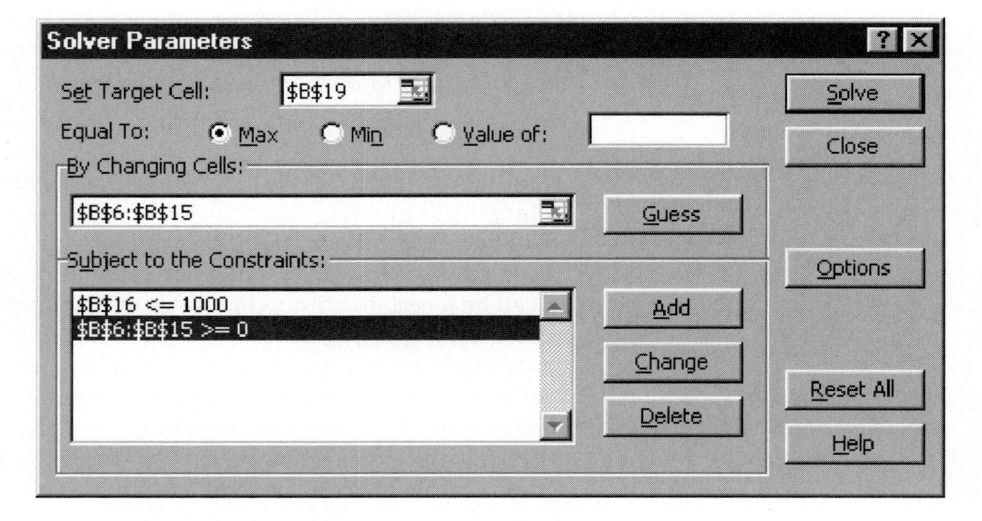

- To maximize the niche breadth of species 1, we want to set Target Cell B19 (B, niche breadth) to a maximum by changing cells B6–B15. You can use your mouse to click on cells for these entries, or you can directly type in the cell references in the appropriate locations. If you do the latter, make sure you type in absolute cell references (e.g., Set Target Cell to $B$19, not B19).
- The spreadsheet will figure out how to change the diets in order to maximize niche breadth, but we will *constrain* the numbers that the spreadsheet uses in the calculations. First select the Add button and add a constraint that the number of foraging attacks for each resource must be greater than or equal to 0. Then select the Add button again and constrain the total number of observations to 1000 (cell B16 must be less than or equal to 1000).
- Once you've entered the constraints, click the Solve button. The program will return a solution, with new values entered automatically in cells B6–B15.

You can either keep the Solver results (which pastes the Solver values into your spreadsheet), or simply interpret the results and cancel the Solver results to return to your original spreadsheet values. (You might want to copy your spreadsheet into a new worksheet if you wish to keep the Solver answers and your original cell entries). Note that you can also find minimums or specify a certain value that you want to be solved.

4. Assuming that species 2 cannot change its resource use, what diet should species 1 consume to minimize niche overlap with species 2? Again, you can use the Solver and set cell B22 to a minimum (and constrain the foraging observations to a total of 365, the original number of observations).
5. Ask an interesting question pertaining to niche overlap or niche breadth. Use your model to answer your question. Provide graphs to support your answer.

LITERATURE CITED

Gotelli, N. 2001. *A Primer of Ecology*, 3rd Edition. Sinauer Associates, Sunderland, MA.

Krebs, C. J. 1999. *Ecological Methodology*, 2nd Edition. Addison-Wesley, New York.

Levins, R. 1968. *Evolution in Changing Environments: Some Theoretical Explorations*. Princeton University Press, Princeton, NJ.

MacArthur, R. and R. Levins. 1967. The limiting similarity, convergence, and divergence of coexisting species. *American Naturalist* 101: 377–385.

Pianka, E. R. 1986. *Ecology and Natural History of Desert Lizards*. Princeton University Press, Princeton, NJ.

23 POPULATION ESTIMATION: MARK-RECAPTURE TECHNIQUES

Objectives

- Simulate the process of mark and recapture of individuals in a closed population.
- Estimate abundance using the Lincoln-Petersen method.
- Perform a Monte Carlo simulation to estimate the accuracy of the Lincoln-Petersen results.
- Determine how the number of individuals marked and number of individuals recaptured affects the precision of the Lincoln-Petersen index.
- Evaluate how emigration and capture probability can bias the Lincoln-Petersen index.

INTRODUCTION

How many moose are in Vermont? What is the population size of breeding black ducks in the Adirondacks? How many jaguars are in the Calakmul Biosphere Reserve in Mexico? How "confident" are we in our estimates? Estimating abundance in animals is a very common procedure for ecologists and land managers. This is because the size of a population can profoundly affect, among other things, its genetic make-up, probability of persistence, and rates of immigration, emigration, birth, and survival.

There are two basic ways of determining population size. The first is an actual "head count" of individuals, or **census**; the second is **estimation of population size through sampling**. The second method is the only option when (as is often the case) counting all individuals is impractical or impossible. There are different strategies for estimating plant and animal population sizes over time. The foremost difference is that animals move from location to location, whereas plants remain rooted in place and are thus often (but not always!) easier to count.

Because most animals are mobile, animal abundance is often estimated through mark-recapture techniques (Lancia et al. 1994). Deer, for example, are often marked with ear tags, and birds can be marked with color-coded bracelets attached to their legs. Marked animals are released and move freely about the population. A follow-up recapture session involves capturing a random sample of individuals from the population. Some individuals will contain markings, some will not. Mark-recapture techniques are based on the notion that the proportion of marked individuals in the second sample should be approximately equal to the proportion of marked animals in the total population. In other words, if you know the number of marked and unmarked individuals captured in the second sampling session, and you know the number marked in the first sampling session, you can estimate the original population size in the first sampling session.

Several different mark-recapture models exist, including the Lincoln-Petersen model, the Schnabel model, and the Jolly-Seber model. Of these, the Lincoln-Petersen method is the simplest, involving only a single marking session and a single recapture session. This procedure was used by C. J. G. Petersen in studies of marine fishes and by F. C. Lincoln in studies of waterfowl populations (Seber 1982). The data in the model include the number of individuals marked in the first sample (M); the total number of individuals that are captured in the second sample (C); and the number of individuals in the second sample that have markings (R). These data are used to estimate the total population size, N, as

$$\frac{N}{M} \approx \frac{C}{R} \qquad \text{Equation 1}$$

Let's assume we are trying to estimate the population size of ladybug beetles in a given area. Equation 1 says that the ratio of the *total* number of ladybugs in the population to the total number of *marked* ladybugs is equal to the ratio of the number of ladybugs in the *sample* to the number of *marked (recaptured)* ladybugs in the sample. We can rearrange Equation 1 to get an estimate, of the total population size:

$$\hat{N} = \frac{CM}{R} \qquad \text{Equation 2}$$

This formula is the Lincoln-Petersen index of population size. In our spreadsheet, we will allow resampling (that is, an individual may be recaptured more than once). In this situation, the following modified index provides a better overall estimate of the population size when multiple trials are conducted:

$$\hat{N} = \frac{M(C+1)}{R+1} \qquad \text{Equation 3}$$

The Lincoln-Petersen estimate assumes that the population is **closed**—that immigration and emigration are negligible and the population does not change in size between the mark and recapture sessions. Other assumptions include:

- The second sample is a random sample.
- Marking does not affect the recapture of individuals.
- Marks are not lost, gained, or overlooked.

The Schnabel model is similar (in theory) to the Lincoln-Petersen method but involves more than one mark and recapture episode. The Jolly-Seber model relaxes the assumption that the population is closed (see Krebs 1999 for an overview of these methods).

Once we have an estimate of population size, it's critical to determine just how *confident* you are in your estimate. After all, you *will* arrive at an estimate, but since all sampling involves error, your estimate is probably off target by some amount. In this exercise, you will use a Monte Carlo simulation to get a feel for the range of values returned by the Lincoln-Petersen index. A **simulation** is any analytical method meant to imitate a real-life system. A **Monte Carlo simulation** is a statistical technique in which a quantity is calculated repeatedly, using randomly selected "what-if" scenarios for each calculation. In a nutshell, the technique uses a data-generating mechanism (such as the random number function in a spreadsheet) to model a process you wish to understand (such as the "behavior" of the Lincoln-Petersen index, when, for example, $M = 20$ and $C = 30$). New samples of simulated data are generated repeatedly, and the results approximate the full range of possible outcomes. The likelihood of each possible result can then be computed. The Monte Carlo technique derives its name from the casinos of Monte Carlo in Monaco, where the major attractions are games of chance and the successful gamblers must constantly calculate the probabilities of multiple possible scenarios in their heads.

PROCEDURES

In this exercise, you'll simulate a mark and recapture of individuals in a population of size 100 (the number you are trying to estimate). You'll calculate the Lincoln-Petersen index of abundance, run a Monte Carlo simulation to see the range of possible outcomes, and examine how the estimate and confidence intervals change as sample effort changes and as assumptions to the model are violated. Once you are an expert at Monte Carlo simulations, you can use the procedure to determine the best strategy for winning money at blackjack and head to Las Vegas (or better yet, Monaco).

As always, save your work frequently to disk.

INSTRUCTIONS	ANNOTATION
A. Set up and mark the model population.	For the sake of this exercise, we will consider a population of 100 individuals. However, you, the field biologist, don't *know* the actual population size is 100—you are trying to estimate it using the mark-recapture technique. You have been granted funding to mark 20 individuals. (We'll explore what happens if you mark fewer or more individuals later in the exercise.)
1. Open a new spreadsheet and set up column headings as shown in Figure 1.	See Figure 1 below.
2. In cell E4, enter the number of individuals you will mark.	Enter the number 20 in cell E4.
3. Enter the letter **m** in cell E5.	The mark you will give to the 20 individuals is the letter **m**. The unmarked individuals will have the letter **u** associated with them.
4. Enter 1 in cells E6 and E7.	The Lincoln-Petersen method assumes that the population is closed (births, deaths, emigration, and immigration are negligible) and that all individuals have the same probability of capture and recapture. The values in cells E6 and E7 will allow us to explore violations of these assumptions. Cell E6 is the probability that an individual will remain in the population. For now it is set to 1 to meet the assumption that the population is closed. If individuals leave the population, either through death or emigration, that probability will decrease. Cell E7 is the probability that an individual will be recaptured, which we will also set to 1. If certain individuals (either marked or unmarked) tend to avoid traps in the recapture session, that probability will decrease. Perhaps they have learned trap locations and have become "trap shy."

	A	B	C	D	E
1	*Population Estimation - Lincoln-Petersen Mark-Recapture Model*				
2					
3					Initial Sampling
4		M = number of new individuals marked =			20
5				Marking =	m
6		probability of remaining in population =			1
7			probability of recapture =		1
8					
9	MARK		RECAPTURE		
10	Individual #	Marking	Individual	Marking	C

Figure 1

5. Set up a linear series from 1 to 100 in cells A11–A110.

Enter 1 in cell A11.
Enter **=1+A11** in cell A12. Copy your formula down to row 110.
This assigns a number to each individual in the population.

6. In cells B11–B110, enter an **IF** formula to mark the first 20 individuals with an **m**, and designate the remainder **u** (unmarked).

Enter the formula **=IF(A11<=$E$4,$E$5,"u")** in cell B11. Copy this formula down to cell B110.
This formula tells the spreadsheet to examine the number in cell A11. If that number is less than or equal to (<=) to value in cell E4 (i.e., **20**), return the marking listed in cell E5 (i.e., **m**); otherwise, return the letter **u**.

7. Save your work.

B. Simulate the recapture of individuals.

Now we have a sample of marked individuals that have been released back into the population, and we can (after a period of time) resample the population and compute the Lincoln-Petersen index. First we will "reshuffle" the population, draw individuals from the population at random, and determine whether the individuals are marked or not.

1. In cell C11, generate a random number between 1 and 100. Copy your formula down to row 110.

Two different formulae can be used to generate a random number between 1 and 100:

- **=RANDBETWEEN(1,100)**
- **=ROUNDUP((RAND()*100),0)**

The **RANDBETWEEN** formula is fairly straightforward. If this function is not available in your spreadsheet package, the second formula will work by generating a random number between 0 and 1 (the **RAND()** portion of the formula), multiplying the number by 100 (***100**) and rounding the result up to **0** decimal places.

2. Use the **AND** function in cell G3.

Enter the formula **=AND(RAND()<=$E$6,RAND()<=$E$7)** in cell G3.
We'll take a moment to learn about the **AND** function, which we'll use as part of the formula in the next step. The **AND** function evaluates conditions you specify and returns the word "true" only if *all* the conditions you specify are true, and the word "false" if *any* of the conditions are not true. It has the syntax **AND(condition1, condition2, . . .)**. The formula in cell G3 generates two random numbers between 0 and 1 (the **RAND()** portion of the formula). The conditions are that the first random number must be less than or equal to the value in cell **$E$6** (the probability of remaining in the population), and that the second random number must be less than or equal to the value in cell **$E$7** (the probability of being captured in the second sampling bout). Since cells E6 and E7 are currently set to 1, both random numbers will be less than or equal to 1, so the program will return the word "true."

Now set cell E6 and E7 to 0.7 and press F9, the calculate key, to see how this formula works. Occasionally, a random number greater than 0.7 will be drawn, and the program will return the word "false." When you are satisfied that you understand how the **AND** function works, return cells E6 and E7 to the value 1 and continue to the next step.

3. In cells D11–D110, enter a formula to determine whether or not a recaptured individual was marked.

Enter the formula **=IF(AND(RAND()<$E$6,RAND()<$E$7), VLOOKUP($C$11:$C$110,$A$11:$B$110,2),".")** in cell D11. Copy this formula down to cell D110.
Now we are ready to determine if the individual that was sampled was marked or not. We also need to determine if the individual that was sampled left the population through death or emigration (cell **$E$6**) and if the individual is trap-shy (cell **$E$7**). The formula in cell D11 is a combination of four functions: **IF**, **AND**, **RAND**, and **VLOOKUP**. Keep in mind that Excel performs the innermost functions first and then moves to the outer functions.

The two inner functions are **RAND()** functions, which draw a random number between 0 and 1. The first random number is compared to the value in cell **$E$6**, which is the probability that the individual remains in the population. The second random number is compared to the probability that the individual is not trap-shy. In order for an individual to be captured in the recapture session, both probabilities need to be considered; this is done with the **AND** function, which will return "true" only if the individual stays in the population and is not trap-shy. Now we are ready for the **IF** function. If the individual remained in the population and was not trap-shy, then Excel moves to the **VLOOKUP** function. However, if the individual either left the population through death or emigration, or was trap-shy, Excel returns a missing value in the cell (**"."**).

The **VLOOKUP** formula searches for a value in the leftmost column of a table and then returns a value in the same row from a column you specify in the table. It has the syntax **VLOOKUP(lookup_value, table_array, col_index_num, range_lookup)**. So, assuming the individual was indeed captured, Excel will look up the value given in column C (the shuffled individuals) in a table given in columns A and B (specifically, cells A11–B110), and will return the value in the second column of the table (**m** or **u**); note that the **range_lookup** parameter is optional, and we are leaving it blank. In other words, assuming the individual is still in the population and can be captured, the **VLOOKUP** formula will find its number in column A and relay its marking from column B.

4. In cells E11–E110, sum two COUNTIF formulae to tally *C*, a running tally of the number of marked (**m**) plus unmarked (**u**) individuals recaptured.

Enter the formula **=COUNTIF($D$11:D11,"u")+COUNTIF($D$11:D11,"m")** in cell E11. Copy the formula down to cell E110.
To calculate C in column E, we count the individuals that are marked and those that are unmarked, then sum the two together. Remember to "anchor" the first reference to cell D11 with dollar signs (absolute reference). Also remember to use quotes around the letters u and m since they are nonnumerical data. Note that when cells E6 and E7 are both set to 1 (i.e., when the population is closed and no individuals learn to evade recapture), this formula will simply produce a linear series from 1 to 100 in column E. When the value in E6 or E7 is less than 1, however, not every capture attempt in column D will result in capturing an individual, so we will need this column to keep track of those that do.

5. Press F9, the calculate key, to simulate recapture outcomes.

C. Calculate and graph the Lincoln-Petersen index.

1. Set up column headings in cells F9–G10 as shown.

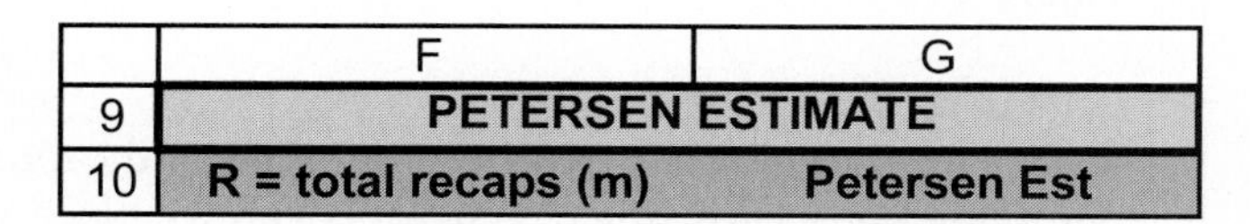

	F	G
9	PETERSEN ESTIMATE	
10	R = total recaps (m)	Petersen Est

Figure 2

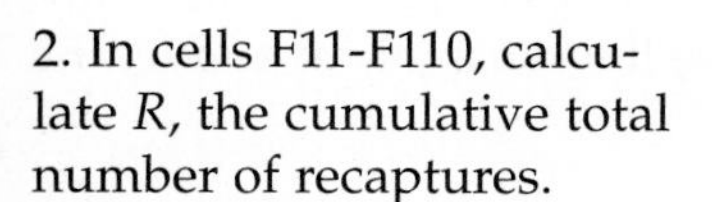

2. In cells F11-F110, calculate *R*, the cumulative total number of recaptures.

To calculate the Lincoln-Petersen index, we need to keep track of *M*, *C*, and *R*. We'll assume that we start to recapture individuals one at a time, and we'll calculate the Lincoln-Petersen index each time a new individual is captured. The number marked, *M*, is given in cell E4. The numbers captured in the second session, *C*, are given in column E. A count of the number recaptured that were marked (*R*) will be tallied in column F. Row 11 simulates our first capture (Figure 3). We need to determine if the individual was marked or not, and then keep a running tally of recaptured individuals as we continue to capture individuals.
Enter the formula **=COUNTIF($D$11:D11,"m")** in cell F11. Copy your formula down to row 110.

	C	D	E	F
10	Individual	Marking	C	R = total recaps
11	12	m	1	1
12	95	u	2	1
13	31	u	3	1
14	2	m	4	2
15	2	m	5	3
16	88	u	6	3

Figure 3

3. Calculate the Petersen estimate in cells G11-G110.

Enter the formula **=($E$4*(E11+1))/(F11+1)** in cell G11. Copy the formula down to cell G110.
This is the spreadsheet version of Equation 3:

$$\hat{N} = \frac{M(C+1)}{R+1}$$

4. Graph the Lincoln-Petersen index as a function of *C*, the number of individuals captured in the second sampling bout.

Use the line graph option and label your axes fully. Your graph should resemble Figure 4.

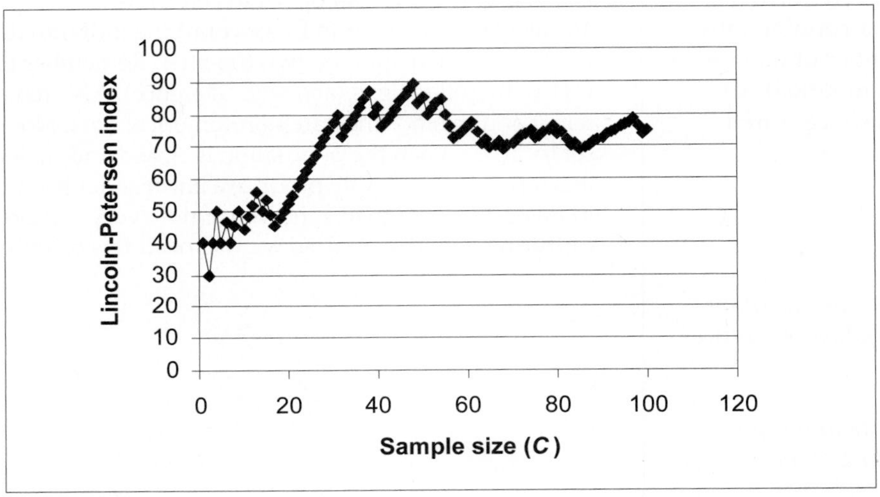

Figure 4

5. Answer questions 1–3 at the end of the exercise before proceeding.

D. Perform a Monte Carlo simulation.

Let's suppose that we mark 20 individuals and capture 20 individuals in the second sampling bout. How much confidence can we place in the resulting Lincoln-Petersen estimate? In this section we will set up a Monte Carlo simulation to see the range of estimates returned by our Lincoln-Petersen index. To do this, we will need to repeat our entire exercise 1000 times, each time generating a new index. Then we will examine how the index "behaves" based on our 1000 trials. We'll write a macro and let the computer do the tedious work for us.

1. Return the value in cell E4 to 20 individuals marked.

2. Set up new column headings in cells I9–O10 as shown in Figure 5.

	I	J	K	L	M	N	O
9	MONTE CARLO SIMULATION						
10	Trial	L-P index	Low 2.5%	High 2.5%	Mean	Summary	

Figure 5

3. Set up a linear series from 1 to 1000 in cells I11–I1010.

Enter 1 in cell I11.
Enter **=1+I11** in cell I12. Copy the formula down to cell I1010.

4. Set the calculation key to manual.

Open Tools | Options | Calculation and select Manual.

5. Develop a macro to run a Monte Carlo simulation.

Bring your spreadsheet macro program into record mode and assign a name and shortcut key (we used the shortcut <Control>+<m>).

Figure 6

If the small Stop Recording toolbar (Figure 6) doesn't automatically appear, open View | Toolbars | Stop Recording. The filled square on the left is the "stop recording" button, which you press when you complete your macro. The button to the right is the **relative reference button**. By default the button is "off," as shown above, which means that your macro records keystrokes as absolute references. Leave the button off *for now* and record the following steps:

- Select cell E10.
- Press F9, the calculate key, to generate new random numbers and hence a new simulation of mark-recapture.
- Open Edit | Find. Enter the number 20 in the box labeled Find What as shown in Figure 7. Select the Search by Columns and Look in Values options. Click the Find Next button, then Close. Excel will move your cursor down to the 20th individual captured.

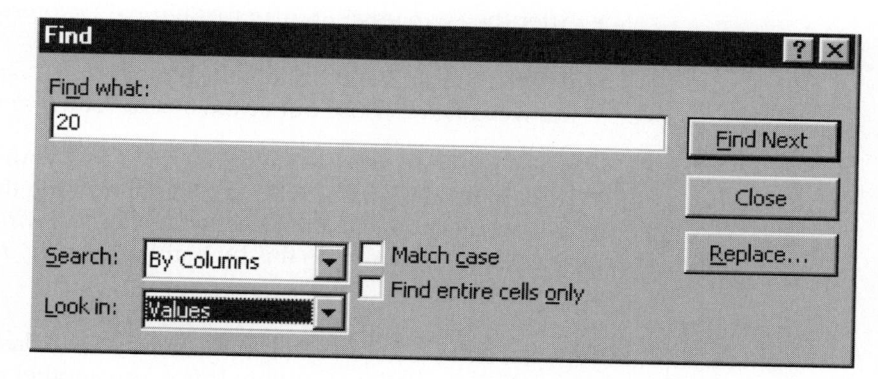

Figure 7

- Press the relative reference button (see Figure 6); it should become a lighter shade when depressed. Excel now assumes that cell references are relative rather than absolute.
- Use the right arrow key to move your cursor two cells to the right. This cell holds the Lincoln-Petersen estimate associated with 20 captured individuals in the second session and a variable number of marked and recaptured individuals.
- Click the relative reference button off.
- Open Edit | Copy.
- Select cell J10.

- Open Edit | Find. Leave the Find What box blank and Search by Columns. Click the Find Next button, then Close.
- Open Edit | Paste Special. Then select the Paste Values option. Press OK.
- Click on the Stop Recording button.

Now when you press your shortcut key 1000 times you will generate 1000 new Lincoln-Petersen indices, each one generated by random numbers and following the parameters established in the model. A simple shortcut outlined in the next step can save you 1000 keystrokes.

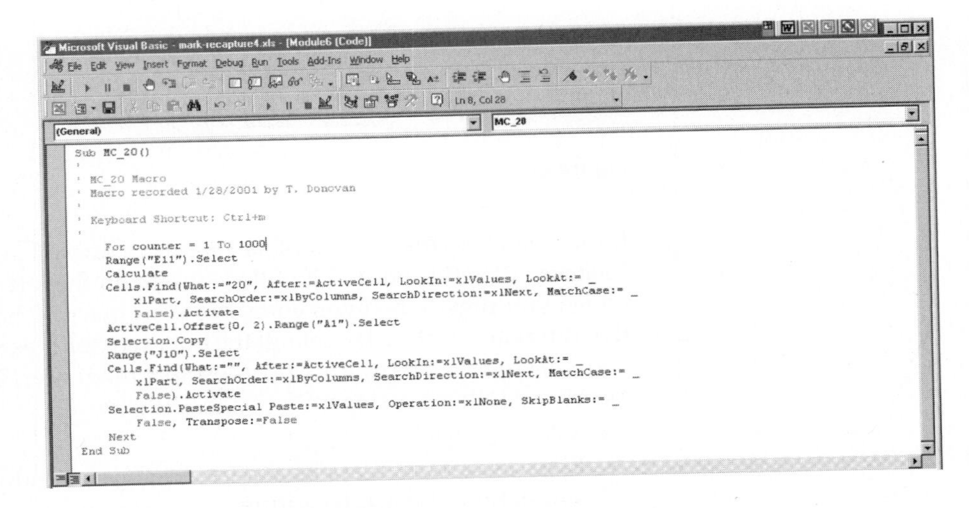

Figure 7

6. (Optional) Edit your macro using the Visual Basic code.

To edit your macro, open Tools | Macro | Macros, and click the Edit button. You should see a box that reveals the Visual Basic Applications code that Excel recorded as you entered your macro (Figure 7)

- After the `Keyboard Shortcut Control+m`, press Return and type in the words **`For counter = 1 to 1000`**
- Before the last line of code, which reads `End Sub`, create a new line and type in the word **`Next`**. Close out of the box to return to your spreadsheet.

Now you press <control>+<m> just once and your new macro, which consists of 1000 different simulations, will run. Before running the macro, you should delete any previous results from column J (otherwise you will wind up with more than 1000 results in this column). You can do this by highlighting any results in this column and pressing the Delete key.

When you press <control>+m, your computer will flash for several minutes as it cranks through the simulation. Caution: If you use another program while the simulation is running, be careful not to copy material to the clipboard—the simulation is making extensive use of the clipboard (through copy and paste), so putting other material there can cause errors.

7. Examine your results from 1000 trials.

Bear in mind that in actual mark-recapture experiments we don't know the total population size—that's what we're trying to estimate. This Monte Carlo simulation allows us to determine, for the special case in which $N = 100$, just how likely the Lincoln-Petersen index is to come up with an "acceptable" estimate. What is acceptable will depend on the purpose of the experiment (see question 4 at the end of the exercise).

When analyzing results, scientists like to be at least 95% certain that a given result is not due to chance. You can use your spreadsheet to see the range of values that the Lincoln-Petersen estimate will return 95% of the time. Since you have 1000 results from your Monte Carlo simulation, the "middle" 950 values represent this range. The remaining 50 values are the 25 highest and 25 lowest Lincoln-Petersen estimates from your simulation. We are interested in determining the 25th highest observation and the 975th highest observation. The **LARGE** function does this: it returns the *k*th largest value in a data set—you specify the data set and what value you want returned.

8. In cell K11, compute the value of the 975th highest estimate.

Enter the formula **=LARGE(J11:J1010,975)** in cell K11.

9. In cell L11, compute the value of the 25th highest estimate.

Enter the formula **=LARGE(J11:J1010,25)** in cell L11.

10. In cell M11, compute the average Lincoln-Petersen index from your simulation.

Enter the formula **=AVERAGE(J11:J1010)** in cell M11.

E. Optional: Generate descriptive statistics on your results.

This step requires that that the Analysis ToolPak be activated. To activate the ToolPak, go to Tools | Add-Ins and click on the ToolPak option, then press OK. To generate descriptive statistics, go to Tools | Data Analysis | Descriptive Statistics. The dialog box in Figure 8 will appear.

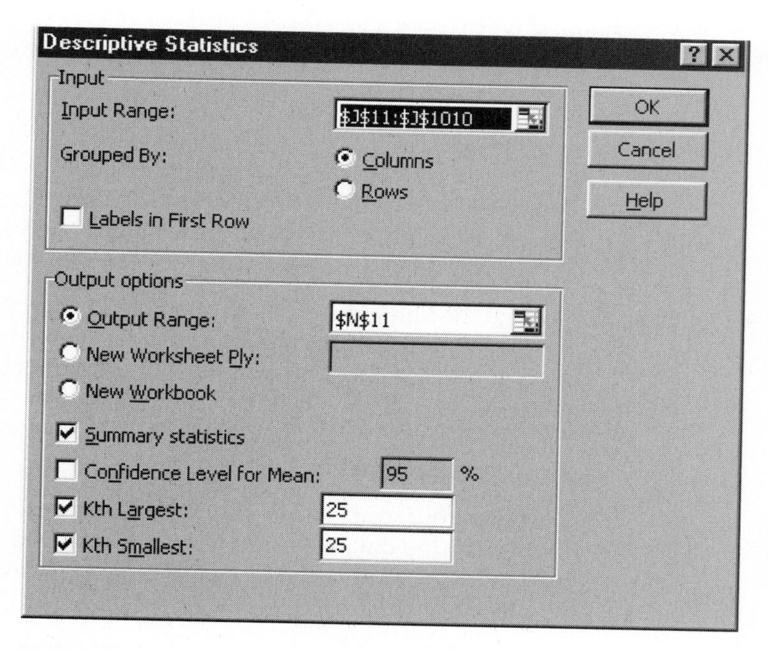

Figure 8

The Input Range will be the results of your 1000 simulations (you can use the relative reference button to enter this). Use **$N$11** as the output range, check the Summary statistics option, and enter **25** as the Kth Largest and Kth Smallest values. Excel will return descriptive statistics in columns N and O, as shown in Figure 9. The 95% confidence intervals are obtained by examining the 25 highest and lowest values. The confidence values should match the values you computed in cells K11 and L11. Our simulation revealed that, for a population where 20 individuals are marked in the first session and 20 individuals are captured in the second session, the Lincoln-Petersen index fell between 46.7 and 210 individuals 95% of the time. Your answer might be slightly different. Remember that the true population size is 100 individuals. You may be able to get better estimates by changing M and/or C.

	N	O
11	*Column1*	
12		
13	Mean	100.3880152
14	Standard Error	1.746782658
15	Median	84
16	Mode	84
17	Standard Deviation	55.23811776
18	Sample Variance	3051.249654
19	Kurtosis	13.9588804
20	Skewness	3.130416209
21	Range	385
22	Minimum	35
23	Maximum	420
24	Sum	100388.0152
25	Count	1000
26	Largest(25)	210
27	Smallest(25)	46.66666667

Figure 9

QUESTIONS

1. Based on your initial setting of $M = 20$, how does C (the number captured in the second sampling bout) affect the Lincoln-Petersen index? Press F9, the calculate key, to run several simulations and get a qualitative feel for the relationship.
2. Change the value in cell E4 to 50, then 70, then 90 to increase the proportion of the population that is initially marked. For each value, press F9 several times to get a general feel for the results. How does this increase in proportion of marked individuals affect the Lincoln-Petersen estimate? What happens to the Lincoln-Petersen estimate when 100 individuals are marked? Use graphs to illustrate your answer.
3. Examine your graph from Part E (the Lincoln-Petersen index as a function of C). How were the data collected to generate such a relationship? Is this a legitimate way to evaluate how the Lincoln-Petersen index changes as C increases? Why or why not?
4. Suppose you are planning to study population fluctuations of a species of frog living in a particular pond, and your initial "guesstimate" is that the pond currently has about 100 frogs living in it. Discuss the value of estimating variations in the population size by marking 20 individuals and recapturing 20 individuals. Try different values for M and C to try to determine an experimental design that will produce an "acceptable" margin of error. Which has a greater effect on the range of results: increasing M or increasing C?

   To change M, simple change the value in cell E4. To change C, you need to edit the macro: Open Tools | Macro | Macros, highlight your macro on the list that appears, and click on the edit button. In the macro editing window that opens, the first "Find What" value represents C, and you can change it to any value up to 100. Remember to clear your results from column J before running the macro each time (or, if you want to keep your previous results, save your spreadsheet with a different name).

5. Set cell E4 equal to 50. How do violations of the assumptions of "closed" population and equal catchability affect the Lincoln-Petersen estimate? Set cells E6 to 0.6 (thus, 40% of the individuals leave the population) and set cell E7 to 0.7 (thus, 30% of the individuals are unlikely to be captured in the second sampling bout for some reason). Assuming that you can recapture 30 individuals ($C = 30$), how do the results of the Monte Carlo simulation change as a result of these violations?

*6. Assume that the population is closed (cell E6 = 1). Assume further that the probability of recapture pertains only to those individuals that were marked in the initial sampling period (perhaps the individuals have learned to avoid traps after being captured earlier). How could the model be modified to reflect this situation?

LITERATURE CITED

Krebs, C. J. 1999. *Ecological Methodology*, 2nd Ed. Benjamin/Cummings. Menlo Park, CA.

Lancia, R. A., J. D. Nichols and K. H. Pollock. 1994. Estimating the number of animals in wildlife populations. *In* T. Bookhout (ed.), *Research and Management Techniques for Wildlife and Habitats*, 5th Ed., pp. 215–253. The Wildlife Society, Bethesda, MD.

Seber, G. A. F. 1982. *The Estimation of Animal Abundance and Related Parameters*, 2nd Ed. Macmillian, New York.

24 SURVIVAL ANALYSIS

Objectives

- Simulate the fates of 25 individuals over a 10-day period.
- Calculate the Kaplan-Meier product limit estimate.
- Graphically analyze the Kaplan-Meier survival curve.
- Assess how sample size affects the Kaplan-Meier estimate.
- Assess how censorship affects the Kaplan-Meier estimate.

Suggested Preliminary Exercise: Life Tables and Survivorship Curves

INTRODUCTION

A population of black bears has been surveyed for 10 years, and ecologists note that the number of bears in the population has declined over this time frame. Why? Changes in numbers of individuals over time can be directly traced back to the population's birth, death, immigration, and emigration rates. The population may have declined because the birth rate dropped, the death rate increased, immigration dropped, or emigration increased. A combination of any or all of these factors may also be responsible for the decline. *Mortality* and its counterpart, *survival*, are keys to the demographic equation for all organisms. How do ecologists estimate these two important parameters? In this exercise we'll explore one method for estimating survival.

In your life table exercise, you tracked the fates of individuals over time, noting how many individuals in the cohort were still alive at each time step, and then calculated the survivorship schedule and survival probabilities from your data. Suppose we followed a cohort of 100 newborns over time, carefully noting when deaths occurred. We start with $S_0 = 100$, count individuals again at the next time step (S_1) and then at time step S_2. Suppose $S_1 = 40$ and $S_2 = 10$. The *survivorship schedule* (see Exercise 12, "Life Tables, Survivorship Curves, and Population Growth") tells us that the probability that an individual will survive *from birth to time x*. Thus, the probability of surviving to age 1 is $S_1/S_0 = 40/100 = 0.4$, and the probability of surviving from birth to age 2 is $S_2/S_0 = 10/100 = 0.1$. **Age-specific survival probabilities**, in contrast, tell us the probability that an individual will survive ***from one age to the next***—such as the probability that an individual alive in time S_1 will be alive at time S_2. In life table calculations, the age-specific survival probability is calculated as $g_x = l_{x+1}/l_x$. In our example, the probability that an individual of age S_1 will survive to age S_2 is $0.10/0.40 = 0.25$. The life table "cohort" analysis is one way of calculating survival. However, this method is not always possible to use, especially if the organisms of interest are long-lived. Fortunately, alternatives for estimating survival exist.

Kaplan-Meier Survival Analysis

When the research question can be posed as "how long does it take until death occurs?" the **Kaplan-Meier survival analysis**, also known as the Kaplan-Meier product limit estimate or the Kaplan-Meier survival curve, can be used to estimate survival. The Kaplan-Meier method (1958) involves tracking the fates of individuals over time and estimating how long it takes for death to occur. The method has been applied broadly to measure how long it takes for *any specific event* to occur—such as the time it takes until death, the time until a cancer patient recovers from a treatment, the time until an infection appears, the time until pollination occurs, and so on.

The Kaplan-Meier method is conceptually similar to life table calculations because you keep track of the number of individuals alive and the number of deaths that occur over intervals of time. Specifically, you count the number of individuals who die at a certain time and divide that number by the number of individuals that are "at risk" (alive and part of the study) at that time. If we do this for each time period in the study, we will be able to compute two survival probabilities: the *conditional survival probability* and the *unconditional survival probability*. We will describe how each is computed with a brief example.

Suppose you initiate a study on beetle mortality and track 20 individuals over 5 days, each day recording the number of deaths and the number of individuals still alive. Let's also suppose that some of your population decides to emigrate out of the population so you can no longer track their fates. The data you collect are:

	A	B	C
1	Day	Emigrants	Deaths
2	1	1	3
3	2	0	4
4	3	1	2
5	4	0	1
6	5	0	2

Now let

t be a particular time period, such as 1 day

d be the number of deaths at time t_i

n be the number of individuals at risk at the beginning of time t_i.

The **conditional survival probability,** P_c, is the probability of surviving to a specific time, given that you survived to the previous time (this is similar to the age-specific survival probabilities in the life table). P_c is computed as

$$P_c = 1 - \frac{d_i}{n_i} \qquad \text{Equation 1}$$

The term d_i / n_i gives the number of individuals that die in time step i divided by the number of individuals still alive *and* still in the population (the number at risk). This is the conditional mortality probability, or the probability that an individual will die during that time step. Since survival can be computed as 1 minus mortality, Equation 1 gives the conditional survival probability.

Because we started with a population of 20 individuals, the number at risk for death at the beginning of day 1 is 20. During that day, 3 individuals died, so the conditional mortality probability is 3/20 = 0.15, and the conditional survival probability is 1 – 0.15 = 0.85. Now let's consider day 2. At the beginning of day 2, there are only 16 individuals at risk. Three individuals died the previous time step, and one left the population through emigration. The individual that left the study is called a **censored observation**.

Individuals that die in the previous time step, as well as censored individuals, cannot be considered at risk, so on day 2 only 16 individuals are at risk. On day 2, 4 deaths occurred, so the conditional mortality probability is 4/16 = 0.25, and the conditional survival probability is 1 – 0.25 = 0.75. The rest of the computations are shown in Figure 1.

	A	B	C	D	E	F
1	Day	Emigrants	Deaths	# at risk	Deaths / at risk	P_c
2	1	1	3	20	= 3/ 20 = 0.15	1 - 0.15 = 0.85
3	2	0	4	= 20 - 3 - 1 = 16	= 4 / 16 = 0.25	1 - 0.25 = 0.75
4	3	1	2	= 16 - 4 - 0 = 12	= 2 / 12 = 0.16	1 - 0.16 = 0.84
5	4	0	1	= 12 - 2 - 1 = 9	= 1 / 9 = 0.11	1 - 0.11 = 0.89
6	5	0	2	= 9 - 1 = 8	= 2 / 8 = 0.25	1 - 0.25 = 0.75

Figure 1

The **unconditional survival probability**, P_u, is the probability of survival from the start of the study *to* a specific time (this is similar to the survivorship schedule in the life table). The unconditional probability is equal to the cumulative product of the conditional probabilities, which is why the Kaplan-Meier method is sometimes called the Kaplan-Meier product limit estimate. The equation can be expressed as

$$P_u = \prod_{j=1}^{i} \left(1 - \frac{d_j}{n_j}\right) \qquad \text{Equation 2}$$

where the Π symbol means "multiply all of the individual conditional probabilities together." The computations are shown in Figure 2.

For day 1, the unconditional survival probability is the same as the conditional survival probability. P_u for day 2 gives the probability that an individual at the start of the study will survive through day 2. This is obtained by multiplying the conditional survival probability for day 1 by day 2, since both conditions must be met in order for an individual to be alive at the end of day 2.

	A	B	C	D	E	F	G
1	Day	Emigrants	Deaths	# at risk	Deaths / at risk	P_c	P_u
2	1	1	3	20	= 3/ 20 = 0.15	1 - 0.15 = 0.85	= 0.85
3	2	0	4	= 20 - 3 - 1 = 16	= 4 / 16 = 0.25	1 - 0.25 = 0.75	= 0.85 * 0.75 = .6375
4	3	1	2	= 16 - 4 - 0 = 12	= 2 / 12 = 0.16	1 - 0.16 = 0.84	= 0.85 * 0.75 * 0.84 = .54
5	4	0	1	= 12 - 2 - 1 = 9	= 1 / 9 = 0.11	1 - 0.11 = 0.89	= 0.85 * 0.75 * 0.84 * 0.89 = .48
6	5	0	2	= 9 - 1 = 8	= 2 / 8 = 0.25	1 - 0.25 = 0.75	= 0.85 * 0.75 * 0.84 * 0.89 * 0.75 = .36

Figure 2

Notice that P_u decreases with each day because the probability of living to a given period must decrease as ever-greater time periods are considered. Sometimes ecologists are interested in expressing P_u as a daily probability. To obtain a daily survival estimate, you would take the appropriate root. For example, $P_u = 0.36$ on day 5 in Figure 2. This gives the probability that an individual will survive through day 5. What would daily survival be to obtain $P_u = 0.36$ on day 5? A daily probability of x would have to yield 0.36 when multiplied by itself once for each day, so $x^5 = 0.36$. By taking the fifth root of 0.36, you could solve for x. The spreadsheet formula is **0.36^(1/5)**.

Kaplan-Meier Survival Curves

The results of the Kaplan-Meier analysis are often graphed; graphs are known as the Kaplan-Meier survival curves (Figure 3). Comparing the survival curves of two different populations can yield insightful information about the timing of deaths in

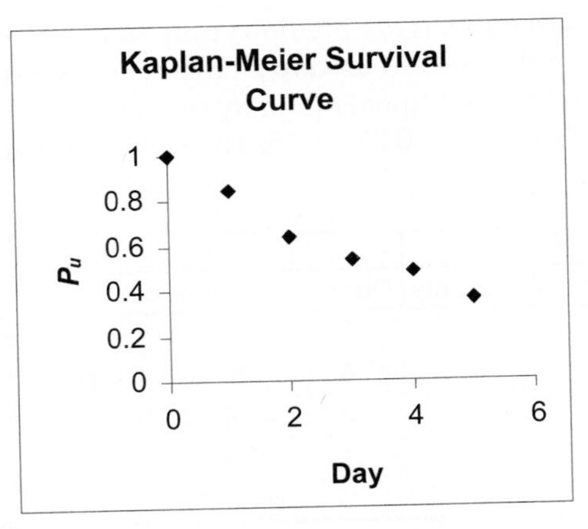

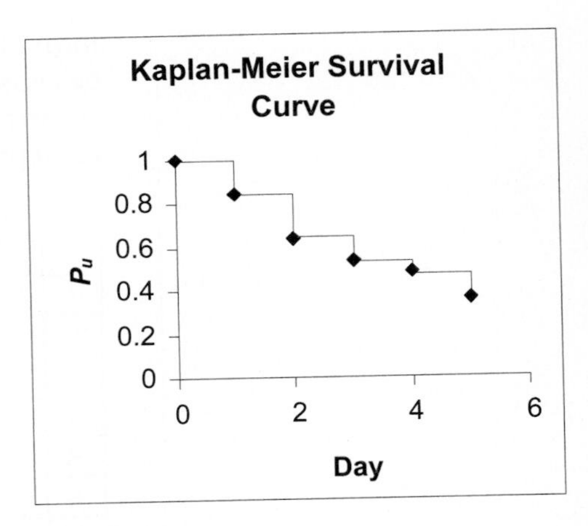

Figure 3 Kaplan-Meier survival curves for a hypothetical population. The unit time is plotted on the x-axis; P_u is plotted on the y-axis. In Kaplan-Meier curves, the raw data are plotted as in the graph on the left, then the data points are connected with horizontal and vertical bars as shown on the right. Large vertical steps downward indicate a large number of deaths in the given time period, while large horizontal steps indicate few deaths have occurred during an interval.

response to different environmental conditions. Often in the literature, you will see the survival curves for two different populations on the same graph so that you can compare the two easily.

PROCEDURES

The method outlined by Kaplan and Meier (1958) is one of the most referenced papers in the field of science, suggesting that is has played an important role in ecology and other sciences since its publication. The goal of this exercise is to set up a spreadsheet model of the Kaplan-Meier product limit estimate, and to learn how censored observations and sample size affect the survival probabilities. As always, save your work frequently to disk.

INSTRUCTIONS

ANNOTATION

A. Set up the model population.

1. Open a new spreadsheet and set up column headings as shown in Figure 4.

We'll track 25 individuals for 10 days and keep track of their fates over time. Row 10 will track Individual 1's fate, Row 11 will track Individual 2's fate, and so on to Row 34.

	A	B	C	D	E	F	G	H	I	J	K
1	*Survival Analysis*										
2											
3	Model Inputs:										
4	Survival =	0.9									
5	Total sample =	25									
6	Prob. of censor =	0.1	0.1	0.1	0.1	0.1	0.1	0.1	0.1	0.1	0.1
7											
8		Day									
9	Individual	1	2	3	4	5	6	7	8	9	10

Figure 4

2. Set up a linear series from 1 to 25 in cells A10–A34.

In cell A10 enter the value 1.
In cell A11 enter the formula **=1+A10**. Copy this formula down to cell A34.

3. In cell B4, enter a value for the probability that an individual will survive each 24-hour period (daily survival).

Enter the value 0.9 in cell B4. In reality, you wouldn't know what this number is; you are using the Kaplan-Meier method to estimate this parameter.

4. Enter the number of individuals in the initial population in cell B5.

Enter the value 25 in cell B5.

5. In cells B6–K6, enter a value for the probability that an individual in the population will be censored on a given day.

Enter the value 0.1 in cells B6–K6.
This is the probability that an individual will leave the study on any given day so that its fate cannot be tracked over time. For now, we set that probability to 0.1 for all days. Later in the exercise you will change these values to determine how censored observations, and the time at which they occur, affect survival probability estimates.

5. Save your work.

B. Simulate fates of individuals over time.

1. In cells B10–B34, enter a formula to assign a fate to each individual for day 1.

In cell B10 enter the formula **=IF(RAND()<$B$6,"C",IF(RAND()>$B$4,"D",1))**. Copy your formula down to row 34.
The formula in B10 will assign a fate to individual 1 on day 1. The individual will be either alive (1), censored (C), or dead (D). The formula contains two **IF** functions and a **RAND** function, so it is a nested formula. Remember that the **IF** function consists of three parts separated by commas. In the first part of the function, you specify a criteria. If the criteria is true, the spreadsheet will do or carry out whatever you specify in the second portion of the function. If the criterion is false, the spreadsheet will carry out what you specify in the third portion of the function. Let's review the B10 formula carefully.

The criterion is that a random number (the **RAND()** portion of the formula) is less than the value in cell B6 (the probability of being censored on day 1). If the criterion is true, the individual is censored and the spreadsheet will return the letter C. If the criterion is false, the individual is not censored, and the second **IF** function will be computed.

The second **IF** function tells the spreadsheet to evaluate whether a random number between 0 and 1 is greater than the value in cell B4—the true (but unknown to you, the researcher) daily survival probability. If the random number is greater than the survival probability, the individual will die (the spreadsheet will return the letter D). If the random number is less than the value in cell B4, the spreadsheet will return the number 1, indicating that the individual survived that day. When you copy your formula down for the 25 individuals in the population, you should see that some individuals die and some become censored. Press F9, the calculate key, to generate new fates for individuals in the population.

2. In cell C10, enter a formula to assign a fate to individual 1 for day 2.

In cell C10 enter the formula **=IF(OR(B10="D",B10="C",B10=""),"",IF(RAND()<$C$6, "C",IF(RAND()>$B$4,"D",1)))**. Don't be intimidated by the length of this formula. If the individual in cell C10 died or was censored on day 1, we want to return a blank cell (i.e., two double quotes). If the individual survived day 1, then we want to know what happened on day 2. The formula in cell C10 is another nested **IF** function. There are multiple criteria, however, in the first **IF** function, and these criteria are given with an **OR** function. The **OR** function is used to evaluate whether the value in cell B10 is "D" or "C" or "". If any one of those three conditions is *true*, the spreadsheet will return a blank, or "". If none of the conditions is true, the individual must have survived day 1, and the second **IF** function is computed; it has the same form as the formula in cell B10, with the spreadsheet again returning a value of "C," "D," or the number 1.

3. Select cell C10, and copy its formula across to cell K10. Modify the formula in each cell to reflect the probability of censorship for the appropriate day.

Double-check your formulae. They should read as follows:

- In cell D10,
  =IF(OR(C10="D",C10="C",C10=""),"",IF(RAND()<$D$6,"C",IF(RAND()>$B$4, "D",1)))
- In cell E10,
  =IF(OR(D10="D",D10="C",D10=""),"",IF(RAND()<$E$6,"C",IF(RAND()>$B$4, "D",1)))
- In cell F10,
  =IF(OR(E10="D",E10="C",E10=""),"",IF(RAND()<$F$6,"C",IF(RAND()>$B$4, "D",1)))

and so on. Your spreadsheet should now resemble Figure 5, although the fates of your individuals will likely be different than that shown.

4. Select cells C10–K10, and copy the formula down to row 34.

5. Save your work.

	A	B	C	D	E	F	G	H	I	J	K
8						Day					
9	Individual	1	2	3	4	5	6	7	8	9	10
10	1	1	D								
11	2	1	1	C							
12	3	C									
13	4	1	1	1	1	1	1	1	1	C	
14	5	1	1	1	1	D					

Figure 5

C. Compute survival probabilities.

The first calculations in the Kaplan-Meier estimate involve counting the number of individuals at risk (still alive) during each day, and to count the number of deaths that occur each day.

1. Set up new headings as shown in Figure 6.

	A	B	C	D	E	F	G	H	I	J	K
35	# at risk										
36	# deaths										
37	# censored										
38	Conditional P_c										
39	Unconditional P_u										
40	Expected survival										
41	Daily survival										

Figure 6

2. In cell B35, enter the number of at-risk individuals in the population on day 1.

Enter 25 in cell B35.
The number at risk on day 1 is 25 because we started with a sample size of 25.

3. In cell B36, enter a formula to count the number of deaths on day 1.

In cell B36 enter the formula **=COUNTIF(B10:B34,"D")**.
The number of deaths on day 1 is the number of D's that appear for the 25 individuals.

4. In cell B37, enter a formula to count the number of censored observations on day 1.

In cell B37 enter the formula **=COUNTIF(B10:B34,"C")**.
The number of censored observations on day 1 is the number of C's that appear for the 25 individuals.

5. In cell B38, enter a formula to compute the *conditional* probability of survival, P_c.

In cell B38, enter the formula **=1-(B36/B35)**.
This is the spreadsheet version of Equation 1:

$$P_c = 1 - \frac{d_i}{n_i}$$

The conditional probability of survival is the probability of survival to a particular time period, *given that you survived to the previous time*. This probability is easy to calculate if you know the number of deaths at a specific time and the number of individuals at risk at that same time. The number of deaths divided by the number at risk gives the conditional probability of mortality, so 1 minus that value is the conditional probability of survival.

6. In cell B39, enter a formula to compute the *unconditional* probability of survival, P_u.

In cell B39 we used the formula **=PRODUCT($B$38:B38)**.
The unconditional probability of survival is the probability of surviving to a particular time. It is calculated in Equation 2 as the cumulative product of the conditional probabilities:

$$P_u = \prod_{j=1}^{i}\left(1 - \frac{d_j}{n_j}\right)$$

7. In cell B40, enter a formula to compute the expected P_u for day 1, given the survival parameter in cell B4.

In cell B40 enter the formula **=$B$4^B9**.
The ^ symbol means raises the value in cell C4 (the survival probability) to the number of days under consideration.

8. In cell B41, enter a formula to compute the actual daily survival for each P_c.

In cell B41 enter the formula **=B39^(1/B9)** to obtain the daily survival estimate for day 1.
Remember that the P_c gives the probability of surviving to a specific time period. To convert the P_c to daily survival probabilities, take the appropriate root. For example, take the third root of P_c for day 3, the seventh root of P_c for day 7, and so on, to obtain the daily survival estimate. To obtain roots in spreadsheets, use the exponent form with the exponent as a fraction.

9. In cell C35, compute the number of individuals at risk for day 2.

In cell C35 enter the formula **=B35-(B36+B37)**.
Remember that the number of individuals at risk are those currently alive and not censored.

10. Select your formulae from steps 3–8 and copy them across to column K.

Your spreadsheet should now look something like Figure 7, but (with the exception of Row 40) your numbers will likely be different.

	A	B	C	D	E	F	G	H	I	J	K
35	# at risk	25	20	14	12	10	6	5	5	5	4
36	# deaths	3	2	0	1	1	0	0	0	1	0
37	# censored	2	4	2	1	3	1	0	0	0	0
38	Conditional P_c	0.88	0.9	1	0.917	0.9	1	1	1	0.8	1
39	Unconditional P_u	0.88	0.792	0.792	0.726	0.653	0.653	0.653	0.653	0.523	0.523
40	Expected survival	0.9	0.81	0.729	0.656	0.59	0.531	0.478	0.43	0.387	0.349

Figure 7

11. Save your work.

D. Create graphs.

1. Graph P_c, P_u, and expected P_u as a function of time. Interpret your graph.

Use the line graph option and label your axes fully.

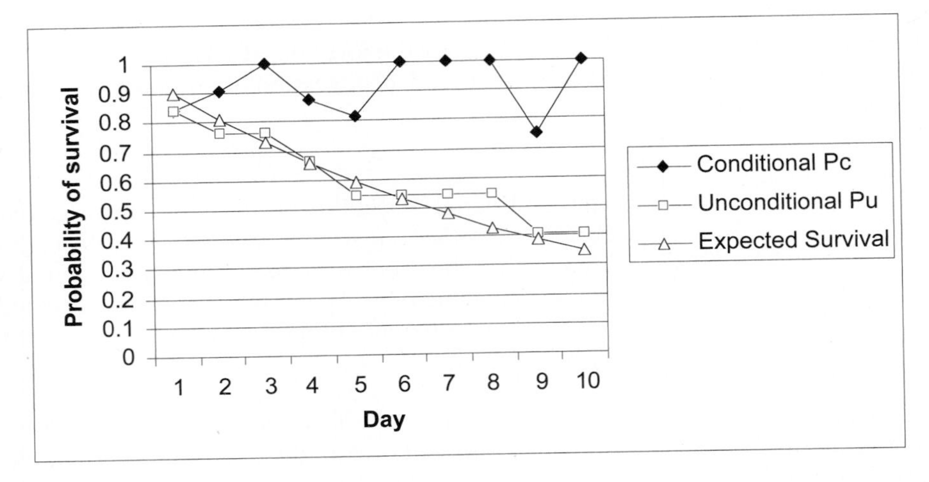

Figure 8

Your graph will look different than the Kaplan-Meier survival curve because the points are connected differently. However, the graphs are interpreted the same way. Note that the expected P_u is a straight line because we set the daily survival probability as a constant over time. Sharp drops in the P_u line indicate more mortality on a given day, and shallow drops in a line indicate fewer deaths occurring during a particular interval. Figure 8 shows few (no) deaths actually occurred from Day 5 to Day 8.

2. Press F9 to generate a new simulation. How do your results appear to change with each new simulation?

Your results should vary from simulation to simulation. This is due to the random number function changing the data set, and it is also due to the fact that our population consists of only 25 individuals (so there is some demographic stochasticity in this model). In order to fully understand how P_c and P_u "behave" over the 10-day period, we need to run several simulations, and track our results. We will do that in the next step.

E. Track 100 simulations.

1. Set up new headings as shown in Figure 9, but extend the trials to 100 (cell M109) and the days to 10 (cell W9).

	M	N	O	P	Q	R
9	**Trial**	**Day 1**	**Day 2**	**Day 3**	**Day 4**	**Day 5**
10	1					
11	2					
12	3					
13	4					
14	5					

Figure 9

2. Record a macro to track P_u for 100 trials, logging your results in cells N10–W109.

Open up the macro function as described in Exercise 2 or your user's manual. Once you have assigned a shortcut and the macro is in Record mode, perform the following steps:

- Select cells B39–K39. Copy.
- Select cell N9. Open Edit | Find.
- Leave the Find What box empty, and search by columns. Select Find Next, then Close. Your cursor should move down to cell N10.

- Open Edit | Paste Special and select the Paste Values option. Click OK.
- Select Tools | Macro | Stop Recording.

Run your macro until 100 trials have been computed.

3. Use the **AVERAGE** function in cells N110–W110 and **STDEV** function in cells N111–W111 to compute the average P_u and standard deviation for the 100 trials.

Your formula for day 1 should be **=AVERAGE(N10:N109)**. This gives the average unconditional probability that an individual will survive past day 1. The standard deviation is computed as **=STDEV(N10:N109)**. You will want to divide this number by 2 for graphing purposes in the next step.

4. Graph the average P_u for each day.

Use the column graph option. Your graph should resemble Figure 10 (without the error bars).

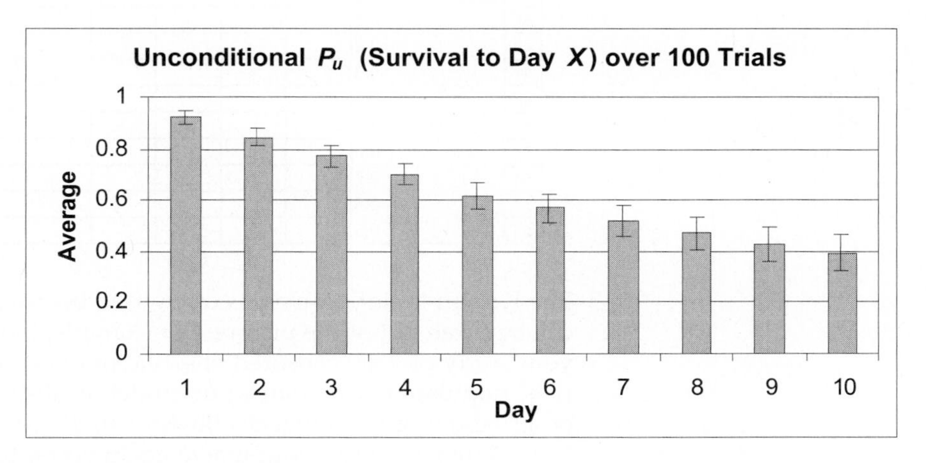

Figure 10

5. Add error bars to your graph. First, divide each standard deviation by 2 in cells N112–W112.

To add error bars, click on the columns in the graph to select them. Then go to Format | Selected Data Series | Y Error Bars. Select the Custom option. Click on the Display Both option. Place your cursor in the box labeled +, then use your mouse to select the standard deviations for your 100 trials divided by 2 (cells N112–W112). Do the same for the box labeled –. Click OK and your graph should be updated.

6. Save your work.

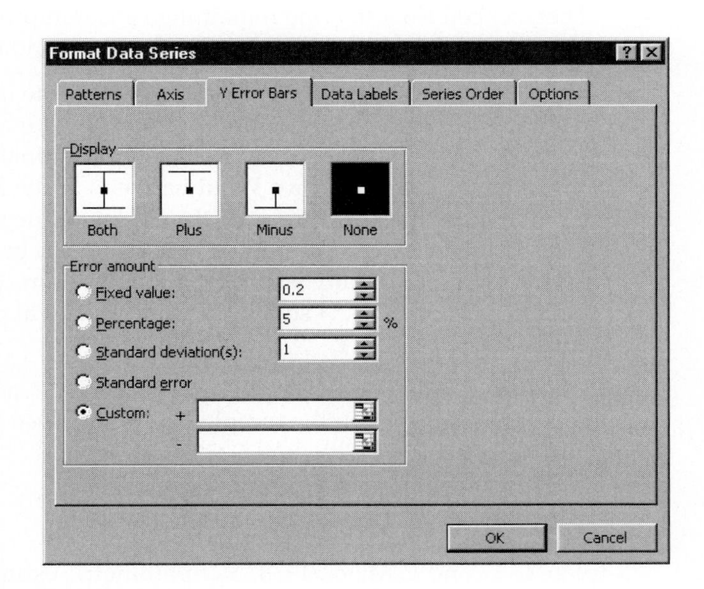

Figure 11

QUESTIONS

1. Interpret the Kaplan-Meier conditional and unconditional probabilities graph (e.g., Figure 8). What do long stretches of slightly sloping or horizontal lines indicate? What do steeply sloping vertical drops indicate?

2. What level of daily survival is needed to ensure that the population will persist for 10 days? Set up your spreadsheet as shown. Enter the expected P_u's for each level of daily survival (given in cells A45–A53). For example, cell B45 should compute P_u for day 1 when the daily survival is 0.1. Under what conditions is a population likely to persist for at least 10 days? Graph your results.

	A	B	C	D	E	F	G	H	I	J	K
44	Daily Survival	Expected P_u									
45	0.1										
46	0.2										
47	0.3										
48	0.4										
49	0.5										
50	0.6										
51	0.7										
52	0.8										
53	0.9										

3. The Kaplan-Meier estimate is often used because "uncooperative" individuals can be taken out of the picture. For example, individuals that fly away from your study plot are censored observations and can be subtracted from your "at risk" population. Compare your model results to a population where censored observations are absent (cells B6–K6 = 0). Erase your macro results (cells N10–W109), then run your macro again under the new conditions. Compare the average P_u and the standard deviations of the trials.

4. Under some conditions, censored observations may occur early in the study, and under some conditions censored observations may occur late in the study. For example, dispersal of individuals out of your study population may occur early or late in the study, depending on the time of year your study is being conducted. Compare how early censorship and late censorship affect P_c and P_u. Set cell B6 = 0.5 to assess early censorship (the remaining cells should be 0). Then set cell K6 = 0.5 (the remaining censorship probabilities should be 0). Describe your results in terms of P_u and its standard deviation.

5. In the spreadsheet model, we simulated the fate of individual's death or survival by linking a random number to a daily survival probability in cell B4. Thus we assumed that for each day, an individual had the same probability of surviving as any other day. What happens to the Kaplan-Meier estimates when survival probabilities vary over the course of the study? Modify your model to include this change and discuss your results in graphical form. For example, establish different daily survival probabilities in cells B4–K4, and adjust the formulae in cells B10–K34 so that the daily survival probability reflects your new entries in cells B4–K4.

*6. (Advanced) How does sample size affect both P_c and P_u? Modify your model and compare results when the sample size is increased from 25 to 50 individuals.

LITERATURE CITED

Kaplan, E. L. and P. Meier. 1958. Nonparametric estimation from incomplete observations. *Journal of the American Statistics Association* 53: 457–481.

25

HABITAT SELECTION

In collaboration with David N. Bonter

Objectives

- Develop a spreadsheet model of ideal-free habitat selection.
- Compare the ideal-free and ideal-despotic habitat selection models.

INTRODUCTION

Imagine it is time for dinner, and you are deciding where to eat this evening. Your options are either ordering pizza or going to the dining hall. You'd prefer pizza, but you know that as soon as the pizza delivery person appears, everyone in the dorm will be interested in getting a piece of your pizza. Although your first choice is pizza, competition for each slice may leave you hungry. On the other hand, you know that there will be plenty to eat at the dining hall. It may not be pizza, but at least you won't be hungry while studying tonight. Which do you choose? Does it depend on how many friends are in the dorm tonight?

Similarly, organisms must routinely choose between habitat patches that present different opportunities for meeting foraging and other resource needs. The choice between the dining hall (suboptimal forage) and pizza delivery (optimal forage) is analogous to the choice between habitat patches, where the number of people in the dorm is the density of organisms within the habitat. Competitors may decrease an organism's intake through interference or by reducing the resources available in a patch through exploitation competition. Facing these circumstances, an organism may do better by moving to a patch with fewer competitors, even if the overall resources are inferior. In other words, if your dorm is crowded tonight with many hungry competitors for pizza, you may reach your daily foraging requirements better by eating in the dining hall!

Ideal-Free Habitat Selection

The **intrinsic** or **basic suitability** of a habitat may depend on factors such as food and predators; some patches are higher in quality than others. Individuals that compete for similar resources can reduce this basic suitability, so that "crowded" habitats may be much lower in actual suitability, even if the basic suitability is high. Thus, even though one habitat may be intrinsically "better" than the other, an organism can do equally well in either habitat, depending on the density of individuals within the habitats. This model of habitat selection is known as **ideal-free**, because individuals are assumed to have full or "ideal" knowledge of what the intrinsic suitabilities of each habitat are, as well as the densities in each habitat, and individuals are "free" to select and enter habitats that will optimize their fitness. Hence, individuals make behavioral decisions based on the behavior of other individuals in the population (Fretwell and Lucas 1970).

Numerous assumptions are usually associated with the ideal-free distribution model.

- Individuals are of identical competitive ability.
- Habitat patches vary in quality.
- Competitors are free to move without costs or constraints.
- Each competitor will move to where its expected gains are highest.
- The value of a patch declines as more individuals exploit that patch.
- Maximum patch suitability occurs when the population density approaches zero.

The model predicts that all competitors will experience equal gains and that the average rate of gain in all habitats is equal. In other words, at equilibrium, no individual should be able to improve its situation by moving to another patch.

Obviously, many of these assumptions are violated in the real world, and we will address some of these assumptions later. But the ideal-free distribution provides a sound place to start our model. Mathematically, we can express the **suitability** of the *i*th habitat as a function of its basic (or intrinsic) suitability, modified by the density of organisms in the habitat:

$$S_i = B_i - f_i(d_i) \qquad \text{Equation 1}$$

where S_i is the realized suitability of habitat i, B_i is the basic (intrinsic) suitability of habitat i, and d_i is the density of organisms in habitat i. The term $f_i(d_i)$ expresses the lowering effect on suitability as a result of an increase in density. When f_i is large, each individual occupying the habitat reduces the basic suitability of the habitat by a large amount.

A hypothetical comparison between the suitability of two habitats is shown in Figure 1.

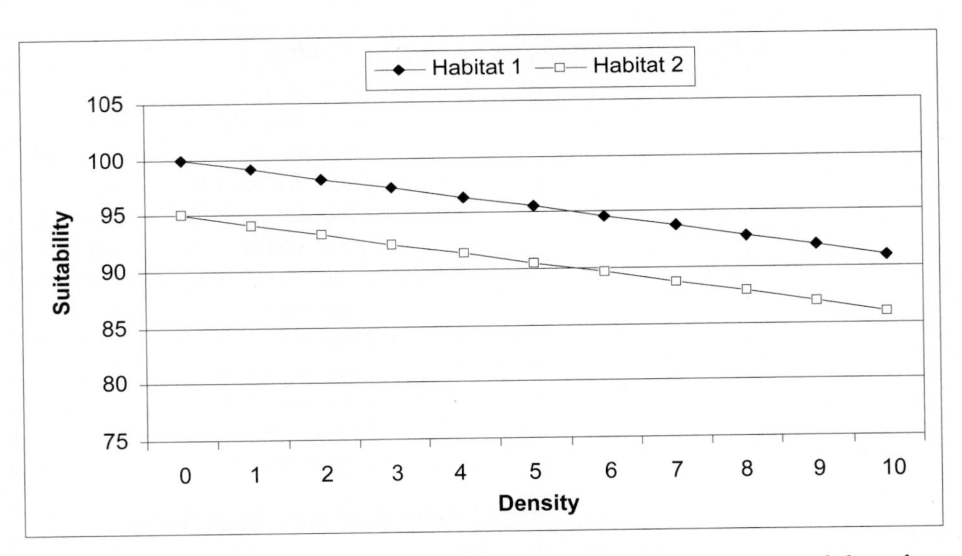

Figure 1 In this example, the basic suitability of habitat 1 is 100 units and that of habitat 2 is 95 units. The amount that each resident lowers suitability, $f_i(d_i)$, is the same for both habitat patches. As individuals begin to colonize the two empty habitats, selecting habitat 1 will maximize their fitness. However, after the first five individuals have established residence in habitat 1, the suitability of this habitat has decreased to be identical to that of habitat 2 (still with 0 occupants). The sixth colonist would do best to colonize habitat 2. This colonist will then reduce the quality of habitat 2 such that habitat 1 will be selected by the seventh colonist, and so on.

Ideal-Despotic Habitat Selection

If individuals are not free to occupy the patch of their choice, we can modify our model of habitat selection and develop the **ideal-despotic** model. In this model, some individuals cannot freely occupy a habitat because other individuals (the "despots," or "dictators") already present in the patch prevent them from colonizing. Thus, for example, decisions of unsettled birds are influenced by the behavior of resident birds—the nonresidents are not always free to select the habitat they want. Mathematically, the lowered suitability of a habitat patch for future colonists due to resident behavior can be expressed by

$$T_i = S_i\,[1 - t(d_i)] \qquad \text{Equation 2}$$

where T_i is the *apparent suitability* of the habitat for the unsettled bird, or how the colonizing individual perceives the quality of habitat i. Equation 2 says that the apparent suitability is equal to the realized, or actual, habitat suitability, S_i (calculated in Equation 1), discounted by a factor that takes into account the density of occupants already present in the habitat (d_i) and the resistance of those occupants to new colonists (t). When $t = 0$, the occupants do not resist new colonists at all, and $T_i = S_i$ (there is no despotism). When $t = 1$, the occupants strongly resist new colonists. As long as $t > 0$, $1 - t(d_i)$ is less than 1, and higher densities mean lowered apparent suitability. The relationship between a site's basic or intrinsic suitability, its suitability when population density is factored in (from the ideal-free model), and its apparent suitability (from the ideal-despotic model) is represented in Figure 2.

Various factors can act to decrease the apparent suitability of a habitat patch. Often an organism will have to choose between habitat patches that differ in predation risk in addition to resource availability. We may think that by adding predation risk to habitat selection considerations, the ideal site will have plentiful resources, few competitors,

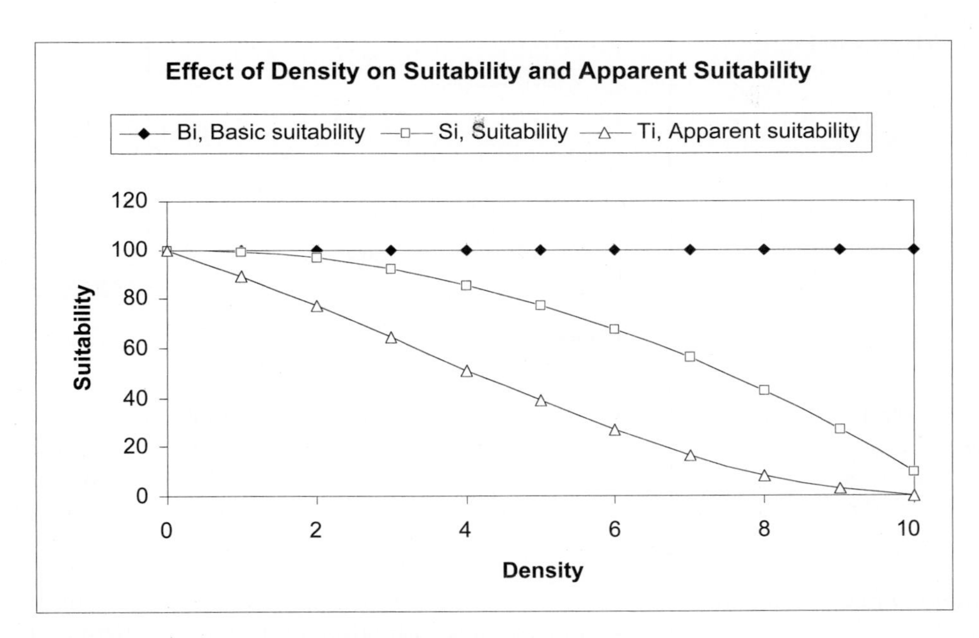

Figure 2 The basic (intrinsic) suitability of a habitat patch is fixed and remains constant regardless of population density. However, this relationship is unlikely to represent conditions in the real world. Basic suitability is often diminished as a function of population density, $f_i(d_i)$, because individuals compete for resources. Here we see that 10 individuals reduce the suitability of the habitat by 90%. If patch residents act to exclude future colonists, the apparent suitability is reduced even further, by $t(d_i)$. In this example, the patch is no longer hospitable to future colonists after 10 individuals have established residence.

and a low predation risk. However, the interrelationships between habitat characteristics may be more complicated. For instance, choosing a patch with numerous conspecifics may reduce predation risk. In this situation, allies in predation avoidance become competitors in resource acquisition. The nature of the relationship between gain and risk with group size will influence which habitat patches are exploited (Moody et al. 1996). Abiotic factors may also impact habitat suitability. Differences in temperature or exposure to wind may produce differential energetic costs in different habitat patches.

PROCEDURES

This exercise presents a simple model that focuses only on a density-dependent decrease in habitat suitability. In this model, competition for resources in "good" patches may result in lower energetic gains due to loss of resources to rivals. In "poor" patches, it may be harder to locate available resources, but less competition may make this choice worthwhile. The ideal-free distribution model often successfully predicts the distribution of organisms in the real world, and has become the basis for more complex models.

In this exercise, you will develop a spreadsheet model of the ideal-free distribution and explore its consequences on habitat selection. You'll also compare the ideal-free model to the ideal-despotic model. As always, save your work frequently to your disk.

INSTRUCTIONS

ANNOTATION

A. Set up an ideal-free model for two-habitats.

1. Open a new spreadsheet and set up column headings as shown in Figure 3.

	A	B	C	D	E	F	G	H	I	J	K	L
1	*Habitat Selection*											
2												
3						Habitat 1	Habitat 2					
4	*Bi* = Basic suitability				===>							
5	*fi* = Lowering effect				===>							
6	*t* = Resistance to settling				===>							
7												
8	HABITAT 1						HABITAT 2					
9	Density	*Bi*	*f * d*	*Si*	*t * d*	*Ti*	Density	*Bi*	*f * d*	*Si*	*t * d*	*Ti*

Figure 3

2. Enter 100 in cell F4 and 95 in cell G4.

We will consider two habitats, habitat 1 and habitat 2.
Habitat 1 has a higher basic suitability than habitat 2. The values entered reflect the basic or intrinsic suitabilities of each habitat. Remember, these basic suitability scores are based on factors such as food abundance, predators, and so on, when the patches are not yet occupied by colonists.

3. Enter 0.9 in cells F5 and G5.

These values represent *f(i)* for the two habitats, or the "lowering effect" of habitat quality of each new individual occupying the habitat. Each individual occupying a habitat will reduce the habitats' quality by this amount.

4. Enter densities from 0–10 for habitat 1 in cells A10–A20

First we'll focus on habitat 1, then we'll repeat the steps for habitat 2 to examine how basic suitability is lowered as more individuals colonize the different habitats.
Enter 0 in cell A10. Enter **=1+A10** in cell A11, and copy this formula down to cell A20.

5. In cells B10–B20, enter a value for the habitat's basic suitability.

The basic suitability (B_i) for habitat 1 is given in cell F4, so enter the value **=$F$4** in cells B10–B20.
This is the suitability of the habitat based on intrinsic qualities such as the amount of food, number of predators, and so on.

6. In cell C10, enter a formula for the lowering effect of density on the suitability of habitat 1.

This is the value in cell $F$5 times the density (given in cell A10) in habitat 1. So, enter the formula **=A10*$F$5** in cell C10 and copy this formula down to cell C20.
The lowering effect, *f*, is a fixed value currently set at 0.9. For any given density, however, the total reduction in suitability is the product of *f* times the density in the habitat.

7. In cell D10, enter a formula to calculate the realized suitability of habitat 1. Copy this formula down to cell D20.

In cell D10, enter the formula **=B10-C10**. Copy this formula down to cell D20.
The suitability of habitat 1, according to the Fretwell-Lucas model (1970), is $S_i = B_i - f_i (d_i)$. Take a good look at this equation. It says that the suitability of a habitat is its intrinsic suitability (cell B10) minus the density of individuals in the patch times the amount that each individual lowers the basic suitability ($f_i \times d_i$) in cell C10.

8. Repeat steps 4–7 to fill out cells G10–J20.

Now we are ready to concentrate on habitat 2. Make sure to reference parameters associated with habitat 2 in cells G4–G5 in your formulae.

9. For habitat 1, make a graph that compares the basic suitability with the actual suitability.

You will be graphing the values in cells A9–B20 and those in cells D9–D20. Use the XY scatter graph option and label your axes fully. Your graph should resemble Figure 4.

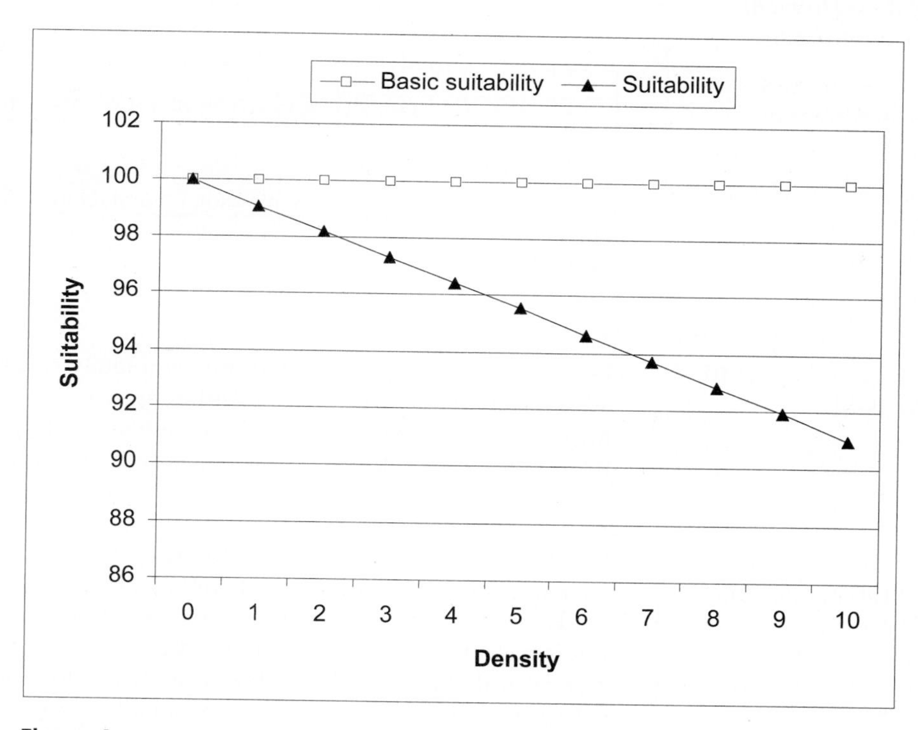

Figure 4

10. Graph the suitabilities of both habitat types as a function of density.

You will be graphing the values in cells A9–A20, D9–D20, and cells J9–J20. Remember to hold down the <Control> key to select cells that are not contiguous. Use the XY scatter graph option and label your axes fully. Your graph should resemble Figure 5.

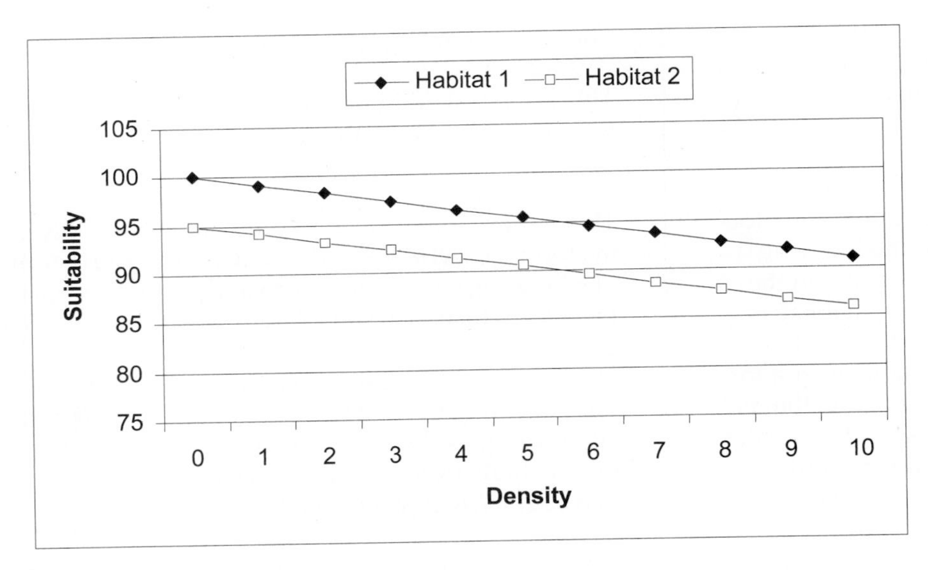

Figure 5

11. Save your work.

B. Simulate settlement patterns of individuals.

1. Set up new column headings as shown in Figure 6.

	M	N	O	P	Q	R
6	HABITAT SELECTION SIMULATION - 10 INDIVIDUALS					
7						
8		Habitat	Habitat 1		Habitat 2	
9	Individual	choice	Running total	Suitability	Running total	Suitability

Figure 6

2. Enter the numbers 1–10 in cells M10–M19.

Let's imagine that both habitats are completely empty; then 10 individuals arrive (not all at once) and have options of settling into habitat 1 or habitat 2. Remember, the goal for individuals is to maximize their success, so they will choose whatever habitat has the highest suitability. In this step you will simulate the decisions of the 10 individuals on your spreadsheet.

3. In cell N10, enter the formula **=IF(F4>G4,1,2)**. In cell O10, enter the formula **=IF(N10=1,1,0)**. In cell Q10, enter the formula **=IF(N10=1,0,1)**.

For the first individual, the decision is easy. It will select the habitat with the greatest basic suitability. We can use an **IF** function in the spreadsheet to return the choice made. An **IF** function returns one value if a condition you specify is true and another value if it is false. It has the syntax **IF(logical_test,value_if_true,value_if_false)**. The formula in cell N10 tells the spreadsheet to examine the contents of cells F4 and G4, the basic suitabilities of the two habitats. If F4 > G4, the spreadsheet will return the number 1 (which indicates habitat 1 was selected); otherwise, it will return the number 2 (which indicates habitat 2 was selected). Use IF functions in cell O10 and Q10 to keep a running total of individuals in habitats 1 and 2.

4. In cell P10, enter the formula **=VLOOKUP (O10,$A$10:$D$20,4)**. Copy this formula down to cell P19.

We need to record the suitabilities of each habitat, depending on what their current densities are. We'll use **VLOOKUP** to do this. The **VLOOKUP** function searches for a value in the leftmost column of a table, and then returns a value in the same row from a column you specify in the table. It has the syntax **VLOOKUP(lookup_value, table_array,col_index_num,range_lookup)**, where **lookup_value** is the value to be found in the first column of the table, **table_array** is the table of information in which the data are looked up, and **col_index_num** is the column in the table that contains the value you want the spreadsheet to return. **Range_lookup** is either true or false (use false for your formula). For example, the formula in cell P10 tells the spreadsheet to look up the value in O10 (which is the running tally of individuals in habitat 1) in the table in cells A10–D20, and return the value associated with the fourth column of the table that is associated with the value listed in O10.

5. Use the **VLOOKUP** formula in cell R10 to return the current suitability of habitat 2 (based on its current occupancy). Copy the formula down to cell R19.

In cell R10 enter the formula **=VLOOKUP(Q10,$G$10:$J$20,4)**.

6. In cell N11, enter the formula **=IF(P10>=R10,1,2)**. Copy this formula down to cell N19.

Now we need to focus on the second individual. The **IF** formula tells the spreadsheet to determine if the value in cell P10 is greater than or equal to (>=) the value in cell R10. If so, the spreadsheet returns a 1 (indicating a selection of habitat 1); otherwise the spreadsheet returns a 2 (indicating a selection of habitat 2).

7. Enter the formula **=COUNTIF($N$10:N11,1)** in cell O11 and the formula **=COUNTIF($N$10:N11,2)** in cell Q11. Copy these formulae down to cells O19 and Q19, respectively.

To keep a running tally of how many individuals are in habitats 1 and 2, we can use the **COUNTIF** function. The **COUNTIF** function counts the number of cells within a range that meet the given criteria. It has the syntax **COUNTIF(range,criteria)**, where **range** is the range of cells from which you want to count cells, and **criteria** is what you want to count. For example, the formula in cell O11 tells the spreadsheet to count how many 1s there are in cells N10–N11.

8. Graph the running population totals of habitat 1 and habitat 2.

You will be graphing the values in cells O9–O19 and cells Q9–Q19. Use a line graph and use the values in cells M10–M19 as your *x*-axis (under the Series tab). Your graph should resemble Figure 7.

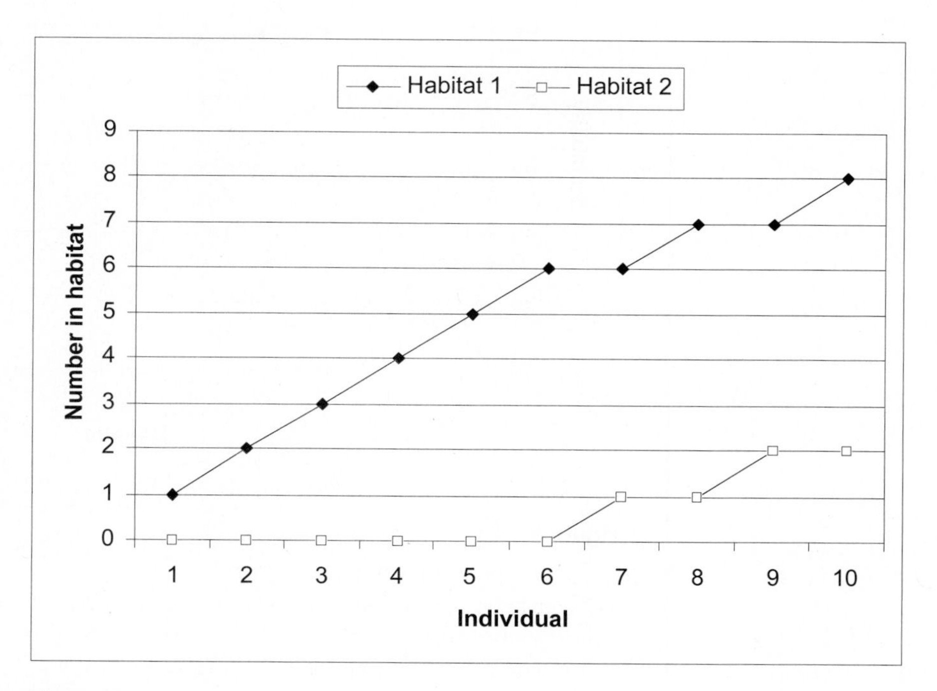

Figure 7

9. Save your work, and answer Questions 1–4 at the end of the exercise before proceeding.

C. Enter parameters for the ideal-despotic model.

Now we will consider the ideal-despotic model of habitat selection, where unsettled individuals are restricted by the "despotic" behavior of already settled individuals. Thus, even though they may "choose" to settle in a particular habitat based on its suitability, the colonists may not successfully settle and hence their success will be lower than expected for that habitat.

1. Enter 0.1 in cells F6 and G6.

The parameter t represents how "resistant" a resident individual is to new colonizers. Its value ranges from 0 to 1, where 0 means no resistance to new settlers and 1 indicates full resistance to new settlers. For now, $t = 0.01$, indicating little resistance. You will be able to change this value later in the exercise.

2. In cell E10, enter the formula **=$F$6*A10**. Copy this formula down to cell E20.

The total resistance of the habitat to new colonizers is a function of how many residents there are in the habitat. Thus, the term $t \times d$ is an indication of the overall resistance to new colonists.

3. In cell F10, calculate the apparent suitability, T, as $S_i[1 - (td_i)]$. Copy your formula down to cell F20.

T is the apparent suitability of a habitat, from the perspective of an individual looking to settle into a habitat. We used the formula **=D10*(1-E10)** in cell F10 to calculate the apparent suitability of habitat 1 when habitat 1 is vacant.

4. Enter formulae in cells K10–L20 for habitat 2.

Now consider the influence of despotic behavior on the apparent suitability of habitat 2. We used the formula **=$G$6*G10** in cell K10 and **=J10*(1–K10)** in cell L10.

5. Graph the suitabilities and apparent suitabilities for habitat 1.

Highlight cells A9–A20, D9–D20, and F9–F20. Use the XY scatter graph option and label your axes fully. Your graph should resemble Figure 8.

6. Save your work, and answer questions 5–7.

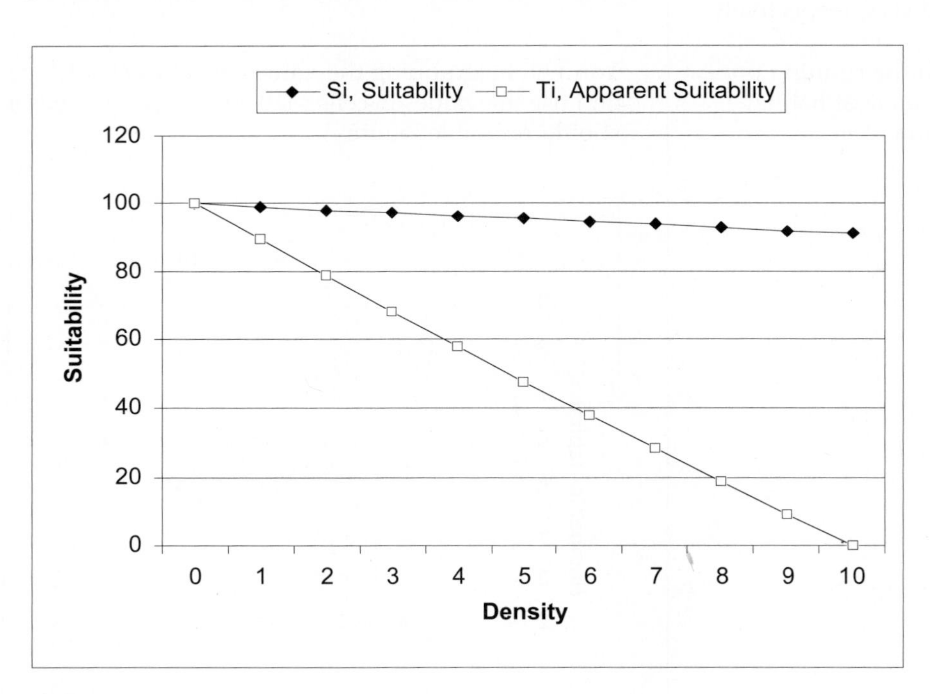

Figure 8

QUESTIONS

1. Based on your graph and the parameters used in the model in Section A of the exercise, if the density of habitat 1 is 3, and the fourth individual is looking for a place to settle, which habitat should it select? What if the density in habitat 1 is 8 and the density in habitat 2 is 0, which habitat should an individual select? When all 10 individuals have settled into their respective habitats, how do the two habitats compare in terms of per capita fitness?
2. In the ideal-free model, how does f affect suitability? Enter various values into the spreadsheet and examine graphical results from Section A of the exercise for habitat 1.
3. Your ideal-free model suggests a linear decline in suitability as density increases. Is this assumption justified? Modify your model so that each additional individual adds more and more of a "penalty" to suitability. For example, each new individual decreases suitability by a squared function of the density [$S_i = B_i - f_i(d_i^2)$]. How does your modification change your basic results?
4. One assumption of the ideal-free model is that all individuals are free to move into any habitat patch. In reality, individuals currently occupying a habitat patch may attempt to prevent others from entering. What influence would these "despots" have on the apparent suitability of a habitat patch? Consider how *you* would modify your model to be an ideal-despotic model. (We will do this in Part C, but your ideas may be better than ours.)
5. In the ideal-despotic model, what effect does t have on habitat suitability? Enter various values in your model and interpret your results.
6. Does the ideal-despotic distribution lead to a condition similar to what we found in the ideal-free model, where T_i is relatively equal in all habitats?
7. Both the ideal-free and the ideal-despotic models assume that individuals have "ideal" knowledge of relative habitat quality. Hypothesize about the effects on habitat selection if this assumption were violated.

LITERATURE CITED

Fretwell, S. D. and H. L. Lucas. 1970. On territorial behavior and other factors influencing habitat distribution in birds. *Acta Biotheoretica* 19: 16–36.

Moody, A.L., A. I. Houston and J. M. McNamara. 1996. Ideal free distributions under predation risk. *Behavioural Ecology and Sociobiology* 38: 131–143.

26

OPTIMAL FORAGING MODELS

In collaboration with David N. Bonter

Objectives

- Develop a spreadsheet model of foraging choices among two prey types, prey 1 and prey 2.
- Determine the conditions in which individuals should be specialists (consume either prey 1 or prey 2) or generalists (consume both prey types).

INTRODUCTION

What are you going to eat for lunch today? Your choices may be many or few, depending on how far you are from various restaurants, how much change you have in your pocket, or whether you packed a lunch from home. The decision of what to eat for most animals is not a matter of luxury, but of survival, and the decisions that organisms make in their selection of food can be strongly shaped by natural selection. Costs and benefits are ultimately calculated in terms of Darwinian fitness (survival and reproduction). In this exercise, we use energy gained from foraging as a surrogate measure of fitness.

Let's suppose that you are enjoying a snack consisting of peanuts (prey 1, still in their shell) and popcorn (prey 2, already popped). Let's further suppose that you are very, very hungry. Which food item will you choose to eat first? When will you stop eating the first food item and switch to the second? Ecologists think about the choices animals make in terms of economic **profitabilities**. Each food item has a benefit associated with it if consumed: energy (E). Each item also has a cost, which includes the time it takes to manipulate the food so that it can be consumed (called **handling time**, h). The "profitability" of a particular food item is E/h.

Should you eat the peanuts or popcorn? Peanuts have more energy per unit than the popcorn, but their handling time can be quite large, especially if the nuts are tightly closed. You, the predator, should eat the peanuts when:

$$E_{\text{peanuts}}/h_{\text{peanuts}} > E_{\text{popcorn}}/h_{\text{popcorn}} \qquad \text{Equation 1}$$

At the beginning of your snack, this is likely to be true. You simply find the peanuts that are cracked half-open, which have lower handling times and can be consumed fairly quickly. Spending time eating popcorn means that you'll be missing the opportunity to consume the more energetically profitable peanuts. But this may not continue to be the case. When should you start eating popcorn? When the gain from eating popcorn is greater than the gain from rejecting the popcorn and searching for the more profitable peanuts. That is, you should eat popcorn when

$$E_{\text{popcorn}} / h_{\text{popcorn}} > E_{\text{peanuts}} / h_{\text{peanuts}} \qquad \text{Equation 2}$$

Even if the search times were equal, you might switch to popcorn when you get to the last of the peanuts, where the nuts that are so tightly sealed that the handling time becomes enormous, sending the profitability of peanuts spiraling downward.

With this analogy in mind, in this exercise we will develop an optimal foraging model for two prey types. We will predict when a predator will specialize in the more profitable prey type, and when it will become a generalist and consume either prey type when encountered. Assuming that we can measure prey value, that handling times are fixed, that prey are recognized instantaneously, and that prey are encountered randomly, we can make a few predictions. First, the most valuable prey item will never be ignored. Second, the lower value prey will be ignored until

$$E_{\text{lower value}}/h_{\text{lower value}} > E_{\text{higher value}}/h_{\text{higher value}}$$

This simple ecological model suggests that foragers should make decisions that "optimize" their energy gain. Our model makes several assumptions in addition to those mentioned above: a single predator has only two choices of prey items; fitness is related to energy gain; and the predator can make "informed" decisions about whether to consume or bypass an encountered prey item.

Specialists and Generalists

In addition to handling time (h), prey **availability** (λ) may be added into the foraging cost portion of Equations 1 and 2. Prey availability ranges between 0 and 1, and the search time is defined as $1/\lambda$ (Figure 1). When the more profitable prey type is common ($\lambda \sim 1$), the search time is low and the predator wastes little energy locating the more profitable prey type. In such cases, it never pays to miss an opportunity to consume that prey type by spending time and energy pursuing or handling the less profitable prey. But as the more profitable prey item becomes less available ($\lambda < 1$), and search time increases nonlinearly. That is, even when E/h remains constant over time, decreasing availability (λ) leads the overall value of the prey to decline. Equation 3 shows how profitability (E/h) is modified to include both search and handling time costs:

$$\frac{E}{\frac{1}{\lambda}+h} \qquad \text{Equation 3}$$

which can also be written as

$$\frac{\lambda E}{1+\lambda h}$$

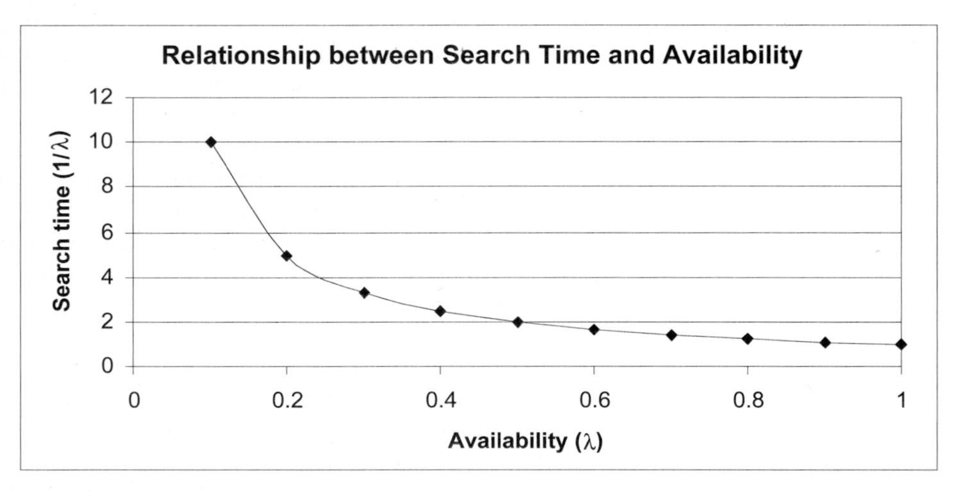

Figure 1 Search time is inversely related to prey availability. When availability is 0, search time is infinite.

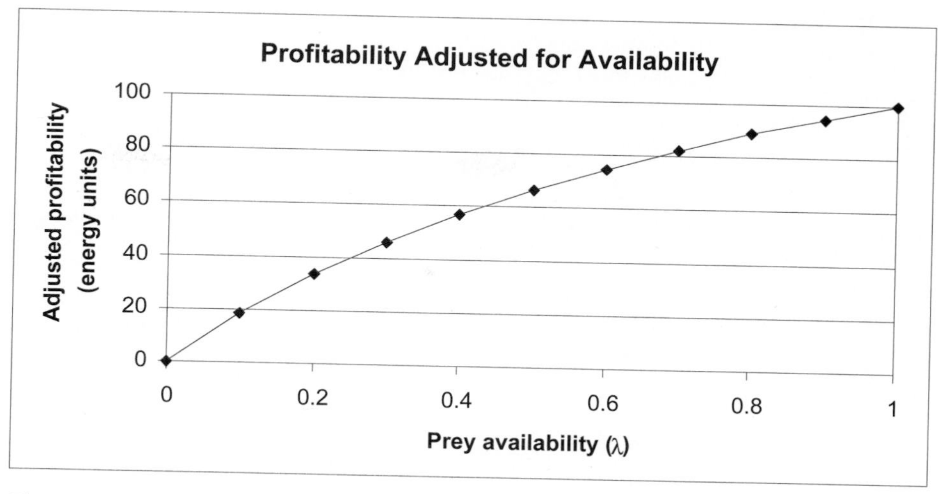

Figure 2 Profitability of a prey item after being adjusted for both handling time (h) and availability (λ) as shown by Equation 3. In this example, we set $E = 200$ and $h = 1$ to illustrate how adjusted profitability decreases sharply as prey availability decreases. This graph would differ if the values of E and h were changed.

if we multiply both the numerator and the denominator by λ. We can see that as availablity (λ) declines in Equation 3, the denominator increases, and profitability declines in a nonlinear manner (Figure 2). Even when $\lambda = 1$, profitability (E/h). is downwardly adusted to included energy involved with locating prey (i.e., search time).

As the more profitable prey type becomes rare, a point is reached where profitabilities of both prey types become roughly equivalent. Consuming the lesser quality prey will provide as much energetic benefit as spending time and searching for the remaining highly profitable items. In order to maximize energy gain per unit time, the predator will specialize on prey type 1 if

$$\frac{\lambda_1 E_1}{1+\lambda_1 h_1} > \frac{\lambda_1 E_1 + \lambda_2 E_2}{1+\lambda_1 h_1 + \lambda_2 h_2} \qquad \text{Equation 4}$$

That is, energetic gain from specializing on prey type 1 alone is greater than that from foraging on both prey types. As long as this inequality is true, the predator will ignore prey 2 and specialize on prey 1. At some point, the decreasing availability of prey 1 will force a change in foraging strategy, and our predator will become a generalist and consume either prey type it encounters. Figure 3 shows that the energetic value of foraging exclusively on prey 1 is higher than generalizing (consuming both prey types) until approximately the sixtieth encounter. At this point, the left side of Equation 4 is no longer greater than the right side. If the predator stays and continues to forage in the habitat patch, it will eventually deplete both prey types as the energy gained per unit time foraging steadily diminishes.

Optimal foraging models lead to a number of predictions (Begon et al. 1986):

- Predators with short handling times compared to search times are likely to be generalists. If the time lost handling less profitable prey items is small, the predator will consume the less profitable prey while continuing to search for preferred prey. Fish consuming aquatic insects may be an example. Once the prey item is located, time spent pursuing, subduing, and consuming the prey is negligible; the largest energetic costs are in finding the prey (search time), and any prey that are located are readily consumed.
- Predators with large handling times relative to search times should be specialists. A large carnivore (a lion, for example) may have negligible search times. Their potential prey (ungulates on an African savannah) are usually all around them. However, catching the prey—the handling time—is energetically expen-

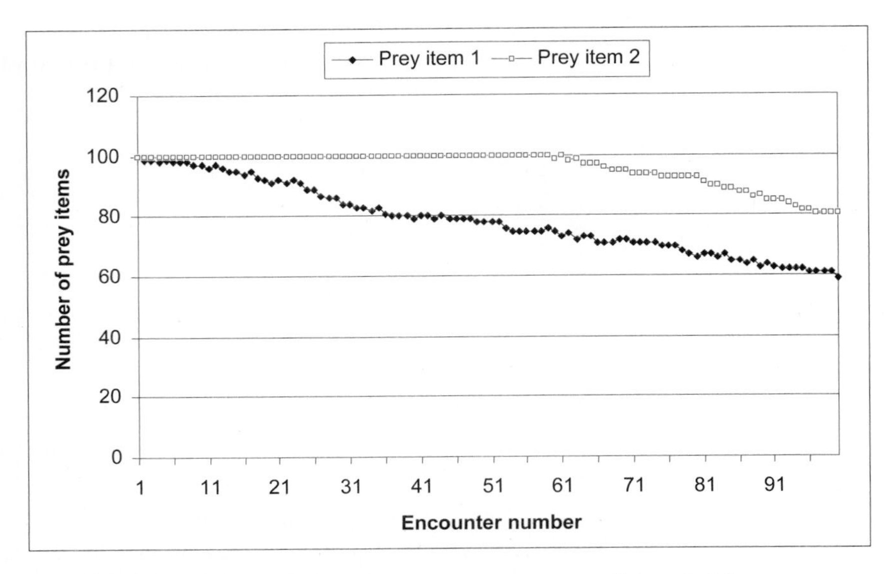

Figure 3 Here, a predator encounters prey types 1 and 2 and either consumes or bypasses them. The energetic value of foraging exclusively on prey 1 is higher than generalizing (consuming both prey types) until just shy of the sixtieth encounter. The overall reward of foraging diminishes as the more valuable prey type is depleted, and then as both prey types are depleted.

sive and often unsuccessful. Therefore, lions typically specialize in those prey items that are easier to handle (i.e., young, old, or sick individuals).

- Predators in unproductive environments are more likely to be generalists than predators in productive environments, as search times are likely to be high. On the other hand, when prey densities are high in a productive environment, the predator will benefit from specialization, as search times are negligible.
- Predators should specialize when profitable food types are common, and generalize when profitable items are rare.
- Predators should discriminate when the differences in profitabilities are great and be indiscriminate when the differences in profitabilities are negligible.

PROCEDURES

In this exercise, we'll see how these predictions are developed mathematically by modeling the conditions under which our energy-maximizing predator should be a specialist or a generalist. As always, save your work frequently to disk.

INSTRUCTIONS

A. Set up the model populations.

1. Open a new spreadsheet and set up column headings as shown in Figure 4.

ANNOTATION

	A	B	C	D	E
1	*Optimal Foraging*		Handling	Initial	Current
2	Prey	Energy (*E*)	time (*h*)	profitability (*E*/*h*)	availability (λ)
3	1	400	3		
4	2	50	1		
5		Number prey 1 =>	100	100	<== Initial prey 1
6		Number prey 2 =>	100	100	<== Initial prey 2
7		Total prey =>			

Figure 4

2. Enter **100** in cells C5–D6.

Let's assume that our predator is foraging in a patch that initially consists of 100 items of prey 1 and 100 items of prey 2. The initial numbers are given in cells D5–D6, and represent the number of each prey present before our forager enters a patch. The current number of prey items is given in cells C5 and C6. The values in cells C5–C6 will decrease as our forager consumes prey.

3. In cell C7, **SUM** the total prey in the patch.

Enter the formula **=SUM(C5:C6)** in cell C7.

4. Enter **400** in cell B3 and **50** in cell B4, and enter **3** in cell C3 and **1** in cell C4.

We need to establish the energy and handling times of each prey type. **In this model, prey 1 will always be more profitable than prey 2.** Let's assume that prey 1 has an energy of 400 calories/individual, and prey 2 has 50 calories/individual. Let's assume, like the peanuts and popcorn, that prey 1 has a slightly larger handling time (3 seconds vs. 1 second) than prey 2.

5. In cells D3 and D4, calculate initial profitability for each prey type.

Enter the formula **=B3/C3** in cell D3.
Enter the formula **=B4/C4** in cell D4.
Profitability is E/h for each prey type. (This profitability is not adjusted for availability.)

6. In cell E3 and E4, calculate the current availabilities (λ) of prey 1 and prey 2.

Enter the formula **=C5/($D$5+$D$6)** in cell E3.
Enter the formula **=C6/($D$5+$D$6)** in cell E4.
The availability λ (type the letter 'l' and change the font to Symbol) is the proportion of current prey type out of the total initial prey.

B. Determine a foraging strategy.

Now we are ready to determine which prey type should be eaten. Since prey 1 is more profitable than prey 2, the choices are whether to consume only prey 1 or to consume both prey 1 and prey 2.

1. Set up new column headings as shown in Figure 5.

	A	B	C
9	Which prey should be consumed?		
10	**Prey 1**	**Either prey**	**Behavior**
11			
12	**Select prey**	1	
13		.	

Figure 5

2. In cells A11 and B11, enter formulae based on Equation 4 to calculate the energy gain from specializing or generalizing.

Enter the formula **=(E3*B3)/(1+E3*C3)** in cell A11.
Enter the formula **=(E3*B3+E4*B4)/(1+E3*C3+E4*C4)** in cell B11.
Recall from Equation 4 that, in order to maximize energy per unit time, the predator specialize on prey type 1 if

$$\frac{\lambda_1 E_1}{1+\lambda_1 h_1} > \frac{\lambda_1 E_1 + \lambda_2 E_2}{1+\lambda_1 h_1 + \lambda_2 h_2}$$

If this inequality is true (the left side of the equation is indeed greater than the right side of the equation), only prey 1 should be consumed. Otherwise, both prey items should be consumed. This equation suggests that there can be a swift switch from being a specialist to being a generalist.

3. In cell C11, enter a formula to determine whether a predator should be a specialist on prey type 1, or a generalist.

In cell C11, enter the formula **=IF(A11>B11,"specialist","generalist")**.
This formula uses an **IF** function to return either the word "specialist" or the word "generalist." The C11 formula examines cell A11. If the value is greater than the value in cell B11, the inequality is true and the predator should specialize on prey type 1; otherwise it should be a generalist. (Given your initial conditions, a specialist strategy should be adopted.)

4. In cell B12 and B13, indicate which prey items will be selected given the foraging strategy employed.

In cell B12, enter the value **1** (because prey item 1 will *always* be taken, whether the predator is a specialist or not).
In cell B13, enter the formula **=IF(C11="generalist",2,".")**.
If only prey 1 is selected, we want the number 1 to appear in cell B12 and a missing value (a period) to appear in cell B13. If both prey are selected, we want the number 1 to appear in cell B12 and the number 2 to appear in cell B13. The **IF** statement in cell B13 returns the value 2 if the forager can consume both prey types, and returns a missing value if the forager is a specialist. Make sure your spreadsheet is working correctly by changing the energy associated with prey 1 (cell B3) from 400 to 200 calories per individual, and press F9 to see your results. Although prey 1 is still more profitable than prey 2, it is now economically most cost effective to consume both prey types, and this should be reflected in cells A11–B13. When you are finished, reset cell B3 to 400.

C. Simulate foraging decisions over time.

Now we'll set up a simulation to see what happens and what kinds of foraging decisions are made as the food in the patch is consumed. Assuming that our predator enters a patch with 100 items of prey 1 and 100 items of prey 2, it should consume prey 1 and bypass prey 2. The forager's first encounter is listed as Encounter X in cell A16.

	A	B	C	D	E	F
15	**Encounter**	**Prey**	**Selected?**	**Prey 1**	**Prey 2**	**Consumed**
16	x	2	no	100	100	0
17						
18						
19						
20	**Encounter**	**Prey**	**Selected?**	**Prey 1**	**Prey 2**	**Consumed**

Figure 6

1. Set up new column headings as shown in Figure 6.

Which prey our forager encounters is given in cell B16, and depends on prey availability, λ. The encounter number will change over time as our predator continues to forage. Cell C15 indicates whether the encountered prey was consumed or bypassed, and cells D16 and E16 indicate how many prey remain in the patch. If the encountered prey was pursued, the energy gained associated with the prey is given in cell F16.

2. In cell B16, enter an **IF** formula to specify whether the forager encounters prey 1 or prey 2.

Enter the formula **=IF(RAND()<C5/C7,$A$3,$A$4)** in cell B16.
This formula simply states that prey items are encountered according to their current proportions in the patch. If the random number (the **RAND()** portion of the formula) is less than the current proportion of prey 1, then the organism encounters prey 1, otherwise it encounters prey 2.

3. In cell C16, enter a formula to determine whether the prey encountered was consumed.

Enter the formula **=IF(OR(B16=$B$12,B16=$B$13),"yes","no")** in cell C16.
Now, although both prey 1 and prey 2 may be encountered, a prudent predator will bypass prey 2, since prey 1 is more profitable. Cell C16 returns the word "yes" if the prey encountered was consumed, and "no" if the prey was bypassed. It is an **IF** formula with an **OR** formula embedded in it. The **OR** portion of the formula—**OR(B16=$B$12,B16=$B$13)**—returns the value "true" if *any* of the arguments specified are true. Thus, if B16 = B12 *or* B16 = B13, the formula returns the value "true." Because the **OR** statement is embedded in an **IF** statement, the spreadsheet returns the word "yes" if either of the **OR** conditions is met, and "no" if neither condition is met.

4. In cells D16 and E16, adjust the number of each prey type remaining in the

Enter the formula **=IF(B16=1,$C$5-1,$C$5)** in cell D16.
Enter the formula **=IF(AND(B16=2,C16="yes"),$C$6-1,$C$6)** in cell E16.
The formulae in cells D16 and E16 reflect a decrease in number of prey 1 and prey 2,

patch (which depends on the decisions of the forager).

respectively, based on whether individuals of each type were consumed. The D16 formula is another **IF** formula: If cell B16 = 1, we know that prey item 1 was selected, so the total number of prey 1 is reduced by 1. The E16 formula is an **IF** formula with an **AND** formula embedded. In this case, cell B16 (the prey encountered) must equal 2 *and* cell C16 must equal "yes" in order for prey 2 to be depleted. Make sure these formulas are working correctly by pressing F9 several times. When prey item 1 is encountered, it should be selected and the total number of prey 1 should be reduced to 99 individuals. If prey 2 is encountered, it should be bypassed and the total number of prey 2 should remain at 100 individuals.

5. In cell F16, enter a formula to calculate the energy gained from a prey item, given that the prey item was selected and consumed.

Enter the formula **=IF(C16="no",0,IF(B16=1,B3,B4))** in cell F16.
This formula tells the spreadsheet to examine cell C16. If cell C16 has the word "no" in it, return a 0; otherwise, run through the second **IF** statement, **IF(B16=1,B3,B4)**. If the prey encountered was prey 1, the energy consumed is given in cell B3. Otherwise, prey item 2 was selected and the energy consumed is given in cell B4.

6. Save your work.

D. Write a macro to simulate foraging over time.

Now we are ready to let our predator continue their foraging in the patch, encountering prey 1 and prey 2 according to their availabilities, which change as the predator forages. The best way to simulate our predator's behavior is to record a macro that repeats the steps in row 16 several times, keeping track of the total number of prey 1 and prey 2 left in the patch.

1. Set up a linear series from 1 to 100 in cells A21–A120.

Enter the number **1** in cell A21.
Enter the formula **=1+A21** in cell A22. Copy this formula down to cell A120.

2. Copy cells B16–F16, and paste the values into cells B21–F21.

You've already simulated the first encounter, so simply paste the values you obtained into the row associated with encounter 1 (when pasting, select Edit | Paste Special | Paste Values). Now you are ready for encounter 2.

3. Set the calculation key to manual.

From the menu, open Tools | Options | Calculation and select Manual.

4. Record a macro to simulate encounters 2–100.

Bring your spreadsheet macro program into record mode and assign a name and shortcut key. Your macro should repeat the steps in row 16 several times, keeping track of the total number of prey 1 and prey 2 left in the patch. Record the following steps in your macro:

- Press F9, to obtain a new random number that will generate which prey type is encountered by the predator.
- Highlight cells B16–F16 and select Edit |Copy.
- Highlight cell B20, then go to Edit | Find. Select Search by Columns (not by rows). Leave the Find What box empty and select Find Next, then Close. Cell B22 should be highlighted.
- Select Edit | Paste Special | Paste Values (not the formulas) and select OK.
- Select cell D16–E16, then select Edit | Copy.
- Highlight cell C5, then select Edit | Paste Special | Paste Values. Make sure to select the Transpose option.
- Stop recording.

Now when you press your shortcut key 99 times you should be able to see how the our predators' foraging decisions changed over the course of time.

5. (Optional) Edit your macro using the Visual Basic code.

You can edit Excel's Visual Basic Editor code to avoid pressing the shortcut key 99 times. Push <Alt>+<F8> and select Edit; the code should appear. After the first line, simply enter the code `For counter = 1 to 100` in the first line of your program. The word `Calculate` should now be the second line of code. At the end of your program, before the words `End Sub` appear, type in a new line of code that reads `Next`. Now when you press your shortcut key just once, the macro will repeat 100 times.

6. Save your work.

E. Create graphs.

1. Graph the prey items remaining as a function of encounter number.

Use the line graph option and label your axes fully. Your graph should resemble Figure 7.

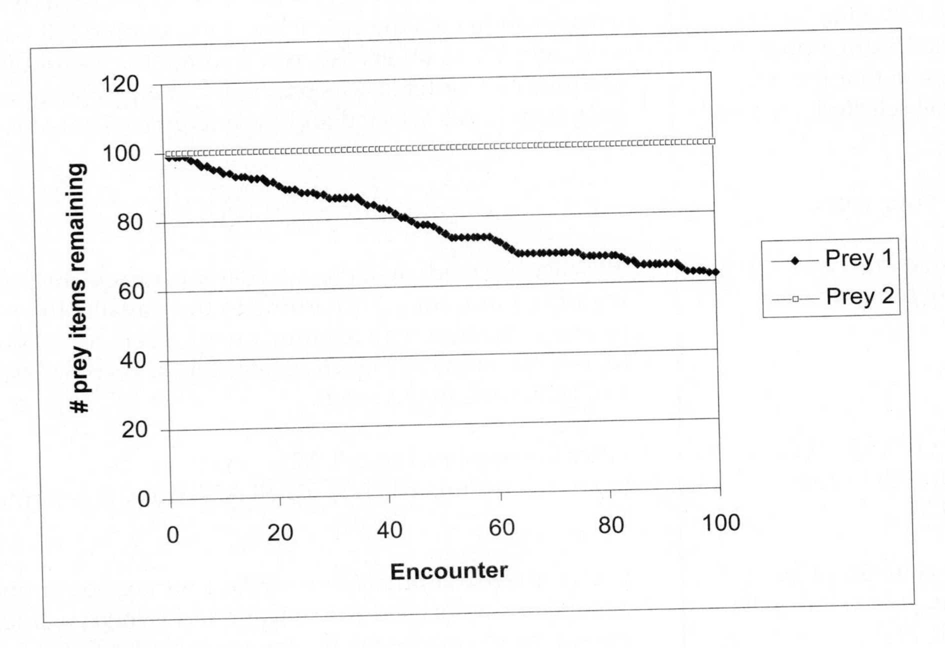

Figure 7

QUESTIONS

1. Interpret the results of your model. Did the forager specialize or generalize? Why?
2. Assuming your answer from Question 1 was "specialize," the forager must have bypassed several food items of the non-preferred prey. What is a major assumption of the model (not explicit in the model) regarding the metabolic costs of our forager while bypassing prey item 2?
3. Change the energy for prey 2 to 75 units (cell B4). Erase your macro results (cells B21–F120), and reset your initial prey abundances to 100 (cells C5–D6). Run your macro again. Interpret your results. Did the forager specialize or generalize? At what point did a change in behavior occur? Why?
4. Examine the availability λ as your simulation progressed. Why does the availability change as the simulation proceeds? How would availability (cells E3 and E4) change if one prey type were very rare, but highly profitable? Set the initial number of prey 1 to 10 individuals (cells C5–D5), and the initial number of prey 2 to 100 individuals (cells C6–D6). How are these differences reflected in availability? In the encounter probability (cell B16)? As prey is consumed, how do these values change?

5. Critically consider some assumptions of this model. Are energy content, handling time and prey availability the only factors that influence foraging decisions? Name other factors.

6. Which parameters drive the outcome of the model most: handling time, energy, or the initial prey availability? Run several trials that vary in 1 parameter (e.g., handling time) while keeping the other two parameters constant. Repeat for the other two parameters. Set up column headings so that you can track your results, and present your results graphically.

LITERATURE CITED

Begon, M., J. L. Harper and C. R. Townsend. 1986. *Ecology: Individuals, Populations and Communities*. Blackwell Scientific Publications, Oxford.

Charnov, E. 1976. Optimal foraging: The marginal value theorem. *Theoretical Population Biology* 9: 129–136.

Krebs, J. R. and N. B. Davies. 1991. *Behavioural Ecology: An Evolutionary Approach*, 3rd Ed. Blackwell Scientific Publications, Oxford.

27

RANGE EXPANSION

Objectives

- Build a spatially explicit model of range expansion by a logistically growing population.
- Model the expansion of a species' range in one and two dimensions.
- Determine how the rate of range expansion relates to population growth and emigration.

Suggested Preliminary Exercise: Logistic Population Models

INTRODUCTION

Species occasionally invade new habitat, often as an intentional or unintentional result of human activities. Invaders usually consist of a few founding individuals, occupying a small area. If the invasion is successful, the invading population grows in numbers and in area occupied—its **range**. Invading species often become pests, displacing or attacking native species, poisoning livestock, or otherwise making nuisances of themselves. It is therefore important to understand how and why a species' range expands. We will focus on two factors that influence range expansion: population growth and emigration rate. You will determine how a population's range expands or contracts, depending on its rates of growth and emigration.

In this exercise, you will treat each cell in the spreadsheet as a patch of habitat, which may house a local population. Each local population may grow or shrink according to its birth and death rates, and it may exchange members with neighboring local populations by emigration and immigration. The number of cells occupied by local populations is the range of the population. The model developed here is loosely based on one in Case (2000).

We begin by assuming that the local population in each cell grows according to a logistic model:

$$N_{t+1} = N_t + (b + b'N_t)N_t - (d + d'N_t)N_t \qquad \text{Equation 1}$$

Here, N_t and N_{t+1} represent the size of the local population at times t and $t + 1$, b represents the per capita birth rate and d the per capita death rate, each when the local population is very small and uncrowded. The symbols b' and d' represent the change in per capita birth and death rates caused by each additional member of the local population. This is the same equation as you used for the logistic model in Exercise 8.

Equation 1 ignores immigration and emigration, which we want to include in our model for this exercise. We could just add terms for immigration and emigration, but the model would rapidly become unwieldy. So let's simplify Equation 1 a bit first. If we multiply out the terms in parentheses, we get

$$N_{t+1} = N_t + bN_t + b'N_t^2 - d\,N_t - d'N_t^2$$

We can rearrange these terms to get

$$N_{t+1} = N_t + bN_t - dN_t + b'N_t^2 - d'N_t^2$$

Factoring gives us

$$N_{t+1} = N_t + (b - d)N_t + (b' - d')N_t^2$$

We can use the symbol R_{max} to represent the population's maximum geometric rate of growth and R_{dd} for the density-dependent reduction in population growth rate—i.e., the amount by which each added member of the population reduces the population's per capita rate of growth. If we define $R_{max} = b - d$ and $R_{dd} = b' - d'$, we can write

$$N_{t+1} = N_t + R_{max}N_t + R_{dd}N_t^2$$

Factoring out N_t once again gives us

$$N_{t+1} = N_t + (R_{max} + R_{dd}N_t)N_t \qquad \text{Equation 2}$$

As the above derivation shows, this model is identical to the logistic model you used in an earlier exercise, but instead of showing per capita birth and death rates and their density-dependent changes explicitly, it combines all that into R_{max} and R_{dd}.

Next, we incorporate emigration out of the cell, symbolized E:

$$N_{t+1} = N_t + (R_{max} + R_{dd}N_t)N_t - (E_{min} + E_{dd}N_t)N_t \qquad \text{Equation 3}$$

Here E_{min} represents the minimum emigration rate and E_{dd} the density-dependent increase in emigration, that is, the amount by which each added member increases the per capita emigration rate from the cell. According to this model, a small proportion of the members of the population emigrate when the population of the cell is small, and the proportion emigrating increases as the population grows (see the exercise, "Metapopulation Dynamics").

These emigrants have to go somewhere, and in this model we will assume they move equally into immediately adjacent cells to the right and left of their natal cell. From the point of view of the population in a cell, new members move in from neighboring cells at rates determined by the sizes of the populations in those neighboring cells, which we simply call Left and Right. We can now write the whole equation for the population of a cell as

$$N_{t+1} = N_t + (R_{max} + R_{dd}N_t)N_t - (E_{min} + R_{dd}N_t)N_t + 0.5(E_{min} + E_{dd}\text{Left}_t)\text{Left}_t + 0.5(E_{min} + E_{dd}\text{Right}_t)\text{Right}_t \qquad \text{Equation 4}$$

In words, the population of each cell grows by reproduction of its own members and loses members by emigration. It also receives members from adjacent cells. The factor of one-half in each of these immigration terms comes from the assumption that half of the emigrants from each cell go to the left, and half go to the right.

Although we have written this equation as a logistic model, you can make it into a geometric model by setting R_{dd} to zero. Notice that this makes $(R_{max} + R_{dd}N_t)N_t = (R_{max} + 0N_t)N_t = R_{max}N_t$, which is our old geometric model (with immigration from neighboring cells added).

Likewise, the model assumes that emigration grows in a density-dependent fashion, but you can make emigration a constant proportion of population size by setting E_{dd} to zero. If you set E_{min} to zero, that represents a situation in which no individuals leave the population when it is very small (when $N_t = 0$, strictly speaking). If you set both E_{min} and E_{dd} to zero, it represents a situation with no emigration at all.

Thus by choosing appropriate values for the model parameters, you can model geometric or logistic population growth, with density-dependent or density-independent emigration, or no emigration at all. You will use the model to find out how a species' range expands under each of these scenarios.

PROCEDURES

First we will model range expansion in one dimension. You might think of one-dimensional habitat as something like a narrow stream, or a narrow riparian zone. Then we will expand the model to two dimensions and see if any of the model predictions change.

As always, save your work frequently to disk.

INSTRUCTIONS

ANNOTATION

A. Set up the one-dimensional model.

1. Open a new spreadsheet and set up titles and column headings as shown in Figure 1. Enter the parameter values shown for R_{max}, R_{dd}, E_{min}, and E_{dd}.

These are all literals, so just select the appropriate cells and type them in.

	A	B	C	D	E	F	G	H	I	J	K	L	M	N	O	P	Q	R
1	*Range Expansion Across Uniform One-Dimensional Habitat*																	
2																		
3			Parameters															
4			R_{max}	0.6	E_{min}	0.5												
5			R_{dd}	-0.01	E_{dd}	0.001												
6																		
7									Local populations									
8	Time		Site A	Site B	Site C	Site D	Site E	Site F	Site G	Site H	Site I	Site J	Site K	Site L	Site M		Total pop	Range
9	0																	
10	1																	
11	2																	
12	3																	

Figure 1

2. Set up a linear series from 0 to 50 in column A.

In cell A9 enter the value 0.
In cell A10 enter the formula **=A9+1**; copy this formula into cells A11–A59.

3. Enter zeros into cells C9–H9, and cells J9 through O9. Enter the value 2 into cell I9.

These are the initial population sizes in each habitat patch, or site. For now, we model a situation in which the population begins with two individuals at the center of the potential range. We call these the "seed population." You can change the initial conditions later.

4. In cell C10, enter a formula into to calculate the size of the population in that cell at time 1.

Enter the formula **=C9+($D$4+$D$5*C9)*C9-($F$4+$F$5*C9)*C9+0.5*($F$4+$F$5*B9)*B9+0.5*($F$4+$F$5*D9)*D9**.
This corresponds to Equation 4:

$$N_{t+1} = N_t + (R_{max} + R_{dd}N_t)N_t - (E_{min} + R_{dd}N_t)N_t + 0.5(E_{min} + E_{dd}\text{Left}_t)\text{Left}_t + 0.5(E_{min} + E_{dd}\text{Right}_t)\text{Right}_t$$

Note that the formula refers to cell B9, which is empty. The spreadsheet treats empty cells as zero values. In effect, we assume that cell B9 is unsuitable habitat, from which no emigrants emerge and within which immigrants die.

5. Copy the formula into cells D10–O10.

Note that the formula in cell O10 refers to cell P9, which is empty. The same interpretation applies here as in the case of cell B9.

6. Copy cells C10–O10 into cells C11–O59.

7. In cell Q9, enter a formula to calculate the total population (in all cells) at time 0. Copy this formula down the column.

Enter the formula **=SUM(C9:P9)**. Copy the formula in cell Q9 and paste it into cells Q10–Q59.

8. In cell R9, enter a formula to calculate the total range of the population at time 0. Copy this formula down the column.

Enter the formula **=COUNTIF(C9:O9,">1")**. Note the quotation marks around >1. Copy this formula into cells R10–R59. This formula tells the spreadsheet to count the number of cells in columns C–O of the current row that contain values greater than one.

We use the cutoff value of 1 rather than 0, because if we use 0 the behavior of the model becomes unrealistic due to the way the spreadsheet handles very small numbers. This cutoff is also more biologically reasonable, because we should not count habitat as occupied until the population there has reached some minimum size. Later, you can try raising the threshold value higher than 1 to see what effect that has.

9. Save your work.

Your one-dimensional model is now complete. You can now use it to graph various aspects of the population's size and range.

B. Graph various aspects of the one-dimensional model.

1. Graph the total population size against time. Edit your graph for readability.

Select cells A8–A59. Hold down the control key (Windows) or the key (Macintosh) and select cells Q8–Q59. Make an XY graph. Your graph should resemble Figure 2.

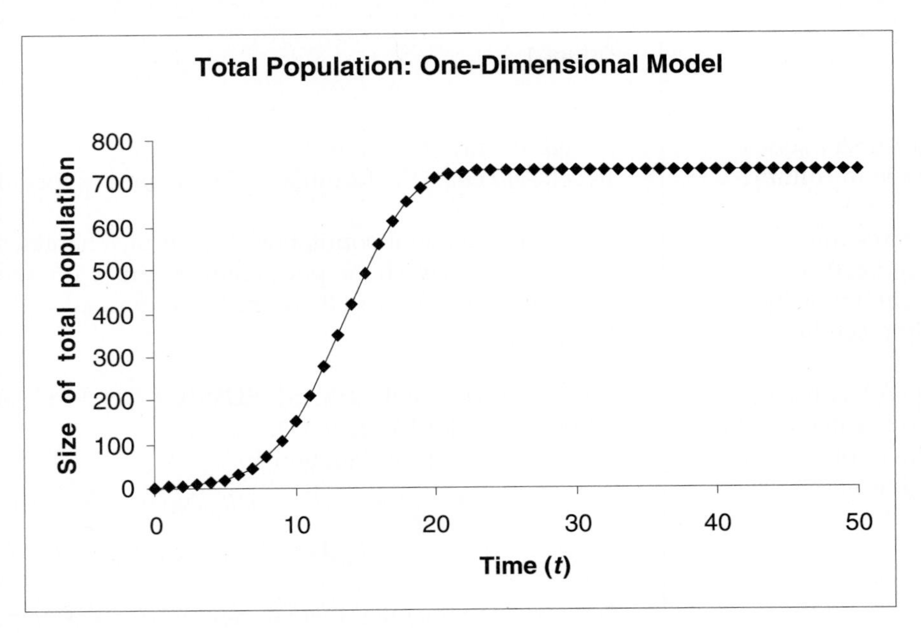

Figure 2

2. Graph the range of the population (number of occupied cells) against time. Edit your graph for readability.

Select cells A8–A59. Hold down the control key (Windows) or the key (Macintosh) and select cells R8–R59. Make an XY graph. Your graph should resemble Figure 3.

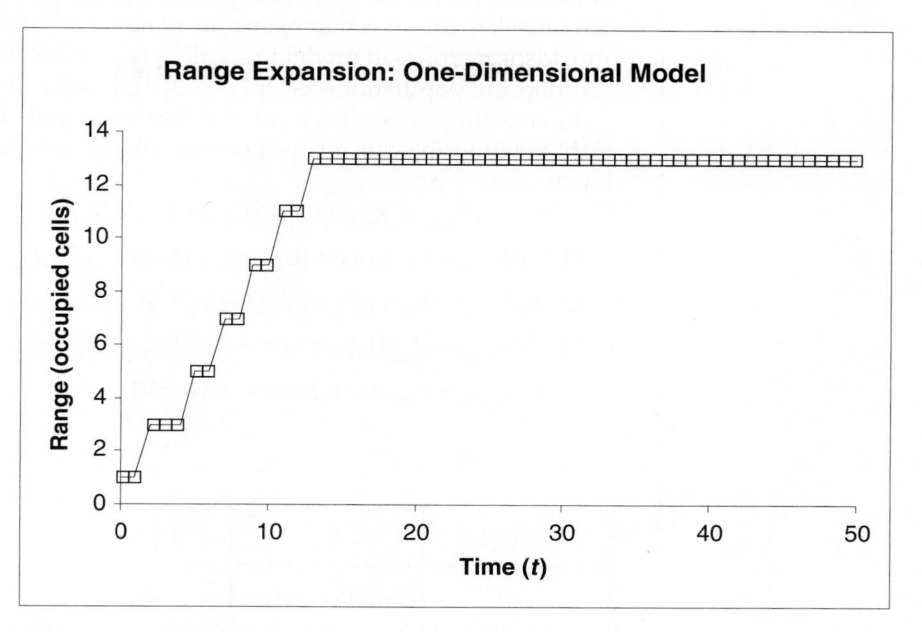

Figure 3

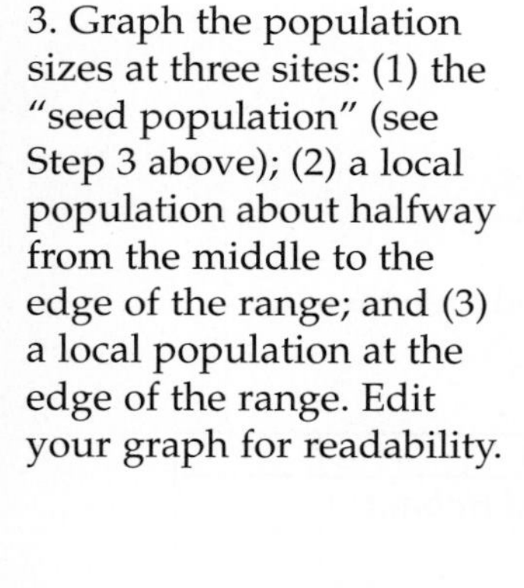
3. Graph the population sizes at three sites: (1) the "seed population" (see Step 3 above); (2) a local population about halfway from the middle to the edge of the range; and (3) a local population at the edge of the range. Edit your graph for readability.

Select cells A8–A59.
Hold down the control key (Windows) or the key (Macintosh), and select cells C8–C59.
Hold down the control key (Windows) or the key (Macintosh), and select cells F8–F59.
Hold down the control key (Windows) or the key (Macintosh), and select cells I8–I59.
Make an XY graph. Your graph should resemble Figure 4.

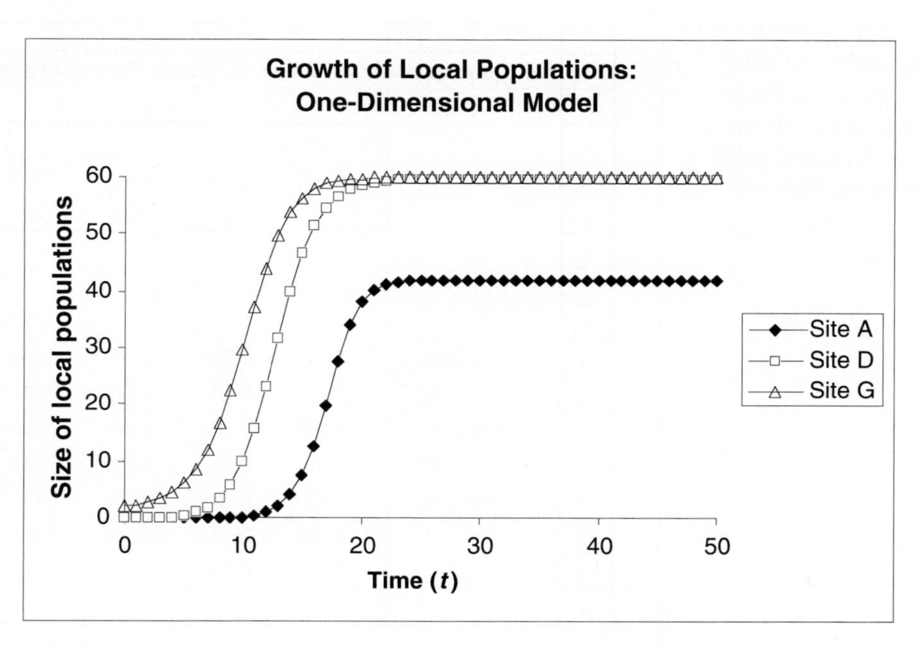

Figure 4

Expanding into Two Dimensions

In the next part of the exercise, we examine range expansion in two dimensions. You can visualize this as a homogeneous plain—perhaps a tract of prairie or forest. In the spreadsheet, you will model this with a two-dimensional grid of cells. We can refer to neighboring populations as UpLeft, Up, UpRight, etc. (Figure 5).

In two dimensions, each cell will lose emigrants to, and receive immigrants from, eight neighboring cells instead of two. We can represent this in an equation similar to Equation 4:

$$N_t = N_{t+1} + (R_{max} + R_{dd}N_t)N_t - (E_{min} + E_{dd}N_t)N_t + (1/8)(E_{min} + E_{dd}\text{UpLeft}_t)\text{UpLeft}_t + (1/8)(E_{min} + E_{dd}\text{Up}_t)\text{Up}_t + (1/8)(E_{min} + E_{dd}\text{UpRight}_t)\text{UpRight}_t + (1/8)(E_{min} + E_{dd}\text{Left}_t)\text{Left}_t + (1/8)(E_{min} + E_{dd}\text{Right}_t)\text{Right}_t + (1/8)(E_{min} + E_{dd}\text{DownLeft}_t)\text{DownLeft}_t + (1/8)(E_{min} + E_{dd}\text{Down}_t)\text{Down}_t + (1/8)(E_{min} + E_{dd}\text{DownRight}_t)\text{DownRight}_t \quad \text{Equation 5}$$

UpLeft	Up	UpRight
Left	Focal Population	Right
DownLeft	Down	DownRight

Figure 5 Eight neighboring populations around a central, focal population.

INSTRUCTIONS

C. Set up the two-dimensional model.

1. Open a new spreadsheet and set up labels in cells A1–F5 as shown in Figure 6. Enter the labels shown in cells A9, A11, and A13. Enter the parameter values shown for R_{max}, R_{dd}, E_{min}, and E_{dd}.

ANNOTATION

These are all literals, so just select the appropriate cells and type them in.

	A	B	C	D	E	F	G	H	I
1	***Range Expansion Across Uniform Two-Dimensional Habitat***								
2									
3			Parameters						
4			R_{max}	0.75	E_{min}	0.5			
5			R_{dd}	-0.1	E_{dd}	0.001			
6									
7									
8									
9	Time		0.00	0.00	0.00	0.00	0.00	0.00	0.00
10	0		0.00	0.00	0.00	0.00	0.00	0.00	0.00
11	Total pop		0.00	0.00	0.00	0.00	0.00	0.00	0.00
12	2.00		0.00	0.00	0.00	2.00	0.00	0.00	0.00
13	Range		0.00	0.00	0.00	0.00	0.00	0.00	0.00
14	1		0.00	0.00	0.00	0.00	0.00	0.00	0.00
15			0.00	0.00	0.00	0.00	0.00	0.00	0.00
16									

Figure 6

2. Set up a two-dimensional matrix of cells to represent the local populations at time 0.	The matrix of cells C9–I15 represent the area of suitable habitat. At time 0, the habitat is empty except for a small seed population in cell F12. Enter the values shown for the local populations at time 0. The easiest way to do this is to enter a value of 0 into cell C9, then copy that and paste it into cells C10–C15. Then copy cells C9–C15 and paste into cells D9–I15. Finally, enter a value of 2 into cell F12.
3. Mark this matrix as representing time 0.	In cell A10, enter the value 0.
4. In cell A12, enter a formula to calculate the total population size.	Enter the formula **=SUM(B9:I15)**.
5. In cell A14, enter a formula to calculate the range of the population.	Enter the formula **=COUNTIF(C9:I15,">1")**.
6. Set up a separate matrix of cells to represent the population at time 1.	Copy cells A9–I15 and paste into cells A17–I23. Change the time-value in cell A18 from 0 to **=A10+1**.
7. Enter formulae to calculate the size of each local population at time 1.	Equation 5 above is the basis for this formula. In cell C17, enter the formula **=C9+($D$4+$D$5*C9)*C9-($F$4+$F$5*C9)*C9+(1/8)*($F$4+$F$5*B8)*B8+(1/8)*($F$4+$F$5*C8)*C8+(1/8)*($F$4+$F$5*D8)*D8+(1/8)*($F$4+$F$5*B9)*B9+(1/8)*($F$4+$F$5*D9)*D9+(1/8)*($F$4+$F$5*B10)*B10+(1/8)*($F$4+$F$5*C10)*C10+(1/8)*($F$4+$F$5*D10)*D10.** Copy this formula into cells C18–C23. Copy cells C17–C23 and paste into cells D17–I23.
8. Copy the entire matrix of cells for time 1 down the spreadsheet to model the spread of the species through time 15.	Copy cells A17–I23. Paste separately into each of the following cells: A25, A33, A41, A49, A57, A65, A73, A81, A89, A97, A105, A113, A124, and A129.
9. Set up titles and column headings as shown in Figure 7.	These are all literals, so just select the appropriate cells and type them in.

	L	M	N	O	P	Q
8	Time	Total pop	Central pop	Medial pop	Edge pop	Range
9	0					
10	1					
11	2					
12	3					

Figure 7

10. Set up a linear series from 0 to 15 in cells L9–L22.	Enter the value 0 in cell L9. In cell L10, enter the formula **=L9+1**. Copy this formula into cells L11–L22.

11. Set up cells M9–M22 to echo the values of total population size at times 0–15.

In cell M9, enter the formula **=A12**.
In cell M10, enter the formula **=A20**.
In cell M11, enter the formula **=A28**.
Continue down the column, incrementing the row address by 8, until you reach cell M22, which should contain the formula **=A132**.

12. Set up cells series in columns N, O, and P to echo the population size of the central population (Central Pop), a local population about halfway to the edge of the suitable habitat (Medial Pop), and a local population at the edge of suitable habitat (Edge Pop).

In cells N9–N22, enter the formulae **=F12** through **=F132**, in steps of 8, as you did for the formula in column M.
Similarly, in cells O9 through O22, enter the formulae **=D10** through **=D130** in steps of 8.
In cells P9 through P22, enter the formulae **=C9** through **= C129** in steps of 8.

13. Set up series in column Q to echo the size of the range at times 0 through 15.

In cells Q9 through Q22, enter the formulae **=A14** through **=A134** in steps of 8.

14. Save your work.

Your spreadsheet is now complete.

D. Graph aspects of the two-dimensional model.

1. Graph total population size against time.

Select cells L9 through M22, and make an XY graph. Edit your graph for readability. It should resemble Figure 8.

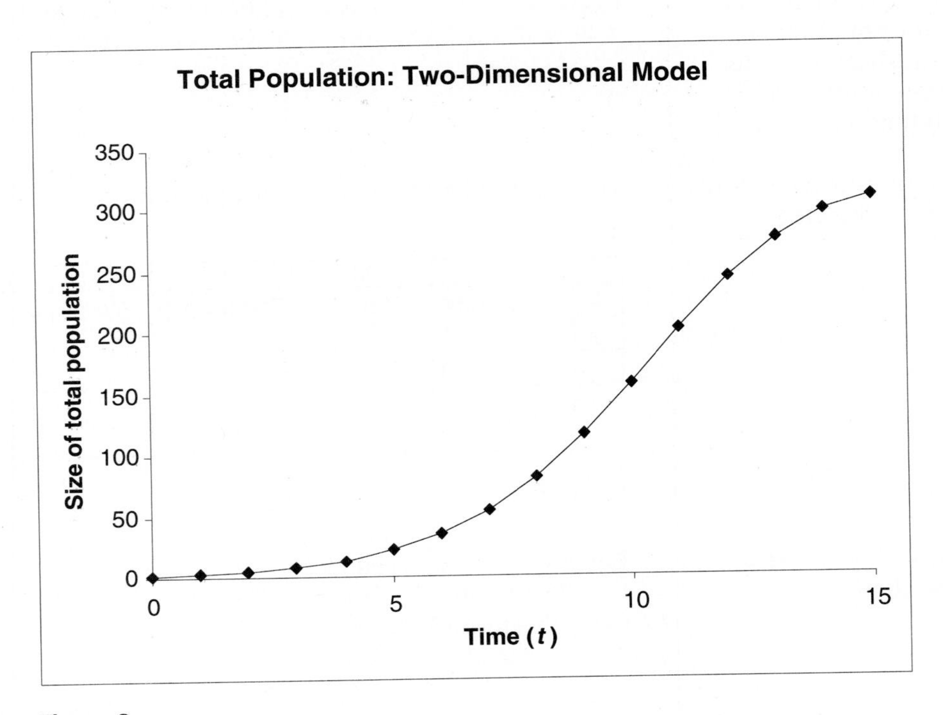

Figure 8

2. Graph the sizes of the central population, medial population, and edge population against time.

Select cells L9 through L22. Hold down the key (Macintosh) or control key (Windows) while selecting cells N9 through P22. Make an XY graph. Edit your graph for readability. It should resemble Figure 9.

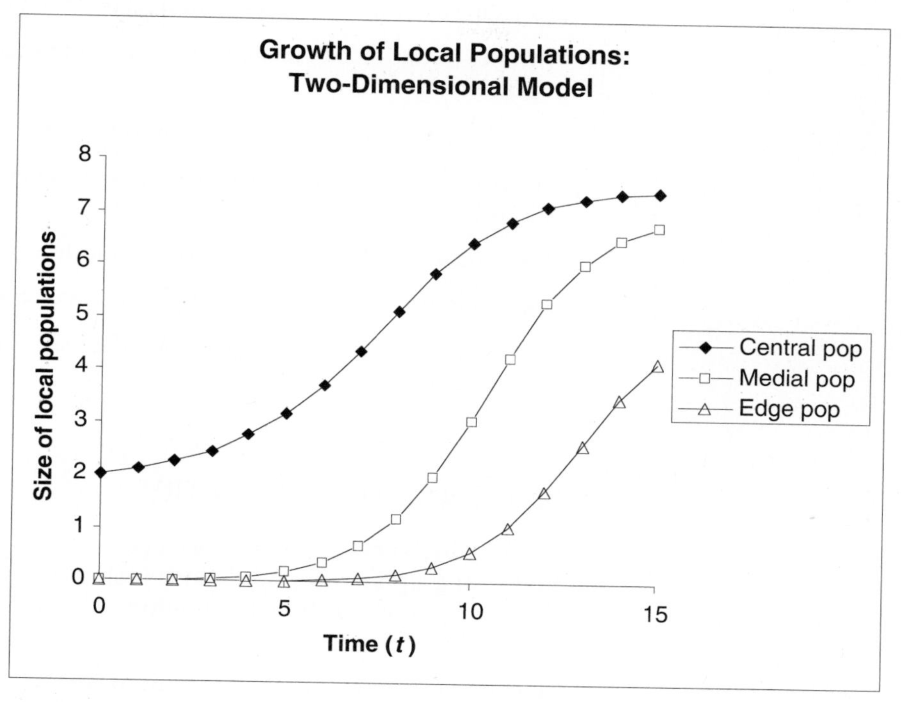

Figure 9

3. Graph the range of the population against time.

Select cells L9 through L22. Hold down the key (Macintosh) or control key (Windows) while selecting cells Q9 through Q22. Make an XY graph. Edit your graph for readability. It should resemble Figure 10.

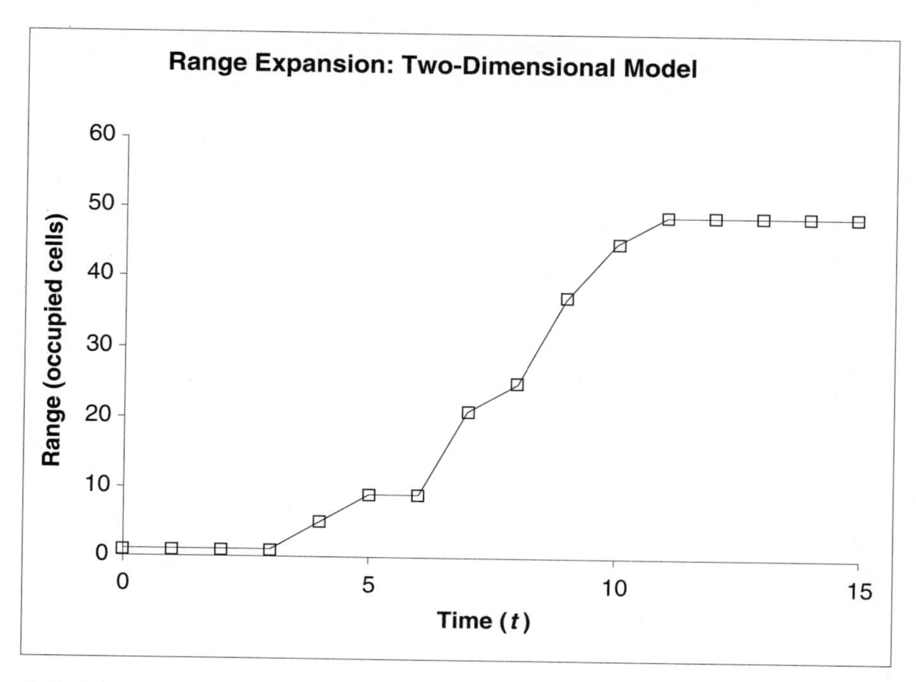

Figure 10

QUESTIONS

Answer questions 1–5 first for the parameters given for the one-dimensional model, and then again using the parameters in the two-dimensional model. (You should find that the two models behave very similarly.)

1. With the parameter values given, how do the local populations and the total population grow?
2. How does the range of this population expand?
3. Can you change parameter values to model geometrically growing local populations?

   Does this affect the predictions of the model?
4. Does changing the emigration parameters change the behavior of the model?
5. How is the rate of range expansion affected by rates of local population growth and emigration?

ADDITIONAL THINGS TO TRY

1. Set R_{max} high enough to produce cyclic or chaotic behavior in the seed population, and graph a few populations (columns) as well as total population. How do the dynamics of these populations compare to dynamics of isolated logistic populations?
2. Start two or more populations at nonzero values (i.e., set up two or more seed populations). Graph each seed population and the total population. What happens when their ranges meet and overlap?

LITERATURE CITED

Case, T. J. 2000. *An Illustrated Guide to Theoretical Ecology*. Oxford University Press, New York.

28 SUCCESSION

Objectives

- Understand the concept of succession and several theories of successional mechanisms.
- Set up a spreadsheet matrix model of succession.
- Use the model to explore predictions of various theories of succession.

INTRODUCTION

Succession is change in community composition at one site over time-scales longer than a year and shorter than millenia. We exclude shorter time-spans because we want to exclude cyclic seasonal changes in abundance, and longer time-scales because we want to exclude evolutionary changes and responses to climate changes.

Succession may occur on newly exposed substrate, such as glacial till or fresh lava, in which case it is called **primary succession**. Succession may also occur on previously vegetated soil, from which much or all of the biota has been removed by some disturbance, such as fire or clear-cutting. In this case, we call it **secondary succession**.

A Markov Chain Model of Succession

Primary and secondary succession often differ in the sequence of organisms that appear at a site and in the mechanisms that determine that sequence. However, we can describe either kind of succession in purely phenomenological terms by specifying transition probabilities from one state of the community to another. The technical term for such a model is a **Markov chain** or **Markov process**. A Markov chain is a sequence of states of a system in which each successive state depends only on the previous state and the transition probabilities between possible states.

To make the concept a bit more concrete, consider algal succession on rock in the intertidal zone of an ocean shore. A storm roils the surf, shifting boulders, and scraping some clean. Let us focus on the surface of one such boulder. After the waters calm, propagules of species A may settle out and begin to grow. Soon thereafter, propagules of species B may settle out on the same rock, compete with species A, and eventually take over the rock. Somewhat later, species C may similarly displace species B. In short, we have a successional sequence of species A $\rightarrow$ B $\rightarrow$ C.

The system here is the community (consisting in this case of a single species) occupying the rock surface. The states of the system are "Occupied by bare rock," "Occupied by species A," "Occupied by species B," and "Occupied by species C." Whatever state the system is in at any given time, there is some probability that it will be in each of the other states one time unit later. These probabilities are the transition probabilities.

We can conveniently represent the states of the system, and the transition probabilities between states, in matrix form (Table 1). By convention, the top row of the matrix lists all possible current states of the system (species occupying the rock) at some time t; the left column lists all possible succeeding states of the system one arbitrary time unit later. The entries in the body of the matrix represent the probabilities of each possible transition from one state to another state or the same state over that time period.

Table 1. Matrix of hypothetical transition probabilities between successional states on a rock in the intertidal zone.

		Species occupying the rock at time t			
		Bare Rock	A	B	C
Species occupying the rock at time $t + 1$	Bare Rock	0.10	0.10	0.10	0.10
	A	0.80	0.75	0.02	0.01
	B	0.06	0.10	0.80	0.04
	C	0.04	0.05	0.08	0.85

According to this matrix, bare rock is unlikely ($p = 0.10$) to remain bare from time t to time $t + 1$. The probability that a bare rock will be colonized by species A in that time is 0.80; that it will be colonized by species B is 0.06; and by species C, 0.04. A patch of rock already occupied by species A is likely to remain so ($p = 0.75$), but there is a 10% chance that it will succeed to species B and a 5% chance that it will succeed to species C. There is also a 10% chance that a new disturbance will remove whatever species currently occupies the rock (note the values of 0.10 in the three right-hand cells of the top row).

Notice that each column of the transition matrix adds to 1. This has to be, because we must account for the fate of all patches that began the interval from t to $t + 1$ in each state.

To apply the transition matrix, we must begin with the number of rocks currently in each stage of succession (i.e., bare rock or occupied by species A, B, or C). These numbers are arranged in a **state vector**, which describes the current state of the system. For example, if we examined our intertidal area at some time and found 70% of the rocks were bare, 20% occupied by species A, 5% occupied by species B, and 5% occupied by species C, we could write that as a state vector $\mathbf{s}_t$

$$\mathbf{s}_t = \begin{bmatrix} 0.70 \\ 0.20 \\ 0.05 \\ 0.05 \end{bmatrix}$$

To predict the number of rocks occupied by each species (or bare) in the future, we multiply the state vector by the transition matrix $\boldsymbol{A}$:

$$\mathbf{A} = \begin{bmatrix} 0.10 & 0.10 & 0.10 & 0.10 \\ 0.80 & 0.75 & 0.02 & 0.01 \\ 0.06 & 0.10 & 0.80 & 0.04 \\ 0.04 & 0.05 & 0.08 & 0.85 \end{bmatrix}$$

That is,

$$\mathbf{s}_{t+1} = \mathbf{A} \times \mathbf{s}_t$$

or in our example,

$$\mathbf{s}_{t+1} = \begin{bmatrix} 0.10 & 0.10 & 0.10 & 0.10 \\ 0.80 & 0.75 & 0.02 & 0.01 \\ 0.06 & 0.10 & 0.80 & 0.04 \\ 0.04 & 0.05 & 0.08 & 0.85 \end{bmatrix} \times \begin{bmatrix} 0.70 \\ 0.20 \\ 0.05 \\ 0.05 \end{bmatrix}$$

We can carry our predictions as far into the future as we wish by iterating this matrix multiplication:

$$\mathbf{s}_{t+1} = \mathbf{A} \times \mathbf{s}_t$$

$$\mathbf{s}_{t+2} = \mathbf{A} \times \mathbf{s}_{t+1}$$

$$\mathbf{s}_{t+3} = \mathbf{A} \times \mathbf{s}_{t+2}$$

$$\mathbf{s}_{t+4} = \mathbf{A} \times \mathbf{s}_{t+3}$$

and so on. (If you are unfamiliar with matrix multiplication, or have forgotten the details, consult the Appendix at the end of this exercise.)

We can ask a variety of interesting questions about long-term model predictions. For example, will the system eventually come to equilibrium? If so, will the equilibrium consist of a single species (a **climax**), or will it consist of a stable mixture of species? If the latter, what will be the proportion of each species? Does the eventual state of the system depend on the initial state vector, or only on the transition probabilities?

It may be tempting to conceive of successional changes not from one species to another but of entire communities. This presupposes that communities in a successional sequence are discrete entities, corresponding to the discrete states of a Markov chain. However, the evidence from field ecology shows that communities are not discrete entities, and that succession is not a change from one discrete community to another, but rather individualistic, species-by-species changes in abundance, presence, and absence. Therefore, to model successional change accurately at the community level requires a species-level model.

We can use a Markov chain model, however, if we keep in mind that we are modeling a continuous process as if it proceeded in discrete steps. That is, we may choose to look at community composition at, say, 50-year intervals. With that much time, community composition may have changed enough to permit us to regard communities as discretely different, despite our knowledge that change over the intervening years was individualistic and continuous.

Whether we think of our model as representing species-by-species replacement or whole-community replacement, the mathematics is the same, only our interpretation changes. Indeed, the model is mathematically identical to a Leslie matrix model of a size-structured or stage-structured population.

PROCEDURES

Connell and Slatyer (1977) described three fundamentally different ways in which succession might proceed. Early-arriving individuals ("pioneers") may change the environment in ways that favor other species at the expense of their own offspring, as for example by casting shade or adding organic matter and other substances to the soil. Connell and Slatyer call this the **facilitation model**. Alternatively, early-arriving individuals may simply hold onto their sites, and the only way other individuals can enter the community is if disturbance removes the site-holders. Connell and Slatyer call this the **inhibition model**. Finally, it is logically conceivable that existing individuals may have no significant influence, either positive or negative, on the establishment of others. Connell and Slatyer call this the **tolerance model**. You can examine the outcome of each of these models with the Markov chain model set up in this exercise.

As always, save your work frequently to disk.

INSTRUCTIONS / ANNOTATION

A. Markov chain model of succession.

1. Open a new spreadsheet and set up titles and column headings as shown in Figure 1

The text items are all literals, so just select the appropriate cells and type them in. The transition probabilities correspond to Table 1.

	A	B	C	D	E
1	***Succession***				
2	A Markov-chain model of community change over time.				
3	Example: Table 1 from Introduction				
4		Transition matrix: A			
5		Bare rock	Species A	Species B	Species C
6	Bare rock	0.10	0.10	0.10	0.10
7	Species A	0.80	0.75	0.02	0.01
8	Species B	0.06	0.10	0.80	0.04
9	Species C	0.04	0.05	0.08	0.85
10	Sum	1.00	1.00	1.00	1.00

Figure 1

2. Enter formulae to sum up each column of transition probabilities.

In cell B10 enter the formula **=SUM(B6:B9)**. Copy this formula into cells C10–E10. You will use these sums to check your transition probabilities when you change them later in the exercise. Remember that each column of the transition matrix must add up to 1.

3. Enter column and row headings shown in Figure 2. Continue the sequence of time values to the right until you reach $t = 20$ in column AB.

	G	H	I	J	K	L
4		State vectors: $s(t)$				
5	Time (t)	0	1	2	3	4
6	Bare rock	0.70				
7	Species A	0.20				
8	Species B	0.05				
9	Species C	0.05				
10	Sum	1.00				

Figure 2

4. Enter the initial state vector.

Enter the values shown in cells H6 through H9.

5. Enter a formula to total the frequencies in the initial state vector.

In cell H10 enter the formula **=SUM(H6:H9)**.
This is another check on your model. State vectors must also add up to 1.

6. Enter a formula to calculate the state vector at time 1.

In cell I6 enter the formula **=$B6*H$6+$C6*H$7+$D6*H$8+$E6*H$9**.
Be careful to use absolute and relative addresses exactly as shown. This allows you to copy the formula into other cells and get correct results. Any deviation from the formula will produce erroneous results.

7. Copy the formula from cell I6 into cells I7 through I9.

8. Copy cells I6 through I9 into cells J6 through AB9

9 Your spreadsheet is complete. Save your work.

10. Graph the proportion of rock surfaces occupied by each species (or bare rock) against time.

Select cells G5 through AB9. Make an XY (Scatterplot) Chart. Edit your graph for readability. It should resemble Figure 3.

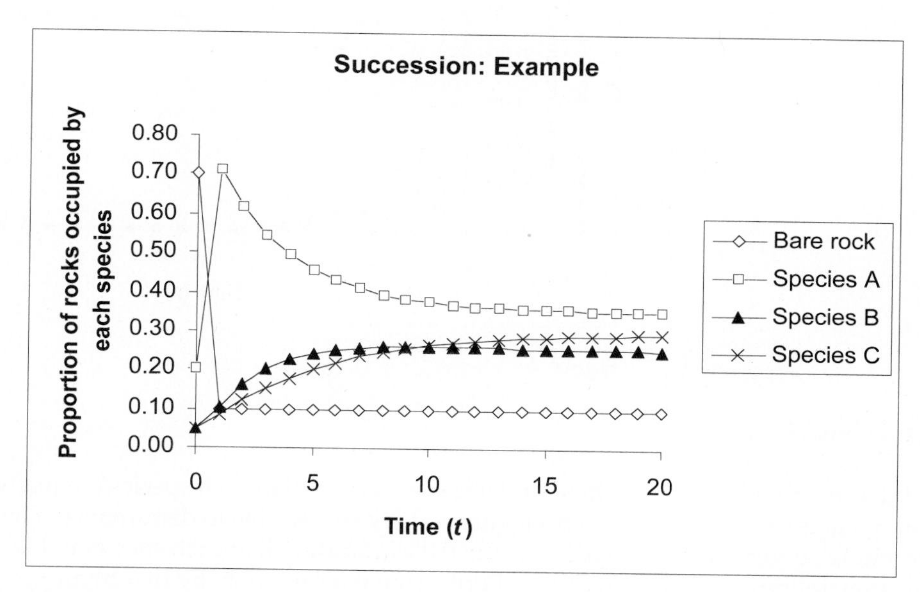

Figure 3

B. Facilitation model.

1. To see the predictions of Connell and Slatyer's (1977) facilitation model of succession, change the transition probabilities in your spreadsheet to those given in Table 2.

These probabilities indicate that bare rock is frequently replaced by species A, species A by species B, and species B by species C. All these species are equally likely to be replaced by bare rock. Species C is unique in that it is almost always replaced by itself, only rarely by bare rock, and never by other species.

Table 2. Transition matrix for the Connell and Slatyer (1977) facilitation model.

	Transition Matrix: Facilitation			
	Bare Rock	Species A	Species B	Species C
Bare Rock	0.10	0.10	0.10	0.01
Species A	0.90	0.10	0.00	0.00
Species B	0.00	0.80	0.10	0.00
Species C	0.00	0.00	0.80	0.99
Sum	1.00	1.00	1.00	1.00

2. Change the initial state vector so that the initial frequency of Bare Rock is 1.00, and all other species have frequencies of 0.00. Graph the results.

Your graph should resemble Figure 4.

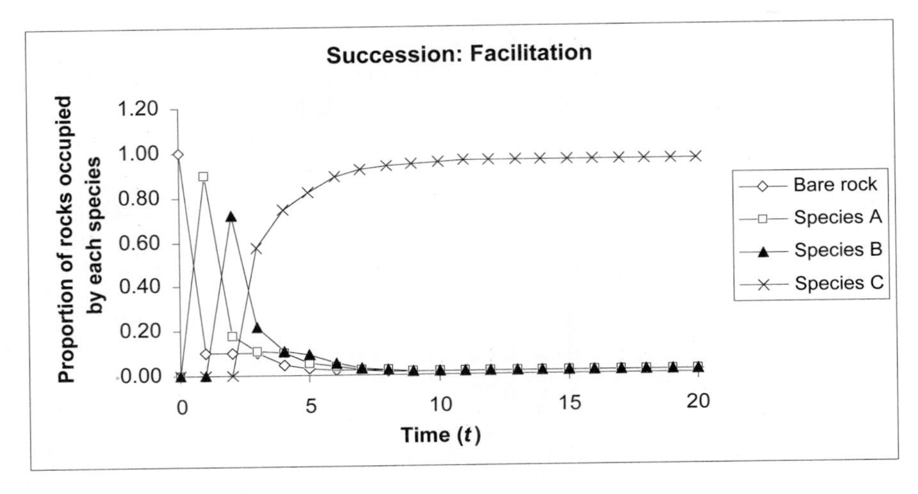

Figure 4

C. Inhibition model.

1. To see the predictions of Connell and Slatyer's inhibition model, enter the transition probabilities given in Table 3 in your spreadsheet.

These probabilities indicate that each species is equally likely to colonize bare rock, and all species are equally susceptible to disturbance. The transition probabilities between species are all 0.00, indicating that each species holds its site and inhibits occupancy by al others. Replacement occurs only by disturbance.

Table 3. Transition matrix for the Connell and Slatyer (1977) inhibition model.

	Transition Matrix: Inhibition			
	Bare Rock	Species A	Species B	Species C
Bare Rock	0.10	0.10	0.10	0.10
Species A	0.30	0.90	0.00	0.00
Species B	0.30	0.00	0.90	0.00
Species C	0.30	0.00	0.00	0.90
Sum	1.00	1.00	1.00	1.00

2. Keep the initial state vector set with the initial frequency of bare rock at 1.00 and all other frequencies at 0.00. Graph the results

Your graph should now resemble Figure 5.

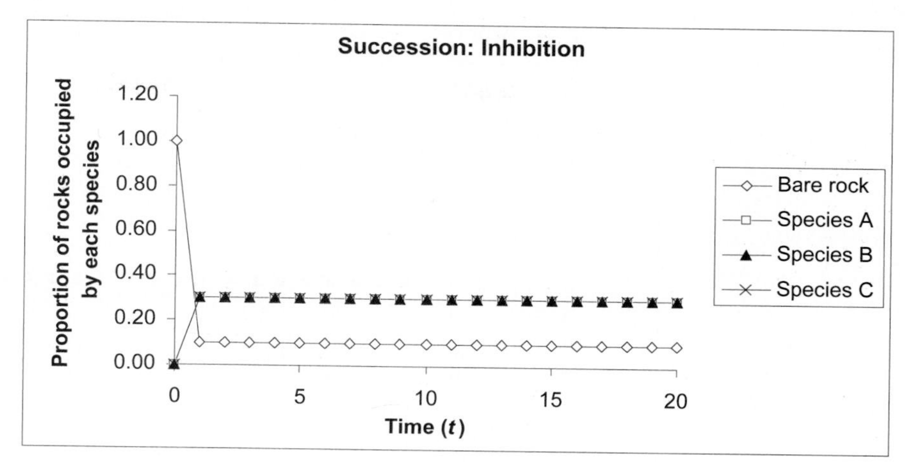

Figure 5

D. Tolerance model.

1. To see the predictions of Connell and Slatyer's tolerance model, enter the transition probabilities given in Table 4.

As you can see, all the transition probabilities are equal. This indicates that any species is equally likely to replace any other, and equally susceptible to disturbance.

Table 4. Transition matrix for the Connell and Slatyer (1977) tolerance model.

	Transition Matrix: Tolerance			
	Bare Rock	Species A	Species B	Species C
Bare Rock	0.25	0.25	0.25	0.25
Species A	0.25	0.25	0.25	0.25
Species B	0.25	0.25	0.25	0.25
Species C	0.25	0.25	0.25	0.25
Sum	1.00	1.00	1.00	1.00

2. Keep the initial state vector set with the initial frequency of bare rock at 1.00 and all other frequencies at 0.00. Graph the results.

Your graph should now resemble Figure 6.

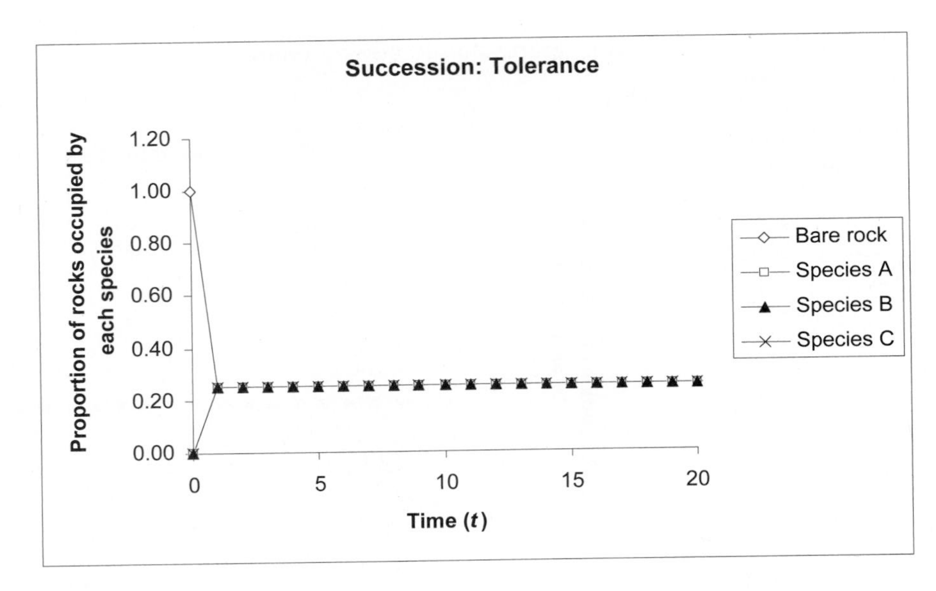

Figure 6

QUESTIONS

1. Will the system eventually come to equilibrium? That is, will the frequencies of rocks occupied by three species and bare rock stop changing?
2. Does the equilibrium consist of a single species occupying all rocks, or is there a stable mixture of species?
3. Are the equilibrium frequencies determined by the initial frequencies (initial state vector), by the transition probabilities, or both?
4. Will any valid transition matrix (valid meaning that the columns each add to 1) result in equilibrium? Or are there valid transition matrices that do not lead to an equilibrium?
5. Describe each of Connell and Slatyer's (1977) models of succession, based on the information in the graphs you produced in Sections B–D of this exercise.

(A) Facilitation model (Figure 4)
(B) Inhibition model (Figure 5)
(C) Tolerance model (Figure 6)

LITERATURE CITED

Connell, J. H. and R. O. Slatyer. 1977. Mechanisms of succession in natural communities and their role in community stability and organization. *American Naturalist* 111: 119–144.

Appendix: MATRIX MULTIPLICATION

A matrix is a rectangular array of numbers characterized by the number of its rows and columns. Matrix **A** below is a 2×3 matrix. A matrix with one row or one column is called a vector. Vector **B** is a 3×1 vector.

$$\mathbf{A} = \begin{bmatrix} 31 & 7 & 23 \\ 11 & 5 & 17 \end{bmatrix} \qquad \mathbf{B} = \begin{bmatrix} 2 \\ 4 \\ 6 \end{bmatrix}$$

Matrices and vectors can only be multiplied by other matrices or vectors if the number columns of the first equals the number of rows of the second. Thus, matrix **A** could be multiplied by vector **B**; that is, $\mathbf{A} \times \mathbf{B}$ is a valid matrix multiplication.

Matrix multiplication is not commutative: that is, $\mathbf{A} \times \mathbf{B} \neq \mathbf{B} \times \mathbf{A}$. Indeed, $\mathbf{B} \times \mathbf{A}$ cannot be done, since the number of columns in **B** does not equal the number of rows in **A**.

Finally, here is how to do $\mathbf{A} \times \mathbf{B}$:

$$\begin{bmatrix} 31 & 7 & 23 \\ 11 & 5 & 17 \end{bmatrix} \times \begin{bmatrix} 2 \\ 4 \\ 6 \end{bmatrix} = \begin{bmatrix} 31 \times 2 + 7 \times 4 + 23 \times 6 \\ 11 \times 2 + 5 \times 4 + 17 \times 6 \end{bmatrix} = \begin{bmatrix} 228 \\ 144 \end{bmatrix}$$

Notice that the resulting matrix (vector in this case) has the same number of rows as the first matrix and the same number of columns as the second.

29 HARDY-WEINBERG EQUILIBRIUM

Objectives

- Understand the Hardy-Weinberg principle and its importance.
- Understand the chi-square test of statistical independence and its use.
- Determine the genotype and allele frequencies for a population of 1000 individuals.
- Use a chi-square test of independence to determine if the population is in Hardy-Weinberg equilibrium.
- Determine the genotypes and allele frequencies of an offspring population.

Suggested Preliminary Exercises: Statistical Distributions; Hypothesis Testing

INTRODUCTION

When you picture all the breeds of dogs in the world—poodles, shepherds, retrievers, spaniels, and so on—it can be hard to believe they are all members of the same species. What accounts for their different appearance and talents, and how do dog breeders match up a male and female of a certain breed to produce prize-winning offspring? The physical and behavioral traits we observe in nature, such as height and weight, are known as the **phenotype**. An individual's phenotype is the product of its **genotype** (genetic make-up), or its environment, or both. In this exercise, we focus on the genetic make-up of a population and how it changes over time. This field of study is known as **population genetics.**

Genes, Alleles, and Genotypes

A **gene**, loosely speaking, is a physical entity that is transmitted from parents to offspring and determines or influences traits (Hartl 2000). In one of the great achievements of the life sciences, Gregor Mendel studied the inheritance of flower color and seed shape in common peas and hypothesized the existence and behavior of such an entity of heredity many years before genes were actually described and shown to exist (Mendel 1866).

The multitude of genes in an organism reside on its chromosomes. A particular gene will be located at the same position, called the **locus** (plural, loci), on the chromosomes of every individual in the populations. In sexually reproducing diploid organisms, individuals have two copies of each gene at a given locus; one copy is inherited paternally (from the father), the other maternally (from the mother). The two copies considered together determine the individual's genotype. Genes can exist in different forms, or states, and these alternative forms are called **alleles**. If the two alleles in an individual are identical, the individual's genotype is said to be **homozygous**. If the two are different, the genotype is **heterozygous**.

Although *individuals* are either homozygous or heterozygous at a particular locus, *populations* are described by their **genotype frequencies** and **allele frequencies**. The word *frequency* in this case means occurrence in a population. To obtain the genotype frequencies of a population, simply count up the number of each kind of genotype in the population and divide by the total number of individuals in the population. For example, if we study a population of 55 individuals, and 8 individuals are A_1A_1, 35 are A_1A_2, and 12 are A_2A_2, the genotype frequencies (f) are

$$f(A_1A_1) = 8/55 = 0.146$$

$$f(A_1A_2) = 35/55 = 0.636$$

$$f(A_2A_2) = 12/55 = 0.218$$

$$\text{Total} = 1.00$$

The total of the genotype frequencies of a population always equals 1.

Allele frequencies, in contrast, describe the proportion of all alleles in the population that are of a specific type (Hartl 2000). For our population of 55 individuals above, there are a total of 110 alleles (of any kind) present in the population (each individual has two copies of a gene, so there are $55 \times 2 = 110$ total alleles in the population). To calculate the allele frequencies of a population, we need to calculate how many alleles are A_1 and how many are A_2. To calculate how many copies are A_1, we count the number of A_1A_1 homozygotes and multiply that number by 2 (each homozygote has two A_1 copies), then add to it the number of A_1A_2 heterozygotes (each heterozygote has a single A_1 copy). The total number of A_1 copies in the population is then divided by the total number of alleles in the population to generate the **allelele frequency**. The total number of A_1 alleles in our example population is thus $(2 \times 8) + (1 \times 35) = 51$. The frequency of A_1 is calculated as $51/(2 \times 55) = 51/110 = 0.464$. Similarly, the total number of A_2 alleles in the population is $(2 \times 12) + (1 \times 35) = 59$, and the frequency of A_2 is $59/(2 \times 55) = 59/110 = 0.536$.

As with genotype frequencies, the total of the allele frequencies of a population always equals 1. By convention, frequencies are designated by letters. If there are only two alleles in the population, these letters are conventionally $\boldsymbol{p}$ and $\boldsymbol{q}$, where p is the frequency of one kind of allele and q is the frequency of the second kind of allele. For genes that have only two alleles,

$$p + q = 1 \qquad \text{Equation 1}$$

If there were more than two kinds of alleles for a particular gene, we would calculate allele frequencies for the other kinds of alleles in the same way. For example, if three alleles were present, A_1, A_2, and A_3, the frequencies would be p (the frequency of the A_1 allele), q (the frequency of the A_2 allele) and r (the frequency of the A_3 allele). No matter how many alleles are present in the population, the frequencies should always add to 1. In this exercise, we will keep things simple and focus on a gene that has only two alleles.

In summary, for a population of N individuals, the number of A_1A_1, A_1A_2, and A_2A_2 genotypes are N_{A1A1}, N_{A1A2}, and N_{A2A2}, respectively. If p represents the frequency of the A_1 allele, and q represents the frequency of the A_2 allele, the estimates of the allele frequencies in the population are

$$f(A_1) = p = (2N_{A1A1} + N_{A1A2})/2N \qquad \text{Equation 2}$$

$$f(A_2) = q = (2N_{A2A2} + N_{A1A2})/2N \qquad \text{Equation 3}$$

The Hardy-Weinberg Principle

Population geneticists are not only interested in the genetic make-up of populations, but also how genotype and allele frequencies *change* from generation to generation. In the broadest sense, **evolution** is defined as the change in allele frequencies in a population over time (Hartl 2000). The Hardy-Weinberg principle, developed by G. H. Hardy and W. Weinberg in 1908, is the foundation for the genetic theory of evolution (Hardy 1908). It is one of the most important concepts that you will learn about in your studies of population biology and evolution.

Broadly stated, the Hardy-Weinberg principle says that given the initial genotype frequencies p and q for two alleles in a population, after a single generation of random mating the genotype frequencies of the offspring will be $p^2{:}2pq{:}q^2$, where p^2 is the frequency of the A_1A_1 genotype, $2pq$ is the frequency of the A_1A_2 genotype, and q^2 is the frequency of the A_2A_2 genotype. The sum of the genotype frequencies, as always, will sum to one; thus,

$$p^2 + 2pq + q^2 = 1 \qquad \text{Equation 4}$$

This equation is the basis of the Hardy-Weinberg principle.

The Hardy-Weinberg principle further predicts that genotype frequencies and allele frequencies will remain constant in any succeeding generations—in other words, the frequencies will be in **equilibrium** (unchanging). For example, in a population with an A_1 allele frequency p of 0.75 and an A_2 allele frequency q of 0.25, in Hardy-Weinberg equilibrium, the genotype frequencies of the population should be:

$$f(A_1A_1) = p^2 = p \times p = 0.75 \times 0.75 = 0.5625$$

$$f(A_1A_2) = 2 \times p \times q = 2 \times 0.75 \times 0.25 = 0.375$$

$$f(A_2A_2) = q^2 = q \times q = 0.25 \times 0.25 = 0.0625$$

Now let's suppose that this founding population mates at random. The Hardy-Weinberg principle tells us that after just one generation of random mating, the genotype frequencies in the next generation will be

$$f(A_1A_1) = p^2 = p \times p = 0.75 \times 0.75 = 0.5625$$

$$f(A_1A_2) = 2 \times p \times q = 2 \times 0.75 \times 0.25 = 0.375$$

$$f(A_2A_2) = q^2 = q \times q = 0.25 \times 0.25 = 0.0625$$

Additionally, the initial allele frequencies will remain at 0.75 and 0.25. These frequencies (allele and genotype) will remain unchanged over time.

The Hardy-Weinberg principle is often called the "null model of evolution" because genotypes and allele frequencies of a population in Hardy-Weinberg equilibrium will remain unchanged over time. That is, populations won't evolve. When populations violate the Hardy-Weinberg predictions, it suggests that some evolutionary force is acting to keep the population out of equilibrium. Let's walk through an example.

Suppose a population is founded by 3,000 A_1A_1 and 1,000 A_2A_2 individuals. From Equation 2, the frequency of the A_1 allele, p, is $(2 \times 3000 + 0)/(2 \times 4000) = 0.75$. Because $p + q$ must equal 1, q must equal $1 - p$, or 0.25. So, since p and q are equal to the values we used above to calculate the equilibrium genotype frequencies, if this population were in Hardy-Weinberg equilibrium, 56% of the population *should* be homozygous A_1A_1, 38% *should* be heterozygous, and 6% *should* be homozygous A_2A_2. But the *actual* genotype frequencies in this population are 75% homozygous A_1A_1 and 25% homozygous

A_2A_2—there are *no* heterozygotes! So this founding population is *not* in Hardy-Weinberg equilibrium.

To determine whether an observed population's deviations from Hardy-Weinberg expectations might be due to random chance, or whether the deviations are so significant that we must conclude, as we did in the preceding example, that the population is not in equilibrium, we perform a statistical test.

The Chi-Square Test of Independence

Once you know the actual allele frequencies *observed* in your population and the genotype frequencies you *expected* to see in an equlibrium population, you have the information to answer the question, "Is the population in fact in a state of Hardy-Weinberg equilibrium?"

When we know the values of what we expected to observe and what we actually observed, a **chi-square (χ^2) test of independence** is commonly used to determine whether the observed values in fact match the expected value (the **null model** or **null hypothesis**) or whether the observed values deviate significantly from what we expect to find (in which case we reject the null model).

Chi-square statistical tests are performed to test hypotheses in all the life and social sciences. The test basically asks whether the differences between observed and expected values could be due to chance. The mathematical basis of the test is the equation

$$\chi^2 = \sum \frac{(O-E)^2}{E} \qquad \text{Equation 5}$$

where O is the observed value, E is the expected value, and Σ means you sum the values for different observations. Hardy-Weinberg genotype frequencies offer a good opportunity to use the chi-square test.

In conducting a χ^2 test of independence, it's useful to set up your data in a table format, where the observed values go in the top row of the table, and the expected values go in row 2. The expected values for each genotype are those predicted by Hardy-Weinberg, computed as $p^2 \times N$, $2pq \times N$, and $q^2 \times N$ for the A_1A_1, A_1A_2, and A_2A_2 genotypes, respectively. If $N = 1000$ individuals and $p = 0.5$ and $q = 0.5$, our expected numbers would be 250 A_1A_1, 500 A_1A_2, and 250 A_2A_2 (Figure 1).

To compute the χ^2 test statistic, we start by computing the difference between the observed and expected numbers for a genotype, square this difference, and then divide by the expected number for that genotype. We do this for the remaining genotypes, and then add the terms together:

$$\chi^2 = \frac{(O_{A1A1} - E_{A1A1})^2}{E_{A1A1}} + \frac{(O_{A1A2} - E_{A1A2})^2}{E_{A1A2}} + \frac{(O_{A2A2} - E_{A2A2})^2}{E_{A2A2}}$$

	J	K	L	M
7			***Parental Population***	
8			**A2A1**	
9		**A1A1**	**A1A2**	**A2A2**
10	**Observed**	258	504	238
11	**Expected**	$p^2 * N = 250$	$2pq * N = 500$	$q^2 * N = 250$

Figure 1 The top row gives the observed genotypes in a population of 1,000 individuals in which both p and $q = 0.5$. The bottom row gives the expected genotype distribution for those values of p and q if the population were in Hardy-Weinberg equilibrium.

The χ^2 test statistic for Figure 1 would be computed as

$$\chi^2 = \frac{(258-250)^2}{250} + \frac{(504-500)^2}{500} + \frac{(238-250)^2}{250} = 0.864$$

D.F. and Critical Value

You now need to see where your computed χ^2 test statistic falls on the **theoretical χ^2 distribution**. If you are familiar with the normal distribution, you know that the mean and standard deviation control the shape and placement of the distribution on the x-axis (see Exercise 3, "Statistical Distributions"). A χ^2 distribution, in contrast, is characterized by a parameter called **degrees of freedom (d.f.)**, which controls the shape of the theoretical χ^2 distribution. The degrees of freedom value is computed as

$$\text{d.f.} = (\text{number of rows minus 1}) \times (\text{number of columns minus 1})$$

or

$$\text{d.f.} = (r-1) \times (c-1) \qquad \text{Equation 6}$$

In Figure 1, we had two rows (observed and expected) and three columns (three kinds of genotypes), so our degrees of freedom = $(2-1) \times (3-1) = 2$.

The mean of a χ^2 distribution is its degrees of freedom, and the mode of a χ^2 distribution is the degrees of freedom minus 2. The distribution has a positive skew, but this skew diminishes as the degrees of freedom increases. Figure 2 shows two χ^2 distributions for different degrees of freedom. The χ^2 distributions in Figure 2 were generated from an infinite number of χ^2 tests performed on data sets where *no effects* were present. In other words, the theoretical χ^2 distribution is a null distribution. Even when no effects are present, however, you can see that, by chance, some χ^2 test statistics are large and appear with a low frequency. Thus, you can get a very large test statistic by chance even when there is no effect.

By convention, we are interested in knowing if our computed χ^2 statistic is larger than 95% of the statistics from the theoretical curve. The 95% value of the theoretical curve's χ^2 statistic is called the **critical χ^2 value**, and at this value, exactly 5% of the test statistics in the χ^2 distribution are greater than this critical value ($\alpha = 0.05$; see Exercise 5, "Hypothesis Testing"). For example, the critical value for a χ^2 distribution with 4 degrees of freedom is 9.49, which means that 5% of the test statistics in the χ^2 distribution are equal to or greater than this value. The critical value for a χ^2 distribution with 10 degrees of freedom is 18.31.

Table 1 gives the critical values for χ^2 distributions with various degrees of freedom when $\alpha = 0.05$ (the "95% confidence level"). Tables of χ^2 critical values for different α values can be found in almost any statistics text. If our computed statistic is *less* than the

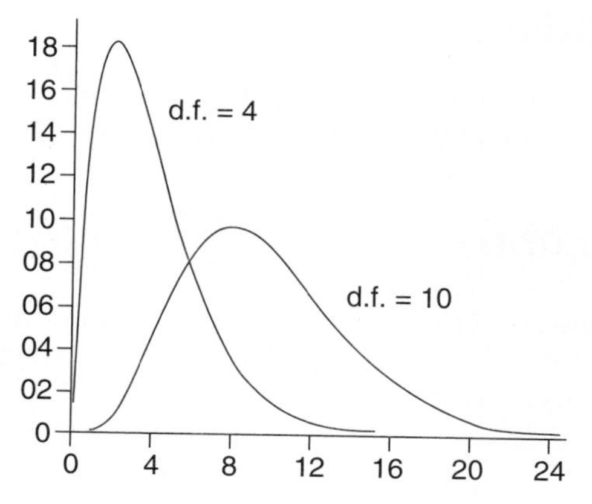

Figure 2 Two χ^2 distributions. Note that the curve steepens (positive skew increases) when the degrees of freedom (d.f.) parameter is smaller.

critical value, we conclude that any difference between our observed and expected values are not **significant**—the difference could be due to chance—and we accept the null hypothesis (i.e., that the population is in Hardy-Weinberg equilibrium). But if our computed statistic is *greater* than the critical value, we conclude that the difference *is* significant, and we reject the null model (i.e., we conclude the population is not in equilibrium).

TABLE 1. Critical values of χ^2 at the 0.05 level of significance (α)

Degrees of freedom	$\alpha = 0.05$	Degrees of freedom	$\alpha = 0.05$
1	3.84	11	19.68
2	5.99	12	21.03
3	7.82	13	22.36
4	9.49	14	23.69
5	11.07	15	25.00
6	12.59	16	26.30
7	14.07	17	27.59
8	15.51	18	28.87
9	16.92	19	30.14
10	18.31	20	31.41

Source: χ^2 values from R. A. Fisher and F. Yates, 1938, *Statistical Tables for Biological, Agricultural, and Medical Research*. Longman Group Ltd., London.

How do you interpret a significant χ^2 test? Interpretation requires that you examine the observed and expected values and determine which genotypes affected the value of the computed χ^2 statistic the most. In general, the larger the deviation between the observed and expected values, the greater the genotype contributed to the χ^2 statistic. In our first example, in which we expected 38% of an equilibrium population would be heterozygotes but in fact observed *no* heterozygotes, the deviation from Hardy-Weinberg expectations is caused primarily by the absence of heterozygotes. You could then proceed to form hypotheses as to *why* there are no heterozygotes.

What forces might keep a population out of Hardy-Weinberg equilibrium? **Evolutionary forces** include natural selection, genetic drift, gene flow, nonrandom mating (inbreeding), and mutation. These forces are introduced in other exercises, but here we will set up the "null model" of a population in Hardy-Weinberg equilibrium.

PROCEDURES

In this exercise, you will develop a spreadsheet model of a single gene with two alleles in population and will explore various properties of Hardy-Weinberg equilibrium.

INSTRUCTIONS	ANNOTATION
A. Set up the model parent population.	Here we are concerned with a single locus, and imagine that this locus has two alleles, A_1 and A_2. Thus, an individual can be homozygous A_1A_1, heterozygous A_1A_2, or homozygous A_2A_2 at the locus.

1. Open a new spreadsheet and set up titles and column headings as shown in Figure 3.

	A	B	C	D	E	F	G	H
1	*Hardy-Weinberg Equilibrium*							
2								
3	Allele	p = A1 =			Calculated	p =		
4	frequencies	q = A2 =			frequencies	q =		
5								
6		Parental		Random	Mom's	Random	Dad's	Offspring
7	Individual	genotype	Gamete	mom	egg	dad	sperm	genotype

Figure 3

2. Set up a linear series from 0 to 999 to represent 1000 individuals in cells A8–A1007.

In cell A8, enter the value 0.
In cell A9, enter **=A8+1**. Copy the formula in cell A9 down to cell 1007 to designate the 1,000 individuals in the population.

3. In cell C3, enter a value for p.

Enter 0.5 in cell C3 to indicate that the frequency of the A_1 allele, or p, is 0.5.

4. In cell C4, enter a formula to compute the value for q.

Enter the formula **=1-$C$3** in cell C4 to designate the frequency of the A_2 allele, or q. Remember that $p + q = 1$.

5. In cells B8–B1007, enter an **IF** formula to assign genotypes to each individual in the population based on the allele frequencies designated in cells C3 and C4.

Enter the formula **=IF(RAND()< $C$3,"A1","A2")& IF(RAND()< $C$3,"A1","A2")** in cell B8. Copy this formula down to cell B1007.
The **IF** formula returns one value if a condition you specify is true, and another value if the condition you specify is false. The **RAND()** part of the formula in cell B8 tells the spreadsheet to choose a random number between 0 and 1. Then, *if* that random number is less than the value designated in cell C3, assign it an allele of A_1; otherwise, assign it a value of A_2. Because there are two alleles for a given locus, you need to repeat the formula again, and then join the alleles obtained from the two **IF** formulas by using the **&** symbol. Once you've obtained genotypes for individual 1, copy this formula down to cell B1007 to obtain genotypes for all 1,000 individuals in the population.

6. Set up new spreadsheet headings as shown in Figure 4.

	J	K	L	M
7	***Parental Population***			
8			**A1A2**	
9		**A1A1**	**A2A1**	**A2A2**
10	**Observed**			
11	**Expected**			
12				
13	Hand-calculated chi-square			
14	Degrees of freedom			
15	Chi test statistic			
16	Spreadsheet-calculated chi-square			
17	Significantly different from H-W prediction?			

Figure 4

7. In cells K10, L10, and M10, use the **COUNTIF** formula to count the number of A_1A_1, A_1A_2, and A_2A_2 genotypes.

The **COUNTIF** formula counts the number of cells within a range that meet the given criteria. It has the syntax **COUNTIF(range,criteria)**, where range is the **range** of cells from which you want to count cells, and **criteria** is what you want to count. We used the formulae:

- Cell K10 **=COUNTIF($B$8:$B$1007,"A1A1")**
- Cell L10 **=COUNTIF($B$8:$B$1007,"A1A2")+COUNTIF($B$8:$B$1007,"A2A1")**
- Cell M10 **=COUNTIF($B$8:$B$1007,"A2A2")**

The formula in cell K10 counts the number of A_1A_1 individuals in cells B8 through B1007. In cell L10, you'll want to count both the A_1A_2 and the A_2A_1 heterozyotes. Your total observations should add to 1000. You can double-check this by entering **=SUM(K10:M10)** in cell N10.

The values from these formulae are your "observed" genotypes, and you'll compare these to the genotypes predicted by Hardy-Weinberg. (Your observed genotypes should be in Hardy-Weinberg equilibrium because of the way you assigned the genotypes. In a natural setting, however, you probably won't know the initial frequencies, but you can count genotypes, and then determine if the organisms are in Hardy-Weinberg equilibrium or not.)

8. In cell G3, enter a formula to calculate the actual frequency of the A_1 allele. In cell G4, enter a formula to calculate the actual frequency of the A_2 allele.

Enter the formula **=(K10*2+L10)/(2*A1007)** in cell G3.
Enter the formula **=1-G3** in cell G4.
Since each individual carries two copies of each gene, your population of 1,000 individuals has 2,000 "gene copies" (alleles) present. To calculate the allele frequency, you simply calculate what proportion of those 2000 alleles are A_1, and what proportion are A_2. The frequency of the A_1 allele is 2 times the number of A_1A_1 genotypes, plus the A_1's from the heterozygotes. The frequency of the A_2 allele is 2 times the number of A_2A_2 genotypes, plus the A_2's from the heterozygotes. Since $p + q = 1$, q can be computed also as $1 - p$. Your estimates of allele frequencies should add to 1.

9. Save your work.

B. Calculate expected genotype frequencies in the parent population.

Now that you have computed the observed allele frequencies, you can calculate the estimated genotype frequencies predicted by Hardy-Weinberg. Remember that if the population is in Hardy-Weinberg equilibrium, the genotype frequencies should be $p^2 + 2pq + q^2$. This means that the number of A_1A_1 genotypes should be $p \times p$ (p^2), the number of A_1A_2 genotypes should be $2 \times p \times q$, and the number of A_2A_2 genotypes should be $q \times q$ (q^2).

1. In cell K11, enter a formula to calculate the expected number of A_1A_1 genotypes, given the p value calculated in cell G3.

Enter the formula **=$G$3^2*1000** in cell K11.
The caret symbol (^) followed by the number 2 indicates that the value should be squared. Thus, we obtained expected number of A_1A_1 genotypes by multiplying $p \times p$, which gives us a proportion, and then multiplied this proportion by 1,000 to give us the number of individuals out of 1,000 that are expected to be A_1A_1 if the population is in Hardy-Weinberg equilibrium.

2. Calculate the expected number of heterozygotes in cell L11.

Enter the formula **=2*$G$3*$G$4*1000** in cell L11.

3. Calculate the expected number of A_2A_2 genotypes in cell M11.

Enter the formula **=$G$4^2*1000** in cell M11.
The expected numbers should add to 1000. You can double-check this by entering **=SUM(K11:M11)** in cell N11.

4. Graph your observed and expected results.

Use a column graph and label your axes fully. Your graph may look a bit different than Figure 5, and that's fine.

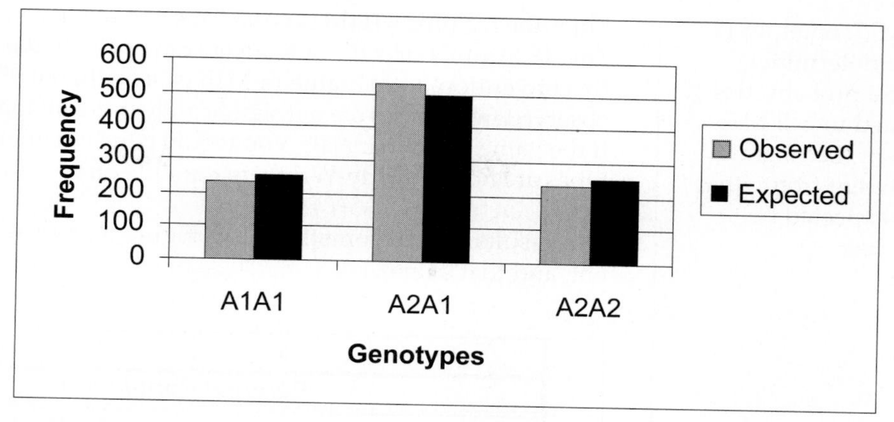

Figure 5

5. Interpret your graph. Does your population appear to be in Hardy-Weinberg equilibrium?

6. Press F9, the calculate key, to generate new random numbers and hence new genotypes. Does your population still appear to be in equilibrium?

C. Calculate chi-square test statistics and probability.

Now you are ready to perform a χ^2 test to verify whether your population's observed genotype frequencies are statistically similar to those predicted by Hardy-Weinberg.

1. In cell M13, enter the formula to calculate your χ^2 test statistic. Refer to Equation 5.

Enter the formula **=(K10-K11)^2/K11+(L10-L11)^2/L11+(M10-M11)^2/M11** in cell M13. This corresponds to Equation 5:

$$\chi^2 = \sum \frac{(O-E)^2}{E}$$

Starting with A_1A_1, we observed 245 individuals and determined that there should be 255 individuals (you may have obtained slightly different numbers than that). Following the chi-square formula, 245 – 255 = 10, 10^2 = 100, 100 divided by 255 = 0.392. Repeat this step for the A_1A_2 and A_2A_2 genotypes. As a final step, add your three calculated values together. This sum is your chi-square (χ^2) test statistic.

2. In cell M14, enter a value for degrees of freedom.

Enter the value 2 in cell M14.
Recall from Equation 6 that the degrees of freedom value is the (number of rows minus 1) × (number of columns minus 1), or $(r - 1) \times (c - 1)$. In our example, we had two rows (observed and expected) and three columns (three kinds of genotypes), so our degrees of freedom = $(2 - 1) \times (3 - 1) = 2$.

3. In cell M15, use the **CHIDIST** function to determine the probability of obtaining your χ^2 statistic.

Enter the formula **=CHIDIST(M13,M14)** in cell M15.
The **CHIDIST** function has the syntax **CHIDIST(x,degrees_freedom)**, where **x** is the test statistic you want to evaluate and **degrees_freedom** is the degrees of freedom for the test. The formula in cell M15 returns the probability of obtaining the test statistic you calculated, given the degrees of freedom—if this probability is less than 0.05, your test statistic exceeds the critical value. If this probability is greater than 0.05, your test statistic is less than the critical value. You can now make an informed decision as to whether your population is in Hardy-Weinberg equilibrium or not.

4. In cell M16, double-check your work by using the **CHITEST** function to calculate your test statistic, degrees of freedom, and probability.

Enter the formula **=CHITEST(K10:M10,K11:M11)** in cell M16.
The **CHITEST** formula returns the test for independence (the probability) when you indicate the observed and expected values from a table. It has the syntax **CHITEST(actual_range,expected_range)**, where actual range is the range of observed data (in your case, cells K10–M10), and expected range is the range of expected data (in your case, cells K11–M11). This number should be very close to what you obtained in cell M15. (If it's not, you did something wrong.)

5. In cell M17, enter an **IF** formula to determine whether the probabilities you obtained in cell M15 is significant (i.e., significantly different from what would be expected by chance alone).

Enter the formula **=IF(M15<0.05,"Yes","No")** in cell M17.
This **IF** formula tells the spreadsheet to evaluate the probability obtained in cell M15. By convention, if the value in M15 is *more* than 0.05, you would conclude that your observed frequencies are not significantly different than those expected by chance alone. If the value is *less* than 0.05, you would conclude that the population's observed genotypes are not in Hardy-Weinberg equilibrium.

Our results looked something like Figure 6 (your results are probably slightly different, and that's fine).

	J	K	L	M
7	**Parental Population**			
8			**A1A2**	
9		**A1A1**	**A2A1**	**A2A2**
10	**Observed**	253	498	249
11	**Expected**	252.004	499.992	248.004
12				
13	Hand-calculated chi-square			0.015872764
14	Degrees of freedom			2
15	Chi test statistic			0.992095028
16	Spreadsheet-calculated chi-square			0.992095028
17	Significantly different from H-W prediction?			No

Figure 6

6. Answer questions 1 and 2 at the end of exercise before proceeding.

D. Simulate random mating to produce the genotypes of the next (F_1) generation.

Now that you have an idea of whether your population of 1,000 is in Hardy-Weinberg equilibrium, we will let your population mate and produce offspring that make up the next generation.

1. In cells C8–C1007, enter a formula to simulate the random assortment of alleles into gametes.

Enter the formula **=IF(RAND()<0.5,RIGHT(B8,2), LEFT(B8,2))** in cell C8. Copy this formula down to cell C1007.
Homozygotes can produce only one kind of gamete, while heterozygotes can produce both A_1 and A_2 gametes. We'll assume that each individual produces a single gamete, and that which of the two possible gametes are actually incorporated into the zygote is randomly determined. The formula in cell C8 tells the spreadsheet to draw a random number between 0 and 1 (the **RAND()** portion of the formula). If the random number is less than 0.5, the program returns the **RIGHT** two characters in cell B8; otherwise, it will return the **LEFT** two characters in cell B8. (You might want to explore the **RIGHT** and **LEFT** functions in more detail.) This formula simulates the random assortment of alleles into gametes that will ultimately fuse with another gamete to form a zygote.

2. In cells D8 and F8, enter a formula to randomly select a male and a female from the population that will mate and produce a zygote.

Enter the formula **=ROUND(RAND ()*1000,0)** in cells D8 and F8. Copy the formula down to cells D1007 and F1007, respectively.
This formula simulates random mating by choosing a random female and random male from our population to mate. The formula tells the spreadsheet to draw a random number between 0 and 1, multiply this number by 1,000, then round it to 0 decimal places. This action will "choose" which individuals will mate. Note that not all individuals in

the population will actually mate, but that each individual has the same probability of mating as every other individual in the population.

3. In columns E and G, enter **VLOOKUP** formulae to determine the gamete contributed by each parent randomly selected in step 2.

In cell E8 enter the formula **=VLOOKUP(D8,$A$8:$C$1007,3)**.Copy this formula down to E1007.

In cell G8, enter the formula **=VLOOKUP(F8,$A$8:$C$1007,3)**. Copy this formula down to G1007.

The formula in cell E8 tells the spreadsheet to look up the value in D8, which is the random mom, from the table A8 through A1007, and return the associated value listed in the third column of the table. In other words, find mom from column A and relay the gamete associated with that mom in column C. The formula in G8 does the same for the random dad.

The **VLOOKUP** function searches for a value in the leftmost column of a table, and then returns a value in the same row from a column you specify in the table. It has the syntax **VLOOKUP(lookup_value,table_array,col_index_num,range_lookup)**, where **lookup_value** is the value to be found in the first column of the table, **table_array** is the table of information in which the data are looked up, and **col_index_num** is the column in the table that contains the value you want to return. **Range_lookup** is either true or false. If **Range_lookup** is not specified, by default it is set to "false," which indicates that an exact match will be found.

4. In cell H8, enter a formula to obtain the genotypes of the zygotes by pairing the egg and sperm alleles contributed by each parent.

Enter the formula **=E8&G8** in cell H8. Copy this formula down to cell H1007.

E. Calculate Hardy-Weinberg statistics for the F_1 generation.

Now you can determine if the offspring generation has genotypes predicted by Hardy-Weinberg. Remember, the Hardy-Weinberg principle holds that whatever the initial genotype frequencies for two alleles may be, after one generation of random mating, the genotype frequencies will be p^2:$2pq$:q^2. Additionally, both the genotype frequencies and the allele frequencies will remain constant in succeeding generations. The observed genotypes are calculated by tallying the different genotypes in cells H8–H1007. The expected genotypes are calculated based on the parental allele frequencies given in cells G3 and G4.

1. Set up new column headings as shown in Figure 7.

	J	K	L	M
20	**Offspring Population**			
21			**A1A2**	
22		**A1A1**	**A2A1**	**A2A2**
23	Observed			
24	Expected			
25				
26	Hand-calculated chi-square test statistic:			
27	Degrees of freedom			
28	Chi test statistic			
29	Spreadsheet-calculated chi-square			
30	Significantly different from H-W prediction?			

Figure 7

2. Enter formulae in cells K23–M24 to calculate observed and expected genotypes of the new generation.

If you've forgotten how to calculate a formula, refer to the formulas you entered for the parents as an aid. Double-check your results:

- K23 **=COUNTIF($H$8:$H$1007,"A1A1")**
- L23 **=COUNTIF($H$8:$H$1007,"A1A2")+COUNTIF($H$8:$H$1007,"A2A1")**
- M23 **=COUNTIF($H$8:$H$1007,"A2A2")**
- K24 **=$G$3^2*1000**
- L24 **=2*$G$3*$G$4*1000**
- M24 **=$G$4^2*1000**

3. Enter formulae in cells M26–M30 to determine if the new generation is in Hardy-Weinberg equilibrium.

You can also simply copy and paste the formulae from the parental population; the program should automatically update your formulae to the new cells (but double-check, just to be sure).

4. Graph your observed and expected results.

QUESTIONS

1. The Hardy-Weinberg model is often used as the "null model" for evolution. That is, when populations are out of Hardy-Weinberg equilibrium, it suggests that some kind of evolutionary process may be acting on the population. What are the assumptions of Hardy-Weinberg?
2. Press F9, the Calculate key, to generate a new set of random numbers, which in turn will generate new genotypes, new allele frequencies and new Hardy-Weinberg test statistics. Press F9 a number of times and track whether the population remains in Hardy-Weinberg equilibrium. Why, on occasion, will the population be out of HW equilibrium?
3. A basic tenet of the Hardy-Weinberg principle is that genotype frequencies of a population can be predicted if you know the allele frequencies. This allows you to answer such questions as *Under what allelic conditions should heterozygotes dominate the population?* In cell C3, modify the frequency of the A_1 allele (the A_2 allele will automatically be calculated). Begin with a frequency of 0, then increase its frequency by 0.1 until the frequency is 1. For *each* incremental value entered, record the expected genotype frequencies of A_1A_1, A_1A_2, and A_2A_2 given in cells K11–M11. (You can simply copy and paste these values into a new section of your spreadsheet, but make sure you use the Paste Values option to paste the expected genotypes.). You spreadsheet might look something like this (but the frequencies will extend a few more rows until the frequency of A_1 is 1:

	O	P	Q	R
13			Expected genotypes	
14	Frequency of A1	A1A1	A1A2	A2A2
15	0	0	0	1000
16	0.1	9	180	817
17	0.2	36	320	658
18	0.3	86	420	498
19	0.4	173	480	341
20	0.5	262	500	239

Make a graph of the relationship between frequency of the A_1 allele (on the x-axis) and the expected numbers of genotypes. Use a line graph, and fully label your axes and give the graph a title. Consider the shapes of each curve, and write a one- or two-sentence description of the major points of the graph.

4. The Hardy-Weinberg principle states that after one generation of random mating, the genotype frequencies should be $p^2{:}2pq{:}q^2$. That is, even if a parental population is out of Hardy-Weinberg equilibrium, it should return to the equilibrium status after just one generation of random mating. Prove this to yourself by modifying the genotypes of the 1,000 individuals listed in column B. Let individuals 0–499 have genotypes A_1A_1; individuals 500–999 have genotypes of A_2A_2. (You'll have to overwrite the formulas in those cells.) Estimate the gene frequencies and determine if this parental population is in Hardy-Weinberg equilibrium. Graph your results, and indicate the chi-square test statistic somewhere on your graph. After one generation of random mating, what are the allele frequencies and genotype frequencies? Is this "new" population in Hardy-Weinberg equilibrium?

LITERATURE CITED

Hardy, G. 1908. Mendelian proportions in a mixed population. *Science* 28: 49–50.

Hartl, D. L. 2000. *A Primer of Population Genetics*, 3rd Edition. Sinauer Associates, Sunderland, MA.

Mendel, G. 1866. Experiments in plant hybridization. Translated and reprinted in J. A. Peters (ed.), 1959. *Classic Papers in Genetics.* Prentice-Hall, Englewod Cliffs, NJ.

30 MULTILOCUS HARDY-WEINBERG AND LINKAGE DISEQUILIBRIUM

Objectives

- Develop a spreadsheet model of allele and genotype frequencies at two loci.
- Examine properties of independent assortment of alleles.
- Use the chi-square test to determine if an offspring population is in Hardy-Weinberg equilibrium.
- Calculate D, the linkage disequilibrium coefficient.
- Graphically determine whether the population is in linkage equilibrium.

Suggested Preliminary Exercise: Hardy-Weinberg Equilibrium

INTRODUCTION

Now that you have been introduced to the Hardy-Weinberg equilibrium principle, it's time to explore the model in greater detail. Recall that this "null model" of evolution specifies algebraically what will happen across generations to the frequencies of alleles and genotypes. The bottom line is that in the absence of natural selection, genetic drift, mutation, and gene flow (and given a population of infinite size where mating is random), allele and genotype frequencies will not change over generations. That is, populations will not evolve. If the allele frequencies for a given locus in a population are given by p and q, the genotype frequencies will be p^2, $2pq$, and q^2 if the population is in Hardy-Weinberg equilibrium.

In a previous exercise, you developed a single-locus model of the Hardy-Weinberg principle for locus A where p_1 was the frequency of the A_1 allele and q_1 was the frequency of the A_2 allele. In reality, organisms may have hundreds of loci on each of their chromosomes, and thus we need to start thinking about evolution at *multiple* loci.

In this exercise, you will learn that when multiple loci are involved, there are two kinds of equilibrium states: one is *Hardy-Weinberg* equilibrium, in which allele frequencies remain constant from generation to generation, and the second is *linkage* equilibrium. You will extend your single-locus model to examine two loci, loci A and B, simultaneously and to discover whether they are in fact in linkage equilibrium.

Hardy-Weinberg Equilibrium for Two Loci

Let's assume that the two alleles at locus B have the frequencies p_2 for the B_1 allele and q_2 for the B_2 allele. Furthermore, let's assume that the locus B is located on a different chromosome than locus A. Since the A and B loci each have only two alleles present in the population, the frequencies for *each* locus (p and q) must add to 1:

$$p_1 + q_1 = 1 \qquad \text{Equation 1}$$

and

$$p_2 + q_2 = 1 \qquad \text{Equation 2}$$

When two loci are considered, the genotype of an organism is characterized by its genotype at both loci, and 9 different genotypes are possible:

$A_1A_1B_1B_1$	$A_1A_1B_1B_2$	$A_1A_1B_2B_2$
$A_1A_2B_1B_1$	$A_1A_2B_1B_2$	$A_1A_2B_2B_2$
$A_2A_2B_1B_1$	$A_2A_2B_1B_2$	$A_2A_2B_2B_2$

Now suppose our hypothetical population mates randomly to produce a new generation of offspring. Individuals produce gametes (sex cells) through the process of meiosis. The end result is an egg or sperm cell that contains a single allele for the A locus and a single allele for the B locus. When an egg and sperm unite via sexual reproduction, the offspring zygote will regain its full complement of alleles. Depending on their genotype, individuals can produce between 1 and 4 different kinds of gametes (called **gamete classes**). The $A_1A_1B_1B_1$ individual can produce only 1 kind of gamete: A_1B_1. The $A_1A_2B_1B_2$ individual can produce 4 kinds of gametes: A_1B_1, A_1B_2, A_2B_1, and A_2B_2. In the space provided in Figure 1, write in the kinds of gametes that each genotype can produce.

Genotype	Gametes
$A_1A_1B_1B_1$	
$A_1A_1B_1B_2$	
$A_1A_1B_2B_2$	
$A_1A_2B_1B_1$	
$A_1A_2B_1B_2$	
$A_1A_2B_2B_2$	
$A_2A_2B_1B_1$	
$A_2A_2B_1B_2$	
$A_2A_2B_2B_2$	

Figure 1

The frequencies of each gamete class (A_1B_1, A_1B_2, A_2B_1, and A_2B_2) in a population depend on the genotype frequencies in the adult population. Thus, the gamete frequencies in the total population must be related in some way to the allele frequencies in the population. Indeed, the frequency of a gamete class is the product of the frequencies of the alleles that make up the gamete (Hartl 2000):

$$\text{Frequency of the } A_1B_1 \text{ gamete} = p_1 \times p_2 \qquad \text{Equation 3}$$

$$\text{Frequency of the } A_1B_2 \text{ gamete} = p_1 \times q_2 \qquad \text{Equation 4}$$

$$\text{Frequency of the } A_2B_1 \text{ gamete} = q_1 \times p_2 \qquad \text{Equation 5}$$

$$\text{Frequency of the } A_2B_2 \text{ gamete} = q_1 \times q_2 \qquad \text{Equation 6}$$

If we assume that p and q are known for each locus, Equations 3–6 allow us to predict the genetic makeup of the offspring population. Let's walk through an example. If our parental population has initial frequencies of $p_1 = 0.5$ and $q_1 = 0.5$ for the first locus, and $p_2 = 0.25$ and $q_2 = 0.75$ for the second locus, the frequencies of the gamete classes are:

$$\text{Frequency of the } A_1B_1 \text{ gamete} = p_1 \times p_2 = 0.5 \times 0.25 = 0.125$$

$$\text{Frequency of the } A_1B_2 \text{ gamete} = p_1 \times q_2 = 0.5 \times 0.75 = 0.375$$

$$\text{Frequency of the } A_2B_1 \text{ gamete} = q_1 \times p_2 = 0.5 \times 0.25 = 0.125$$

$$\text{Frequency of the } A_2B_2 \text{ gamete} = q_1 \times q_2 = 0.5 \times 0.75 = 0.375$$

Note that the sum of the gamete frequencies is 1, as it should be. Now that we know what the gamete frequencies are, we can predict the genotype frequencies of the offspring population by multiplying the probability that two gamete types will join to form a zygote. For example, an $A_1A_1B_1B_1$ genotype in the offspring population is the result of combining an A_1B_1 egg with a A_1B_1 sperm. The frequency of this genotype should be $0.125 \times 0.125 = 0.015625$ in the offspring population.

Because the gamete frequencies are related to the allele frequencies in the parental population, a second way of predicting the genotype frequencies of the offspring populations is to multiply their *independent* allele probabilities together. For example, if we want to estimate the proportion of $A_1A_1B_1B_1$ individuals in the next generation, we would multiply the probability that the offspring would inherit two A_1 alleles,

$$\text{Probability} = p_1 \times p_1 = p_1^2$$

by the probability of inheriting two B_1 alleles, or

$$\text{Probability} = p_2 \times p_2 = p_2^2$$

In our example, the proportion of $A_1A_1B_1B_1$ individuals is expected to be $(0.5 \times 0.5) \times (0.25 \times 0.25) = 0.015625$, or about 1.5% of the population. This is the same answer obtained by the gamete probability method. As a second example, if we want to estimate the proportion of $A_1A_2B_2B_2$ individuals in the population or in the next generation, we would multiply the probability of being heterozygous at the A locus ($2 \times p_1 \times q_1$, or $2 \times 0.5 \times 0.5$) by the probability of being homozygous B_2B_2 at the B locus ($q_2 \times q_2$, or 0.75×0.75). This generates a probability of $(2 \times 0.5 \times 0.5) \times (0.75 \times 0.75)$, which is 0.28125, or about 28% of the population. It's really that simple ... or is it?

Linkage Disequilibrium

A key assumption in calculating Hardy-Weinberg frequencies for two or more loci is that the loci are independent of each other. Essentially, this means that if you know what genotype the organism has at the first locus, you can't necessarily predict what its genotype will be at the second locus beyond what Hardy-Weinberg predicts. Knowing that an individual is A_1A_1 at the first locus doesn't tell us what the genotype at the second locus will be. Given that $p_2 = 0.25$ and $q_2 = 0.75$, Hardy-Weinberg tells us it has a 0.0625 chance of being B_1B_1 at the second locus, a 0.375 chance of being B_1B_2 at the second locus, and 0.5625 chance of being B_2B_2 at the second locus. Note that these frequency probabilities for this locus would be the same regardless of the genotype at the first locus. When alleles at different loci associate independently (at random), they are said to be in **linkage equilibrium**.

Sometimes, however, the two loci are *not* independent. For example, the A_1 allele may always associate with the B_1 allele and the A_2 allele with the B_2 allele. When this happens, the population is said to be in **linkage disequilibrium**. Linkage disequilibrium means, for example, that the different B genotypes are not distributed randomly among the different A genotypes and that, generally speaking, if you know the genotype at the A locus, you have a good idea of what the genotype at the B locus will be. Figure 2, for instance, shows that the B_1B_1 genotype occurs more commonly with the A_1A_1 genotype, and the B_2B_2 genotype occurs more frequently with the A_2A_2 genotypes.

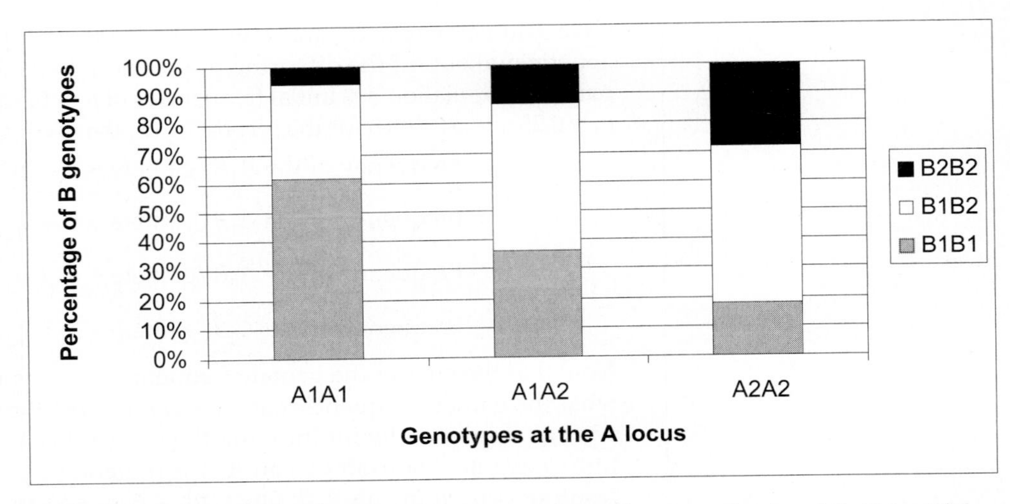

Figure 2 The bar graph shows linkage disequilibrium between the *A* and *B* alleles. If the population were in linkage equilibrium, the three different *B* genotypes would be distributed more or less equally among the three genotypes of the *A* allele.

Linkage disequilibrium can occur when the two loci are physically linked, meaning that they must be located close to each other on the same chromosome. During meiosis, the two alleles on the same chromosome may tend to segregate into the same gamete because of this physical linkage (Be aware, however, that not all alleles on the same chromosome are physically linked.)

Alleles can also be associated with each other because they are *coadapted*. **Coadaptation** is a beneficial interaction between alleles at different loci. For instance, if the A_1 and B_1 allele "work well" together to benefit an organism in its environment, they are said to be coadapted. Ayala (1982) gives this analogy to illustrate coadaption of alleles at different loci:

> A successful performance by a symphony orchestra requires not only that each player know how to play his instrument (a gene must be able to function), but also that he master his part in the piece being performed (a gene type must interact well with the other genes). A violinist playing his part for Beethoven's Sixth Symphony while the rest of the orchestra was playing Ravel's *Bolero* would be cacophonic.

Linkage disequilibrium can be quantified as the difference between the probability that A_1B_1 gametes unite with A_2B_2 gametes (these are called the **coupling gametes**) and the probability that A_1B_2 and A_2B_1 gametes unite (the **repulsion gametes**). The linkage disequilibrium coefficient is

$$D = G_{A1B1}G_{A2B2} - G_{A1B2}G_{A2B1} \qquad \text{Equation 7}$$

where G_{A1B1} is the frequency of the A_1B_1 gamete, G_{A2B2} is the frequency of the A_2B_2 gamete, etc. The value of *D* ranges from 0 to 0.25. When the two alleles associate randomly, *D* will be 0. If the alleles are not randomly associated, *D* will increase. Assuming that the *A* and *B* loci are situated on different chromosomes, and assuming that the population mates at random without natural selection, gene flow, or mutation, the "level" of linkage disequilibrium breaks down with every passing generation. Unlike the single-locus Hardy-Weinberg model, which demonstrated that populations that are out of equilibrium go back into equilibrium after a single generation, several generations may be required for a population that is in linkage disequilibrium to acquire low levels of *D*.

PROCEDURES

The spreadsheet model you are about to develop is intended to give you some insights into how allele and genotype frequencies change over time when multiple loci are considered, and to help you determine whether or not a population is in linkage equilibrium. In this exercise, you will set up a population of 1000 individuals with a specified genotype frequency, let them mate at random, and then examine the genotype and allele frequencies of the offspring population. You will also calculate D, the linkage disequilibrium coefficient, and graphically determine whether populations are in linkage equilibrium. The approach in assigning genotypes to individuals in the population will be different than in the single-locus Hardy-Weinberg exercise, so that you can easily see how linkage disequilibrium works.

As always, save your work frequently to disk.

INSTRUCTIONS / ANNOTATION

A. Set up the model population.

1. Open a new spreadsheet and set up headings as shown in Figure 3.

	A	B	C	D
1	***Multilocus Hardy-Weinberg***			
2				
3	**Genotype**	**Frequency**	**Tally count**	
4			0	
5	A1A1B1B1	0		
6	A1A1B1B2	0		
7	A1A1B2B2	0		
8	A1A2B1B1	0		
9	A1A2B1B2	1		
10	A1A2B2B2	0		
11	A2A2B1B1	0		
12	A2A2B1B2	0		
13	A2A2B2B2	0		**<= This number MUST = 1**
14				
15				
16	**Individual**	**Random #**	**Genotype**	**Gamete**

Figure 3

2. In cells B5–B13, enter genotype frequency values shown.

Cells B5–B13 give the genotype frequencies for the population. Enter the number 1 in cell B9, and 0s in the remaining cells. This indicates that our population will consist solely of $A_1A_2B_1B_2$ genotypes. Later in the exercise you will modify the values in these cells. Remember that the sum of the genotype frequencies in the population must equal 1.

3. In cells C4–C13, enter formulae to keep a running tally of the total genotype frequencies.

Enter the number 0 in cell C4.
Enter the formula **=B5** in cell C5.
Enter the formula **=SUM($B$5:B6)** in cell C6 and copy this formula down to cell C13. Cell C5 gives the running tally of genotype frequencies when only the first genotype, $A_1A_1B_1B_1$, has been considered. When you use the **SUM** function in cell C6 and copy the formula down to cell C13, it keeps a running tally of the genotype frequencies in your total population. Note that **$B$5** is an absolute reference, whereas the other cells

are relative references. This "anchors" cell B5 in the **SUM** so that the tally is a running tally. If cell C13 does not equal 1, it means that cells B5–B13 don't add to 1. If so, make the necessary adjustments.

	A	B	C	D
3	Genotype	Frequency	Tally count	
4			0	
5	A1A1B1B1	0	0	
6	A1A1B1B2	0	0	
7	A1A1B2B2	0	0	
8	A1A2B1B1	0	0	
9	A1A2B1B2	1	1	
10	A1A2B2B2	0	1	
11	A2A2B1B1	0	1	
12	A2A2B1B2	0	1	
13	A2A2B2B2	0	1	<= This number MUST = 1

Figure 4

Your spreadsheet should now look like Figure 4. This tally will allow you to assign genotypes to individuals in a later step, and will help you determine quickly if your genotype frequencies add to 1.

4. Save your work prior to assigning genotypes to individuals in the next step.

5. In cells A17–A1016, set up a linear series from 1 to 1000.

Enter 1 in cell A17.
Enter the formula **=A17+1** in cell A18. Copy this formula down to cell A1016.
Your population now consists of 1000 individuals.

6. In cells B17–B1016, generate a random number between 0 and 1.

Enter the formula **=RAND()** in cell B17 and copy this formula down to cell B1016. When you press F9, the calculate key, the spreadsheet will generate new random numbers that will be used to assign a genotype to individuals in the population.

7. In cells C17–C1016, enter a formula to assign a genotype to each individual

Enter the formula **=LOOKUP(B17,$C$4:$C$13,$A$5:$A$13)** in cell C17. Copy this formula down to cell C1016.
Here we use the **LOOKUP** function to assign genotypes based on the random number generated for each individuals, the frequencies you entered in cells B5–B13, and the tally of genotype frequencies in cells C4–C13. The function looks up a value (**B17**) in a vector that you specify (cells **$C$4:$C$13**) and returns a genotype for the individual given in the vector **$A$5:$A$13**. (Remember that a *vector* is a single row or column of values.) The **LOOKUP** function is handy because if it can't find the exact lookup value (the random number given in cell B17), it matches the largest value in lookup vector (cells $C$4:$C$13) that is less than or equal to **lookup_value**. The result is that genotypes are assigned to individuals in approximately the proportions that you specified.

Examine your first 10 genotypes. They should all be $A_1A_2B_1B_2$ if the **LOOKUP** function worked properly. To see how the function works, change cells B5 and B13 to 0.5, and set cell B9 to 0. (Remember that the final tally of genotype frequencies must equal 1 in cell C13.) Now examine the genotypes of your first 10 individuals. The genotypes should be either $A_1A_1B_1B_1$ or $A_2A_2B_2B_2$. When you feel you have a handle on how this function works, return cells B5 and B13 to 0, and return cell B9 to 1.

8. Save your work.

B. Calculate allele frequencies, and determine gamete probabilities.

1. Set up new column headings as shown in Figure 5.

	E	F	G	H
3	Allele frequencies			
4	Locus 1		Locus 2	
5	p_1 = A1 =		p_2 = B1 =	
6	q_1 = A2 =		q_2 = B2 =	
7				
8	Gamete probabilities			
9	A1B1	A1B2	A2B1	A2B2
10				
11				

Figure 5

2. Enter formulae in cells F5–F6 and H5–H6 to calculate the allele frequencies for the two loci.

In cell F5 enter the formula **=(COUNTIF(C17:C1016,"A1A1*")*2+COUNTIF(C17:C1016,"A1A2*"))/(2*A1016)**.
In cell F6 enter the formula **=1-F5**.
In cell H5 enter the formula
=(COUNTIF(C17:C1016,"*B1B1")*2+COUNTIF(C17:C1016,"*B1B2"))/(2*A1016).
In cell H6 enter the formula **=1-H5**.
Recall from your first Hardy-Weinberg exercise that the frequencies of the A_1 and A_2 alleles are

$$\text{Frequency } (A_1) = 2N_{A1A1} + N_{A1A2})/2N$$

$$\text{Frequency } (A_2) = 2N_{A2A2} + N_{A1A2})/2N$$

There are 1000 individuals in the population, so the denominator will be 2000, which means that there are 2000 total "gene copies" present in the population. To obtain the frequency of the A_1 allele, we need to know how many of those gene copies are A_1. Since this locus has only two alleles, the remainder of the gene copies will carry allele A_2, so its frequency can be obtained by subtraction.

The * in the **COUNTIF** formulae is a "wildcard" that represents one or more unspecified characters. The F5 formula, for example, tells the spreadsheet to search for and count the number of A_1A_1 individuals regardless of what the remaining text in the cell is. Similarly, the H5 formula tells the spreadsheet to search for and count the number of B_1B_1 individuals regardless of what their genotype was at the A locus.

3. In cells E10–H10, enter formulae to calculate the expected gamete proportions.

Enter the following formulae:
Cell E10 =**F5*H5**.
Cell F10 **=F5*H6**.
Cell G10 **=F6*H5**.
Cell H10 **=F6*H6**.

These formulae correspond to Equations 3–6. Gametes contain a single allele for the A locus and a single allele for the B locus. There are four possible gamete combinations: A_1B_1, A_1B_2, A_2B_1, and A_2B_2. The expected proportions of each combination are calculated by multiplying the appropriate allele frequencies together. For example, the expected proportion of A_1B_1 gametes in the population is the product of the A_1 allele frequency times the B_1 allele frequency.

4. In cell I10, enter a formula to sum the gamete probabilities.

Enter the formula **=SUM(E10:H10)** in cell I10.
The sum of the gamete probabilities will always be 1.

5. In cell D15, enter a formula using the **MID** function to generate a gamete type for each individual.

In cell D15 enter the formula **=MID(C17,1,2)**.
The **MID** function has the syntax **MID(text,start_num,num_chars)**. The formula in cell D15 tells the spreadsheet to examine the text in cell C17 and, starting with the first character, return 2 characters. If the formula were =MID(C17,3,2), the spreadsheet would examine the text in cell C17 and would return 2 characters starting with the *third* character.
In the next step, the **MID** function will allow us to generate a single gamete (selected randomly from the possible gametes that can be produced by an individual) for each individual in the population. If an individual is selected for mating, this gamete will be incorporated into the offspring's gene pool. The gamete will contain either the first allele (A_1) or the second allele (A_2) for the A locus, and either the first allele (B_1) or second allele (B_2) for the B locus.

6. In cell D17–D1016, enter a combination of the **RAND()** and **MID** functions to generate a random gamete for each individual.

In cell D17 enter the formula **=IF(RAND()<0.5,MID(C17,1,2),MID(C17,3,2))&IF(RAND()<0.5,MID(C17,5,2),MID(C17,7,2))**. Copy this formula down to cell D1016.

The first part of this formula (to the left of the &) generates the *A* allele in the gamete, and the second part (to the right of the &) generates the *B* allele. The first part draws a random number between 0 and 1; if this random number is less than 0.5, the spreadsheet returns the first and second values from cell C17; otherwise, it returns the third and fourth values from C17. The second part of the formula draws a random number, and returns the fifth and sixth values from C17 or returns the seventh and eighth values. Joining the two parts with the & symbol results in a gamete for the individual.

7. In cells E11–H11, use the **COUNTIF** formula to calculate the observed gamete frequencies.

Enter the following formulae:
Cell E11 **=COUNTIF($D$17:$D$1016,E9)/1000**.
Cell F11 **=COUNTIF($D$17:$D$1016,F9)/1000**.
Cell G11 **=COUNTIF($D$17:$D$1016,G9)/1000**.
Cell H11 **=COUNTIF($D$17:$D$1016,H9)/1000**.

8. Save your work.

Note that when you press F9, the calculate key, new random numbers are generated. This action generates new genotypes, and also generates a new gamete for each individual in the population.

C. Simulate sexual reproduction.

1. Set up new column headings as shown in Figure 6.

	E	F	G	H	I	J	K	L	M
15	Random	Random	Random	Random	Offspring genotype				
16	mom	mom's egg	dad	dad's sperm	Locus A		Locus B		Genotype

Figure 6

2. Use the **RAND()** and **VLOOKUP** functions to select random parents and lookup their gametes as you did in the Hardy-Weinberg equilibrium exercise.

In cells E17–E1016 and cells G17–G1016 you can enter either one of the follow formulae:
=ROUNDUP(RAND()*1000,0)
=RANDBETWEEN(1,1000)
In cells F17–F1016 enter the formula **=VLOOKUP(E17,$A$17:$D$1016,4)**.
In cells H17–H1016 enter the formula **=VLOOKUP(G17,$A$17:$D$1016,4)**.
Refer to Exercise 29, "Hardy-Weinberg Equilibrium," if needed. Your spreadsheet should look similar to Figure 7, although your numbers will be different.

	E	F	G	H
15	Random	Random	Random	Random
16	mom	mom's egg	dad	dad's sperm
17	654	A2B1	528	A1B2
18	367	A1B1	568	A2B2
19	175	A2B2	70	A2B1
20	643	A2B2	692	A2B1
21	956	A1B2	488	A1B2

Figure 7

In Figure 7, the first random Mom was individual 654, and the first random Dad was individual 528. Since the population has a genotype frequency of $A_1A_2B_1B_2 = 1$, all individuals in the population have the genotype $A_1A_2B_1B_2$. This type of individual can produce four kinds of gametes: A_1B_1, A_1B_2, A_2B_1, and A_2B_2. Although four different kinds of gametes can be produced, a single randomly chosen gamete from an individual will fuse with a gamete from another individual, producing a zygote. Mom 654 has a gamete A_2B_1, while Dad 528 has a sperm gamete A_1B_2. The zygote offspring from this union will have the genotype $A_1A_2B_1B_2$. The next few steps will generate the genotypes of the offspring.

3. In cells I17–I1016 enter a formula to determine the offspring's genotype at the *A* locus.

Enter the formula **=LEFT(F17,2)&LEFT(H17,2)** in cell I17. Copy this formula down to cell I1016.
Offspring 1 in cell I17 will inherit one *A* allele from its mother and one *A* allele from its father. The formula in cell I17 takes the left two characters from cell F17 and combines them with the left two characters from cell H17.

4. In cells K17–K1016 enter a formula in cell K17 to determine the offspring's genotype at the *B* locus.

Enter the formula **=RIGHT(F17,2)&RIGHT(H17,2)** in cell K17. Copy this formula down to cell K1016.

5. In cells J17–J1012 and L17–L1012, enter a formula to adjust the genotypes so that all heterozygotes are described as either A_1A_2 or B_1B_2 (not A_2A_1 or B_2B_1).

Enter the formula **=IF(I17="A2A1","A1A2",I17)** in cell J17 and copy it down to cell J1012.
Enter the formula **=IF(K17="B2B1","B1B2",K17)** in cell L17 and copy it down to cell L1012.
This step is necessary because an A_1A_2 heterozygote is the same thing as an A_2A_1 heterozygote, but the spreadsheet "interprets" them as being different.

6. In cells M17–M1016 enter a formula to determine the genotype of each offspring.

Enter the formula **=J17&L17** in cell M17. Copy your formula down to cell M1016.
The genotype of the offspring is the combination of genotypes at the *A* and *B* loci.

7. Save your work.

D. Determine if the population is in Hardy-Weinberg equilibrium and linkage equilibrium.

1. Set up new column headings as shown in Figure 8.

	J	K	L	M
1		**Observed genotype frequencies**		
2		**A1A1**	**A1A2**	**A2A2**
3	**B1B1**			
4	**B1B2**			
5	**B2B2**			
6		**Expected genotype frequencies**		
7		**A1A1**	**A1A2**	**A2A2**
8	**B1B1**			
9	**B1B2**			
10	**B2B2**			
11	**Chi-square test *P* value =**			
12	**Hardy-Weinberg equilibrium?**			
13	**Linkage disequilibrium coefficient = *D* =**			

Figure 8

2. Enter formulae in cells K3–M5 to calculate the observed genotype frequencies in the offspring population.

Double-check your results. Your frequencies should add to 1.

Enter the following formulae:

- Cell K3 **=COUNTIF($M$17:$M$1016,"A1A1B1B1")/1000**
- Cell K4 **=COUNTIF($M$17:$M$1016,"A1A1B1B2")/1000**
- Cell K5 **=COUNTIF($M$17:$M$1016,"A1A1B2B2")/1000**
- Cell L3 **=COUNTIF($M$17:$M$1016,"A1A2B1B1")/1000**
- Cell L4 **=COUNTIF($M$17:$M$1016,"A1A2B1B2")/1000**
- Cell L5 **=COUNTIF($M$17:$M$1016,"A1A2B2B2")/1000**
- Cell M3 **=COUNTIF($M$17:$M$1016,"A2A2B1B1")/1000**
- Cell M4 **=COUNTIF($M$17:$M$1016,"A2A2B1B2")/1000**
- Cell M5 **=COUNTIF($M$17:$M$1016,"A2A2B2B2")/1000**

3. Enter formulae in cells K8–M10 to calculate the expected genotype frequencies in the offspring population.

Double-check your results. Your frequencies should add to 1.

Remember that you can calculate the expected genotype frequencies of the offspring in either one of two ways:

- Multiply the expected gamete frequencies *or*
- Multiply the allele frequencies in the adult population

Both methods should both yield the same results.

If you calculate the expected frequencies based on expected gamete frequencies in the adult population, remember to calculate the variety of ways in which gametes from Mom and Dad can combine. For example, if the frequency of an offspring genotype of $A_1A_1B_1B_2$ can be generated in two ways: Mom's egg can be A_1B_1 and Dad's sperm can be A_1B_2, or Mom's egg can be A_1B_2 and Dad's sperm can be A_1B_1. Both possibilities need to be accounted for to generate correct offspring genotype frequencies. Enter the following formulae:

- Cell K8 **=E10*E10**
- Cell K9 **=E10*F10+F10*E10**
- Cell K10 **=F10*F10**
- Cell L8 **=E10*G10+G10*E10**
- Cell L9 **=E10*H10+H10*E10+F10*G10+G10*F10**
- Cell L10 **=F10*H10+H10*F10**
- Cell M8 **=G10*G10**
- Cell M9 **=G10*H10+H10*G10**
- Cell M10 **=H10*H10**

If you calculate the expected frequencies based on allele frequencies in the parental population, enter the following formulae:

- Cell K8 **=F5*F5*H5*H5**
- Cell K9 **=F5*F5*2*H5*H6**
- Cell K10 **=F5*F5*H6*H6**
- Cell L8 **=2*F5*F6*H5*H5**
- Cell L9 **=2*F5*F6*2*H5*H6**
- Cell L10 **=2*F5*F6*H6*H6**
- Cell M8 **=F6*F6*H5*H5**
- Cell M9 **=F6*F6*2*H5*H6**
- Cell M10 **=F6*F6*H6*H6**

4. Save your work.

5. In cell M11, conduct a chi-square test on observed and expected frequencies.

Enter the formula **=CHITEST(K3:M5,K8:M10)** in cell M11.
Refer to Exercise 29, on "Hardy-Weinberg Equilibrium," for the information on this test and its interpretation.

6. In cell M12, enter a formula to answer "yes" or "no" to the question "Is the population in Hardy-Weinberg equilibrium?"

Enter the formula **=IF(M11>0.05,"yes","no")** in cell M12.

7. In cell M13, enter a formula to calculate D, the linkage disequilibrium coefficient.

In cell M13 enter the formula **=E11*H11-F11*G11**.
Equation 7 gave the formula for the disequilibrium coefficient D as

$$D = G_{A1B1}G_{A2B2} - G_{A1B2}G_{A2B1}$$

where G represents the frequency of the different kinds of gametes observed in the population. Remember that D ranges between 0 and 0.25. When the population is in linkage equilibrium, $D = 0$. Your result for this exercise should be very close to 0, indicating that your population *is* in linkage equilibrium.

E. Create graphs.

1. Create a column graph of the genotypes observed in the offspring population. Label your axes fully.

Select cells J2–M5. Use the bar graph option and label your axes fully. Your graph should resemble Figure 9, although your frequencies will likely be a bit different than the ones shown.

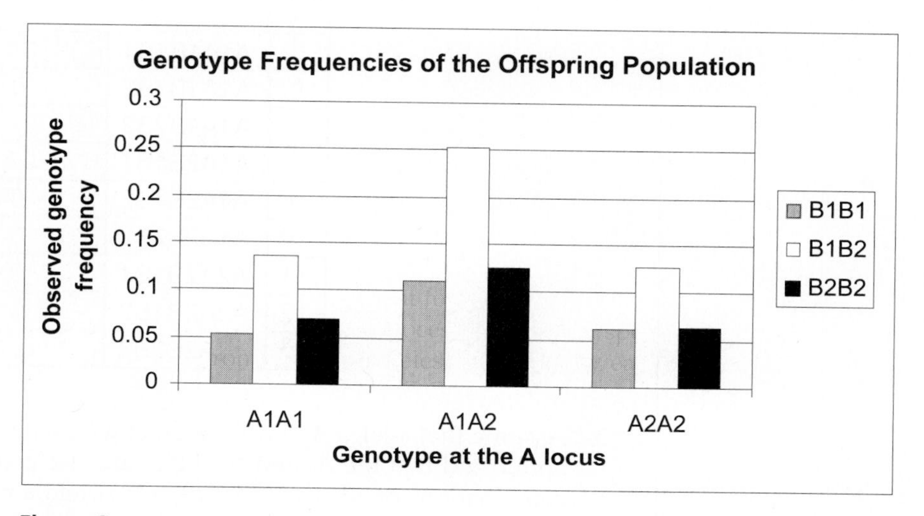

Figure 9

2. Graphically determine whether the various B genotypes are distributed more or less equally among the various A genotypes.

3. Save your work.

Select cells J2–M5 again. Create a new bar graph and choose the 100% stacked column option. Your graph should resemble Figure 10. This graph breaks down the percentage of each B genotype within each A genotype. Since the percentages are relatively equal, this population is in linkage equilibrium.

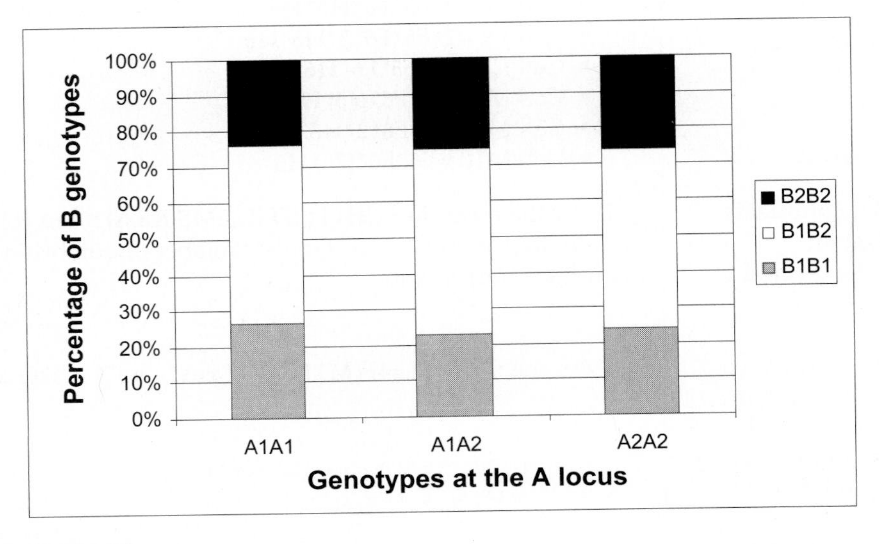

Figure 10

QUESTIONS

1. Interpret the graph you generated in the very last step. In particular, comment on whether the frequencies of B_1B_1, B_1B_2, and B_2B_2 are proportionately the same for A_1A_1, A_1A_2, and A_2A_2 individuals. Is your population in linkage equilibrium? Why or why not?

2. Alter allele frequencies as shown below. Update your graphs and calculate D. Comment on whether the frequencies of B_1B_1, B_1B_2, and B_2B_2 are the same for all of the A_1A_1, A_1A_2, and A_2A_2 individuals.

	A	B
3	Genotype	Frequency
4		
5	A1A1B1B1	0.1
6	A1A1B1B2	0
7	A1A1B2B2	0.1
8	A1A2B1B1	0.7
9	A1A2B1B2	0
10	A1A2B2B2	0
11	A2A2B1B1	0
12	A2A2B1B2	0
13	A2A2B2B2	0.1

3. Assume that alleles $A_1$and B_1 interact well with each other and thus are coadapted, and that the A_2 and B_2 alleles are also coadapted. Assume also that other combinations of alleles ($A_1A_2B_1B_1$, etc.) yield a poorly adapted phenotype. In

this case, $A_1A_1B_1B_1$ and $A_2A_2B_2B_2$ individuals will dominate the population. Alter allele frequencies so that $A_1A_1B_1B_1 = 0.5$, and $A_2A_2B_2B_2 = 0.5$. Modify values in cells B5 and B13, and set the remaining genotype frequencies to 0. Is the parental population in Hardy-Weinberg equilibrium? Is the offspring population in Hardy-Weinberg equilibrium? What is *D*? Graph your results and interpret *D*.

4. If your offspring population from question 3 were to reproduce, how would *D* change over time? How does the frequency of the A_1 allele (p_1) and the frequency of the B_1 (p_2) allele change over time? Simulate the reproduction of individuals over three generations. Set up column headings as shown in the figure below. Start with the genotype frequencies shown for generation 1. Enter 0.5 in cells B5 and B13. Set the remaining genotypes to 0. Calculate D, p_1, and p_2. Record this information in cells V15–V17. Examine the genotypes of the offspring. Enter those genotype frequencies in cells W5–W13 (as shown in the figure below; your numbers will be slightly different). Enter them again in cells B5–B13. Calculate D, p_1, and p_2 for the second generation. Record your results in cells W15–W17. Repeat the process for generation 3. For each generation, examine the 100% column graph (as in Figure 9). Graphically show how D, p_1, and p_2 change over generations.

	U	V	W	X
4		**Generation 1**	**Generation 2**	**Generation 3**
5	A1A1B1B1	0.5	0.266	
6	A1A1B1B2	0	0	
7	A1A1B2B2	0	0	
8	A1A2B1B1	0	0	
9	A1A2B1B2	0	0.494	
10	A1A2B2B2	0	0	
11	A2A2B1B1	0	0	
12	A2A2B1B2	0	0	
13	A2A2B2B2	0.5	0.24	
14				
15	*D* =	0.25		
16	A1 =	0.5		
17	B1 =	0		

LITERATURE CITED

Ayala, F. J. 1982. *Population and Evolutionary Genetics: A Primer.* Benjamin/Cummings, Menlo Park, CA.

Hartl, D. L. 2000. *A Primer of Population Genetics*, 3rd Edition. Sinauer Associates, Sunderland, MA.

31 MEASURES OF GENETIC DIVERSITY

Objectives

- Estimate allele frequencies from a sample of individuals using the maximum likelihood formulation.
- Determine polymorphism for a population, P.
- Determine heterozygosity for a population, H.
- Evaluate how sample size affects estimates of allele frequency, polymorphism, and heterozygosity.

Suggested Preliminary Exercise: Hardy-Weinberg Equilibrium

INTRODUCTION

The amount of genetic variation on earth is astounding. Think of the genetic programming that creates first a larva, then a caterpillar, then a cocoon, and finally an adult butterfly. Or think of the programming that created a single Sequoia tree, and then the different kinds of programming that created an entire forest of Sequoias. Or marvel at the programming required to create you inside your mother's womb. Who would have thought that a mere four molecules—adenine, thymine, cytosine, and guanine, the bases of the genetic code—could be arranged in such a multitude of ways to produce the astonishing variation found among the organisms, living and extinct, that have called the earth home.

The total genetic variation existing on earth today can be "partitioned" or "organized" into four different levels: variation among species; variation among populations of a species; variation among individuals within a population; and variation within a single individual (Hunter 1996). The genetic differences among species such as Sequoias, butterflies, and humans clearly accounts for a large chunk of the total genetic diversity. But populations and individuals of the same species differ in their genetic makeup too. For example, a population of garter snakes living near Lake Ontario may have a very different genetic make-up than a population of the same species of snake living in the Ozark Mountains. Even within a single population, individuals can be quite variable, although they can also be genetically very similar to one another. And within an individual—you, for instance—some portion of the total genome is heterozygous (two different alleles of a gene are present at a locus), and some portion of the genome is homozygous (the two alleles at a locus are both the same). The diversity within any individual can be great or small, depending on how many gene loci are heterozygous. It is important to realize that diversity is measured as a continuum from little or no diversity to very high levels of diversity.

How is genetic diversity measured in populations? Typically, a sample of individuals is obtained from the population and the genotype of each individual is determined using one of several methods (e.g., protein electrophoresis or DNA sequencing). From there, allele frequencies can be estimated, and two other measures of genetic diversity—polymorphism and heterozygosity—can be measured (Hartl 2000).

Let's illustrate these measures with an example. Suppose you sample five individuals of mice from a nearby farm field. For two loci, you obtain the genotypes shown in the table.

Individual	Locus A Genotype	Locus B Genotype
1	A1A1	B1B1
2	A1A2	B1B2
3	A1A2	B1B3
4	A1A1	B1B1
5	A2A2	B1B1

Based on your sample, there are two "alleles" present at the *A* locus (A_1 and A_2) and three alleles present at the *B* locus (B_1, B_2, B_3). For the *A* locus, the frequency of the A_1 allele is 0.6 because 6 of the 10 total alleles (5 individuals, each with two alleles) at this locus are A_1. Likewise, the frequency of the A_2 allele is 0.4. For the *B* locus, the frequency of the B_1 allele is 0.8, the frequency of the B_2 allele is 0.1, and the frequency of the B_3 allele is 0.1. Note that the sum of the frequencies for any locus must equal 1. By sampling five individuals from the population and deriving allele frequency estimates, you are hoping that the five individuals sampled reflect the greater population of mice that live in the field but were not sampled. But does the greater population of field mice really have these frequencies? If we sampled five additional mice, our allele frequency estimates might change. And they might continue to change until every single mouse in the field population is sampled; at that point we could calculate (as opposed to estimate) the true allele frequency of the mouse population.

Estimating Polymorphism and Heterozygosity

Sampling, by nature, involves some error. But we can estimate what the *most likely* allele frequencies are in the greater population, given the size of our sample. The procedure to estimate the frequencies is called **maximum likelihood formulation**. And we can make a statement about how accurate our estimates are by calculating the variance of the estimates themselves.

If we assume that the genetic system of the *A* and *B* alleles is one of co-dominance, the **maximum likelihood estimate** of p (the frequency of the A_1 allele) is

$$\hat{p} = \frac{0.5 \times N_{A_1A_2} + N_{A_1A_1}}{N} \quad \text{Equation 1}$$

and the variance in $\hat{p}$ is

$$V(\hat{p}) = \frac{p(1-p)}{2N} \quad \text{Equation 2}$$

Equation 1 should look familiar to you. Using these formulae, the maximum likelihood estimate of the A_1 allele is 0.6, and the variance is .024. The frequency of the A_2 allele, q, can be similarly calculated.

Once we have estimated the allele frequencies in the population, we can estimate another useful measure of genetic diversity, **polymorphism**, *P*. The word "polymorphism" literally means "many forms." It follows that *P* measures whether a locus contains many different forms of a gene (i.e., alleles), or whether a locus contains few forms or even just one allele. In our example above, the A locus has two alleles (A_1 and A_2), while the B locus has three (B_1, B_2, and B_3). Both loci are polymorphic. Since 2 loci out of 2 loci sampled (A locus and B locus) each have different kinds alleles, $P = 2/2 = 1$. On the other hand, if all five individuals were B_1B_1 genotypes at locus B, the B locus would be monomorphic (literally, "one form"), and so 1 of 2 loci examined would be polymorphic, and P would equal 0.5. Thus, P can be defined as

$$P = \text{Number polymorphic loci/Total number loci evaluated} \qquad \text{Equation 3}$$

In a large population, almost all loci will have more than one allele (Hartl 2000), so if we consider a polymorphism to be any locus that has more than one allele, the value of P will never be very far from 1. To make P more meaningful, a locus is usually considered to be polymorphic only if the *frequency of the most common allele* is less than some arbitrary threshold, usually 0.95 (Ayala 1982). Sample size is therefore a key issue in estimating P. Suppose, for example, that we are examining the C locus in a population and the first four individuals all have the genotype C_1C_1, but the fifth has the genotype C_1C_2. Of the ten alleles we have sampled so far, all but one are C_1, so our estimate of the frequency of C_1 is 9/10, or 0.9. On the basis of this very small sample we would conclude that the C locus is polymorphic. If we continue to sample and find that the next 45 individuals are all C_1C_1, however, we need to reconsider—now we've sampled 100 alleles (from 50 individuals in all), and 99 of them are C_1, so our new estimate of the frequency of C_1 is 0.99. It's beginning to look as if the C_2 allele is less common than our initial sample of five individuals suggested, and the C locus may actually not be polymorphic (if we use a frequency of 0.95 as the cutoff in our definition). A larger sample size yet would give us greater confidence in our results.

Another useful measure of genetic diversity is **heterozygosity**, H, which measures the percentage of genes at which the *average individual* is heterozygous. In our example, individual 1 is homozygous at both the A and B locus, so its heterozygosity is 0 out of 2 loci = 0. Individual 2 is heterozygous at both the A and B locus, so its heterozygosity is 2 out of 2 loci examined = 1. The average individual heterozygosity for these two individuals is then the average of individual 1 and individual 2, so $H = 0.5$. In mathematical terms, **average heterozygosity** is calculated as

$$\hat{H} = \frac{1}{Nm}\sum_{i=1}^{N}\sum_{j=1}^{m} H_{ij} \qquad \text{Equation 4}$$

and the variance in $\hat{H}$ is

$$V(\hat{H}) = \frac{\hat{H}(1-\hat{H})}{Nm} \qquad \text{Equation 5}$$

where N is the sample size and m is the number of loci examined. You'll see clearly how these formulae function as you work through the exercise.

PROCEDURES

In this exercise, you'll learn how to estimate allele frequencies using the maximum likelihood formulation, and you will learn how to calculate P, H and $\hat{H}$. We'll examine only four loci (A, B, C, and D) and we will assume that each locus has only two alleles present in the population. We'll also assume that you are sampling individuals one at a time from a very large population and can identify the genotypes of each individual at the different loci. You'll examine how your estimates of allele frequencies, P, and $\hat{H}$ change as new individuals are sampled and sample size increases. As always, save your work frequently to disk.

INSTRUCTIONS	ANNOTATION

A. Set up the hypothetical population.

1. Open a new spreadsheet and set up headings as shown in Figure 1.

	A	B	C	D	E	F	G	H
1	***Measures of Genetic Variation***			Two alleles per locus, 4 loci evaluated				
2								
3		**Genotypes**						
4		**A1A1**	**B1B1**	**C1C1**	**D1D1**			
5		**A1A2**	**B1B2**	**C1C2**	**D1D2**		**Polymorphism criteria:**	
6		**A2A1**	**B2B1**	**C2C1**	**D2D1**		**0.95**	**0.05**
7		**A2A2**	**B2B2**	**C2C2**	**D2D2**			
8								
9	**Frequency:**	**A1**	**B1**	**C1**	**D1**			
10								
11		**A2**	**B2**	**C2**	**D2**			
12								

Figure 1

2. In rows 10 and 12, assign true allele frequences to a very large hypothetical population. We will try to estimate these frequencies by sampling individuals from the population.

Enter 0.5 in cells B10–E10.
Enter **=1-B10** in cell B12, and copy this formula across to cell E12.
The values in cells B10–E10 and B12–E12 represent the true allele frequencies of a very large (infinite) population from which we'll sample individuals and estimate allele frequencies, *P*, and *H*. To begin, we'll let the true frequencies of each allele for each locus be 0.5. Remember, the sum of the allele frequencies for a given locus must equal 1. The values in cells B10–E10 can be modified directly as you go through the exercise (cells B12–E12 will automatically be updated).

3. Set up spreadsheet headings as shown in Figure 2.

	A	B	C	D	E
14		**Genotype**			
15	**Individual**	**A Locus**	**B Locus**	**C Locus**	**D Locus**

Figure 2

4. Set up a linear series from 1 to 100 in cells A16–A115.

Enter 1 in cell A16.
In cell A17, enter the formula **=1+A16**.
Copy the formula down to cell A115.
We will sample 100 individuals from this large population and determine the genotypes of each individual. We will then assume that individuals are sampled in order (from 1 to 100), and will then estimate the allele frequencies, polymorphism, and heterozygosity as new individuals are included in the total sample.

5. Assign genotypes at the *A* locus to each individual in the population, based on the allele frequencies designated in Step 2.

In cell B16, enter the formula **=IF(RAND()<$B$10,$B$9,$B$11)&IF(RAND()<$B$10,$B$9,$B$11)**.
This formula will assign genotypes based on the allele frequencies that we designated in cells B10 and B12. The **IF** formula in cell B16 is used to determine the genotype of individual 1. The first part of the formula in cell B16 tells the spreadsheet to choose a random number between 0 and 1 (the **RAND()** portion of the formula), and if that random number is less than the value designated in cell B10, then return the value in cell B9 (A_1); otherwise, return the value in cell B11 (A_2). All individuals have two alleles for

a given locus, so you need to repeat the formula again, and then join the two alleles obtained from the two **IF** formulas by using the **&** symbol.

Once you've obtained genotypes for individual 1, copy this formula down to cell B115 to obtain genotypes for all 100 individuals in the population. Note that when you press F9, the calculate key, the spreadsheet generates a new random number, and hence a new genotype.

6. Enter formulae in cells C16–E16 to generate genotypes for individual 1 at the *B*, *C*, and *D* loci.

Enter the formulae:

- C16 **=IF(RAND()<$C$10,$C$9,$C$11)&IF(RAND()<$C$10,$C$9,$C$11)**
- D16 **=IF(RAND()<$D$10,$D$9,$D$11)&IF(RAND()<$D$10,$D$9,$D$11)**
- E16 **=IF(RAND()<$E$10,$E$9,$E$11)&IF(RAND()<$E$10,$E$9,$E$11)**

7. Copy cells B16–E16 down to row 115.

When you copy the formula down, note that the genotypes are assigned based on the random numbers and the allele frequencies in row 10, and the allele designations in rows 9 and 11. These formulae require absolute cell references (with row and columns preceded by $ signs) so that when the formulae are copied down to individual 100, the spreadsheet will go back to the appropriate, fixed, cells in assigning genotypes to individuals.

8. Save your work.

B. Calculate likelihood estimators.

1. Set up spreadsheet headings as shown in Figure 3.

	F	G	H	I
14	**Estimator**			
15	***p* (hat)**	***r* (hat)**	***t* (hat)**	***v* (hat)**

Figure 3

2. In cell F16, enter a formula to estimate the frequency of the A_1 allele of our population (this will be a maximum likelihood formula based on Equation 1).

We'll let $\hat{p}$ estimate the frequency of the A_1 allele, $\hat{r}$ be the estimate of the B_1 allele frequency, $\hat{t}$ be the estimate of the C_1 allele frequency, and $\hat{v}$ be the estimate of the D_1 allele frequency. Enter the formula **=(COUNTIF($B$16:B16,$B$4)+COUNTIF($B$16:B16,$B$5) *0.5+COUNTIF($B$16:B16,$B$6)*0.5)/$A16** in cell F16. This represents Equation 1, the formula for estimating the frequency of an allele in a population:

$$\hat{p} = \frac{0.5 \times N_{A_1A_2} + N_{A_1A_1}}{N}$$

The first step is to tally the number of A_1A_1 homozygotes and the number of A_1A_2 heterozygotes. The tally of heterozygotes is then multiplied by 0.5. The sum is divided by the number of individuals sampled, *N*. The formula in cell F16 does this with the **COUNTIF** function. The formula in cell F16 counts the number of A_1A_1 homozygotes (cell $B$4) in the range of cells $B$16–B16, then counts the number of A_1A_2 heterozygotes in the same range and multiplies this number by 0.5, then counts the number of A_2A_1 heterozygotes and multiplies this number by 0.5. (Remember that a heterozygote can be either A_1A_2 or A_1A_2 in your spreadsheet.) The sum of these numbers is bracketed by parentheses so that the total is divided by *N*, the sample size. In this case, the sample size is 1, given in cell A16. Note the use of absolute and relative references. This will allow you to copy your formula down to cell F115 while updating *N* and the range of cells to be counted.

3. Enter formulae in cells G16–I16 to compute the estimated allele frequencies of the B_1, C_1, and D_1 alleles.

In cell G16, enter the formula **=(COUNTIF($C$16:C16,$C$4)+COUNTIF($C$16:C16,$C$5)* 0.5+COUNTIF($C$16:C16,$C$6)*0.5)/$A16.**
In cell H16, enter the formula **=(COUNTIF($D$16:D16,$D$4)+COUNTIF($D$16:D16,$D$5)* 0.5+ COUNTIF($D$16:D16,$D$6)*0.5)/$A16.**

In cell I16, enter the formula**=(COUNTIF($E$16:E16,$E$4)+COUNTIF($E$16:E16,$E$5)*0.5+ COUNTIF($E$16:E16,$E$6)*0.5)/$A16.**

4. Select cells F16–I16 and copy their formulae down to row 115.

5. Graph the estimated allele frequencies as a function of sample size. Set the y-axis scale to range between 0 and 1.

Use the line graph option and label your axes fully. Your graph should resemble Figure 4.

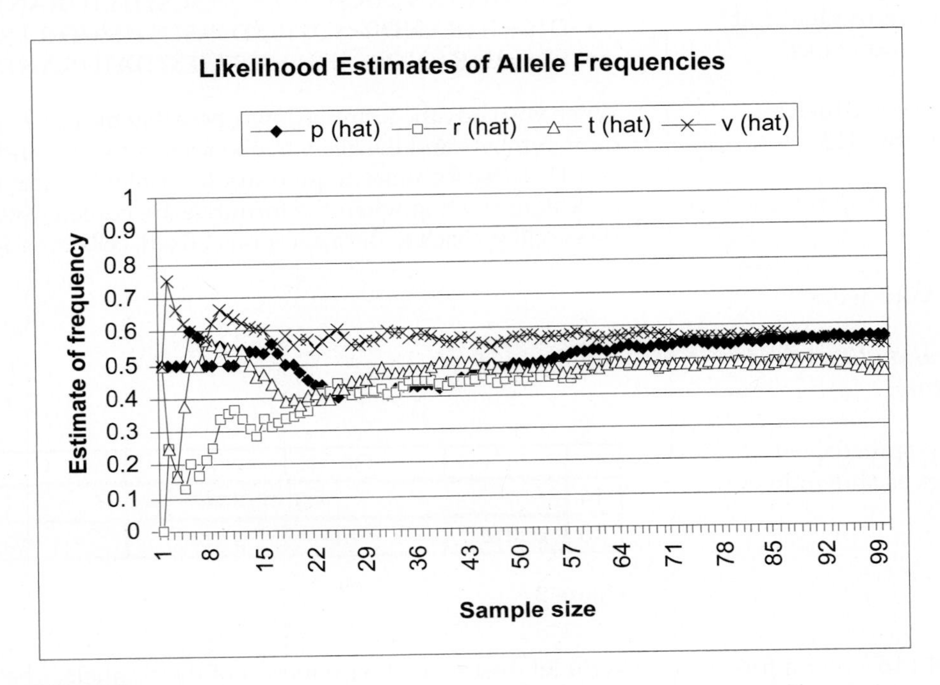

Figure 4

6. Press F9 to generate new random numbers, and hence new genotypes.

How closely do your samples reflect the allele frequencies given in rows 10 and 12? Examine your graph carefully and write a one or two sentence summary of the major results.

7. Save your work.

C. Estimate polymorphism, P.

1. Enter new spreadsheet headings as shown in Figure 5.

	J	K	L	M	N
14			Polymorphism		
15	A Locus?	B Locus?	C Locus?	D Locus?	P

Figure 5

2. In cell G6, enter the criterion parameter for polymorphism.

Now we are ready to estimate polymorphism. To begin, our criterion will be 0.95, so enter 0.95 in cell G6.

3. Enter **=1-G6** in cell H6.

Remember, a gene locus is polymorphic if the frequency of the *most common allele* is less than the criterion. Another way of saying this is that a locus is considered monomorphic if *any* of the alleles at that locus has a frequency >0.95. Thus, if either the A_1 or A_2 allele has a frequency of greater than 0.95, the locus is monomorphic. Concentrating on just the A_1 allele, the *A* locus is polymorphic if the A_1 allele has a frequency of <0.95 (which means that the A_1 allele frequency is <.95) or >0.05 (which means that the A_2 allele frequency is <.95). Otherwise, it is monomorphic.

4. Determine whether the locus is polymorphic (1) or monomorphic (0).

In cell J16, enter the formula **=IF(OR(F16>$G$6,(F16<$H$6)),0,1)**.
We have already calculated the estimated allele frequencies for our population. We'll examine these estimates to determine whether or not the locus is polymorphic. The formula in cell J16 evaluates individual 1. Based on this single sample, if the value in cell F16 is either greater than the criterion in cell G6 or less than the criterion in cell H6, we will consider the locus to be monomorphic (0). Otherwise, it is considered to be polymorphic (1). The **OR** part of this formula—**OR(F16>$G$6,(F16<$H$6)**—allows us to evaluate both conditions; if either one is true the spreadsheet will return the number 0. If both criteria are false, the spreadsheet will return the number 1.

5. Enter formulae in cells K16–M16 to determine the polymorphism at the *B*, *C*, and *D* loci.

Select cell J16, and copy its formula across to cell M16, or enter the following:
In cell K16, enter the formula **=IF(OR(G16>$G$6,(G16<$H$6)),0,1).**
In cell L16, enter the formula **=IF(OR(H16>$G$6,(H16<$H$6)),0,1).**
In cell M16, enter the formula **=IF(OR(I16>$G$6,(I16<$H$6)),0,1).**

6. In cell N16, compute the average *P* for individual 1.

Enter the formula **=AVERAGE(J16:M16).**

7. Select cells J16–N16, and copy their formulae down to row 115.

Keep in mind that although the average polymorphism appears to be calculated for each individual, column A really gives the sample size from the population. The allele frequency estimates are based on all of the samples up to and including the individual sampled, so the *P* estimates are really estimates that change as individuals are added to the sample. Also keep in mind that since only four loci have been evaluated, *P* can take on only five values: 0, 0.25, 0.5, 0.75, and 1, where 0/4, 1/4, 2/4, 3/4, or 4/4 loci are polymorphic.

8. Graph *P* as a function of sample size. Set the *y*-axis scale to range between 0 and 1.

Use the line graph option and label your axes fully. Your graph should resemble Figure 5.

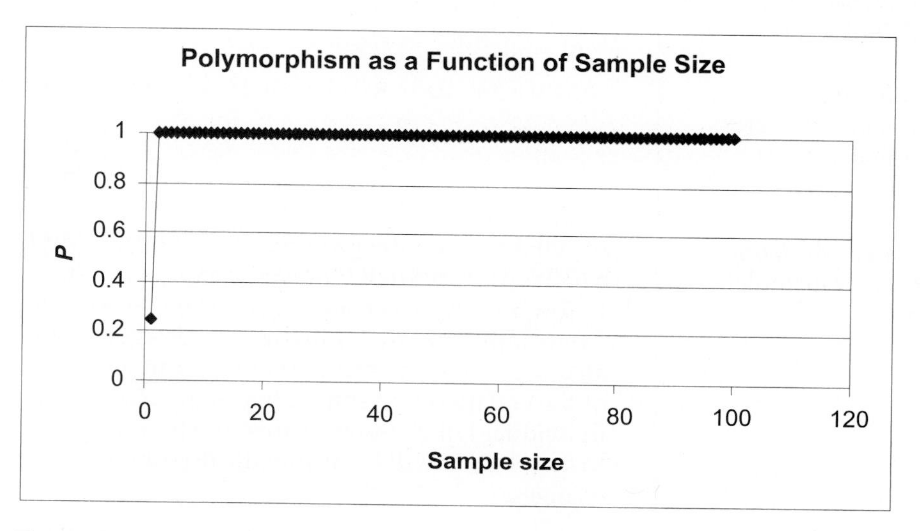

Figure 5

9. Press F9 several times and examine how P changes as the sampled individuals change in genotypes.

Since all loci have allele frequencies around 0.5 (for large enough sample sizes), P should equal 1, indicating that all four loci are polymorphic.

10. Save your work.

D. Estimate heterozygosity, H.

1. Enter new spreadsheet headings as shown in Figure 6.

Remember that heterozygosity has two components: within individuals (H) and among (or across) individuals ($\hat{H}$). Columns O through R tackle the within individual component. Column S uses that information to calculate the among-individuals component.

	O	P	Q	R	S
14			Heterozygosity		
15	A Locus?	B Locus?	C Locus?	D Locus?	H (hat)

Figure 6

2. Determine the heterozygosity of locus A for each individual.

In cell O16, enter the formula **=IF(OR(B16=$B$5,B16=$B$6),1,0).** Within an individual, heterozygosity is the proportion of loci that are heterozygous. The O16 formula examines the A locus for individual 1 and returns a 1 if the individual is heterozygous at that locus, and a 0 if it is homozygous at that loci. An **OR** formula is used because either A_1A_2 or A_2A_1 heterozygotes should be counted. Copy this formula down to row 115 to determine the heterozygosity of the A locus for each individual in the sample.

3. Enter formulae in cells P16–R16 to compute heterozygosity for each individual in the sample at the B, C, and D loci.

In cell P16, enter the formula **=IF(OR(C16=$C$5,C16=$C$6),1,0)**
In cell Q16, enter the formula **=IF(OR(D16=$D$5,D16=$D$6),1,0)**
In cell R16, enter the formula **=IF(OR(E16=$E$5,E16=$E$6),1,0)**

4. Determine $\hat{H}$, the average heterozygosity across all individuals.

Now we are ready to calculate $\hat{H}$, which is calculated with Equation 4:

$$\hat{H} = \frac{1}{Nm}\sum_{i=1}^{N}\sum_{j=1}^{m} H_{ij}$$

5. Select cells O16–S16, and copy their formulae down to row 115.

In cell S16, enter the formula **=1/(4*A16)*SUM($O$16:R16)**. The formula **=AVERAGE($0$16:R16)** gives the same result.
In row 16, we are considering $\hat{H}$ when the sample size consists of a single individual. Our sample size, N, is 1 in this row, designated by cell A16. The number of loci evaluated, m, is 4. So the first part of the formula is easy to take care of. For the second part of the equation (the summation signs, Σ), we simply need to sum the 0's and 1's for individual 1, then multiply this sum by $1/Nm$, or $1/4$. As you copy this formula down to row 115, $\hat{H}$ will be automatically updated as a running estimate as sample size changes.

6. Graph $\hat{H}$ as a function of sample size.

Use the line graph option and label your axes fully. Your graph should resemble Figure 7.

7. Press F9 and evaluate how changes in sampling affect your estimates.

8. Save your work.

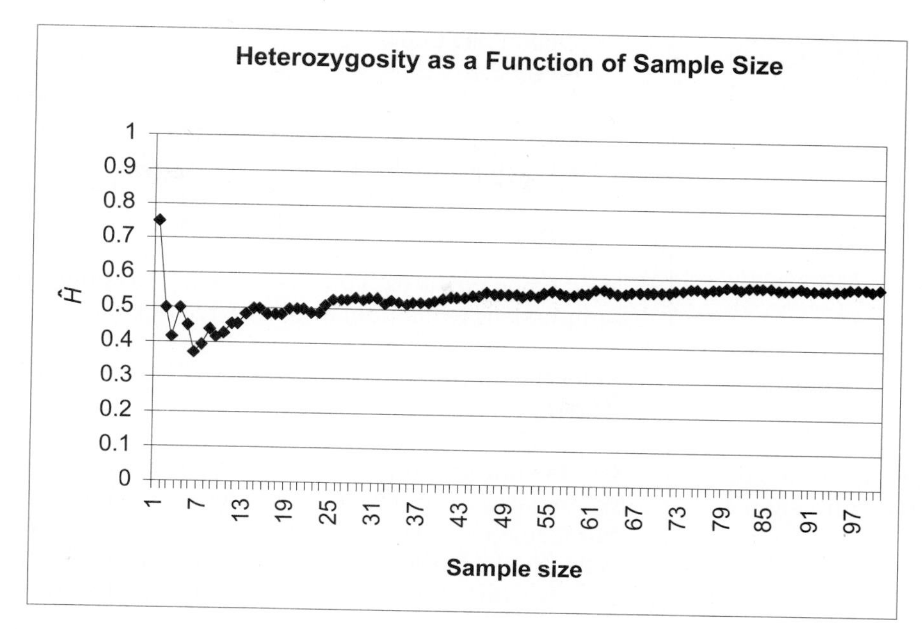

Figure 7

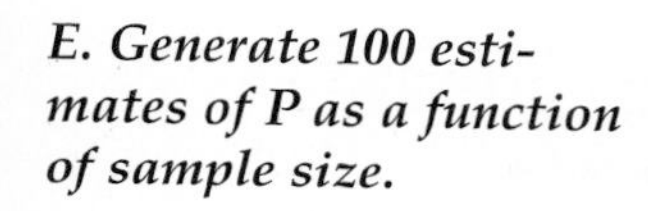

E. Generate 100 estimates of P as a function of sample size.

By now you should have noticed that when you press F9, the calculate key, all of your results, including your graphs, change. This is because the genotypes of individuals change when a new random number is generated. Although you can get a "feel" for how estimates change as sample size increases by pressing F9 a number of times and examining the graphs, quantitative approaches are usually used. How can you therefore assess how sample size affects your estimates of P, $\hat{H}$, and $\hat{p}$, when your results keep changing? In order to determine how sample size affects these estimates, we need to press F9 many times (say 100), and compute the average estimate. This is called a **Monte Carlo simulation**. We will do this in the next step for P; you may wish to evaluate other metrics as well.

1. Set up new spreadsheet headings as shown, in Figure 8, but extend the trials down to 100 (cell U115), and extend the sample size out to 100 (in increments of 5, cell AO15).

	U	V	W	X	Y	Z
14		Sample size				
15	Trial	5	10	15	20	25
16	1	1	1	1	1	1
17	2	1	1	1	1	1
18	3	1	1	1	1	1
19	4	1	1	1	1	1
20	5	1	1	1	1	1

Figure 8

2. Write a macro to record estimates of P for different sample sizes, tracking your results for 100 trials.

See Exercise 2, "Spreadsheet Functions and Macros," for information on how to record a macro. When you are in the Record Macro mode, assign a name (e.g., Trials) and a shortcut key (e.g., <Control>+t) to your macro. Then record the following steps:

- Press F9, the Calculate key, to generate new genotypes for the population.
- Highlight cell N20 (the P estimate for a sample size of 5).
- Press down the <Control> key, and P estimates for sample sizes 10 (N25), 15 (N30), up to N(115).

- Open Edit | Copy.
- Select cell V15.
- Open Edit | Find. A dialog box will appear. Leave the Find What box blank, search by columns and values. Select Find Next, and then Close.
- Open Edit | Paste Special, and select the Paste Values and Transpose options. Click OK. Your results should be pasted into row 16.
- Open Tools | Macro | Stop Recording.

Now when you press your shortcut key 100 times, your estimates of P under different sample sizes will automatically be recorded.

3. Compute the average P in row 116.

Enter the formula **=AVERAGE(V16:V115)** in cell V116. Copy this formula over to cell AO116.

4. Graph your results, the average P as a function of sample size.

Use the line graph option and under the Series tab, select cells V15–AO15 as Category (x) axis labels. Your graph should look like Figure 9. Perhaps this figure is a bit boring, but it suggests that when the frequencies at all four loci are 0.5 for each allele (set in cells B10–E10), the estimate of P is insensitive to sample size. You will see that this is not the case when there are rare alleles at a locus.

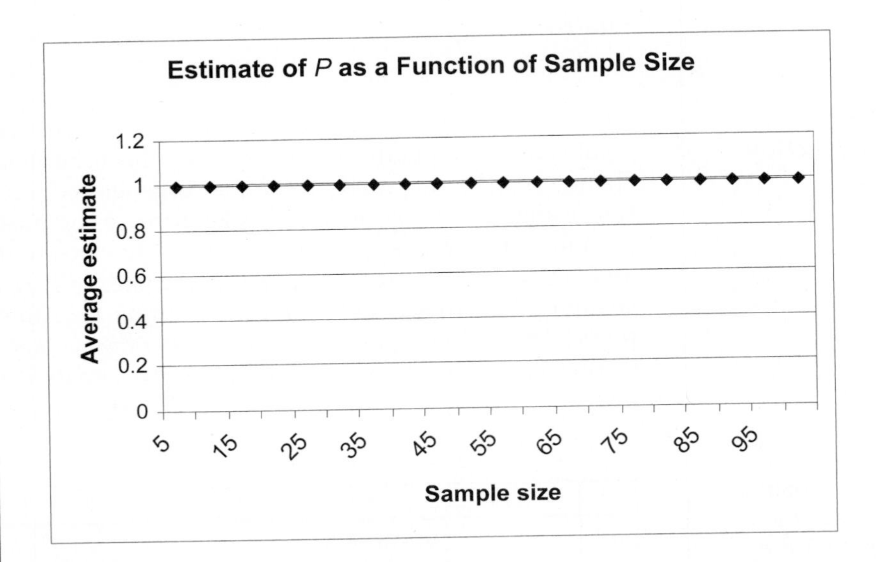

Figure 9

5. Examine the visual basic for application code to learn how to modify your macro for other metrics.

You can edit your macro to examine other metrics ($\hat{p}$, H) by making some slight modifications. (You can also just record a brand new macro if the idea of editing the code of a current macro does not appeal to you).

Open Tools | Macro, then select the macro Trials and Edit. You should now see the Visual Basic for Applications code that the spreadsheet "wrote" when you went through your keystrokes. Read through the code. It should make some sense to you, since it is simply a record of which cells you selected, copied, and pasted. We added two sentences to our code: `For counter = 1 to 100` was added after the fourth line (a keyboard shortcut) and the word `Next` was typed into the second to the last line of the code (before the last line, `ENDSUB`) so that when the macro is run, all 100 trials are completed. In this macro, estimates of polymorphism (P) are given in column N. If you manually replace the letter N with the letter F in all of the appropriate places, your macro can be used to evaluate how $\hat{p}$ or other estimates change as a function of sample size.

QUESTIONS

1. Examine your estimates of P as a function of sample size (last step). How do the allele frequencies affect your result? Set cell B5 to 0.05 (cell B6 should be updated to 0.95). Erase your macro results (cells V16–AO115), and then run your macro again. Your graphs should automatically be updated. Interpret your results.
2. Change polymorphism criteria from 0.95 to some other value, such as 0.9. How does the criteria affect the polymorphism estimate?
3. Which measure is a better indicator of genetic diversity for your population, P or H? Why is it useful to have multiple measures of diversity?
4. Add a fifth and sixth allele to your spreadsheet model. How does increasing the number of alleles affect polymorphism and heterozygosity estimates? If you were given additional funds to evaluate additional loci, would these dollars be well spent? Use graphs to illustrate your answer.

*5. (Advanced) Our model is based on a co-dominant allele system, but several other kinds of genetic systems are possible. Modify your model to estimate allele frequencies in a system where one allele is dominant over the other. Compare your results in terms of maximum likelihood estimators, polymorphism, and heterozygosity.

LITERATURE CITED

Ayala, F. 1982. *Population and Evolutionary Genetics*. Benjamin Cummings, Menlo Park, CA.

Hartl, D. L. 2000. *A Primer of Population Genetics*, 3rd Edition. Sinauer Associates, Sunderland, MA.

Hunter, M. L. Jr. 1996. *Fundamentals of Conservation Biology*. Blackwell Science, Inc., Cambridge, MA.

32 NATURAL SELECTION AND FITNESS

Objectives

- Mathematically define absolute fitness, relative fitness, and the selection coefficient.
- Predict the course of evolution by natural selection from any given initial allele condition, using the formula

$$p_{t+1} = \frac{W_{11}p_t^2 + W_{12}p_t q_t}{(W_{11}p_t^2 + W_{12}2p_t q_t + Wq_t^2)}$$

- Predict the change in population size over time, using the formula

$$N_{t+1} = (W_{11}p_t^2 + W_{12}2p_t q_t + W_{22}q_t^2) \times N_t$$

- Develop a spreadsheet model of a population of 100 individuals that undergo natural selection and track genotypes through time.

Suggested Preliminary Exercises: Geometric and Exponential Population Models; Hardy-Weinberg Equilibrium

INTRODUCTION

Evolutionary biologists are interested in how genotypes and allele frequencies will change over time. **Natural selection** takes place in a population when different genotypes have different probabilities of survival or different abilities to reproduce (Roughgarden 1998). That is, genotypes themselves have growth rates, where "fit" genotypes increase in the population relative to "less fit" or "unfit" genotypes. Stated more succinctly, dN/dt varies among genotypes when natural selection is acting on a population (Wilson and Bossert 1971). Because natural selection affects the growth rates of genotypes, it can profoundly affect how allele frequencies change from one generation to the next. One of the assumptions of the Hardy-Weinberg principle is that natural selection does not act on the population. In this exercise, you'll explore how violating this assumption affects the evolution of a population.

Let's start with a quick review of the Hardy-Weinberg principle. You might recall that if there are only two alleles at a given locus, A_1 and A_2, the frequencies of the alleles are symbolized by p and q, where p is the frequency of the first allele (A_1)

and q is the frequency of the second allele (A_2). Recall further that, for genes with only two alleles,

$$p + q = 1 \quad \text{Equation 1}$$

Assume that the A locus has allele frequencies of $p = 0.6$ and $q = 0.4$. Given these frequencies, the Hardy-Weinberg principle allows us to predict the genotype frequencies of a population, assuming that the population is large, that mating occurs at random, and that there is no gene flow, natural selection, or mutation acting on the population. The predicted genotypes of a population in Hardy-Weinberg equilibrium are $p^2{:}2pq{:}q^2$, where p^2 is the frequency of the A_1A_1 genotype, $2pq$ is the frequency of the heterozygous genotype (A_1A_2 and A_2A_1), and q^2 is the frequency of the A_2A_2 genotype. The sum of the genotype frequencies will be 1. In this example, a population in Hardy-Weinberg equilibrium will have roughly the following genotype frequencies:

- frequency $(A_1A_1) = p^2 = p \times p = 0.6 \times 0.6 = 0.36$, or 36% of the population will be A_1A_1.
- frequency $(A_1A_2) = 2 \times p \times q = 2 \times 0.6 \times 0.4 = 0.48$, or 48% of the population will be A_1A_2.
- frequency $(A_2A_2) = q^2 = 0.4 \times 0.4 = 0.16$, or 16% of the population will be A_2A_2.

Note that the genotype frequencies add to 1:

$$p^2 + 2pq + q^2 = 1 \quad \text{Equation 2}$$

The numbers of *individuals* of each genotype that are expected in the population can be calculated by multiplying the genotype frequencies by the population size, N.

$$\text{Number of } A_1A_1 \text{ individuals} = p^2 \times N$$

$$\text{Number of } A_1A_2 \text{ individuals} = 2pq \times N \quad \text{Equation 3}$$

$$\text{Number of } A_2A_2 \text{ individuals} = q^2 \times N$$

If our population consists of 400 individuals, for example, $0.36 \times 400 = 144$ individuals are expected to be A_1A_1, $0.48 \times 400 = 192$ individuals are expected to be A_1A_2, and $0.16 \times 400 = 64$ individuals are expected to be A_2A_2.

Natural Selection

When natural selection is at work on a population, the genotype frequencies may not match the frequencies predicted by Hardy-Weinberg. If some genotypes are more likely to survive than others, the genotype frequencies in the population will be altered. In turn, the allele frequencies of the population may also change.

Consider a population of 100 individuals that consists of 25 A_1A_1 individuals, 50 A_1A_2 individuals, and 25 A_2A_2 individuals. Given the numbers of individuals of each genotype, the allele frequencies can be calculated and are $p = 0.5$ and $q = 0.5$. With these frequencies, $p^2 \times N = 0.5 \times 0.5 \times 100 = 25$ individuals are expected to be A_1A_1, $2pq \times N = 2 \times 0.5 \times 0.5 \times 100 = 50$ individuals are expected to be A_1A_2, and $q^2 \times N = 0.5 \times 0.5 \times 100 = 25$ individuals are expected to be A_2A_2. Because the observed genotype frequencies equal the expected genotype frequencies, the population is in Hardy-Weinberg equilibrium.

Now let's consider what happens to the population when natural selection acts on it. In this exercise we will assume that our population has discrete, nonoverlapping generations, in which individuals start out as zygotes, reach sexual maturity, reproduce, and then immediately die. The probability of surviving to sexual maturity (adulthood) is given by the letter $\boldsymbol{l}$. Given that individuals survive to reproductive age, the number of gametes than an adult contributes to the next generation's gene pool is given by $\boldsymbol{2m}$. (The reason m is multiplied by 2 will become clear later on.) The life cycle of such an organism is depicted in Figure 1.

Let's assume that the A_2A_2 genotype has a low probability—say, 0.2—of surviving to reproductive age. If in fact only 20% of the A_2A_2 genotypes survive and *all* of the A_1A_1

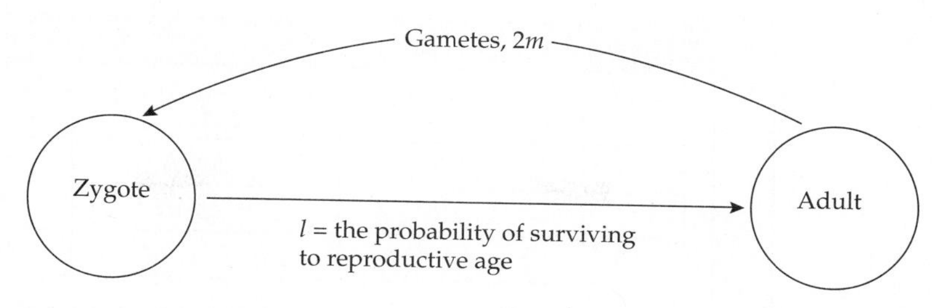

Figure 1

and A_1A_2 genotypes survive, the genotype frequencies of the adult population will be 25 A_1A_1, 50 A_1A_2, and 5 A_2A_2 (because 20 of the A_2A_2 individuals died). A graph of the genotype numbers before and after selection is shown in Figure 2.

Not only has natural selection altered the genotype frequencies, but the allele frequencies have consequently been altered as well. After selection, $p = 0.625$ and $q = 0.375$. Because p and q have changed, it might be tempting to conclude that the population has evolved. However, evolution is a change is allele frequencies *across* generations; so far we have examined the effects of selection *within* a generation. In order to determine the effects of natural selection on evolution, we must calculate p and q in the *next* generation, which depends on both the survival and the reproduction of the different genotypes.

To determine what p and q will be in the next generation, we will utilize the notation outlined by Roughgarden (1998) to follow the progress of a set of individuals from the zygote stage until death, keeping track of how many individuals of each genotype survive to sexual maturity (adulthood) and how many of the total gametes produced by each genotype make it into the next generation's gene pool (Table 1). The starting number of individuals (zygotes) of various genotypes in the population is shown in row 1 of Table 1. This is the Hardy-Weinberg genotype frequency multiplied by the total number of individuals in the population (Equation 3). The probability that a zygote of a given genotype will survive to sexual maturity (adulthood) is denoted by the letter l. The subscript after the letter l indicates the survival probability for a specific genotype; thus, l_{12} is the probability that an A_1A_2 genotype will survive to adulthood. The number of adults of a particular genotype can then be computed as the probability of surviving to adulthood multiplied by the number of zygotes of that genotype. This value appears in row 2 of Table 1.

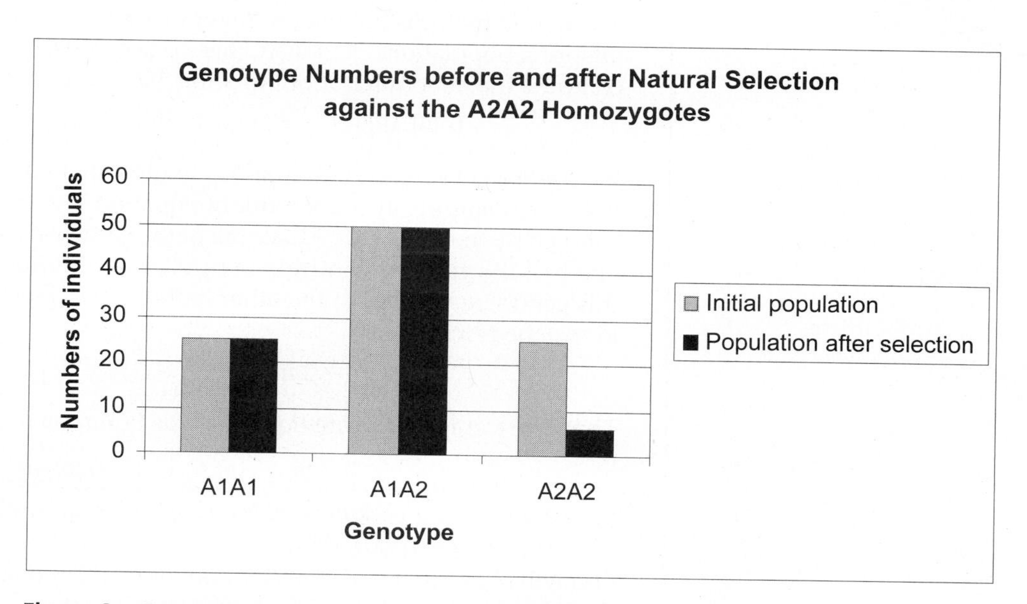

Figure 2 Genotype numbers before and after natural selection against the A_2A_2 homozygote.

TABLE 1.

	A_1A_2	A_1A_2	A_2A_2
1	p^2N	$2pqN$	q^2N
2	$l_{11}p^2N$	$l_{12}2pqN$	$l_{22}q^2N$
3	$2m_{11}l_{11}p^2N$	$2m_{12}l_{12}2qpN$	$2m_{22}l_{22}q^2N$

The number of gametes that are produced per individual of a specified genotype that actually become incorporated into the next generation's gene pool is $2m$: m represents one-half the gametes produced per individual. The total number of gametes from a single genotype in next year's gene pool is $2m$ multiplied by the probability of survival and by the number of individuals of that genotype in the population. This value appears in row 3 of Table 1.

To be clear, let's walk through an example. If $p = 0.5$ and there are 100 individuals in the population, then there would be $p^2N = 25$ A_1A_1, $2pqN = 50$ A_1A_2, and $q^2N = 25$ A_2A_2 zygotes in the population. If $l_{11} = 1$, $l_{12} = 1$, and $l_{22} = 0.2$, all of the A_1A_1 and A_1A_2 zygotes would reach adulthood, but only $0.2 \times 25 = 5$ A_2A_2 zygotes would reach adulthood. If $2m = 3$ for all genotypes, then the number of gametes contributed to the next generation is $3 \times 25 = 75$ gametes for A_1A_1 individuals, $3 \times 50 = 150$ gametes for A_1A_2 individuals, and $3 \times 5 = 15$ gametes for A_2A_2 adults. Thus, in total the next generation consists of 75 + 150 + 15 gametes (240 total), which translates to 120 zygotes in the next generation. Thus, given information in Table 1, you can compute directly how each genotype will impact the gene pool in the next generation.

Absolute and Relative Fitness

We can also be more general in our computations. The frequency of the A_1 allele, p, at time $t + 1$ is

$$p_{t+1} = \frac{A_1 \text{ alleles}_{t+1}}{\text{Total alleles}_{t+1}} \qquad \text{Equation 4}$$

The denominator is the total number of alleles or "gene copies" at the A locus for the offspring population. Obviously, these copies are from the parent's gametes, so you can compute the denominator of Equation 3 as the sum of the bottom row in Table 1:

$$\text{Total alleles}_{t+1} = 2m_{11}l_{11}\, p_t^2N + 2m_{12}l_{12}\, 2p_tq_tN + 2m_{22}l_{22}q_t^2N \qquad \text{Equation 5}$$

We use the subscript t with p_t and q_t to indicate that these represent the frequencies at time t. To compute the numerator of Equation 4, we need to count up the gametes contributed by the A_1A_1 individuals (all of the gametes from this genotype will be A_1), plus one-half the gametes contributed by A_1A_2 individuals (only half of the gametes from this genotype will be A_1; the other half of the gametes will be A_2). Thus, the numerator can be rewritten as

$$2m_{11}l_{11}\, p_t^2N_t + (1/2)(2m_{12}l_{12}2p_tq_tN_t) \qquad \text{Equation 6}$$

Thus, we can now rewrite Equation 4 as Equation 6 divided by Equation 5:

$$p_{t+1} = \frac{2m_{11}l_{11}p_t^2N_t + (1/2)2m_{12}l_{12}2p_tq_tN_t}{2m_{11}l_{11}p_t^2N_t + 2m_{12}l_{12}2p_tq_tN_t + 2m_{22}l_{22}q_t^2N_t} \qquad \text{Equation 7}$$

You will notice that you can factor out both a 2 and an N from both the numerator and the denominator, which cancel out and give

$$p_{t+1} = \frac{m_{11}l_{11}p_t^2 + m_{12}l_{12}p_tq_t}{m_{11}l_{11}p_t^2 + m_{12}l_{12}2p_tq_t + m_{22}l_{22}q_t^2} \qquad \text{Equation 8}$$

Thus, although $2m$ is the number of gametes contributed by an individual to the next generation's gene pool, our computations, once simplified, express each individual's contribution to the next generation's gene pool as m. To simplify things even further, we can combine both the survival probabilities and gamete contributions of a genotype into a single value, capital W:

- $W_{11} = m_{11}l_{11}$
- $W_{12} = m_{12}l_{12}$
- $W_{22} = m_{22}l_{22}$

and by substitution

$$p_{t+1} = \frac{W_{11}p_t^2 + W_{12}p_tq_t}{W_{11}p_t^2 + W_{12}2p_tq_t + W_{22}q_t^2} \qquad \text{Equation 9}$$

Equation 9 is a fundamental formula in evolutionary biology. It was derived in the 1920s by R. A. Fisher, J. B. S. Haldane, and S. Wright. The value W is the **absolute fitness** of a genotype. Knowing W provides information on a genotype's survival probability and its reproductive contribution to the next generation's gene pool (Roughgarden 1998). Accordingly, "fitness" has both survival and reproductive components. Absolute fitness is sometimes designated as λ because it is the finite rate of increase for a particular genotype. Thus, when $W > 1$, the genotype is increasing over time; when $W < 1$, the genotype decreases over time; and when $W = 1$, the genotype remains stable over time. In a broad sense, absolute fitness can be formally defined as *the average per capita lifetime contribution of individuals of that genotype to the population after one or more generations* (Futuyma 1998).

By convention, W is "scaled" such that the genotype with the largest W has the value 1; this scaled value is its **relative fitness**. Relative fitness is designated by a lowercase w, and is computed by

$$w_{ij} = W_{ij}/W_{max} \qquad \text{Equation 10}$$

For instance, assume that the following W's depict the absolute fitnesses of genotypes in the population:

- $W_{11} = 2$
- $W_{12} = 1$
- $W_{22} = 0.4$

The A_1A_1 genotype has the largest absolute fitness, and so we establish the **relative fitness** of this genotype as the standard genotype (the denominator of Equation 10) with which other genotype fitnesses will be compared:

- $w_{11} = W_{11}/W_{11} = 1$
- $w_{12} = W_{12}/W_{11} = 1/2 = 0.5$
- $w_{22} = W_{22}/W_{11} = 0.4 / 2 = 0.2$

The relative fitness values can be interpreted as the growth rate of a genotype relative to the fastest growing genotype. Thus, the A_1A_2 genotype grows at one-half the rate as the A_1A_1 genotype, and the A_2A_2 genotype is growing at 1/5 the rate of the A_1A_1 genotype.

The expression $1 - w$ is called the **selection coefficient** and indicates the degree to which natural selection selects "against" a genotype. Evolutionary modelers often use the relative fitness calculation and selection coefficients rather than the absolute fitnesses, because then the exact numbers of individuals of each genotype in the population do not need to be known. However, in this exercise you will track the fates of 100 individuals over time and will therefore be able to compute absolute fitnesses without difficulty.

The use of absolute fitness over relative fitness has another advantage: Because the number of gametes that each genotype contributes to the next generation is known, the population size of the next generation can also be determined. Refer again to Equation 4:

$$p_{t+1} = \frac{A_1 \text{ alleles}_{t+1}}{\text{Total alleles}_{t+1}}$$

The denominator gives the total number of alleles that will be incorporated into the next generation's gene pool. Since we are talking about a diploid organism, the total number of individuals in the next generation is simply the total number of alleles in $t + 1$, multiplied by 0.5.

$$N_{t+1} = 0.5 \times \text{total alleles at } t + 1 \qquad \text{Equation 11}$$

Remember that the total number of alleles at $t + 1$ is the sum of the bottom row in the table, given in Equation 5:

$$\text{Total alleles}_{t+1} = 2m_{11}l_{11}\, p_t^2 N + 2m_{12}l_{12}\, 2p_t q_t N + 2m_{22}l_{22} q_t^2 N$$

Multiply Equation 5 by 0.5, then replace the m_{ij}'s and l_{ij}'s with W_{ij}'s, and we are left with the formula

$$N_{t+1} = (W_{11}p_t^2 + W_{12}2p_t q_t + W_{22}q_t^2) \times N_t \qquad \text{Equation 12}$$

Hopefully, Equation 12 has a form that is familiar to you.

In Exercise 7, "Geometric and Exponential Population Models," we developed a model with the form

$$N_{t+1} = \lambda \times N_t \qquad \text{Equation 13}$$

Thus, the term $W_{11}p_t^2 + W_{12}2p_t q_t + W_{22}q_t^2$ in Equation 12 is the same thing as λ in Equation 13, the finite rate of increase for the population. This should not be too surprising, since fitness is the growth rate of the various genotypes over time. It is computed by summing the *W's* for each genoytpe, weighting each *W* by the frequency of each genotype (given by Hardy-Weinberg) in the population.

PROCEDURES

In this exercise, you'll set up a spreadsheet model of a population of 100 individuals and subject the population to various selective forces. Your population will consist of individuals that reproduce sexually during a discrete time period and then die (think of an annual plant whose seeds are viable only until the following year). The ultimate goal of the model is to predict the allele frequencies p and q at time $t + 1$ given their initial state at time t, and to predict the new population size as well. As always, save your work frequently to disk.

INSTRUCTIONS

ANNOTATION

A. Set up the model parameters.

We'll consider a population of 100 zygotes of varying genotypes and track their fates to adulthood.

1. Open a new spreadsheet and set up column headings as shown in Figure 3.

	A	B	C	D	E	F
1	*Natural Selection and Fitness*					
2			Tally			
3	Genotypes	# of individuals (zygotes)	0			
4	A1A1	25				
5	A1A2	50				
6	A2A2	25		<== this number MUST total 100		

Figure 3

2. Enter numbers in cells B4–B6 as shown.

To begin, we will have the population consist of 25 A_1A_1 homozygotes, 50 A_1A_2 heterozygotes, and 25 A_2A_2 homozygotes.

3. Enter 0 in cell C3.

Cells C3–C6 will keep track of the total number of individuals by "tallying" the numbers in cells B4–B6. This tally will be used to assign genotypes to individuals in a few steps.

4. Enter **=SUM($B$4:B4)** in cell C4. Copy this formula down to cell C6.

The spreadsheet should return the number 25 in cell C4. Your result in cell C6 should be 100, indicating that the population consists of 100 individuals. Later in the model, you will be free to change the genotype composition of the 100 individuals in cells B4–B6, but you'll want to make sure that cell C6 totals 100.

5. Set up new headings as shown in Figure 4.

	A	B	C	D	E
8			**A1A1**	**A1A2**	**A2A2**
9	**Initial genotype frequencies =**		0.25	0.5	0.25
10	**Probability of genotype survival = *l* =**		1	1	0.2
11	**Half the # of gametes in next gen. = *m* =**		2	2	2
12	**Absolute fitness = *W* = *l* * *m* =**		2	2	0.4

Figure 4

6. Calculate the initial genotype frequencies in cells C9–E9.

In cell C9, enter the formula **=B4/C6.**
In cell D9, enter the formula **=B5/C6.**
In cell E9, enter the formula **=B6/C6.**
The frequency of the various genotypes is simply the number of individuals of a given genotype divided by the total number of individuals in the population.

7. Enter values in cells C10–E10 as shown in Figure 4.

Cells C10–E10 give the viability (survival) fitness component, or the probability of surviving to reproduction. (Make up a hypothetical situation in which the A_2A_2 genotypes are selected against; perhaps their phenotype is more susceptible to being eaten by an introduced herbivore.) A survival probability for A_2A_2 genotypes of 0.2 means that each individual has a 20% probability of surviving to reproductive maturity.

8. Enter values in cells C11–E11 as shown in Figure 4.

For now, let's assume that each genotype that reaches sexual maturity will contribute an equal number of gametes to the next generation (that is, fitness is not affected by reproductive potential). Let m be one-half the number of gametes that a sexually reproducing individual will contribute to the next generation. Since m is the same for all genotypes, each individual (regardless of its genotype) will contribute roughly the same number of gametes per adult to the next generation as any other individual (given that individuals reach adulthood). Note that we don't care *how* these gametes recombine in the next population, only that they *are present* and available for counting when we calculate the p's and q's in the next generation.

9. Compute the absolute fitness, W, in cells C12–E12.

In cell C12, enter the formula **=C10*C11.**
In cell D12, enter the formula **=D10*D11.**
In cell E12, enter the formula **=E10*E11.**
Recall that the absolute fitness, w, is equal to $l \times m$.

10. Save your work.

B. Simulate the survival and reproduction of the 100 individuals in the population.

1. Set up column headings as shown in Figure 5.

	A	B	C	D	E	F	G
14			**Fitness components**		**Surviving**	**Next generation gametes**	
15	**Individual**	**Zygote genotype**	**Survival**	**Reproduction**	**genotypes**	**A1 gametes**	**A2 gametes**

Figure 5

2. Set up a linear series from 0 to 99 in cells A16–A115.

Enter the value 0 in cell A16.
Enter = **1+A16** in cell A17. Copy this formula down to cell A115.

3. Enter a **LOOKUP** formula to assign a genotype to each of the 100 individuals in the population. Copy the formula down to cell B115.

In cell B16, enter the formula **=LOOKUP(A16,$C$3:$C$6,$A$4:$A$6)**.
The **LOOKUP** function will allow us to assign genotypes according to the numbers you entered in cells B4–B6. The vector form of the **LOOKUP** function looks in a one-row or one-column range (known as a vector) for a value and returns a value from the same position in another one-row or one-column range. For instance, the formula in cell B16 tells the spreadsheet to look up the individual's number given in cell *A*16 in the vector C3–C6 (the genotype tallies), and return the appropriate genotype in cells A4–A6. If **LOOKUP** can't find an exact match (for instance, individual 37 cannot be found because the number 37 is not part of the "tally"), LOOKUP returns the genotype that is associated with a number in the tally less than 37. Thus, individuals 0–24 are assigned the genotype listed in cell A4, individuals 25–74 are assigned the genotype listed in cell A5, and individuals 75–99 are assigned the genotype listed in cell A6. The result is that the genotypes are assigned exactly the way you specified in cells B4–B6.

4. In cell C16, enter an **HLOOKUP** formula to calculate the survival probability of each zygote in the population and list the survival probability of its genotype in column C. Copy the formula down to cell C115.

In cell C16, enter the formula **=HLOOKUP(B16,$C$8:$E$12,3,FALSE)**.
We need the spreadsheet to examine individual 0's genotype in cell B16, look up its survival probability in the table in cells C8–E12, and return that probability to cell C16. The **HLOOKUP** function can be used for this purpose. The **HLOOKUP** formula searches for a value in the top row of a table, and then returns a value in the same column from a row you specify in the table. The **HLOOKUP** formula has the form **HLOOKUP(lookup_value,table_array,row_index_num,range_lookup)**, where **lookup_value** is the value to be found in the first row of the table (in our case, we want to look up the individual's genotype in cell B16); **table_array** is a table of information in which data is looked up (in our case, we want to look up information in the table consisting of cells C8–E12); **row_index_num** is the row number in **table_array** from which the matching value will be returned (in our case, we want to return the value associated with survival probabilities, which is the third row in the table). The word **FALSE** tells the program that you require an exact match in the table.

Select cell C15, select the **HLOOKUP** function, and follow the prompts to create your formula. Copy your formula down to record survivorship probabilities for the remaining 99 individuals in the population.

5. In cell D16, use the **HLOOKUP** function to return the gamete contributions for each individual in the population. Copy your formula down to cell D115.

We used the formula **=HLOOKUP(B16,$C$8:$E$11,4,FALSE)**.
Your spreadsheet should now look something like Figure 6.

	A	B	C	D	E
8			**A1A1**	**A1A2**	**A2A2**
9	**Initial genotype frequencies =**		0.25	0.5	0.25
10	**Probability of genotype survival = *l* =**		1	1	0.2
11	**Half the # of gametes in next gen. = *m* =**		2	2	2
12	**Absolute fitness = *W* = *l* * *m* =**		2	2	0.4
13					
14			**Fitness components**		**Surviving**
15	**Individual**	**Zygote genotype**	**Survival**	**Reproduction**	**genotypes**
16	0	A1A1	1	2	
17	1	A1A1	1	2	
18	2	A1A1	1	2	

Figure 6

6. In cell E16, enter a formula to determine which zygotes survive to adulthood. Copy your formula down to cell E115.

In cell E16 enter the formula **=IF(RAND()<C16,B16,".")**.
Remember, the survival probabilities indicate the probability that an individual will survive to reproductive maturity (adulthood). In cell E16, we need a formula that will randomly determine whether individual 0 will survive to adulthood or not, based on the survival probability given in cell C16. The formula in cell E16 uses an **IF** formula to accomplish this task. The formula draws a random number between 0 and 1 (the **RAND()** portion of the formula). If the random number is less than the survival probability given in cell C16, the spreadsheet returns the value in cell B16 (the genotype of the zygote, or shall we now say the genotype of the adult). If the random number is greater than the survival probability, however, the individual died and a period (which designates a missing value) is returned instead. So far, we know which individuals survived to reproduce.

7. In cell F16, enter a formula to count how many gametes the surviving individuals actually contribute to the next generation.

In cell F16, enter the formula **=IF(E16="A1A1",D16*2,IF(E16="A1A2",D16,"."))**.
We will keep track of the A_1 gametes in column F and A_2 gametes in column G. The formula in cell F16 is *two* nested IF functions. The first part of the formula, **IF(E16="A1A1",D16*2**, tells the spreadsheet to examine cell E16, and if cell E16 is an A_1A_1 genotype, to multiply cell D16 by 2 (remember that cell D16 is one-half the gametes contributed, so when this number is multiplied by 2 it is the total number of gametes that an individual of genotype A_1A_1 contributes to the next generation).
However, if cell E16 is *not* genotype A_1A_1, the spreadsheet walks through the second **IF** statement, **IF(E16="A1A2",D16,".")**. This states that if the genotype is A_1A_2, then return the value in cell D16; otherwise return a missing value. Remember that the gametes produced by A_1A_2 genotypes include both A_1 gametes and A_2 gametes in approximately equal numbers, so that half of an individual's gametes are A_1 and half are A_2. Therefore, to count the A_1 gametes from heterozygotes, only half the gamete contribution can be tallied in column F, which is simply m.

8. In cell G16, enter a formula to compute the number of A_2 gametes contributed to the next generation by each surviving individual.

We entered the formula **=IF(E16="A2A2",D16*2,IF(E16="A1A2",D16,"."))**.

9. Copy the formulae in cells F16–G16 down to cells F115–G115.

Your spreadsheet should look something like Figure 7.

10. Save your work.

	F	G
14	**Next generation gametes**	
15	**A1 gametes**	**A2 gametes**
16	4	.
17	4	.
18	4	.
19	4	.
20	4	.
21	4	.
22	4	.
23	4	.

Figure 7

C. Calculate selection statistics.

1. Set up new column headings as shown in Figure 8.

	H	I	J	K	L	M	N
8			A1A1	A1A2	A2A2		
9	Number of zygotes =>						
10	Number of adults =>						
11	Absolute fitness = *W* =>					Largest *W* =	
12	Relative fitness = *w* =>					Average *w* =	
13	Selection coefficient = *S* =>						

Figure 8

2. Enter formulae to count the initial number of zygotes in the population in cells J9–L9 and the number of adults in cells J10–L10.

We entered the following formulae, although you may have come up with other methods for counting individuals:

- J9 **=B4**
- K9 **=B5**
- L9 **=B6**
- J10 **=COUNTIF($E$16:$E$115,J8)**
- K10 **=COUNTIF($E$16:$E$115,K8)**
- L10 **=COUNTIF($E$16:$E$115,L8)**

3. Enter formulae to recompute the absolute fitnesses of each genotype in cells J11–L11.

In cell J11, enter the formula **=C12**.
In cell K11, enter the formula **=D12**.
In cell L11, enter the formula **=E12**.

4. Use the **LARGE** formula in cell N11 to determine the largest absolute fitness of the three genotypes.

The **LARGE** function returns the largest (or second largest, or third largest, etc.) value in a data set. We entered **=LARGE(J11:L11,1)**, where cells J11 and L11 give the data set, and the number 1 at the end of the formula indicates that we want the largest value returned (as opposed to the second largest or third largest value).

5. In cells J12–L12, enter formula to compute the relative fitness, symbolized with a lowercase w, for each genotype.

In cell J12, enter the formula **=J11/$N$11**.
In cell K12, enter the formula **=K11/$N$11**.
In cell L12, enter the formula **=L11/$N$11**.
The *relative fitness* of each genotype is the fitness of each genotype relative to the fittest genotype in the population. In cell N11, you've calculated the largest of the W values in the population. This represents the "fittest" genotype in the population, and all other genotypes will be assigned fitness values relative to this genotype. Relative fitness, w, can be obtained for each genotype by dividing the genotype's absolute fitness (W) by the largest absolute fitness (W_{max}).

6. Calculate the selection coefficient, S, as $1 - w$ for each of the genotypes in cells J13–L13.

In cell J13, enter the formula **=1-J12**.
In cell K13, enter the formula **=1-K12**.
In cell L13, enter the formula **=1-L12**.
Another useful characterization of the strength of natural selection against a genotype is the selection coefficient, S. S is simply $1 - w$, and indicates the relative *decrease* of a genotype due to selection. A high S indicates that a genotype was selected against, while a low S indicates that it was not selected against.

7. Save your work.

D. Make graphs of the selection statistics.

1. Graph the numbers of zygotes and breeding adults for each genotype (cells I8–L10).

Use a column graph and label your axes fully. Your graph should resemble Figure 9 (although the number of **A2A2** adults may differ from our graph.

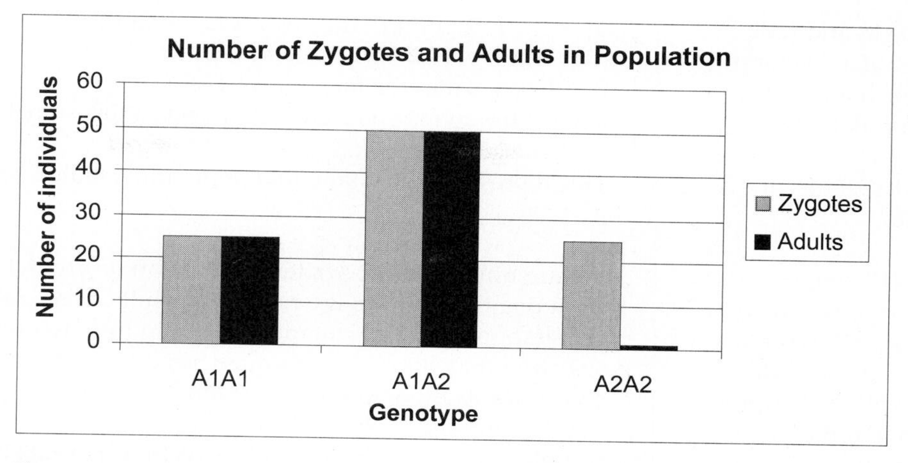

Figure 9

2. Graph W (absolute fitness), w (relative fitness), and S (the selection coefficient) for each genotype (cells I11–L13). Select the Series tab as you make your chart, and select cells J8–L8 as the Category (x) axis labels.

Use a column graph and label your axes fully. Your graph should resemble Figure 10.

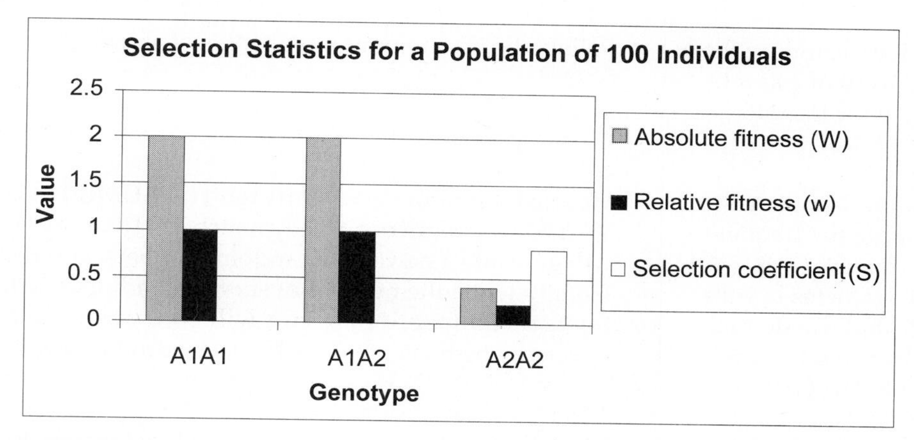

Figure 10

3. Answer Question 1 at the end of this excercise before proceeding.

E. Project allele frequencies and population numbers to next generation.

1. Set up new column headings as shown in Figure 11.

	I	J	K	L
16		**A1 allele**	**A2 allele**	
17	**Time Step**	***p***	***q***	***N***
18	**1**			
19	**2**			
20	**3**			
21	**4**			
22	**5**			
23	**6**			
24	**7**			
25	**8**			
26	**9**			
27	**10**			

Figure 11

2. In cells J18 and K18, enter formulae to compute p and q for the initial zygote population.

In cell J18, enter the formula **=(COUNTIF(B16:B115,"*A1A1*")*2+COUNTIF(B16:B115,"A1A2"))/(2*C6)**.
In cell K18, enter the formula **=1-J18**.
Refer to the exercise on Hardy-Weinberg equilibrium if you are rusty on the computations.

3. In cell L18, enter the formula **=C6**.

This represents the total initial population, tallied in cell C6.

4. In cell J19, enter a formula to compute the new frequency of the A_1 allele, p. Copy your formula down to cell J27. Refer to Equation 9 in the Introduction.

We are now ready to write an equation to predict the change in allele frequencies from one time step to the next as a result of natural selection. Remember that selection happens *within* generations, but in this step we will now consider how natural selection may alter allele frequencies *between* generations. That is, how populations *evolve* as a result of natural selection.

In cell J19, enter the formula **=(($J$11*J18^2)+($K$11*J18*K18))/(($J$11*J18^2)+($K$11*2*J18*K18)+($L$11*K18^2))**.
This corresponds to Equation 9,

$$p_{t+1} = \frac{W_{11}p_t^2 + W_{12}p_t q_t}{W_{11}p_t^2 + W_{12}2p_t q_t + W_{22}q_t^2}$$

5. In cell K19, compute the new frequency of q as **=1-J19**. Copy your formula down to cell K27.

Follow the equation outlined in Step 4 above.

6. For comparison, in cell H19 compute the frequency of the A_1 allele by counting the A_1 gametes in cells F16–F115, and divide that number by the total gametes (in cells F16–G115).

We used the formula **=SUM(F16:F115)/SUM(F16:G115)**.
Your results might not exactly match cell J19. Why? Press F9, the calculate key, and you will generate a new set of random numbers, and hence a new set of adults in column E. Only when the number of surviving adults exactly equals the survival probability times the number of zygotes will your answer match cell J19. This is because you selected which zygotes would reach adulthood with a random number function.

7. In cell L19, enter a formula to compute N_{t+1}. Copy your formula down to cell L27. Refer to Equation 12 in the Introduction.

We will now use the p's, q's, and absolute fitnesses to calculate the new population size in cell L19.
We used the formula **=($J$11*J18^2+$K$11*2*J18*K18+$L$11*K18^2)*L18**. This corresponds to Equation 12:

$$N_{t+1} = (W_{11}p_t^2 + W_{12}2p_t q_t + W_{22}q_t^2) \times N_t$$

Your spreadsheet should now look something like Figure 12.

	I	J	K	L
16		A1 allele	A2 allele	
17	Time Step	*p*	*q*	*N*
18	1	0.5	0.5	100
19	2	0.625	0.375	160
20	3	0.7042254	0.2957746	284
21	4	0.7572203	0.2427797	528.247887
22	5	0.7946929	0.2053071	1006.67819
23	6	0.8224257	0.1775743	1945.46438
24	7	0.8437092	0.1562908	3792.77583
25	8	0.8605251	0.1394749	7437.31905
26	9	0.8741288	0.1258712	14643.1502
27	10	0.8853505	0.1146495	28915.1013

Figure 12

8. For comparison, in cell M19 compute N_{t+1} as the sum of the gametes in cells F16–G115 divided by 2.

We used the formula **=SUM(F16:G115)/2** to compute the number of individuals (zygotes) in time step 2. Your results may not exactly match cell L19 because of the random number function used to determine which genotypes survived.

9. Graph p, q, and N as a function of time.

Use the XY scattergraph and label your axes fully. Your graph should resemble Figure 13. To create a secondary axis on the graph so that the frequencies are shown on the right axis and the number of individuals is on the left axis, double-click on the data in the graph that depicts p or q. A dialog box will appear. Click on the Axis tab, then select Secondary axis. Repeat for the other allele. To label the new axis, select the chart, then go to Chart | Chart Options | Titles and type in the labels for the primary y-axis (Number of individuals) and secondary y-axis (Frequency).

10. Save your work.

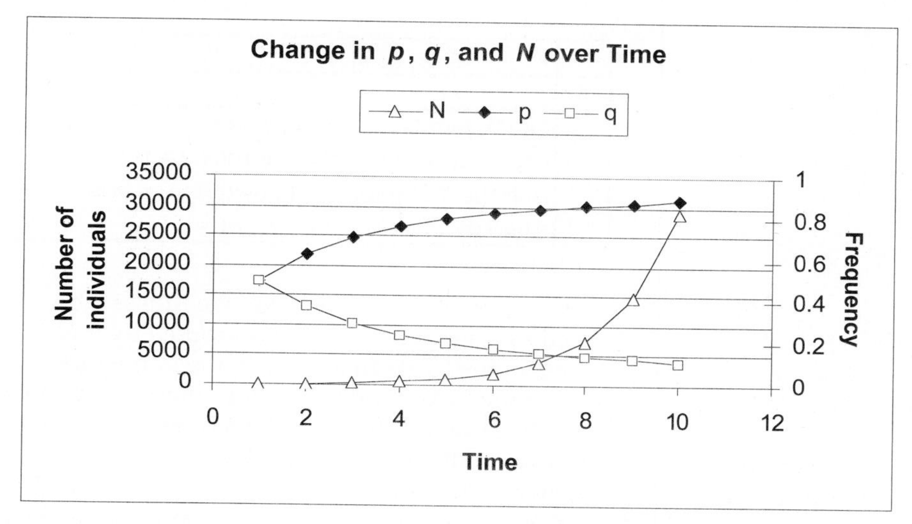

Figure 13

QUESTIONS

1. From your graphs in Section D of the exercise, describe the population in terms of natural selection *within* a generation (Figure 9). Describe the population in terms of W, w, and S (Figure 10).
2. In your model, you've selected against the A_2 homozygote. Yet the A_2 allele persists in the population, even after 10 years of constant selection. Extend your model to 100 years. At what frequency does the A_2 allele appear to stabilize? Why does the A_2 allele persist?
3. Modify your absolute fitness parameters by increasing the gamete contribution of the A_2A_2 genotype to 10 in cell E11. Examine your graph of relative fitness and selection coefficients. How did this change affect your population? What will happen to the frequency of the A_1 and A_2 allele over time?
4. Which affects the genetic rate of change in the population (change in A_1 allele from time step 1 to time step 2), relative fitness or absolute fitness? (Keep in mind that our calculations are based on absolute fitnesses.) Modify cells C10–E11 to answer this question. Enter the values shown below into your model:

   Note that the absolute fitnesses have been changed, but the relative fitnesses (given in cells J12 and L12) remain the same. How does changing the absolute

	A	B	C	D	E
8			A1A1	A1A2	A2A2
9	Initial genotype frequencies =		0.25	0.5	0.25
10	Probability of genotype survival = *l* =		1	1	0.4
11	Half the # of gametes in next gen. = *m* =		4	4	2
12	Absolute fitness = *W* = *l* * *m* =		4	4	0.8

fitness in the manner described affect p and q in the next generation? Modify your model so that the relative fitnesses are altered. How does changing the relative fitness affect p and q in the next generation?

5. Set up new entries as follows:

	A	B	C	D	E
8			A1A1	A1A2	A2A2
9	Initial genotype frequencies =		0.25	0.5	0.25
10	Probability of genotype survival = *l* =		0.6	0.2	0.4
11	Half the # of gametes in next gen. = *m* =		4	4	2
12	Absolute fitness = *W* = *l* * *m* =		2.4	0.8	0.8

Compute the weighted average fitness (absolute fitness) in cell F12. The weighted average fitness can be computed by multiplying the absolute fitness of each genotype by its frequency in the population, and then summing these values together. Now compute λ for your population as N_{t+1}/N_t (given in cells L19 and L18). How do these numbers compare? Change some of your parameters in your model to see if your relationship holds no matter what parameters you change in your model. Why calculate a weighted average, rather than simply the average to predict population growth? Why is absolute fitness (weighted) used as indication of population growth rather than relative fitness?

6. Modify your absolute fitness parameters by selecting against the heterozygotes (absolute fitness = 0). Enter survival and reproductives values for the A_1A_1 and A_2A_2 homozygotes such that their absolute fitnesses are > 0 but equal in value. Change the genotype make-up of the initial population in the following manner:

How do p and q change over time? Next, change your values in cells B4–B6 as shown:

	A	B
3	Genotypes	# of individuals (zygotes)
4	A1A1	30
5	A1A2	50
6	A2A2	20

Update and graph your results. What happens to allele frequencies over time when $A_1 > 0.5$? When $A_1 < 0.5$? Explain your results.

	A	B
3	Genotypes	# of individuals (zygotes)
4	A1A1	20
5	A1A2	50
6	A2A2	30

7. Modify your absolute fitness parameters by selecting for the heterozygotes. Enter survival and reproductive values for the A_1A_1 and A_2A_2 homozygotes that result in an absolute fitness of 0, and values for the heterozygote > 0 as shown:

   How does selection for the heterozygote affect p and q over time?

	A	B	C	D	E
8			A1A1	A1A2	A2A2
9	**Initial genotype frequencies =**		0.25	0.5	0.25
10	**Probability of genotype survival = *l* =**		0	1	0
11	**Half the # of gametes in next gen. = *m* =**		4	3	4
12	**Absolute fitness = *W* = *l* * *m* =**		0	3	0

8. *(Advanced). Modify your model to include frequency dependent selection (the selection of a genotype depends on the frequency of the genotype in the population).

9. *(Advanced). Although you've entered survival and reproductive values for each genotype, these values remain fixed in your model. In reality, survival and reproductive rates are stochastic in nature. Modify your model to incorporate this element of stochasticity.

LITERATURE CITED

Futuyma, D. 1998. *Evolutionary Biology*, 3rd Edition. Sinauer Associates, Sunderland, MA.

Roughgarden, J. 1998. *Primer of Ecological Theory*. Prentice-Hall, Upper Saddle River, NJ.

Wilson, E. O., and W. H. Bossert. 1971. *A Primer of Population Biology*. Sinauer Associates, Inc. Sunderland, MA.

33 ADAPTATION: PERSISTENCE IN A CHANGING ENVIRONMENT

In collaboration with Mary Puterbaugh

Objectives

- Consider how recombination and natural selection can lead to new phenotypes.
- Develop a spreadsheet model of allele and genotype frequencies at three loci.
- Examine how the abruptness of an environmental change affects the ability of a population to adapt to that change.
- Consider how genetic factors (recombination, genetic diversity, and number of genes) influence the likelihood of extinction in a finite population experiencing selective pressure.

Suggested Preliminary Exercise: Hardy-Weinberg Equilibrium

INTRODUCTION

We hear a lot these days about global warming. Global climate change is not a new phenomenon—over its history, the earth has been warmer than it is today, and also much, much colder. But one of the concerns biologists have about the current warming trend is that, because it is occurring so rapidly, many populations will not be able to respond to the changes.

For many organisms even a small increase in environmental temperature can spell the difference between life and death. Estuarine marine organisms, for example, may have to adapt quickly to rising sea levels in order to persist over time. Species that cannot adapt quickly will go extinct, while species that are able to adapt will persist. What factors govern whether a population persists through a period of environmental change? Population size is obviously one answer. But we also should consider whether enough *heritable genetic variation* is present to allow the population to respond to selective pressures. Such variation arises either through mutation or recombination. This exercise will illustrate the process of recombination, an important force for evolution as we understand it.

Recombination is the process by which a sperm or an egg randomly receives one allele from a pair of alleles possessed by each parent. Suppose your mother has the genotype $A_1A_1B_1B_1C_1C_1$ for the A, B, and C loci, and your father has the genotype $A_2A_2B_2B_2C_2C_2$. You must have the genotype $A_1A_2B_1B_2C_1C_2$, because each of your parents produced only one type of allele at those loci, and you inherited one allele from each parent for each locus. In your case, however your **gametes** (eggs or sperm) randomly receive either an A_1 or A_2 allele, a B_1 or B_2 allele, and a

C_1 or C_2 allele during meiosis. Your gametes thus have the potential to carry any one of the following *nine* genotypes: $A_1B_1C_1$, $A_1B_2C_1$, $A_1B_2C_2$, $A_2B_1C_2$, $A_2B_1C_1$, $A_1B_1C_2$, $A_2B_2C_1$, or $A_2B_2C_2$. Your mother could produce only $A_1B_1C_1$ eggs, but the alleles you inherited from your father *recombined* with hers to create genotypes (yours!) that weren't present in the previous generation.

Recombination has a strong influence on the genotypes of offspring, especially for traits that are controlled by multiple genes. For example, beak size in birds is a heritable trait, and many different genes probably act together to determine beak size for an individual bird. When many genes affect the expression of a single trait, it is called a **polygenic trait**. Many traits are polygenic. In the simplest case, each locus makes a contribution to the expressed trait. For example, three different loci (*A*, *B*, and *C*) might contribute to beak size. If an individual inherits an A_1, B_1, or C_1 allele from its parents, it "inherits" a 1-mm contribution to beak size. If it inherits an A_2, B_2, or C_2 allele from its parents, it "inherits" a 2-mm contribution to beak size. Thus, $A_1A_1B_1B_1C_1C_1$ individuals have the smallest beaks (6 mm), while $A_2A_2B_2B_2C_2C_2$ individuals have the largest beaks (12 mm). Individuals that are heterozygous at either gene have intermediate-sized beaks (e.g., $A_1A_2B_1B_2C_1C_2$ genotypes have 9-mm beaks). The loci, then, act **additively** to determine the phenotype. Because several loci contribute to beak size, the population will tend to exhibit **continuous variation** in beak size, with beaks ranging from 6 mm to 12 mm.

The environment may play a large role in determining which genotype combinations are "best suited" in terms of survival and reproduction. For example, large beak size in one of Darwin's finches (*Geospiza fortis*) may be favored in drought years, but small beak sizes may be favored in wet years (Grant and Grant 1993). In other words, certain genotype combinations are favored under drought conditions, while other combinations are favored under wet conditions. Imagine for a moment that the frequencies of the alleles A_2, B_2, and C_2 (the alleles that produce larger beaks) are initially low in a given population. This means that A_2A_2 individuals will be rare, as will B_2B_2 and C_2C_2 individuals. The probability that random mating and recombination will produce an individual with the genotype $A_2A_2B_2B_2C_2C_2$ may be so small that this genotype may never occur in the population. If natural selection favors larger beaks, however, the frequencies of the A_2, B_2, and C_2 alleles in the population will increase, and recombination may occasionally produce individuals with the $A_2A_2B_2B_2C_2C_2$ genotype.

Experiments with corn and fruit flies have demonstrated dramatic changes in phenotype that are probably the result of selection and recombination. In a famous experiment, Clayton and Robertson (1957) started out with a population of fruit flies and counted the bristles on the abdomen of each fly. They found that the number of bristles varied from 30 to 50. Over many generations, Clayton and Robertson consistently took the flies with the highest number of bristles and mated them. After 35 generations, all of the flies had between 60 and 110 bristles—phenotypes that didn't even occur in the original population!

Perhaps some novel mutation arose that increased bristle number, but it is more likely that changes in the frequencies of existing alleles led to the changes in bristle number: If bristle number is polygenic—controlled by several different genes—and if alleles that produce higher bristle numbers are rare, then the probability may be very small that recombination will produce an individual with more than 50 bristles. But by selecting *against* individuals with low bristle numbers (or *for* individuals with high bristle numbers), Clayton and Robertson increased the frequencies of the alleles that produce high bristle numbers, and thus increased the probability that recombination would result in individuals with more than 50 bristles. After 35 generations, the frequencies of alleles that result in high bristle numbers were high enough that recombination occasionally produced individuals with 110 bristles. On the other hand, the frequencies of alleles that produce low bristle numbers decreased, making it very unlikely that recombination could produce an individual with fewer than 60 bristles.

PROCEDURES

From an evolutionary perspective, key questions include "How much genetic variation is needed for a population to persist through a period of rapid environmental change?" and "How does variation in environmental conditions affect the ability of a population to respond?" In this exercise, you'll set up a spreadsheet model to answer these questions. We will consider a single trait (beak size, which determines drought resistance) and the allele frequencies at different loci that influence beak size. To begin, the initial allele frequencies will be determined by you, the modeler. The population at the beginning of the first year will consist of 500 adults with beak sizes determined by the allele frequencies you input. This population will then experience the environmental conditions for that year, again determined by the modeler. Certain individuals will survive to reproduce, while others will not. Those that survive will go on to reproduce at the end of the year. Since beak size is a heritable trait, the "new" population will have beak sizes that reflect the genetic composition of the survivors. You will follow the population for 5 years, during which you can alter environmental conditions (dry, mild, and wet) and alter the phenotypes that survive. If in any given year no individuals survive, the population has gone extinct. If you are pinched for time, you may model just 3 generations.

The goal of the exercise is to explore how much genetic variation is needed for a population to adapt to environmental change, and to explore how variation in environmental conditions affects the genetic diversity of populations. As always, save your work frequently to disk.

INSTRUCTIONS / ANNOTATION

A. Set up the model population.

1. Open a new spreadsheet and set up column headings as shown in Figure 1.

	A	B	C	D	E	F	G	H
1	*Adaptation*							
2								
3	Environmental condition:		Dry	Mild	Wet			
4	Select phenotypes above:		11	8	0			
5								
6								
7			Allele frequencies of surviving parents					
8	Phenotype contribution =>		1	2	1	2	1	2
9	Year	Condition	A1	A2	B1	B2	C1	C2
10	Initial		0.8	0.2	0.8	0.2	0.8	0.2
11	1	Wet						
12	2	Mild						
13	3	Mild						
14	4	Dry						
15	5	Mild						

Figure 1

2. In cells C4–E4, enter 11, 8, and 0 respectively as shown.

We'll consider three types of environmental conditions: dry, mild, and wet. Each condition favors different beak-length phenotypes. If a year is wet, individuals with a beak size of greater than 0 will survive. If there is a severe drought (dry conditions), only individuals with beak sizes greater than 11 will survive.

3. Enter the environmental conditions shown in cells B11–B15.

For now, year 1 will be a wet year, years 2, 3, and 5 will be mild, and year 4 will be a dry year. You will be able to manipulate these environmental conditions later in the exercise.

4. Enter the initial allele frequencies of the population shown in cells C10–H10.

We'll track the allele frequencies of three loci (*A*, *B*, and *C*) over a 5-year period. At each locus, there are just two alleles; their frequencies are *p* and *q*. In year 1, we'll start with allele frequencies of roughly $A_1 = B_1 = C_1 = 0.8$. Because only two alleles are present at each locus, the frequencies of the A_2, B_2, and C_2 alleles must be $(1 - p)$, which is 0.2.

5. Enter the phenotypic contributions of each allele as shown in cells C8–H8.

These contributions ultimately determine what an individual's phenotype will be. For example, the number 1 entered in cell C8 designates that individuals with the A_1 allele inherit a 1-mm contribution to beak size. The number 2 in cell D8 specifies that individuals with the A_2 allele inherit a 2-mm contribution to beak size. With the phenotypic contributions given, the genotype $A_1A_2B_1B_2C_2C_2$ has a phenotype of 1 + 2 + 1 + 2 + 2 + 2, or 10 mm.

6. Save your work.

B. Track the population through year 1.

1. Set up the new column headings shown in Figure 2, but extend and repeat your column headings to 5 years.

Repeat the column headings **Genotype**, **Phenotype**, **Survive?** and **Phenotype** for years 2 – 5 in columns F through U.

	A	B	C	D	E
18				Year 1	
19	**Individual**	**Genotype**	**Phenotype**	**Survive?**	**Phenotype**

Figure 2

2. In cells A20–A519, establish a population of 500 individuals.

Enter the number 1 in cell A20.
In cell A21, enter **=A20+1**. Copy this formula down to cell A519.

3. In cell B20, enter a formula to generate a genotype for individual 1, and copy the formula down to obtain genotypes for the remaining individuals in the population.

Now we will assign genotypes to individuals at the beginning of year 1. These genotypes depend on the allele frequencies of breeders from the previous year, listed as "initial" frequencies. Only some of these genotypes will actually survive to breed at the end of the year. You might review the formulas used in the Hardy-Weinberg exercise.

Enter the formula **=IF(RAND()<$C$10,"A1","A2")&IF(RAND()<$C$10,"A1","A2")&**
IF(RAND()<$E$10,"B1","B2")&IF(RAND()<$E$10,"B1","B2")&
IF(RAND()<$G$10,"C1","C2")&IF(RAND()<$G$10,"C1","C2") in cell B20.
Copy this formula down the column.

Each individual will have two alleles at each of the three loci (*A*, *B*, and *C*); the three loci are joined with the & symbol. (In the above rendition, the formula for each allele is on a separate line; your formula will be entered as a unit, with no spaces around the ampersands). Let's go over the formula for the *A* locus: Have the spreadsheet generate a random number. If this number is less than the allele frequency for the A_1 allele given in cell C10, return an A_1 allele; if the random number is greater than the allele frequency of the A_1 allele given in cell C10, return an A_2. Use the analogous procedure to generate the second allele at the *A* locus, and then to obtain the *B* and *C* alleles.

4. In cell C20, enter a formula to generate phenotypes for each individual. Copy your formula down

Enter the formula **=LOOKUP(MID(B20,1,2),$C$9:$H$9,$C$8:$H$8)+**
LOOKUP(MID(B20,3,2),$C$9:$H$9,$C$8:$H$8)+
LOOKUP(MID(B20,5,2),$C$9:$H$9,$C$8:$H$8)+
LOOKUP(MID(B20,7,2),$C$9:$H$9,$C$8:$H$8)+

to obtain phenotypes for the remaining individuals in the population.

LOOKUP(MID(B20,9,2),$C$9:$H$9,$C$8:$H$8)+
LOOKUP(MID(B20,11,2),$C$9:$H$9,$C$8:$H$8) in cell C20 (there should be no spaces when you enter the formula). Copy this formula down the column.

We used two functions to generate phenotypes: the **LOOKUP** and **MID** functions. The **MID** function returns a specific number of characters from a text string, starting at the position you specify. It has the syntax **MID(text,start_num,num_chars)**, where **text** is the text string containing the characters you want to extract, **start_num** is the position of the first character you want to extract in text, and **num_chars** is the number of characters you want to extract. The first character in text has **start_num 1**, and so on.

For example, **=MID(B20,1,2)** tells the spreadsheet to examine the genotype in cell B20, start with the first character in the genotype, and return two characters. If your genotype in cell B20 is **A1**A1B1B1C1C1, the **MID** function will return the portion of the genotype that is bolded. Similarly, the formula **=MID(B20,5,2)** will examine the genotype in cell B20, start with the fifth character in the genotype, and return two characters (the program will return "B1").

The **LOOKUP** formula returns a value either from a one-row or one-column range or from an array. The **LOOKUP** function has two syntax forms: vector and array. We will use the vector form, which looks in a one-row or one-column range (the vector) for a value and returns a value from the same position in a second one-row or one-column range. It has the syntax **LOOKUP(lookup_value,lookup_vector,result_vector)**, where **lookup_value** is a value the function searches for in the first vector, **lookup_vector** is a range that contains only one row or one column, and **result_vector** is the value that the spreadsheet returns from the same position in a row or column that is adjacent to the lookup vector. For example, **=LOOKUP("A1",C9:H9,C8:H8)** finds the value A_1 in the vector C9–H9 and returns the phenotype contribution associated with that allele.

We have combined **LOOKUP** and **MID** formulae to generate a phenotype. For example, **=LOOKUP(MID(B20,1,2),$C$9:$H$9,$C$8:$H$8)** uses the **MID** formula to determine the first allele in the *A* locus (either A_1 or A_2), finds this value in cells C9–H9, and returns the associated phenotype contribution listed in cells C8–H8. You can add several of these kinds of formulae together to generate a final phenotype. It produces a very long formula that looks intimidating at first, but is really quite simple once you work through it.

5. In cell D20, enter a formula to determine whether individual 1 survived the conditions associated with year 1. Copy the formula down for the remaining individuals in the population.

Enter the formula **=IF(C20>LOOKUP($B$11,$C$3:$E$3,$C$4:$E$4),B20,"")** in cell D20. Copy this formula down the column.
We want to know whether an individual survives to reproduce, given the environmental condition for year 1 (cell B11) and the beak size required to survive the environment for year 1 (listed in cells C4–E4). The formula simply tells the spreadsheet to look up year 1's condition in cell B11, locate that condition in cells C3–E3, and return the minimum phenotype required for survival listed in cells C4–E4. **IF** the individual has a phenotype greater than necessary for survival, return the individual's *genotype*; otherwise, return a blank cell (indicated by the two sets of quotation marks). Year 1 is a wet condition, and hence all genotypes will survive.

6. In cell E20, enter a formula that returns the individual's phenotype if it survived.

Enter the formula **=IF(D20="","",C20)** in cell E20. Copy this formula down for the remaining 499 individuals in the population.

7. Set up new headings as shown in Figure 3.

	I	J
8	Number	Mean
9	surviving	phenotype
10		
11		
12		
13		
14		
15		

Figure 3

8. In cell I11, enter a formula to count the number of survivors in year 1. These individuals will produce offspring for the next generation.

Enter the formula **=COUNTIF(D20:D519,"A*")** in cell I11.
The **COUNTIF** formula counts the number of cells within a range that meet the given criteria. The formula above tells the spreadsheet to examine cells D19–D518 and to count any cell that begins with an *A*. The * following the *A* is a wild card, indicating that it doesn't matter what text follows the *A*. Since only surviving individuals have a genotype listed, the formula will count only those individuals that survived.

9. In cell J11, use the **AVERAGE** function to calculate the mean phenotype of the survivors.

Enter the formula **=AVERAGE(E20:E519)** in cell J11.

10. Enter formulae in cells C11–H11 to compute allele frequencies of the surviving adults.

We entered the following formulae:

- Cell C11 **=(2*COUNTIF($D$20:$D$519,"A1A1*")+ COUNTIF($D$20:$D$519,"A1A2*")+ COUNTIF($D$20:$D$519,"A2A1*"))/(2*$I$11)**
- Cell D11 **=1-C11**
- Cell E11 **=(2*COUNTIF($D$20:$D$519,"*B1B1*")+ COUNTIF($D$20:$D$519,"*B1B2*")+ COUNTIF($D$20:$D$519,"*B2B1*"))/(2*$I$11)**
- Cell F11 **=1-E11**
- Cell G11 **=(2*COUNTIF($D$20:$D$519,"*C1C1")+ COUNTIF($D$20:$D$519,"*C1C2")+ COUNTIF($D$20:$D$519,"*C2C1"))/(2*$I$11)**
- Cell H11 **=1-G11**

You have entered similar formulae in your Hardy-Weinberg exercise. Remember the trick of using the * wild card character. For example, when we used the **COUNTIF** formula to count the number of **A1A1*** individuals, it counted all individuals with the A_1A_1 genotype, regardless of their genotypes at the *B* or *C* locus. The same principle applies to the *B* (***B1B1***) and *C* (***C1C1**) genotypes.

Since these individuals survived to breed, they will determine the genotypes of individuals at the beginning of year 2.

11. Save your work.

Your spreadsheet should now look something like Figure 4. Your numbers will be a bit different in Row 11, and that's fine.

	A	B	C	D	E	F	G	H
7			Allele frequencies of surviving parents					
8	Phenotype contribution =>		1	2	1	2	1	2
9	Year	Condition	A1	A2	B1	B2	C1	C2
10	Initial		0.8	0.2	0.8	0.2	0.8	0.2
11	1	Wet	0.80	0.20	0.78	0.22	0.81	0.19

Figure 4

C. Track the population for year 2.

1. In cells F20–F519, enter a formula to generate a genotype for each individual (offspring), given the allele frequencies listed in cells C11–H11.

The headings in Figure 5 should already be in place. You can simply repeat the step you completed for year 1 to complete column F.
Enter the formula **=IF(RAND()<$C$11,"A1","A2")&IF(RAND()<$C$11,"A1","A2")& IF(RAND()<$E$11,"B1","B2")&IF(RAND()<$E$11,"B1","B2")& IF(RAND()<$G$11,"C1","C2")&IF(RAND()<$G$11,"C1","C2")** in cell F19. Copy this formula down to row F519.

	F	G	H	I
18		Year 2		
19	Genotype	Phenotype	Survive?	Phenotype

Figure 5

2. Select cell C20, and copy it to cell G20.

This will determine the phenotypes of the 500 individuals that are present in the population at the beginning of year 2.

3. Enter a formula in cell H20 to determine if individual 1 survives to breed in year 2.

Refer back to the formula used in year 1. We entered the formula **=IF(G20>LOOKUP($B$12,$C$3:$E$3,$C$4:$E$4),F20,"")**.
This formula looks up the conditions associated with year 2 and returns the phenotype of individuals whose beak sizes are large enough to survive the environmental conditions for year 2.

4. Select cell E20, and copy it to cell I20.

The formula in cell E20 returns the phenotype of individuals that survive to breed.

5. Enter a formula in cell I12 to count the number of survivors in year 2.

Enter the formula **=COUNTIF(H20:H519,"A*")** in cell I12.

6. Enter a formula in cell J12 to determine the average phenotype of survivors in year 2.

Enter the formula **=AVERAGE(I20:I518)** in cell J12.

7. Enter formulae in cells C12–H12 to compute the allele frequencies of survivors for year 2.

As you did for year 1, compute the allele frequencies for the population that survives to breed in year 2. These frequencies will be used to assign genotypes to individuals (offspring) in year 3.

- Cell C12 **=(2*COUNTIF($H$20:$H$519,"A1A1*")+ COUNTIF($H$20:$H$519,"A1A2*")+ COUNTIF($H$20:$H$519,"A2A1*"))/(2*$I$12)**
- Cell D12 **=1-C12**
- Cell E12 **=(2*COUNTIF($H$20:$H$519,"*B1B1*")+ COUNTIF($H$20:$H$519,"*B1B2*")+ COUNTIF($H$20:$H$519,"*B2B1*"))/(2*$I$12)**
- Cell F12 **=1-E12**
- Cell G12 **=(2*COUNTIF($H$20:$H$519,"*C1C1")+ COUNTIF($H$20:$H$519,"*C1C2")+ COUNTIF($H$20:$H$519,"*C2C1"))/(2*$I$12)**
- Cell H12 **=1-G12**

8. Save your work.

Note that when you press F9, the calculate key, the spreadsheet generates new genotypes, and hence a new set of survivors and frequencies.

9. Repeat steps 1–8 to obtain results for each of years 3–5 in cells J20–U519.

D. Create graphs.

1. Graph the frequencies of each allele over time.

Use the line graph option and label your axes fully. Your graph should resemble Figure 6.

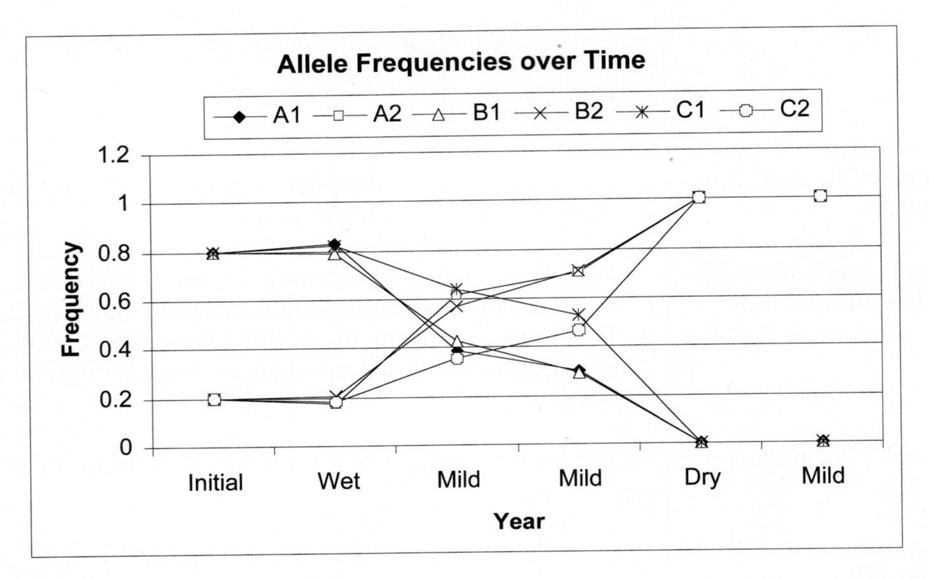

Figure 6

2. Graph the numbers of survivors over the 5-year period.

Your graph should resemble Figure 7.

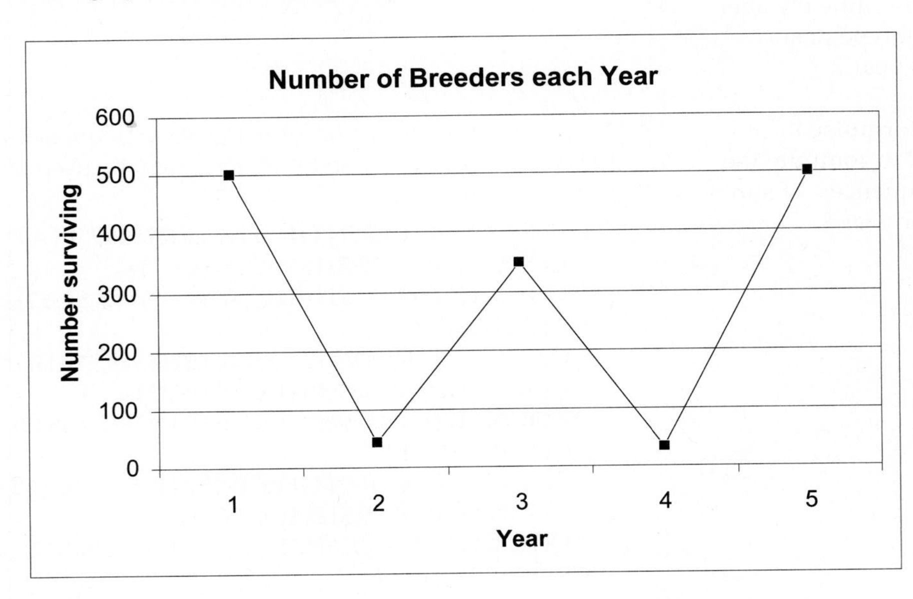

Figure 7

QUESTIONS

1. Hit the F9 key 20 times, keep track of the values in cells I11–I15, and count how many times the population goes extinct. (Under these conditions it will probably never go extinct.) In what percentage of the 20 trials did the population go extinct at any time during the 5-year period? This is the *extinction rate* for the situation in which year 1 is wet, years 2, 3, and 5 have a mild drought, and year 4 is dry (drought conditions). Change cell B12 (year 2) to DRY instead of MILD. Again hit the F9 key 20 times. In what percentage of the 20 trials did the population go extinct? This is the extinction rate for the situation in which the change in precipitation occurred more abruptly. Relate the extinction rate to the genetic variation and phenotypic variation in the population.
2. How do starting initial allele frequencies affect how the population adapts to abrupt changes in environmental conditions?
3. What if initial frequency of the C_2 allele was zero? Would the population ever be able to adapt to a harsh drought? Explain how genetic diversity is important to adaptation.
4. Is the following statement true or false? Explain. "The population had the genetic diversity to adapt, but could not adapt because the environmental change occurred too abruptly."

*5. (Advanced) Explore the model by modifying the trait size needed for survival (cells C4–E4), initial allele frequencies, and the environmental conditions experienced in years 1–5. Provide an interesting observation in terms of adaptation as a result of your exploration.

LITERATURE CITED

Clayton, G. A. and A. Robertson. 1957. An experimental check on quantitative genetical theory. II. Long-term effects of selection. *Journal of Genetics* 55: 152–170.

Grant, B. R. and P. R. Grant. 1993. Evolution of Darwin's finches caused by a rare climatic event. *Proceedings of the Royal Society of London (B)* 251:111–117.

34 GENE FLOW AND POPULATION STRUCTURE

Objectives

- Model two subpopulations that exchange individuals through gene flow.
- Determine equilibrium allele frequencies as a result of gene flow.
- Calculate H (heterozygosity) statistics for the population.
- Calculate F statistics for the population.
- Determine how H, F, and allele frequencies change over time as a result of gene flow.

Suggested Preliminary Exercise: Hardy-Weinberg Equilibrium

INTRODUCTION

Think about a favorite plant or animal species, and consider how it is distributed across the earth. Are the individuals all in one place, or are individuals scattered in their distribution? Most of the earth's species have distributions that are "patchy" in some way. In other words, the greater population is subdivided into smaller units or subpopulations. For example, a species of fish may have a subdivided distribution if individuals inhabit a number of different lakes. Similarly, maple forests may be patchily distributed within a mosaic of farm land, resulting in a number of subpopulations. Even dandelions in a lawn may have distinct patches to which individuals belong. But does this "subdivision" in distribution suggest that the species is made up of several "subpopulations," each with an independent evolutionary trajectory? Or does the species "behave" as a single, panmictic population, where individuals can mix freely in spite of the patchiness? Or perhaps the population is somewhat subdivided, where individuals from one location can mix (breed) with individuals from other locations, but not as freely as a single panmictic population because they are spatially separated from each other.

These questions concerning **gene flow** and **population structure** are important from the perspectives of evolution, ecology, and conservation. A population is "structured" if the individuals that make up the greater, overall population are subdivided spatially, and hence random mating among individuals in the greater population is limited. The degree to which populations are structured depends in large part on the amount of gene flow— the migration of individuals between subpopulations, with subsequent breeding—that takes place between the subdivided populations (or subpopulations). If there is little or no gene flow, then each subpopulation evolves independently of the other. In contrast, if there is substantial gene flow, the structure in the population breaks down because sufficient genetic mixing has occurred. Gene flow is therefore a homogenizing force that causes allele frequencies in subdivided populations to converge (Wilson and Bossert 1971).

Allele Frequencies in Subpopulations

Let's consider gene locus A in two subpopulations. To keep things simple, we'll assume locus A exists in two forms, or **alleles**, A_1 and A_2. Let's assume that subpopulation 1 has an A_1 allele frequency, p_1, of 0.7, while subpopulation 2 has an A_1 allele frequency of $p_2 = 0.2$. Let's now let the two subpopulations exchange individuals through migration, where m is the migration rate of individuals *into* a subpopulation. The individuals that make up the population that did not migrate in are called **residents**, and the resident population is designated as $1 - m$. If $m > 0$, then after a single generation of mixing, p_1 in subpopulation 1 will be changed; subpopulation 1 now consists of some portion of individuals that remained within subpopulation 1, plus some portion of individuals that migrated from subpopulation 2 into subpopulation 1. Mathematically, the new frequency of allele A_1 is designated as p_1', and

$$p_1' = (1 - m)p_1 + mp_2 \qquad \text{Equation 1}$$

Equation 1 says that the new frequency of allele A_1 will have two components: $(1 - m)p_1$, which represents the proportion of subpopulation 1 that does not emigrate times the frequency of A_1 in subpopulation 1 before migration, and mp_2, which represents the proportion of immigrants from subpopulation 2 times the frequency of A_1 in subpopulation 2.

$$\Delta p = p_1' - p_1 \qquad \text{Equation 2}$$

Substituting p_1' from Equation 1 into the Equation 2, we get

$$\Delta p = (1 - m)p_1 + mp_2 - p_1 = p_1 - mp_1 + mp_2 - p_1$$

The p_1s drop out of the equation, and we can factor out $-m$ from the remaining terms to get

$$\Delta p = -m(p_1 - p_2) \qquad \text{Equation 3}$$

Equation 3 says that a change in allele frequency of a recipient population (subpopulation 1) due to migration is a function of the migration rate, as well as of the difference in the allele frequency between the migrants and the recipient population. If the migration rates remain constant over time, eventually the two subpopulations will have exactly the same allele frequencies (Figure 1; Wilson and Bossert 1971).

H and F Statistics

When two populations have reached the same allele frequencies, the larger population will appear to be unstructured. Or is it? Structure depends not only on allele frequencies but also how the A_1 and A_2 alleles are distributed among individuals. Therefore, we must also consider genotype frequencies in the subpopulations.

In many species, especially animals, individuals carry two copies of most genes, one from each parent. Let's assume that subpopulation 1 consists of 5 individuals with genotypes A_1A_1, A_1A_1, A_1A_2, A_2A_2, A_2A_2, and that subpopulation 2 consists of 5 individuals with genotypes A_1A_2, A_1A_2, A_1A_2, A_1A_2, A_1A_2. The subpopulations have identical frequencies of the A_1 allele, $p = 0.5$, but the two subpopulations have quite different levels of **heterozygosity**. Most of the individuals in subpopulation 1 are homozygotes—they carry either two copies of A_1 or two copies of A_2; but the individuals in subpopulation 2 are heterozygotes and each of them carries one copy each of allele A_1 and A_2. So allele frequency alone does not tell us everything about a population's structure. The level of structure depends on levels of heterozygosity in the subpopulations, as well as the level of heterozygosity in the greater population.

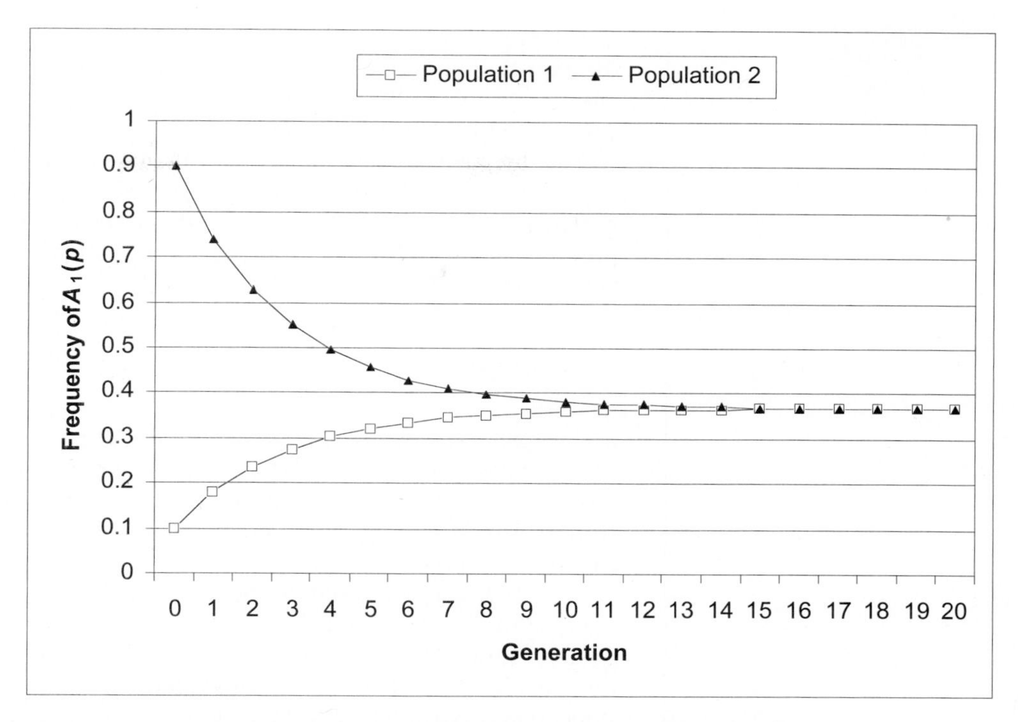

Figure 1 Two subpopulations with different initial frequencies of allele A_1 exchange individuals at two different rates (the migration rate, m). As individuals move between the two populations, the frequency of A_1 in subpopulation 1 approaches that in subpopulation 2, and they eventually become equal.

Why is heterozygosity used to estimate structure? And how is the degree of structuring measured through heterozygosity statistics? Two measures are commonly used, H and F (Hartl 2000).

H is a measure of heterozygosity; it is used to measure structure because individuals within subdivided populations are likely to inbreed due to small population sizes, which typically results in decreased heterozygosity (see Exercise 41/24, "Inbreeding and Outbreeding"). Thus, if there is no gene flow between subpopulations, each subpopulation will (theoretically) have more homozygotes (A_1A_1 or A_2A_2) than predicted by Hardy-Weinberg.

The statistic $\boldsymbol{H_i}$ measures the *observed* level of heterozygosity in a subpopulation For example, 1 of 5 individuals in subpopulation 1 from our previous example were heterozygotes while 5/5 individuals in subpopulation 2 were heterozygotes. This measure is averaged across subpopulations, and can be interpreted as the average heterozygosity of an individual in a subpopulation, or the proportion of the genome that is heterozygous within an individual. For example, H for subpopulation 1 equals $1/5 = 0.2$. H for subpopulation 2 equals $5/5 = 1.0$. The average of the two H scores $= 0.6 = H_i$.

The observed levels of heterozygosity in subpopulations are compared to two other measures of heterozygosity, $\boldsymbol{H_s}$ and $\boldsymbol{H_t}$. H_s is the *expected* level of heterozygosity in a subpopulation if the subpopulation is randomly mating as predicted by Hardy-Weinberg. This measure is also averaged across subpopulations. Returning to our example, both subpopulations have allele frequencies $p = 0.5$ and $q = 0.5$. If each subpopulation were in Hardy-Weinberg equilibrium, we would expect the genotype frequency of heterozygotes to be $2 \times 0.5 \times 0.5 = 0.5$. This number is averaged for the two subpopulations to give us H_s: $(0.5 + 0.5)/2 - 0.5$. Thus, in our example, $H_i = 0.6$ and $H_s = 0.5$. This means that the observed levels of heterozygotes are, on average, higher than what is expected for a population in Hardy-Weinberg equilibrium.H_t is the expected level of heterozygosity that should be observed in the subpopulations if the greater population (subpopulation 1 and subpopulation 2) were really a single, randomly mating, panmictic population. If our subpopulations were really a single, panmictic population, the expected genotype frequency of heterozygotes would be $2 \times p \times q$, where p and q are the *averages* of the subpopulation allele frequencies (Hartl, 2000). In out example, $p = q = 0.5$ for both subpopulations, so the equation is $2 \times 0.5 \times 0.5 = 0.5$.

The three H statistics are used to calculate ***F*** statistics, which are common measures of population subdivision and inbreeding; F is sometimes referred to as the **inbreeding coefficient**. The F statistics use the different H statistics to reveal different things about population subdivision. $\boldsymbol{F_{is}}$ compares observed and expected heterozygosities *within* a subpopulation. It is calculated as

$$F_{is} = \frac{H_s - H_i}{H_s} \qquad \text{Equation 4}$$

and suggests the level of inbreeding at the subpopulation level. Thus, F_{is} is often called the *inbreeding coefficient within subpopulations*. The numerator reveals how much the heterozygosity observed in the subpopulations differs, on average, from what is expected from Hardy-Weinberg. For mathematical reasons, this difference is then "adjusted" by the expected level.

When H_i is approximately the same as H_s, the deviation from Hardy-Weinberg is small, and F_{is} is close to 0, suggesting that observed and expected levels of heterozygosity within subpopulations are close in value. When H_i is much different than H_s, F_{is} deviates from 0. When F_{is} is positive, fewer heterozygotes are observed in subpopulations than predicted by Hardy-Weinberg. When F_{is} is negative, more heterozygotes are observed in the subpopulation than predicted by Hardy-Weinberg. F_{is} is usually large in self-fertilizing (inbred) species.

$\boldsymbol{F_{it}}$ also measures inbreeding, but is concerned with how individuals (H_i) deviate, on average, from the heterozygosity of the larger population (H_t). It is calculated as

$$F_{it} = \frac{H_t - H_i}{H_t} \qquad \text{Equation 5}$$

Thus, it calculates a level of *inbreeding at the total population level*. When H_i is similar to H_t, the observed heterozygosities in subpopulations are close to what is predicted as if the population were really a single large, panmictic population, and F_{it} is 0. When H_i is much different than H_t, F_{it} deviates from 0. When F_{it} is positive, fewer heterozygotes are observed in subpopulations than predicted by Hardy-Weinberg. When F_{it} is negative, more heterozygotes are observed in the subpopulation than predicted by Hardy-Weinberg. These differences can be caused by both inbreeding and by genetic drift, both of which reduce heterozygosity in a subpopulation. Thus, F_{it} measures the amount of inbreeding due to the combined effects of nonrandom mating within subpopulations and to random genetic drift among subpopulations.

$\boldsymbol{F_{st}}$ is a measure of *nonrandom mating among or between subpopulations* relative to the total population, and hence this statistic is often used to indirectly measure the amount of population subdivision. It is calculated as

$$F_{st} = \frac{H_t - H_s}{H_t} \qquad \text{Equation 6}$$

F_{st} is a measure of the genetic differentiation of subpopulations and is always positive. The formula "compares" two expected values from Hardy-Weinberg calculations. The numerator in the formula measures the difference in H_t (the average of the expected heterozygosity in the total population) and H_s (H_s is the average *expected* heterozygosity within the subpopulations). F_{st} is not concerned with individual subpopulations, so it measures the reduction in heterozygosity due to factors other than inbreeding (such as genetic drift). When population subdivision is great, the difference between the values in the numerator increases, F_{st} takes on a high value.

PROCEDURES

The H and F statistics can be confusing until you sit down and work through the math. The purpose of this exercise is to set up a model of two subpopulations of equal size that interact through migration. You'll enter observed genotype frequencies, then calculate gene frequencies and how these frequencies change over time. You'll also calculate and interpret the H and F statistics as gene flow occurs between the two populations. As the simulation progresses, you'll be able to see how the H and F statistics change as the two subpopulations become homogenized, and you'll interpret what the statistics mean.

As always, save your work frequently to disk.

INSTRUCTIONS

ANNOTATION

A. Set up the spreadsheet.

1. Open a new spreadsheet and set up headings as shown in Figure 2.

	A	B	C	D	E	F	G	H
1	***Gene Flow and Population Structure***							
2								
3				**Parameters**		**Genotype frequencies**		
4			***N***	***m***	***r***	**A1A1**	**A1A2**	**A2A2**
5		**Subpopulation 1:**	100	0	1	0.36	0.48	0.16
6		**Subpopulation 2:**	100	0	1	0.04	0.32	0.64

Figure 2

2. Enter N and m subpopulation parameters as shown.

We'll consider a general model of gene flow and population structure that focuses on a single locus, the A locus. We'll start with two subpopulations, 1 and 2, that each consist of N individuals; we designate N as 100 in cells C5 and C6. In this exercise, N will be the same for both populations.

The migration rate, m, ranges between 0 and 1 and is the proportion of the population that migrates from one subpopulation to the other. The value in cell D5 gives the migration rate *into* subpopulation 1 (from subpopulation 2). The value in cell D6 gives the migration rate *into* subpopulation 2 (from subpopulation 1). To begin the exercise, we'll consider two subpopulations where the migration rate between them is 0. We'll modify m later in the exercise.

3. Enter a formula to calculate the value of r (the proportion of each subpopulation that are residents as opposed to migrants).

Enter **=1-D5** in cell E5 and **=1-D6** in cell E6.
The total subpopulation consists of migrants that move into the population plus the residents that remain in the population, so the sum of m (the migration rate) and r (resident population proportion) is equal to 1.

4. Enter the observed genotype frequencies for each subpopulation in cells F5–H6 as shown in Figure 2.

For the purpose of this exercise, we'll assume that you have the ability to determine the genotype of each individual in the subpopulations, and can then calculate the proportion of A_1A_1, A_1A_2, and A_2A_2 genotypes. The current values in cells F5–H6 indicate that both subpopulations are in Hardy-Weinberg equilibrium. (Prove this to yourself before you continue). You will be able to manipulate the observed genotype proportions later in the exercise (i.e., you can model populations that are not in Hardy-Weinberg equilibrium).

5. Sum the genotype frequencies for each subpopulation in cells I5 and I6.

Enter the formula **=SUM(F5:H5)** in cell I5 and **=SUM(F6:H6)** for subpopulation 2. These equations are used to ensure that the genotype frequencies for each subpopulation sum to 1. If the frequencies don't sum to 1, change the observed genotype frequencies so that they sum to 1.

6. Save your work.

B. Set up the general model of gene flow.

1. Set up new headings as shown in Figure 3.

	A	B	C	D	E	F	G
10		Observed allele frequencies					
11			Subpop 1			Subpop 2	
12	Generation	A1	A2	Delta A2	A1	A2	Delta A2

Figure 3

2. Set up a linear series from 0 to 50 in cells A13–A63.

We'll calculate the allele frequencies in our two subpopulations over a 50-generation period. Year 0 will represent the initial conditions in terms of allele frequencies.

3. In cell B13 and C13, enter formulae to calculate the initial frequencies of the A_1 and A_2 alleles in subpopulation 1, respectively.

Remember that a population of 100 individuals has 200 "gene copies" or "total alleles" present. (Each individual has 2 copies). We just need to know how many of those are A_1 alleles, and how many are A_2 alleles. Homozygote A_1A_1 individuals carry two of the A_1 alleles, and heterozygotes carry 1 A_1 allele.
Enter the formula **=(2*F5*C5+G5*C5)/(2*C5)** in cell B13.
Enter the formula **=1-B13** in cell C13.

4. In cells E13 and F13, enter formulae to calculate the starting frequencies of the A_1 and A_2 alleles in subpopulation 2.

Enter the formula **=(2*F6*C6+G6*C6)/(2*C6)** in cell E13.
Enter the formula **=1-E13** in cell F13.

5. Enter formulae in cells B14 and C14 to calculate the allele frequencies of subpopulation 1, given the migration and resident parameters.

Remember that the frequencies in the next time step can be computed as

$$p_{1,t+1} = (1-m)p_1 + mp_2$$

We used the formula **=\$E\$5*C13+\$D\$5*F13** in cell C14 to calculate the frequency of the A_2 allele, and then calculated A_1 as $1-q$ in cell B14 **(=1-C14)**.
Make sure you understand the C14 formula. It says that the frequency of the A_2 allele in subpopulation 1 in year 1 depends on two factors: (1) the frequency of the A_2 allele in the resident population (**\$E\$5*C13**), and (2) frequency of the A_2 allele in the immigrants (**\$D\$5*F13**).

6. Calculate the change in the frequency of the A_2 allele (ΔA_2) in cell D14.

We used the formula **=C14-C13**. (You can make a delta symbol, Δ, by typing in a capital D, and then changing the font to Symbol.)

7. Calculate the allele frequencies and change in the A_2 allele frequency in subpopulation 2 for year 1.

Enter the following formulae:

- E14 **=1-F14**
- F14 **=\$E\$6*F13+\$D\$6*C13**
- G14 **=F14-F13**

8. Select cells B14–G14 and copy their formulae down to row 63.

9. Save your work.

C. Make graphs.

1. Graph the frequency of the A_1 allele over time.

Use the line graph option and label your axes fully. Your graph should look something like Figure 4. (We have graphed only the first 15 generations for clarity.)

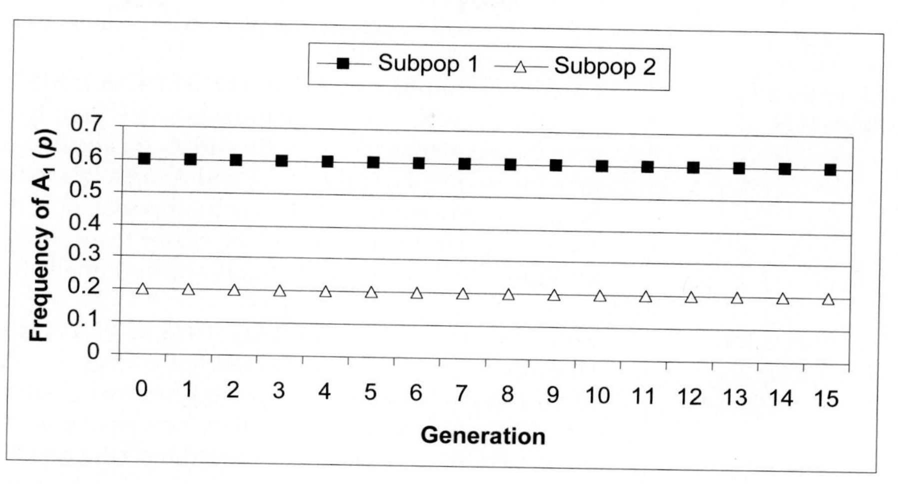

Figure 4

2. Change the migration rate for your two populations (choose any rate between 0 and 1), and construct a new graph of allele frequencies over time.

We generated the graph in Figure 5 by changing the migration rate for subpopulation 1 from 0 to 0.2.

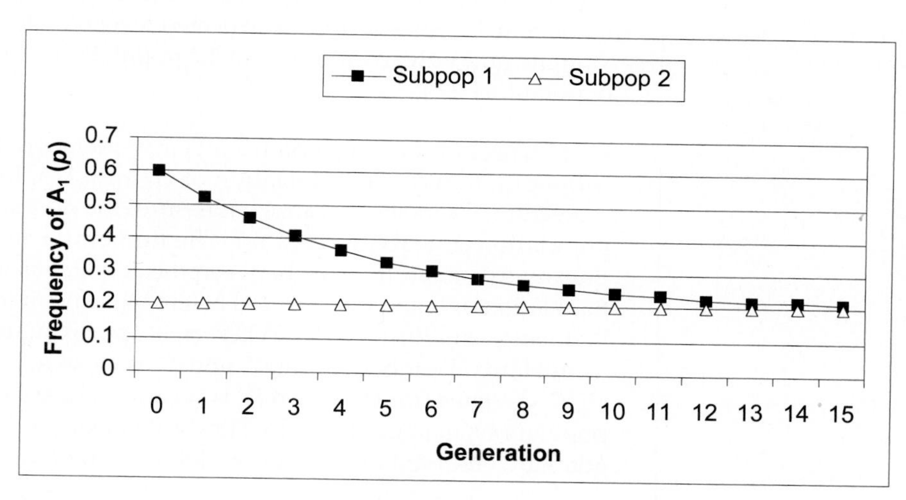

Figure 5

3. Save your work, and answer questions 1–3 at the end of the exercise.

D. Calculate H and F statistics.

1. Set up new headings as shown in Figure 6.

	H	I	J	K	L	M
11	***H* Statistics**			***F* Statistics**		
12	H_i	H_s	H_t	F_{is}	F_{it}	F_{st}

Figure 6

2. In cell H13, enter a formula to calculate H_i.

Enter the formula **=AVERAGE($G$5:$G$6)** in cell H13.

H_i is the average observed heterozygosity within a total population. Thus, we take the average of cells G5 and G6, which are the frequencies of heterozygotes in subpoplation 1 and subpopulation 2. Keep in mind that by making cells G5–G6 absolute references, you are forcing the heterozygote proportions to remain constant over time—this will affect the calculation of *F* statistics later in the exercise.

3. In cell I13, enter a formula to calculate H_s.

Enter the formula **=AVERAGE(2*B13*C13,2*E13*F13)** in cell I13.

H_s is the average *expected* heterozygosity within the subpopulations. Cell B13 and C13 give the frequency of the A_1 (p) and A_2 (q) allele in subpopulation 1. Cells E13 and F13 give the frequency of the A_1 (p) and A_2 (q) allele for subpopulation 2. The Hardy-Weinberg principle tells us that, for each subpopulation, the expected heterozygote frequency is $2 \times p \times q$. The formula in I13 tells Excel to multiply $2 \times p \times q$ for subpopulation 1, then multiply $2 \times p \times q$ for subpopulation 2, and finally to average these two values together.

4. In cell J13, enter a formula to calculate H_t.

Enter the formula **=2*AVERAGE(B13,E13)*AVERAGE(C13,F13)** in cell J13.

H_t is the average of the expected heterozygosity in the total population. H_t is similar to H_s, but it's the average expected heterozygosity for the population at large. Therefore, first we calculate an overall p, then an overall q, and then multiply by 2. The result tells us what heterozygosity should be if the two subpopulations were one panmictic population.

5. In cell K13, enter a formula to calculate F_{is}.

Enter the formula **=(I13-H13)/I13** in cell K13.

Now that we have the *H* statistics calculated, the *F* statistics are fairly straightforward. The *F* statistics compare the different levels of heterozygosities to reveal how the population is structured. All three *F* statistics (F_{is}, F_{it}, F_{st}) have H_t or H_s as the denominator, which "adjusts" for the expected level of heterozygosity if the population were a single randomly mating, panmictic population (H_t) or randomly mating subdivided populations (H_s).

F_{is} measures of the deviation from Hardy-Weinberg heterozygote proportions within subpopulations (or the deviation of H_i from H_s). Remember that F_{is} also called the inbreeding coefficient because it measures the decrease in heterozygosity *within* a subpopulation (due to inbreeding). The numerator in the equation $F_{is} = (H_s - H_i) / H_s$ thus reveals the difference between the actual, observed heterozygosities in the subpopulations (H_i) and the expected heterozygosities if the subpopulations were in Hardy-Weinberg equilibrium (H_s). When H_i is approximately the same as H_s, the deviation from Hardy-Weinberg is small, and F_{is} is close to 0. When H_i is much different than H_s, F_{is} deviates from 0. When F_{is} is positive, fewer heterozygotes are observed in subpopulations than predicted by Hardy-Weinberg. When F_{is} is negative, more heterozygotes are observed in the subpopulation than predicted by Hardy-Weinberg.

6. In cell L13, enter a formula to calculate F_{it}.

Enter the formula **=(J13-H13)/J13** in cell L13.

F_{it} measures the total inbreeding coefficient. It measures the deviations of observed heterozygosities within subpopulations from Hardy-Weinberg proportions of the total population (or the deviation of H_i from H_t). The equation for calculating F_{it} is $F_{it} = (H_t - H_i)/H_t$. When H_i is similar to H_t, the observed heterozygosities in subpopulations are close to what is predicted as if the population were really one large, panmictic population, and F_{it} is 0. Thus, F_{it} measures the amount of inbreeding due to the combined effects of nonrandom mating within subpopulation and to random genetic drift among subpopulations. When H_i is much different than H_t, F_{it} deviates from 0. When F_{it} is positive, fewer heterozygotes are observed in subpopulations than predicted by Hardy-Weinberg. When F_{it} is negative, more heterozygotes are observed in the subpopulation than predicted by Hardy-Weinberg.

7. In cell M13, enter a formula to calculate F_{st}.

Enter the formula **=(J13-I13)/J13** in cell M13.
F_{st} is a measure of the genetic differentiation of subpopulations and is always positive. The formula "compares" two expected values from Hardy-Weinberg calculations. The numerator in the formula $F_{st} = (H_t - H_s)/H_t$ measures the difference in H_t (the average of the expected heterozygosity in the total population) and H_s (H_s is the average *expected* heterozygosity within the subpopulations). Thus, F_{st} is the amount of "inbreeding" due solely to population subdivision (i.e., due to genetic drift). When inbreeding due to subdivision is great, the difference between the values in the numerator increases, and F_{st} takes on a high value.

8. Select cells H13–M13, and copy their formulae down to row 63.

At this time, you might want to play around with your model parameters and contemplate the meaning of the H and F statistics in Generation 0. Then consider the statistics as gene flow occurs in subsequent generations.

9. Save your work.

E. Create graphs.

1. Set the migration rate to 0, and graph the H statistics and allele frequencies as a function of time. Use the line graph option and label your axes fully.

Interpret your graph. Your graph should resemble Figure 7.

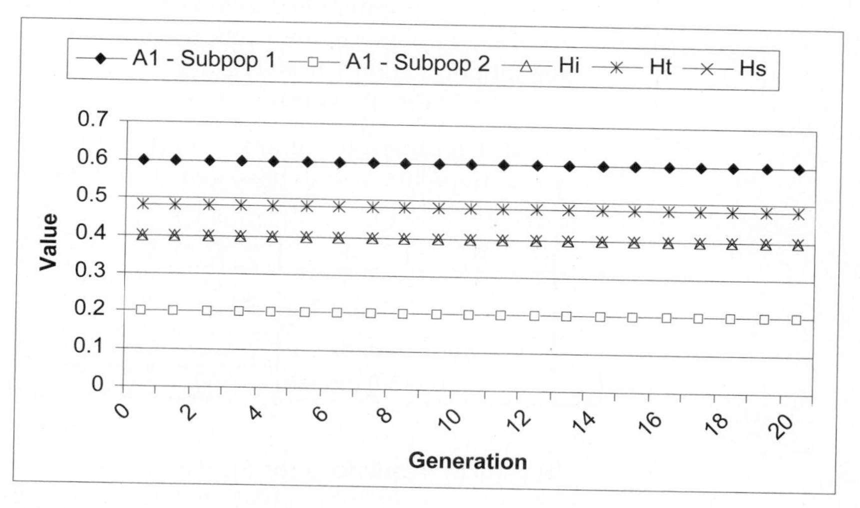

Figure 7

2. Graph the F statistics and allele frequencies as a function of time.

3. Save your work.

Your graph should resemble Figure 8. Interpret your graph.

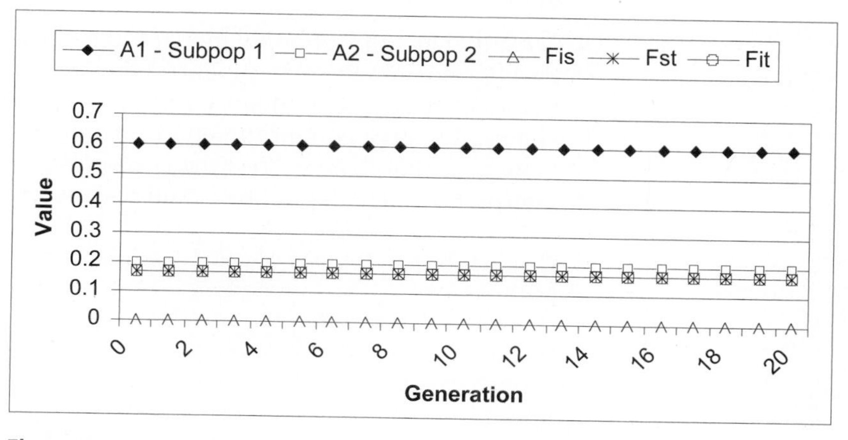

Figure 8

QUESTIONS

1.Enter the following values in your spreadsheet:

	A	B	C	D	E	F	G	H
3				Parameters		Genotype frequencies		
4			*N*	*m*	*r*	A1A1	A1A2	A2A2
5		Subpopulation 1:	100	0	1	0.25	0.5	0.25
6		Subpopulation 2:	100	0	1	0.09	0.42	0.49

Change cell D5 by increments of 0.1. What is the equilibrium allele frequencies for subdivided populations with gene flow? How does changing m determine the point in time is equilibrium reached?

2. How do allele frequencies change in the two populations in an island model (gene flow is uni-directional) compared to a general model in which gene flow is bi-directional? Set m for subpopulation 1 to 0 to indicate that subpopulation 1 is a mainland that sends out emigrants but does not receive immigrants. Set $m = 0.5$ for subpopulation 2 to indicate that subpopulation 2 is an island that receives immigrants from subpopulation 1. Graph your results. Then change m for subpopulation 1 from 0 to 1 in increments of 0.1. How do the two models compare? How do your results change if m for subpopulation 2 is changed?

3. What determines the amount of time to reach equilibrium frequencies in subdivided populations that have gene flow? Set up population genotypes as shown.

	A	B	C	D	E	F	G	H
3				Parameters		Genotype frequencies		
4			*N*	*m*	*r*	A1A1	A1A2	A2A2
5		Subpopulation 1:	100	0.1	0.9	0.83	0.16	0.01
6		Subpopulation 2:	100	0.1	0.9	0.01	0.16	0.83

The allele frequencies for the subpopulations are $p = 0.91$ for subpopulation 1 and $p = 0.09$ for subpopulation 2. Keeping m fixed at 0.1 for both subpopulations, change the intial genotype frequencies (the allele frequencies will also be altered). How does change in initial genotype frequency (and allele frequency) affect the amount of time until equilibrium is achieved?

Return your spreadsheet to its initial settings (Figure 2) and continue to Part D in the exercise.

4. Set m to 0 in both subpopulations, and enter genotype frequencies in cells F5–H6 so that both subpopulations are in Hardy-Weinberg equilibrium, and have identical allele frequencies. (In the exercise both subpopulations were in Hardy-Weinberg equilbrium and had different allele frequencies within them.) How does this change affect the H and F statistics? Graph the results and fully interpret the meaning of the H and F statistics.

5. Set m as 0 values for both subpopulations, then enter genotype frequencies in cells F5–H6 so that at least one subpopulation is out of Hardy-Weinberg equilibrium. For example, you might enter values as shown:

	A	B	C	D	E	F	G	H
3				Parameters		Genotype frequencies		
4			*N*	*m*	*r*	A1A1	A1A2	A2A2
5		Subpopulation 1:	100	0	1	0.5	0	0.5
6		Subpopulation 2:	100	0	1	0.04	0.32	0.64

How do H and F statistics reflect structure? How did F_{is} change? Is it positive or negative? Is it large or small? Explain why you obtained the F_{is} value that you did. What does this tell you about the populations? (Remember that the genotype frequencies will remain out of Hardy-Weinberg equilibrium over time because of the formula entered in cell H13.)

6. For this question, you will ignore the genotype frequencies given in rows 5 and 6, and directly enter the initial allele frequencies for subpopulations in cells B13–F13. (We'll assume the genotypes are in Hardy-Weinberg proportions.) Start with $p = 0.6$ for subpopulation 1 and $p = 0.5$ for subpopulation 2. Record the F statistics for that generation. Then let $p = 0.8$ in supopulation 1 and $p = 0.2$ in subpopulation 2, and record the F statistics. Then let $p = 0.9$ in subpopulation 1 and subpopulation 2, and record the F statistics. How did the F statistics change as the two subpopulations became more differentiated (allele frequencies diverged)? Which F statistic changed the most? Why?

LITERATURE CITED

Hartl, D. 2000. *A Primer of Population Genetics*, Third Edition. Sinauer Associates, Sunderland, MA.

Wilson, E. O., and W. H. Bossert. 1971. *A Primer of Population Biology*. Sinauer Associates, Sunderland, MA.

35 LIFE HISTORY TRADE-OFFS

Objectives

- Develop a spreadsheet model of annual versus perennial life history strategies for plants.
- Determine how adult survival and offspring survival affect the breeding success of plants.
- Evaluate how trade-offs in reproduction and survival affect population growth.
- For a given environment, determine the life history schedule that maximizes growth.

Suggested Preliminary Exercises: Age-Structured Matrix Models; Life Tables and Survivorship Curves

INTRODUCTION

A sockeye salmon (*Oncorhynchus nerka*) is born in an Alaskan stream. It migrates to the ocean and spends several years there while it grows to reproductive size, and then journeys back to its natal stream to spawn. It lays hundreds of eggs (few of which will survive to reproductive age) and then dies. Foxglove (*Digitalis purpurea*) is a plant that flowers when it reaches a critical size (usually 2 years after it germinates), produces hundreds of seeds, and normally dies after setting seed. Human beings (*Homo sapiens*) have a typical life span of more than 65 years and can produce offspring when they are teenagers. Female humans typically produce a single offspring in each reproductive bout (multiple births, even twins, are relatively rare) and provide more than a dozen years of care for their young. These examples describe the life history of various species. If you've worked through a life table exercise, you've essentially charted an organism's life history.

Ecologists describe a species in terms of its reproductive life history. Life history schedules address the following questions:

- At what age does reproduction start?
- How many offspring are typically produced in a single reproductive bout?
- How many reproductive bouts does an organism have in its lifetime?
- Does number of offspring produced vary with the adult's age?

Species that reproduce only once during their life have a **semelparous** life history strategy. Salmon are examples of semelparous species. The fecundity schedule for such an organism would have zero for all age brackets except the age at which the reproduction occurs. Semelparous species can be early reproducers (produce offspring in their first year of life, such as many annual plants), or late reproducers (produce offspring after their first year of life, such as salmon). In contrast, **iteroparous** species reproduce several times in a lifetime. Maple trees, humans, and sea turtles are examples of iteroparous species.

To begin our discussion of life histories, let's assume that a hypothetical species has two age classes and that its life history can be shown with a Leslie matrix. Let's also assume that the second age class is a composite age class consisting of individuals of age 2 and any older individuals. This Leslie matrix has the form

$$A = \begin{bmatrix} F_1 & F_2 \\ P_1 & P_2 \end{bmatrix}$$

Remember that the top row of the Leslie matrix gives the fertility (F) of age class 1 and age class 2+ (which is a composite of 2-year-olds plus any older individuals). Let's assume that $F_1 = F_2 = 10$ individuals per individual per year. The bottom row of the Leslie matrix gives the survival probabilities, P. The left entry is the probability that an individual in age class 1 will survive to age class 2+, and the right entry is the probability that an individual in age class 2+ will survive to live additional years, and remain in the 2+ age class. Let's assume that these parameters are 0.3 and 0.4, respectively. If we describe life histories generally in terms of early reproduction versus late reproduction and semelaparous versus iteroparous, we arrive at four life history strategies and their associated Leslie matrices (Table 1).

Table 1. Four Life History Strategies and their Associated Leslie Matrices

The top row of each matrix gives the fertility of age classes 1 and 2+, F_1 and F_2, respectively. The lower row of each matrix gives the respective survival probability for each class, P_1 and P_2. The left-hand column represents age class 1, the right-hand column age class 2+.

	Semelparous	**Iteroparous**
Early reproduction	$A = \begin{bmatrix} 10 & 0 \\ 0 & 0 \end{bmatrix}$	$A = \begin{bmatrix} 10 & 10 \\ 0.3 & 0.4 \end{bmatrix}$
Delayed reproduction	$A = \begin{bmatrix} 0 & 10 \\ 0 & 0 \end{bmatrix}$	$A = \begin{bmatrix} 0 & 10 \\ 0.3 & 0.4 \end{bmatrix}$

Trade-Offs between Reproduction and Survival

Ideally, a species would reproduce as often as possible, have as many young as possible to maximize lifetime reproduction, and live forever. But is it that simple? An organism has a finite amount of energy to allocate to survival and reproduction. Energy allocated to reproduction means that less energy may be allocated to growth or maintenance (i.e., tasks that enhance survival). This creates a **trade-off** between present reproduction and survival, since organisms cannot maximize both. If size confers a significant survival advantage, for example, an organism may maximize its growth at the expense of reproducing until it reaches a critical size (Silvertown and Dodd 1999). And individuals that invest heavily in early reproduction may have poor survivorship later in life (Gotelli 2001). Figure 1 shows such a trade-off. The x-axis gives the proportional effort invested in reproduction, ranging between 0 and 1. The y-axis gives the survival rate, adjusted for the reproductive effort. When the proportional reproductive

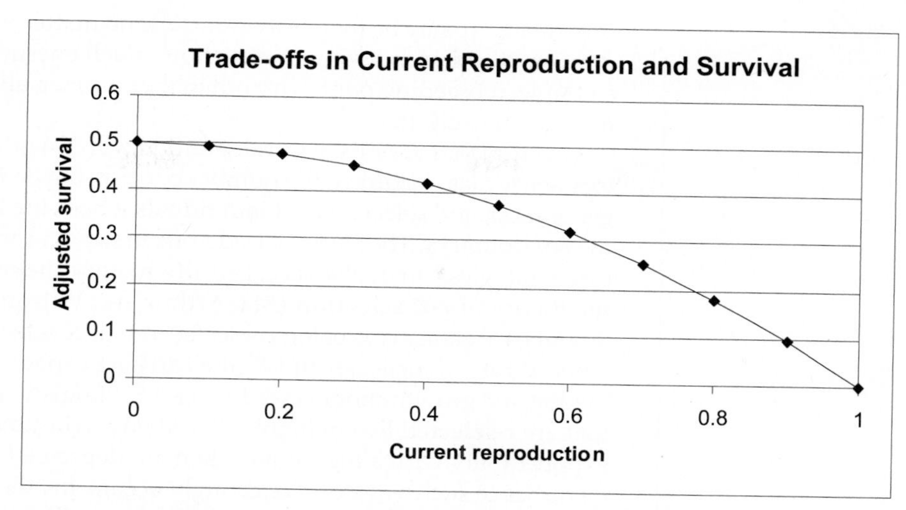

Figure 1 The shape of this curve indicates that virtually any energy devoted to reproduction will negatively impact survival; this species has a "high cost of reproduction" since the curve slopes downward.

effort is 0, no energy is devoted to current reproduction, and survival is determined by the intrinsic qualities of the environment in which the organism lives. In Figure 1, the survival rate is 0.5 even when individuals do not reproduce. When reproductive effort is greater than 0, it has a negative impact on survival, and the nature of this impact depends on the shape of the curve. When the effort is 1, all energy is devoted towards current reproduction, and survival becomes 0. Figure 1 has a fairly steep slope, which suggests that there is a "high cost of reproduction" in this environment. A high cost of current reproduction negatively impacts survival, which in turn affects future population size and hence future reproduction.

Figure 2 also shows trade-offs between survival and reproduction. However, survival is not decreased until almost all energy is devoted towards current reproduction. This environment would be considered a "low cost of reproduction" environment. Such environments may be so benign that resources are available for both survival and reproduction (survival is high no matter how much energy is devoted to reproducing). Or,

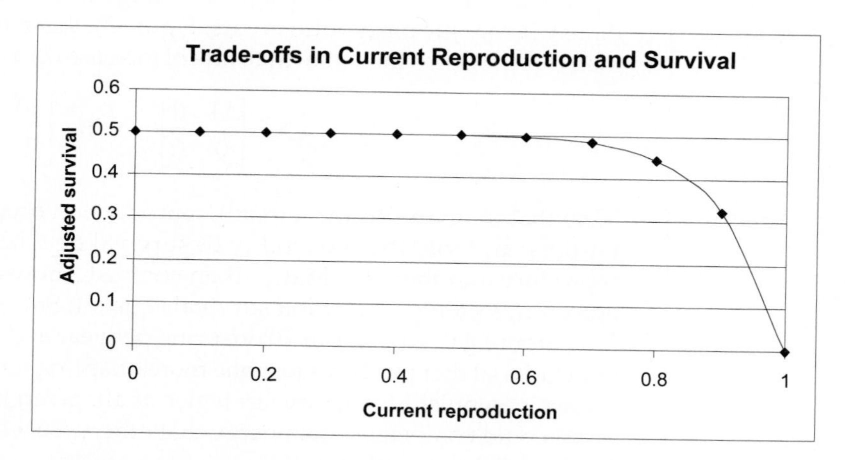

Figure 2 The shape of this curve indicates that energy expended on reproduction has little impact on survival unless almost all of an individual's energy is devoted to reproduction. This species has a "low cost of reproduction."

conversely, it may be that individuals die no matter how much effort is put into reproduction (survival is low no matter how much energy is devoted to reproducing). For example, if breeding ponds dry out in the summer, all the adults die regardless of their reproductive effort.

Given such trade-offs, natural selection will "favor" those individuals whose life history schedules maximize the number of offspring an individual contributes to the next generation, and select against individuals whose life histories are less compatible with the environment. The study of trade-offs in survival and reproduction, and how life history strategies can evolve, is called **life history theory**. One such life history theory is the theory of ***r-K*** **selection** (MacArthur and Wilson 1967; Pianka 1970). This theory describes organisms as being r-selected versus K-selected, where the terms r (the instantaneous rate of increase) and K (the carrying capacity of the environment) come from the logistic growth model (see Exercise 8, "Logistic Population Models"). Organisms that are ***r*-selected** live in highly disturbed environments, tend to increase in numbers exponentially with a high r, and then are depressed dramatically in numbers when a disturbance such as a storm or drought occurs. In other words, their growth is governed by r (or λ) until a disturbance occurs. Such populations rarely approach K and intraspecific competition has a negligible impact on growth rates. Because the future is uncertain in terms of resources, these organisms tend to breed early in life, are semelparous, and have large clutches.

In contrast, organisms that live in more stable, competitive environments are called ***K*-selected** species because their population numbers tend to be stable over time and exist at levels near the carrying capacity of the environment. Intraspecific competition is great for such species. These organisms tend to bear fewer offspring later in life and are iteroparous, because this schedule gives young a competitive advantage to survive in a competitive environment. A summary of how life history attributes are expected to vary for r and K selected species is given in Table 2.

Cole's Paradox

Even before r-K selection theory was formulated, Lamont Cole (1954) wondered about how life histories evolve in plant species. An **annual** plant is one that reproduces in its first year and then dies. Thus, an annual has a semelparous reproductive strategy. A **perennial** plant may also reproduce in its first year, but survives into future years and reproduces each year thereafter; thus it has an iteroparous reproductive strategy.

Cole realized that an annual strategy could achieve the same growth rate (λ) as a perennial strategy, where a perennial is immortal (never dies; survival = 1), as long as the annual can reproduce just one more offspring per year than the perennial. If we assume a population is censused with a prebreeding census (all individuals are counted immediately before the birth pulse occurs, Figure 3), this means that an annual with a Leslie matrix **A** produces the same finite rate of increase (λ) as a perennial with Leslie matrix **B**:

$$\mathbf{A} = \begin{bmatrix} 11 & 0 \\ 0 & 0 \end{bmatrix} \qquad \mathbf{B} = \begin{bmatrix} 10 & 10 \\ 1 & 1 \end{bmatrix}$$

Note that in matrix **A** (the annual), reproduction occurs in only one age class (semelparous), and that the probability of survival beyond age class 1 is 0, so individuals reproduce and then die. Matrix **B**, in contrast, shows reproduction occurring in both age groups (iteroparous), and survival equals 1. Since the two matrices yield the same λ, a perennial that produces 10 offspring per year and lives forever has the same fitness as an annual that produces just one more offspring and then dies. Cole wondered why we see perennial life history strategies at all, given that just a bit more reproductive effort could compensate for energy that otherwise would be devoted to survival. This is called **Cole's paradox**.

The key to understanding Cole's paradox is to realize that in a matrix model, the fertility rates for each age class (F_i for age class i) are the birth rates (b_i) adjusted for survival (see Exercise 13, "Age-Structured Matrix Models"). Figure 3 illustrates this using

Table 2. Summary of *r*- versus *K*-Selected Life History Strategies

r-Selected	*K*-Selected
Reproduce early, since disturbance is frequent and unpredictable; those individuals that wait to reproduce may die before reproduction occurs.	**Reproduce later**, since individuals that reproduce early are likely to have smaller, less competitive offspring.
Produce many offspring per reproductive bout. Saving energy for future reproduction is fruitless if the probability of mortality in the future is uncertain.	**Produce fewer offspring** per reproductive bout, since fewer offspring with parental care are more likely to survive in a competitive environment than many offspring with no parental care.
Produce small offspring, because if there is a finite amount of energy that can be used for reproduction, more offspring can be produced if each offspring is small.	**Produce large offspring**, since smaller offspring will not be able to compete and survive in competitive environments as well as larger offspring.
Smaller adults. Because individuals breed at an early age, breeding individuals may be smaller on average than *K*-selected species.	**Larger adults** able to produce larger offspring.
Tendency is to reproduce once, then die. Allocating energy for future reproduction in an uncertain environment may lead to fewer offspring overall.	**Tendency to reproduce repeatedly**, because only one or few offspring are produced per reproductive bout and require care. Repeated reproduction allows more total offspring to be produced, spread out over the reproductive portion of the life cycle.
Type III survivorship curve. Because the environment is unpredictable, and because offspring are small, survivorship is low for young and intermediate ages.	**Type I survivorship curve**. Because the young are large and competitive, there is high survivorship of young and intermediate ages, then a drop-off as old age sets in.

Summarized from Begon et al. (1986).

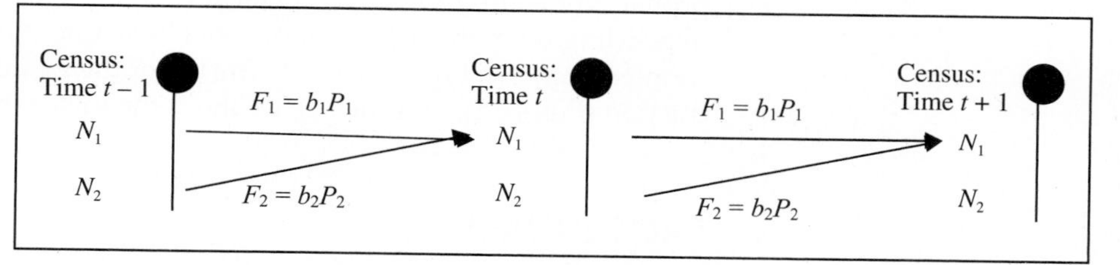

Figure 3 In this hypothetical population, the number of individuals of each age class (N_1 and N_2) is counted during the census, and a birth pulse (filled circles) occurs just after the census. Offspring are produced according to the birth rate (b_1 or b_2). Both age classes contribute individuals to age class 1 in the next year. However, in order for these young to be counted in the population as 1-year-olds (and to reproduce) in the next time step, they must survive almost a full year, until the next birth pulse. Thus, the fertilities are multiplied by the probability that an individual will survive to reproduce the following year (P_1 or P_2). The resulting adjusted fertility (b_iP_i) gives the number of offspring produced per individual that will survive and be counted in the next time step (Caswell 1989).

a hypothetical population with two age classes, censused over a 3-year period (time $t - 1$, time t, and time $t + 1$). Cole's paradox relies on the unlikely assumption that *all* individuals born in year t will survive to year $t + 1$ (i.e., $P_1 = 1$).

Model Development

In this exercise, you will set up a matrix model of Cole's paradox, and will explore the conditions that lead to iteroparity, semelparity, early reproduction, and late reproduction. Our model will take the form of a Leslie matrix model, but will include something that Cole did not consider: trade-offs in survival and reproduction. The standard matrix model has the form

$$\begin{bmatrix} F_1 & F_2 \\ P_1 & P_2 \end{bmatrix} \times \begin{bmatrix} N_{1(t)} \\ N_{2(t)} \end{bmatrix} = \begin{bmatrix} N_{1(t+1)} \\ N_{2(t+1)} \end{bmatrix} \qquad \text{Equation 1}$$

where F_1 and F_2 are the fertility rates of 1-year-olds and 2-year-olds, respectively, and P_1 and P_2 are the survival rates. P_1 gives the probability that an individual in the first age class will survive to the second age class. P_2 gives the probability that an individual in the second age class will survive but remain in age class 2+. The model multiplies the matrix of fertilities and survivals by the number of individuals in each age class at time t to give the number of individuals in each age or stage class at time $t + 1$. For example, the number of individuals in age/stage 1 at time $t + 1$ ($N_{1(t+1)}$) is computed as

$$N_{1(t+1)} = F_1 \times N_{1(t)} + F_2 \times N_{2(t)}$$

The number of individuals in age/stage 2 at time $t + 1$ is computed as

$$N_{2(t+1)} = P_1 \times N_{1(t)} + P_2 \times N_{2(t)}$$

We will modify the standard matrix model by adding terms to the F_i and P_i elements in the Leslie matrix, which control trade-offs in survival and reproduction (after Cooch and Ricklefs 1994).

$$\begin{bmatrix} F_1 \times E & F_2 \times E \\ P_1 \times (1 - E^z) & P_2 \times (1 - E^z) \end{bmatrix} \times \begin{bmatrix} N_{1(t)} \\ N_{2(t)} \end{bmatrix} = \begin{bmatrix} N_{1(t+1)} \\ N_{2(t+1)} \end{bmatrix} \qquad \text{Equation 2}$$

The term E gives the proportional effort of energy allocated towards current reproduction, and ranges from 0 to 1. Thus, the fertility rates are multiplied by E in the modified Leslie matrix. When $E = 1$, all energy is allocated toward current reproduction, so individuals reproduce with fertility rates in the standard model. As E decreases, the current fertility rate decreases proportionately. The trade-off between current reproduction and survival is reflected in the second row of the Leslie matrix. Each survival probability is multiplied by the term $(1 - E^z)$. The survival probabilities are adjusted depending on both E (the proportional investment into reproduction) and z (the environment's cost of reproduction). The lower the value of z, the higher the cost of reproduction (Figure 1), and the higher the z, the lower the cost of reproduction (Figure 2).

PROCEDURES

With this background in mind, let us begin with the model. The goal of the model is to explore how λ, the finite rate of increase, can be maximized given trade-offs in survival and reproduction, and to think about the kinds of environments that promote early versus late reproduction, and semelparous versus iteroparous reproduction. If you are rusty on Leslie matrices, refer back to Exercise 13 before you begin. As always, save your work frequently to disk.

INSTRUCTIONS	ANNOTATION

A. Model Cole's paradox.

1. Open a new spreadsheet and set up headings as shown in Figure 4.

	A	B	C	D	E
1	*Life History Trade-offs*				
2					
3					
4					
5		Age			
6		1	2+	*n*	
7		11	0	10	
8	A =	0	0	0	
9					
10	Numbers over time - no trade-offs				
11	Time	Age 1	Age 2	Total pop	λ

Figure 4

2. Enter the parameter values shown in cells B7–C8.

We will consider a plant that has just two age/stage classes. The matrix of cells

$$\begin{bmatrix} 11 & 0 \\ 0 & 0 \end{bmatrix}$$

is the life history for an annual plant. Each plant in the first year of life produces 11 offspring, and then dies. P_1, the probability that a individual in age class 1 will move to age class 2, is 0. Thus, F_2 and P_2 are also 0.

3. Enter the initial vector of abundances as shown in cells D7 and D8.

The initial vector of abundances

$$\begin{bmatrix} 10 \\ 0 \end{bmatrix}$$

gives the starting number of individuals in age class 1 and age class 2+, respectively.

4. Set up a linear series from 0 to 50 in cells A12–A62.

Enter 0 in cell A12.
Enter **=1+A12** in cell A13. Copy cell A13 down to cell A62.
This will allow us to track the dynamics of this plant species over 50 years.

5. Enter formulae in cells B12 and C12 to link the number of individuals in age classes 1 and 2 to the vector of abundances in cells D7 and D8.

Enter the formula **=D7** in cell B12.
Enter the formula **=D8** in cell C12.

6. Enter a formula in cell D12 to compute the total population size at time 0.

Enter the formula **=SUM(B12:C12)** in cell D12.

7. In cell E12, Compute λ for year 0 as $N_{(0)}/N_{(1)}$.

Enter the formula **=D13/D12** in cell E12. Your result will not make sense until you have computed the total population size in year 1.

8. Enter formulae in cells B13 and C13 to compute the number of individuals in age class 1 and 2 in year 1.

Enter the formula **=$B$7*B12+$C$7*C12** in cell B13 to compute the number of individuals in age class 1 in year 1.
Enter the formula **=$B$8*B12+$C$8*C12** in cell C13 to compute the number of individuals in age class 2+ in year 1.
Remember, the matrix calculations are

$$\begin{bmatrix} a & b \\ c & d \end{bmatrix} \times \begin{bmatrix} x \\ y \end{bmatrix} = \begin{bmatrix} ax+by \\ cx+dy \end{bmatrix}$$

9. Copy cells B13–C13 down to cells B62–C62.

10. Copy cells D12–E12 down to cells D62–E62.

This completes the 50-year projection of your population. Your spreadsheet should now resemble Figure 5.

	A	B	C	D	E
10	Numbers over time - no trade-offs				
11	Time	Age 1	Age 2	Total pop	λ
12	0	10	0	10	11
13	1	110	0	110	11
14	2	1210	0	1210	11
15	3	13310	0	13310	11
16	4	146410	0	146410	11
17	5	1610510	0	1610510	11
18	6	17715610	0	17715610	11
19	7	194871710	0	194871710	11
20	8	2143588810	0	2143588810	11
21	9	2.3579E+10	0	2.3579E+10	11
22	10	2.5937E+11	0	2.5937E+11	11

Figure 5

11. Graph the population numbers over time.

Use the scattergraph option, and label your axes fully. Your graph should resemble Figure 6.

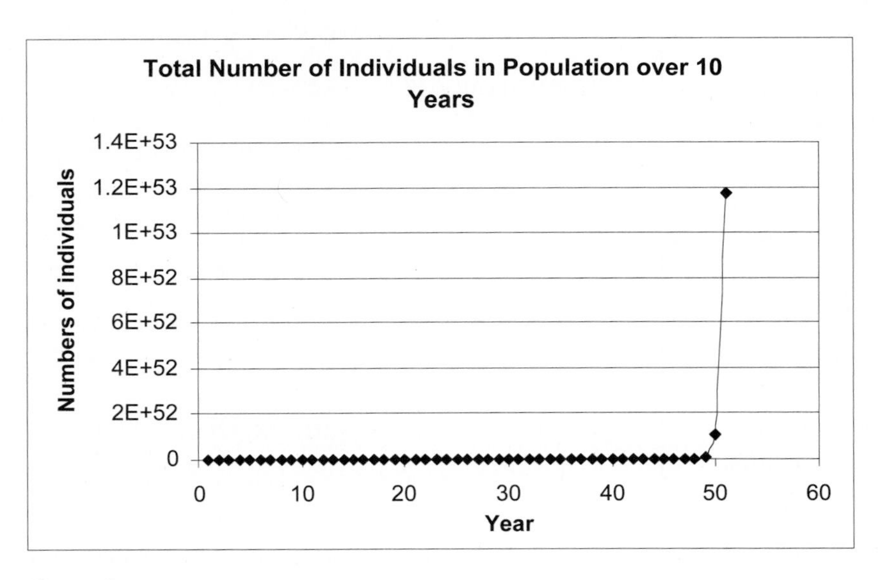

Figure 6

12. Change the matrix entries as shown in Figure 7.

	A	B	C	D
5		Age		
6		1	2+	*n*
7		10	10	10
8	**A** =	1	1	0

Figure 7

13. Update your projections (this may be done automatically, or by pressing F9).

The matrix entries now suggest an everlasting perennial. All individuals produce 10 offspring per year, and survival from age class 1 to age class 2+ is 1. Additionally, all individuals in age class 2+ survive with a probability of 1 to the next age class, and then survive with a probability of 1 to the next age class (and so on).

14. Answer Questions 1–3 at the end of this exercise.

15. Save your work.

B. Establish trade-offs for survival versus reproduction.

Now we will add trade-offs between survival and current reproduction into the model.

1. Set up new headings as shown in Figure 8.

We will let *E* be a proportional reproductive effort. If *E* is 1, then the organism reproduces at fertility rates given in the original Leslie matrix. If *E* is 0, then current reproduction is 0 times the fertility rates in the Leslie matrix. If *E* is any value between 0.1 and 0.9, that number is multiplied by the fertility rates in the Leslie matrix. Thus, *E* "brakes" the fertility rates by a proportional amount.

	K	L
6	**Trade-off parameters**	
7	***E*** =	
8	***z*** =	2
9	***P*** =	0.5
10	***E***	**Adjusted survival**
11	0	
12	0.1	
13	0.2	
14	0.3	
15	0.4	
16	0.5	
17	0.6	
18	0.7	
19	0.8	
20	0.9	
21	1	

Figure 8

2. Enter 0.5 in cell L9.

P is a generic survival value, or the probability that an organism of age *x* will survive to the next time step. For now $P = 0.5$. You will be able to change this shortly.

3. Enter 2 in cell L8.

The variable *z* controls the cost of reproduction in an environment. For our purposes, we will let *z* range between 0 and 20. The higher the *z*, the lower the cost of reproduction, and the lower the *z*, the higher the cost of reproduction in a given environment. Currently $z = 2$, so the cost of reproduction is high. You will be able to see how *z* and *P* affect trade-offs in survival and reproduction shortly.

4. In cells L11–L21, enter an equation to compute adjusted survival for a given level of *E*. Refer to Equation 2.

Enter the formula **=$L$9*(1-K11^$L$8)** in cell L11. Copy the formula down to cell L21. The adjusted survival can be computed as $P \times (1 - E^z)$.

5. Graph the adjusted survival as a function of *E*.

Use the scatterplot option and label your axes fully. Your graph should resemble Figure 9.

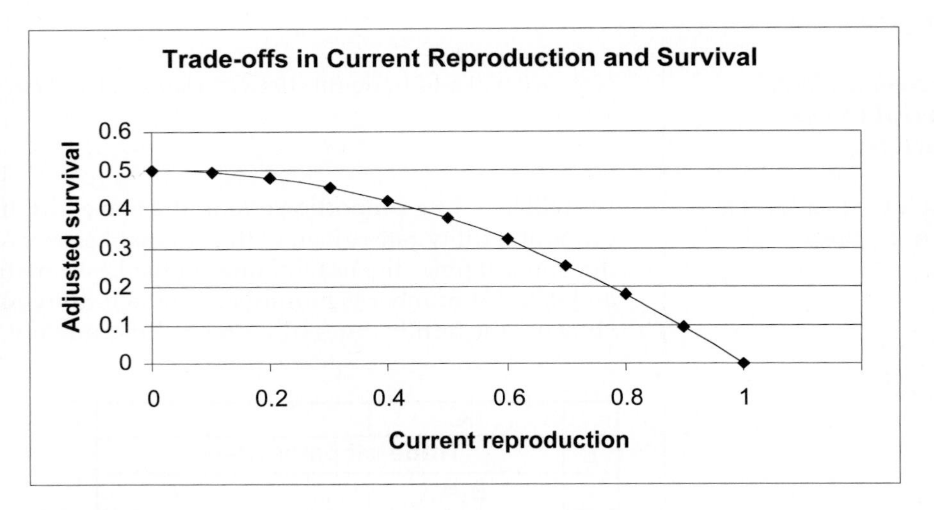

Figure 9

6. Interpret your graph fully.

Keep in mind that this figure is for $z = 2$ and $P = 0.5$. This figure will change as you modify *z* and *P* in the next step. When current reproductive effort is 1 (100%), survival becomes 0 because all energy is devoted to reproduction. When current reproduction is 0, adjusted survival is at 0.5, the baseline survival value. In between, as current reproduction effort is increased, the adjusted survival probability decreases rather abruptly. This is the trade-off between energy allocated to survival and energy allocated to reproduction. We will incorporate this trade-off into the matrix model in Part C.

7. Increase the value of *z* to 20 by units of 5, and interpret your final graph ($z = 20$).

Your graph should resemble Figure 10.
You should see that as *z* increases, the cost of reproduction is lessened. When *z* is 20, there is still a trade-off between survival and reproduction, but survival is adjusted only when reproductive effort is close to 100% effort ($E = 1$). Habitats with high *z*'s are low cost of reproduction habitats.

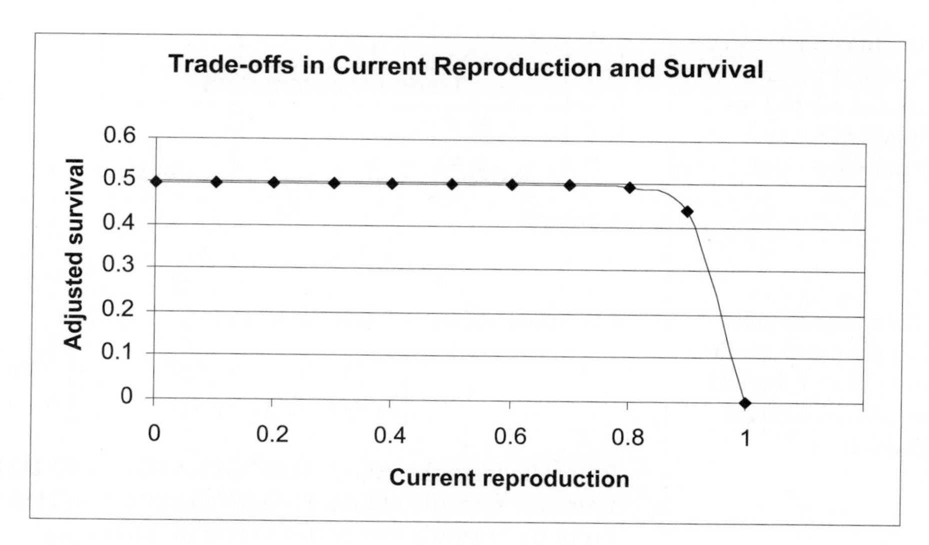

Figure 10

8. Save your work.

Now that you have a handle on how E and z function, the next step is to incorporate E and z in the matrix model.

C. Set up the trade-off model.

1. Set up new headings as shown in Figure 11.

	F	G	H	I	J
1			Trade-off Model		
2					
3					
4					
5		Age			
6		**1**	**2+**	***n***	
7		11	0	10	
8	**A =**	0	0	10	
9					
10			Numbers over time - with trade-offs		
11	**Time**	**Age 1**	**Age 2**	**Total population**	λ

Figure 11

2. Set up a time series from 0 to 50 in cells F12–F62.

Enter 0 in cell F12.
Enter the formula **=1+F12** in cell F13. Copy this formula down to cell F62.

3. Enter a formula in cells G12 and H13 that links to the initial vector of abundances in cells I7 and I8.

Enter the formula **=I7** in cell G12.
Enter the formula **=I8** in cell H12.

4. Enter a formula in cell I12 to compute the total population size. Compute λ in cell J12.

Enter the formula **=SUM(G12:H12)** in cell I12.
Enter the formula **=I13/I12** in cell J12.

5. Enter 0.5 in cell L7 and 2 in cell L8. *E* and *z* establish the cost of reproduction on survival for the trade-off matrix model.

	K	L
6	Trade-off parameters	
7	E =	0.5
8	z =	2

Figure 12

6. Enter formulae in cells G13–J13 to project population sizes in year 1, including trade-offs in survival and reproduction.

Our trade-off matrix has the form

$$\begin{bmatrix} F_1 \times E & F_2 \times E \\ S_1 \times (1-E^z) & S_2 \times (1-E^z) \end{bmatrix} \times \begin{bmatrix} N_{1(t)} \\ N_{2(t)} \end{bmatrix} = \begin{bmatrix} N_{1(t+1)} \\ N_{2(t+1)} \end{bmatrix}$$

Enter the formula **=$G$7*$L$7*G12+$H$7*$L$7*H12** in cell G13.
Enter the formula **=$G$8*(1-$L$7^$L$8)*G12+$H$8*(1-$L$7^$L$8)*H12** in cell H13.
Enter the formula **=SUM(G13:H13)** in cell I13.
Enter the formula **=I14/I13** in cell J13.

7. Copy cells G13–J13 down to row 62.

8. Graph the population size over time.

Use the scatterplot option and label your axes fully. Your graph should resemble Figure 13. What is the asymptotic λ for your model? (This is a key model output that will be compared to the models.)

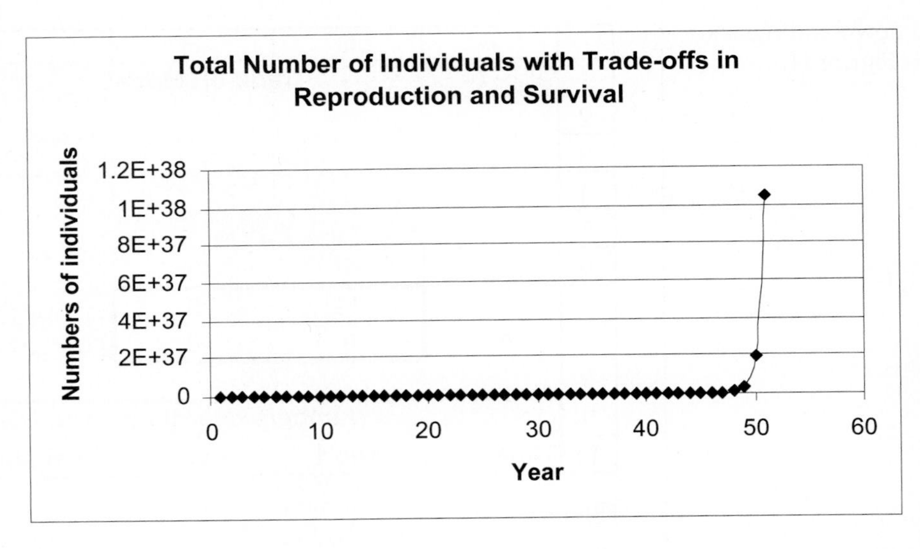

Figure 13

9. Save your work and answer questions 4–9.

QUESTIONS

1. What is Cole's paradox? Which of the two strategies (annual or perennial) is the fittest in this environment? Try entering other fertilities in the Leslie matrix so that the annual has 1 more offspring than the perennial. Does Cole's paradox still hold?

2. In modeling Cole's paradox, we set adult survival of perennials to 1 so that a perennial never dies. What is another major assumption of Cole's paradox regarding the fertility rates of the annual life history strategy?

3. To understand Cole's paradox more fully, it's helpful to break apart the F entry for age class 1 into its components, b_1 and P_1, where b_1 is the per capita birth rate of one-year-old females and P_1 is the probability that an offspring produced will survive to be counted as a one-year-old in the next census. Set up column headings as shown, and enter a formula in cell B7 to compute F_1 as cell B3*B4, or b_1P_1. When the probability of juvenile survival decreases, how much must b_1 increase (cell B3) to match the λ of the everlasting perennial? Track your results for $P_1 = 0.1$ to 1 in increments of 0.1, display your results graphically, and interpret your results.

	A	B	C
3	b_1 =	11	
4	P_1 =	1	
5		**Age**	
6		**1**	**2+**
7		11	0
8	**A** =	0	0

4. In question 3, you addressed what b_1 must be for an annual to match the growth rate of an everlasting perennial when juvenile survivorship (P_1) is not 1. Now let's focus on what happens when a perennial is not immortal, and consider trade-offs between current reproduction and future survival. In the trade-off model, which of the strategies below will yield the highest asymptotic growth rate, λ: the annual matrix, **A**, or the perennial matrix, **B**? Explain your results.

$$\mathbf{A} = \begin{bmatrix} 11 & 0 \\ 0 & 0 \end{bmatrix} \qquad \mathbf{B} = \begin{bmatrix} 10 & 10 \\ 1 & 1 \end{bmatrix}$$

5. How does changing z in Question 4 affect the asymptotic growth rate, λ, for the annual? For the perennial?

6. Set up spreadsheet parameter values as shown below. Is this a low or a high cost-of-reproduction environment? Assuming a hypothetical organism that can produce 100 offspring maximum per year, what kind of reproduction schedule (early versus late, iteroparous versus semelparous) will maximize λ? Given your results, how can adjustments to E affect which life history strategy will be most fit?

	F	G	H	I	J	K	L
5		**Age**					
6		**1**	**2+**	*n*		**Trade-off parameters**	
7				10		*E* =	0.9
8	**A** =	0.1	0.1	10		*z* =	20
9						*P* =	0.1

7. Change the survival rates to 0.9 in your matrix. Would an early semelparous or early iteroparous strategy be favored under these conditions? Why?

	F	G	H	I	J	K	L
5		**Age**					
6		**1**	**2+**	*n*		**Trade-off parameters**	
7				10		*E* =	0.9
8	**A** =	0.9	0.9	10		*z* =	20
9						*P* =	0.9

8. Consider another environment and a different organism. Set up your spreadsheet as shown below. Assuming that your organism can produce a maximum of 5 offspring per year, what kind of reproductive schedule will maximize λ?

	F	G	H	I	J	K	L
5		Age					
6		**1**	**2+**	*n*		**Trade-off parameters**	
7				10		*E* =	0.5
8	A =	0.9	0.9	10		*z* =	1
9						*P* =	0.9

9. Suppose an organism's life history can be described with the Leslie matrix shown below. What level of E will produce the highest λ? Explain your result in detail in terms of trade-offs in reproduction and survival. What level of E would produce the highest λ if cells G8–H8 = 0.9? Explain.

	F	G	H	I	J	K	L
5		Age					
6		**1**	**2+**	*n*		**Trade-off parameters**	
7		1	5	10		*E* =	0.6
8	A =	0.4	0.4	10		*z* =	1
9						*P* =	0.4

10. Two marine bivalves, *Mercenaria mercenaria* and *Gemma gemma*, live in the same habitat. However, their reproductive strategies are very different. *M. mercenaria* is a broadcast spawner, meaning that male and female adults release eggs and sperm into the water column where external fertilization takes place, and the larvae undergo planktonic development. *G. gemma* is a brooder, meaning females retain their eggs and fertilization is internal. The offspring undergo direct development within the female. *G. gemma* produce small broods during the reproductive season, while *M. mercenaria* releases thousands of gametes into the water column. Surprisingly, both species enjoy similar reproductive success. Let's assume that in each reproductive season *G. gemma* will successfully rear 25 offspring that survive to be counted as N_1 individuals, and *M. mercenaria* will release 4000 gametes, all of which will be fertilized. Assuming equal costs of reproduction, what must the survival rate of *M. mercenaria* offspring (P_1) be in order to equal the reproductive output *G. gemma*?

*11. (Advanced) Some organisms have life histories that cannot be described as either r or K. "Bet-hedging" is a strategy that is predicted to evolve in environments that have unpredictable disturbances that increase the mortality of young, but not adults. If young are produced all at once, and it turns out to be a bad year, then an adult's fitness is 0. But if young are spread out across different generations, fitness may be increased by producing at least some young in some years when conditions are good. Add an element of stochasticity to your model that affects juvenile survival rate either by letting F_1 vary stochastically, or by splitting apart F into its components b_1 and P_1 (as in Question 3) and letting each component vary. Adjust your model, then examine the life history conditions that are needed to maximize λ.

LITERATURE CITED

Begon, M., L. Harper and C. R. Townsend. 1986. *Ecology.* Blackwell Scientific, Cambridge, MA.

Caswell, H. 2001. *Matrix Population Models*, 2nd Edition. Sinauer Associates, Sunderland, MA.

Cole, L. 1954. The population consequences of life history phenomena. *Quarterly Review of Biology* 29: 103-137.

Cooch, E., and R. Ricklefs. 1994. Do variable environments significantly influence optimal reproductive effort in birds? *Oikos* 69: 447–459.

Gotelli, N. 2001. *A Primer of Ecology*, 3rd Edition. Sinauer Associates, Sunderland, MA.

MacArthur, R. H. and E. O. Wilson. 1967. *The Theory of Island Biogeography*. Princeton University Press, Princeton, NJ.

Pianka, E. R. 1970. On *r*- and *K*-selection. *American Naturalist* 104: 592–597.

Silvertown, J. and M. Dodd. 1999. The demographic cost of reproduction and its consequences in balsam fir (*Abies balsamea*). *American Naturalist* 154: 321–332.

36

HERITABILITY

In collaboration with Mary Puterbaugh and Larry Lawson

Objectives

- Understand the concept of heritability.
- Differentiate between broad-sense heritability and narrow-sense heritability.
- Learn different methods for computing heritability.
- Understand the conditions that lead to high heritability and low heritability.

INTRODUCTION

Can you think of a physical trait that makes you different from your brother or sister? You may be taller than your sibling or have darker skin or have a different hair color. Can you think of a trait in which you and your sibling are similar? Were either of these traits inherited from your parents, or were they controlled more by environmental factors?

Most people have a good general concept of **heritability**. Surprisingly, the strict scientific definition of heritability is a much more difficult concept to grasp than our everyday use of the word. This is partly because heritability has a theoretical definition that is impossible to directly measure in the field, and there are several different ways to estimate heritability in practice (e.g., twin studies, breeding experiments, offspring-parent regressions, and selection experiments). These different ways of estimating heritability have assumptions. As such, it is not uncommon that two different methods of estimating heritability might lead to quite different values even in the same population in the same environment.

Possibly the most important key to understanding the scientific definition of heritability is to realize that the trait itself is almost completely unimportant to the definition of heritability. Rather, it is the *variation* in the trait that is important. If you repeatedly remind yourself that heritability is defined by the variation in a trait and not by the trait itself, you will avoid falling into many pitfalls with your understanding of the term.

The Theoretical Definition of Heritability

Imagine that you take a black-and-white photograph of people you know and you "score" the darkness of their hair with a single value. The lightest-haired people would receive a zero and the darkest-haired people would receive a 100. Everyone else would receive values between these. You could describe the variation among individuals by calculating the **variance**, a common statistic that you are likely to have calculated in your science, math, or statistics courses. This statistic is (approximately) the average squared deviation from the mean, and we calculate it to measure the amount of variation in a collection of observations. For a sample taken from a population, variance (abbreviated V in this exercise) is calculated as

$$V = \frac{\sum (X_i - \overline{X})^2}{N-1} \qquad \text{Equation 1}$$

N is used when the computations are for a population, and $N - 1$ is used when the computations are for a sample of the population.

For a set of observations, the variance is easily computed with a spreadsheet function. Individuals vary in their hair color for at least two different reasons. One is that they inherited different kinds of genes for hair color, and the other reason is that they've experienced different environments. For example, hair color may depend on a chemical environment (a hair dye or bleach), or on time spent (or not spent) exposed to the sun. Theoretically, the variance in hair color (abbreviated V_p; the "p" subscript comes from term "phenotype") can be divided into the variance that is due to genetic differences among individuals (V_g) and the variance due to differences among the environments of the individuals (V_e) Thus,

$$V_p = V_g + V_e \qquad \text{Equation 2}$$

Heritability (abbreviated here as h^2) in a strict genetic sense is the proportion of total phenotypic variance in a trait that is explained by genetic differences among individuals. Theoretically, heritability can vary between 0 and 1.

$$h^2 = V_g/V_p \qquad \text{Equation 3}$$

Let us look more closely at the V_g and V_e components of total variation. How do we determine the deviations from which these variance components are calculated? If you could take the mean hair color of the population and then ask how much a *particular individual* differs from that *population* mean due to particular alleles it has, and then how much that individual differed from the population mean due to its environment, you could express these deviations with quantities called G and E, respectively, for each individual in the population (Hartl 2000). G represents a deviation of that individual's phenotype from the population mean (μ) due to the particular genotype that individual has, and E represents the deviation of that individual's phenotype from the population mean that is due the environment in which the individual was raised. Once you had a G and E for every individual, the variance in the G and E are the phenotypic variance due to environmental and genotypic effects, respectively. The variance in G would be calculated as

$$V_g = \sum \frac{G^2}{N}$$

and the variance in E would be calculated as

$$V_e = \sum \frac{E^2}{N}$$

Note that capital letters G and E are used for individuals. Provided that individuals were randomly occurring in different environments, $V_g + V_e$ would equal V_p as in Equation 2. Furthermore, you can now define the phenotype of each *individual* in a population in a particular environment (Equation 4). In this equation, the P stands for the phenotype of the particular individual. In theory, all the G's in the population should add to zero, and likewise all the E's should add to zero. Notice that if you took the variance of each variable in the equation below, you would recreate Equation 2 because the variance in μ is zero.

$$P = \mu + G + E \qquad \text{Equation 4}$$

One of the advantages of using modeling is that it can allow you to investigate a process that is not directly measurable in reality. In this exercise, you will construct a population and define the G's and E's, two variables that can not be directly measured in real life, so that you can investigate many aspects of the definition of heritability that are virtually impossible to investigate any other way.

Types of Genetic Variation

Before reviewing some of the practical methods of measuring heritability, it is useful to briefly discuss what types of genetic variation exist. The V_g that you have just reviewed above can also be partitioned into two parts: the phenotypic variances owing to additive genotypic effects (V_a) and the phenotypic variances owing to non-additive genotypic effects (V_{na}). Thus,

$$V_g = V_a + V_{na} \qquad \text{Equation 5}$$

In the exercise to follow, we will assume that $V_{na} = 0$. We will do this by constructing individuals with genotypes for two genes (A and B). For these genes, there will be only two alleles (a "1" allele and a "2" allele, each with a frequency of 0.5). Each allele will have a given affect on the phenotype of the individual regardless of what other allele occurs at that gene and regardless of what alleles occur at other genes. In other words, an A_1 allele will always be worth "+1" units from the mean in terms of your phenotype, an A_2 will be worth "–1"; a B_1 will be worth "+1" and a B_2 will be worth "–1". Thus an individual who is $A_1A_2B_2B_2$ will differ from the mean population phenotype by –2 units because the deviation of this individual's phenotype from the mean is $1 - 1 - 1 - 1 = -2$. In real life, the A_1 and A_2 alleles might interact—for example, A_2 might be dominant over A_1 (in which case the A_1 allele would be worth nothing in the presence of A_2). Likewise, it is not uncommon for **epistasis** to occur, meaning that the effect of an allele at the B gene depends on what alleles are at the A gene.

Thus, in the scientific literature, there are two types of heritability: **broad-sense heritability** (V_g/V_p where V_g includes the nonadditive component) and **narrow-sense heritability** (V_a/V_p where the numerator is only the additive component of genetic variance). In this exercise, all V_g is additive, so the broad- and narrow-sense heritabilities are the same. Narrow-sense heritability is a more useful measure of heritablity as it is the variance in a population that will respond predictably to selection. In the next exercise on quantitative genetics, you will see how heritability is related to a response to selection.

Practical Methods of Estimating Heritability

How does one go about estimating heritability if you cannot measure V_g and V_e directly? Probably the most conceptually simple way is to compare offspring to their parents. The more closely the offspring's phenotype is predicted by their parents' appearances, the more the variation among individuals in a population is due to genetic variation. Specifically, you can measure the trait in an offspring and graph it against the mean of the trait in the two parents (the **midparent trait value**; Figure 1). The slope

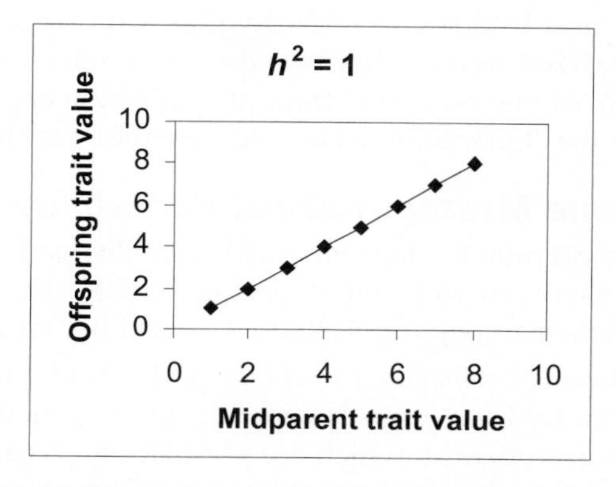

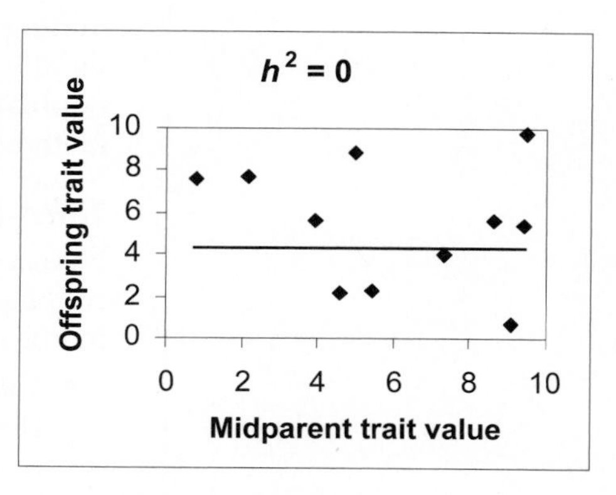

Figure 1 Parent-offspring regressions showing high (left) and low (right) heritability.

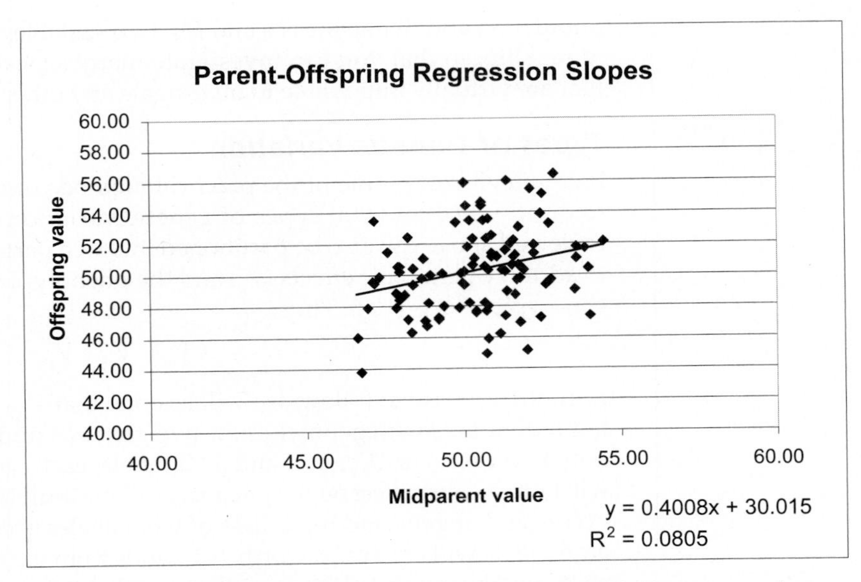

Figure 2 A parent-offspring regression in which heritability is somewhere between 0 and 1. The graph illustrates the typical scatter that you might find around a regression line.

of that plot of offspring values against midparent values is exactly narrow-sense heritability. In this exercise you will see that the slope really does accurately estimate the heritability that you can also calculate as V_g/V_p, if you know V_g.

When an offspring's trait is perfectly matched to the average of its two parents, h^2 = 1 (Figure 1, left). Small parents will have small offspring, and large parents will have large offspring. The slope of the line is 1, and $h^2 = 1$. When an offspring's trait cannot be predicted by the traits of its parents, $h^2 = 0$ (Figure 1, right). Parents of any size can have offspring of any size. In this case, the slope of the regression line is 0, and $h^2 = 0$.

In many cases, the parent-offspring graph for a given trait might look like Figure 2. This graph shows a tendency for larger parents to have larger offspring and for smaller parents to have smaller offspring, but there is substantial scatter. This suggests that h^2 would fall between 0 and 1.

There are other ways to measure heritability that we will not explore in this exercise. One commonly used method in human studies is to investigate twins. This method is based on the idea that monozygotic (identical) twins are more similar genetically than dizygotic (non-identical) twins. Other methods of estimating heritability involve estimating V_g and V_e through carefully planned breeding experiments (Falconer 1989). Finally, **realized heritability** (the degree to which a trait responds to selection in a population) can be estimated through a selection experiment. You will investigate this method in the "Quantitative Genetics" exercise that follows.

Take-Home Messages about Heritability

It is easy to get mired in the details of heritability and forget the big picture. If you recall, we began this exercise by emphasizing that heritability in a scientific sense is defined by the *variation* among individuals. This fact has two important consequences:

- Variation is a *population level trait* and is undefined at the level of an individual.
- Heritability is *not fixed*. It depends on the genetic variation in a population and the environment in which the population occurs. In other words, a population with exactly the same genetic composition as another population can have a different heritability if the two populations are in different environments.

Likewise, if the genetic composition of a population changes, even if the environment stays the same, estimates of heritability for that population will also change.

Let us return for a moment to the hair color example. Suppose cloning had advanced to the point that we could clone all the people you knew and split them into two groups. If we prevent the clones in one group from going out in the sun or using any hair dyes or bleaches, we might be able to eliminate most of variation among individuals in hair color that is due to environment (i.e., we could reduce V_e). If we allow the clones in the other group to go out in the sun and color their hair as they please, the variation in hair color due to environment will be greater, and heritability will be lower. To reiterate, even though the two populations would be identical genetically, the heritability would be different! Perhaps you can begin to see why heritability in a strict scientific sense has some nuances that make it quite different from the way we use the term in everyday conversation.

PROCEDURES

In this exercise, you will explore the theoretical definition of heritability. At the same time you will see that the practical method of constructing a regression of offspring against midparent values can also be used to estimate heritability. Two consequences of the theoretical definition of heritability (that heritability is a population level trait, and that it depends on both the genetic composition and environment of the population) will also be illustrated.

As always, save your work frequently to disk.

INSTRUCTIONS	ANNOTATION

A. Set up the population parameters.

1. Open a new spreadsheet as shown in Figure 3. Enter the values shown in cells B6–E6.

	A	B	C	D	E
1	***Heritability***				
2					
3			**Model inputs**		
4					
5	**Genotype**	**A1**	**A2**	**B1**	**B2**
6	*G*	1	-1	1	-1
7	Freq.	0.5	0.5	0.5	0.5
8					
9				**Population traits**	
10				**Parents**	**Offspring**
11		Average phenotype =		50.00	50.00
12		Environmental heterogeneity =		0.01	0.01

Figure 3

We will assume that genes at two loci control the trait, the *A* locus and the *B* locus. Thus, we are dealing with a **polygenic trait**. We will also assume that each locus has just two alleles, A_1 and A_2, and B_1 and B_2. In the simplest case, each allele makes a contribution to the expressed trait. For example, if an individual inherits an A_1 or B_1 allele from its parents, it "inherits" a +1 unit contribution in the trait size. If it inherits an A_2 or B_2 allele from its parents, it "inherits" –1 units in the trait size. Thus, $A_1A_1B_1B_1$ individuals will have a +4 phenotype, $A_2A_2B_2B_2$ genotypes will have a –4 phenotype. Because $A_1A_2B_1B_2$ (heterozygotes) will have a phenotype of 0, they are the "standard" upon which other genotypes are compared. Note that since two loci contribute to the trait size, the population will tend to exhibit **continuous variation** in trait size, ranging between –4 and +4 units.

2. Enter **0.5** in cells B7–E7.

Cells B7–E7 give the frequencies of each allele. Remember that the frequencies must add to 1. You will be able to change these frequencies later in the exercise. You may wish to enter the formula **=1-B7** in cell C7 and **=1-D7** in cell E7.

3. Enter **50** in cells D11–E11.

This represents the average phenotype of the parental population. In our example, the parents are currently located in an environment and have genotypes that confer, on average, 50 units to trait size (cell D11).

4. Enter **0.01** in cells D12–E12.

Cells D12–E12 set how variable the environmental conditions are for the parental and offspring populations, respectively. Each individual will experience its own set of environmental conditions that will cause its phenotype to deviate from the population's average phenotype, μ. In this model, a very low score (standard deviation) such as 0.01 suggests that the deviation from the mean phenotype due to the environment is very low—in other words, most individuals occupy the same kind of environment. High numbers, such as 10 or greater, suggest that individuals experience dramatically different environments; some will be located in low-quality environments, some in high-quality environments, and some will be found in an "average" environment. If our environmental conditions can be described with a normal distribution, and cell D12 is set to 0.1, then approximately 68% of the adults in the population experience environments that alter their phenotypes by 0.01 units, and 95% of the individuals will experience environments that alter their phenotypes by 0.02 units (±2 standard deviations) from the population mean.

5. Save your work.

B. Set up the parental population.

1. Set up new headings as shown in Figure 4.

Now we will generate genotypes and phenotypes for a population of individuals (parents) who will then mate and produce offspring. Columns B–E will focus on the first parent, and columns F–I will focus on the second parent of each pair.

	A	B	C	D	E	F	G	H	I	J
16	PARENTAL POPULATION									
17	Individual	Genotype parent 1	*G* parent 1	*E* parent 1	Phenotype parent 1	Genotype parent 2	*G* parent 2	*E* parent 2	Phenotype parent 2	Midparent value
18										

Figure 4

2. Generate a linear series from 1 to 1000 in cells A19–A1018.

Enter 1 in cell A19.
Enter **=1+A19** in cell A20. Copy the formula down to cell A1018.

3. In cells B19–B1018, enter a formula to generate a genotype for parent 1 based on the allele frequencies in cells B7–E7.

In cell B19, enter the formula **=IF(RAND()<$B$7,"A1","A2")&IF(RAND()<$B$7,"A1","A2")&IF(RAND()<$D$7,"B1","B2")&IF(RAND()<$D$7,"B1","B2")**. Copy the formula down to cell B1018.
This formula follows the nested formula used in the Hardy-Weinberg exercise. It uses the & function to join the results of 4 separate **IF** functions together, because each individual requires four alleles (two *A* alleles and two *B* alleles) to make up its geno-

type. Each **IF** function draws a random number between 0 and 1 (the **RAND()** portion). For the *A* locus, if the random number is less than the frequency of the A_1 allele given in cell B7, the individual gets an A_1 allele; otherwise it gets an A_2 allele.

4. In cells C19–C1018, use a **LOOKUP** formula to generate a trait size for individual 1, based on individual 1's genotype (cell B19) and the contribution of each allele to trait size (cells B6–E6).

Enter the formula **=LOOKUP(MID(B19,1,2),$B$5:$E$5,$B$6:$E$6)+ LOOKUP(MID(B19,3,2),$B$5:$E$5,$B$6:$E$6)+ LOOKUP(MID(B19,5,2),$B$5:$E$5,$B$6:$E$6)+ LOOKUP(MID(B19,7,2),$B$5:$E$5,$B$6:$E$6)** in cell C19. Copy the formula down to cell C1018.

This long formula is really quite simple; it is just four **LOOKUP** equations added together. The first part of the formula, **=LOOKUP(MID(B19,1,2),$B$5:$E$5,$B$6:$E$6)**, is a nested function because within the **LOOKUP** function is the **MID** function. The **LOOKUP** function looks up the value given by the function **MID(B19,1,2)**. This function examines the first *A* allele for individual 1, which will be either A_1 or A_2. It examines the text in cell B19 (individual 1's genotype), and starting with the first character, returns two characters from the text given in cell B19. The result will be either A_1 or A_2. The program then returns to the lookup function, finds this value in the range of cells B5–E5, and returns the number associated with the appropriate value in cells B6–E6. When this procedure is done for each of the alleles in individual 1's genotype, and the results are added together, it generates the genetic contribution to trait size.

Double-check your results. You should be able to examine a genotype and make sure that the function is generating the proper trait size. Technically, this computation provides the contribution to the phenotype for individuals, rather than the *deviation* of the genotype from the population mean phenotype, which is the correct computation of *G*. Since p and $q = 0.5$ for both loci, the average trait should in fact be 0, so the *G*'s represent deviations from this mean and also the phenotypic contribution.

5. In cells D19–D1018, enter a **NORMINV** function to obtain the deviation in trait size for individual 1 as determined by individual 1's environment.

Enter the formula **=NORMINV(RAND(),0,$D$12)** in cell D19. Copy the formula down to cell D1018.

Remember that the **NORMINV** function draws a random cumulative probability from a distribution whose mean and standard deviation are specified, and then converts that probability to an actual number from the distribution. Here we are interested in how much individual 1 deviates from the average phenotype because of the environment in which it lives, so the mean of 0 and the standard deviation from cell D12 is used. The result shows, generally speaking, what kind of environment each individual is located in. For example, Figure 5 shows that individual 1 has a genotype of $A_2A_2B_2B_2$ and so the genetic contribution to a trait is –4 units (it is 4 units smaller than the heterozygous "standard" in terms of genetic trait size). But this individual is located in an environment that is somewhat better in quality than average (deviation = 0.01 in cell D19). Its phenotype (computed in the next step) will be the genetic trait, plus the environmental deviation, plus the average phenotype of the population. In contrast, individual 4 is 2 units larger than an $A_1A_2B_1B_2$ heterozygote, but it is located in an average environment (0.00), so the deviation in its phenotype is not due to the environment.

	A	B	C	D	E	F
17–18	Individual	Genotype parent 1	*G* parent 1	*E* parent 1	Phenotype parent 1	Genotype parent 2
19	1	A2A2B2B2	-4	0.01	46.01	A1A1B2B2
20	2	A2A2B2B1	-2	0.02	48.02	A2A1B2B2
21	3	A1A2B2B1	0	-0.01	49.99	A1A1B2B2
22	4	A2A1B1B1	2	0.00	52.00	A2A1B2B1

Figure 5

6. Enter a formula in cells E19–E1018 to generate the phenotype for individual 1 (parent 1).

Enter the formula **=$D$11+D19+C19** in cell E19. Copy the formula down to cell E1018. This formula is the spreadsheet version of the Equation 4, $P = u + G + E$.

7. Enter formulae in cells F19–I19 to generate *G*, *E*, and *P* (steps 3–6) for the *second* parent. Copy your formulae down to row 1018.

The formulae are the same except that cell references should refer to columns F, G, H, and I. We entered the formulae

- F19 **=IF(RAND()<$B$7,"A1","A2")&IF(RAND()<$B$7,"A1","A2")&IF(RAND()<$D$7,"B1","B2")&IF(RAND()<$D$7,"B1","B2")**
- G19 **=LOOKUP(MID(F19,1,2),$B$5:$E$5,$B$6:$E$6)+LOOKUP(MID(F19,3,2),$B$5:$E$5,$B$6:$E$6)+LOOKUP(MID(F19,5,2),$B$5:$E$5,$B$6:$E$6)+LOOKUP(MID(F19,7,2),$B$5:$E$5,$B$6:$E$6)**
- H19 **=NORMINV(RAND(),0,$D$12)**
- I19 **=$D$11+H19+G19**

8. In cells J19–J1018, enter a formula to compute the average phenotype for parent 1 and parent 2 (the **midparent value**).

Enter the formula **=AVERAGE(E19,I19)** in cell J19. Copy the formula down to J1018. Review your entries to this point to make sure you fully comprehend the model thus far.

9. Save your work.

C. Generate offspring.

1. Set up new headings as shown in Figure 6.

	K	L	M	N
16	**OFFSPRING POPULATION**			
17	**Genotype**	***G***	***E***	***P***
18	**offspring**	**offspring**	**offspring**	**offspring**

Figure 6

2. In cells K19–K1018, enter formulae to randomly obtain an *A* and a *B* allele from each parent to generate a zygote.

Enter the formula **=IF(RAND()<0.5,MID(B19,1,2),MID(B19,3,2))&IF(RAND()<0.5,MID(F19,1,2),MID(F19,3,2))&IF(RAND()<0.5,MID(B19,5,2),MID(B19,7,2))&IF(RAND()<0.5,MID(F19,5,2),MID(F19,7,2))** in cell K19. Copy the formula down to cell K1018.

With this formula, we simulate gamete formation and independent assortment so that each parent contributes a single *A* allele and a single *B* allele. The alleles from both parents are then joined with the **&** function to specify the offspring's genotype. The first portion of the formula, **=IF(RAND()<0.5,MID(B19,1,2),MID(B19,3,2))** , specifies the *A* allele for parent 1. If a random number is less than 0.5, parent 1 will contribute the first *A* allele listed in its genotype (given by the first and second characters in cell B19). Otherwise, parent 1 will contribute the second *A* allele listed in its genotype (given by the third and fourth characters in cell B19). The second **IF** function concentrates on the *A* allele for parent 1. The third and fourth **IF** functions concentrate on the *B* allele contributions from parents 1 and 2, respectively.

3. Enter formulae in cells L19–N19 to generate *G*, *E*, and *P* for offspring 1. Copy your formula down to row 1018. Be sure to reference cells E11–E12 in your formulae.

Double-check your results:

- L19 **=LOOKUP(MID(K19,1,2),$B$5:$E$5,$B$6:$E$6)+LOOKUP(MID(K19,3,2),$B$5:$E$5,$B$6:$E$6)+LOOKUP(MID(K19,5,2),$B$5:$E$5,$B$6:$E$6)+LOOKUP(MID(K19,7,2),$B$5:$E$5,$B$6:$E$6)**
- M19 **=NORMINV(RAND(),0,$E$12)**
- N19 **=$E$11+M19+L19**

4. Save your work.

D. Obtain frequencies and make graphs.

1. Set up new headings as shown in Figure 7.

	P	Q	R
17		Frequency	
18	Bins	Parents	Offspring
19	40		
20	41		
21	42		
22	43		
23	44		
24	45		
25	46		
26	47		
27	48		
28	49		
29	50		
30	51		
31	52		
32	53		
33	54		
34	55		
35	56		
36	57		
37	58		
38	59		
39	60		

Figure 7

2. In cells Q19–Q39, use the **FREQUENCY** function to generate frequencies of average parent phenotypes.

Remember that the frequency function is an array function, so must be entered differently than a standard function. (Refer to Exercise 2 for information on how to use array functions.) This function computes the frequency of the average parent phenotype (cells J19–J1018) and uses cells P19–P39 as bins. When you are finished, your formula in cells Q19–Q30 should read **{=FREQUENCY(J19:J1018,P19:P39)}**.

3. In cells R19–R39, use the **FREQUENCY** function to generate frequencies of average offspring phenotypes.

For offspring, the formula in cell R19–R39 should read **{=FREQUENCY(N19:N1018,P19:P39)}**.

4. Make a frequency histogram of midparent and offspring phenotypes.

Use the column graph option and label your axes fully. Your graph should resemble Figure 8.

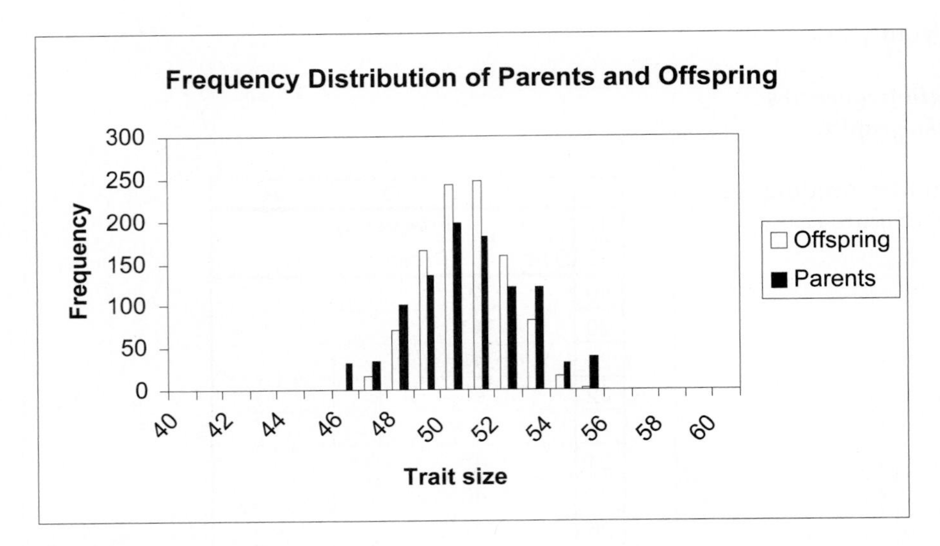

Figure 8

5. Graph the midparent versus the offspring trait size. Use the scattergraph option and add the regression equation to the graph. Adjust your axes so that the each axis ranges from 40–60 units in trait size.

Your graph should resemble Figure 9. To add the trendline, select the Chart Menu, then go to Chart | Add Trendline. Select the Linear option, then click on the Options tab. Select Display equation on the chart.

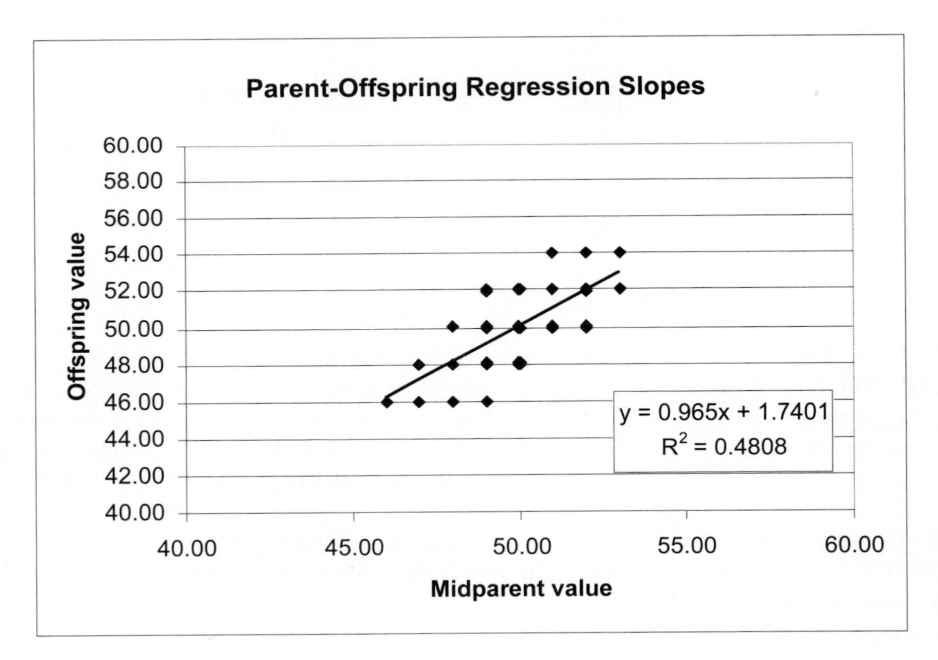

Figure 9

6. Save your work.

E. Compute heritability statistics.

The heritability statistics are based on offspring traits, as well as on parent-offspring regressions.

1. Set up new column headings as shown in Figure 10.

	G	H	I
3	Model outputs		
4			
5	Offspring	V_e	
6	Variance	V_g	
7	Values	V_e+V_g	
8		V_p	
9	Heritability	V_g/V_p	
10		Slope	
11			

Figure 10

2. In cell I5, use the **VAR** function to compute the variance in offspring's environmental conditions.

Enter the formula **=VAR(M19:M1018)** in cell I5.

3. In cell I6, use the **VAR** function to compute the variance in offspring's genetic traits.

Enter the formula **=VAR(L19:L1018)** in cell I6.

4. In cell I7, add $V_e + V_g$.

Enter the formula **=I6+I5** in cell I7.

5. In cell H8, use the **VAR** function to compute the total phenotypic variation in offspring.

Enter the formula **=VAR(N19:N1018)** in cell H8.

6. In cell I9, compute heritability as V_g/V_p.

Enter the formula **=I6/I8** in cell I9.

7. In cell I10, compute heritability as the slope of the parent-offspring regression.

Enter the formula **=SLOPE(N19:N1018,J19:J1018)** in cell I10.

8. Save your work.

QUESTIONS

1. Why do you suppose that the slope is sometimes not exactly equal to V_g/V_p?
2. How does the mean affect heritability? Using the initial conditions you entered upon setting up the spreadsheet model (top of next page) hit the F9 key several times to examine the heritability. Your heritability measures as shown in cells I9 and I10 should be very close to 1. Now, change the mean of the parental and offspring population to 10 (change cells D11 and E11 both to 10). Hit the F9 key several more times. Leave cell D11 as 10 but change cell E11 to 50. Again hit the F9 key and observe the effect on the heritability estimates. You may want to examine the parent-offspring regression as well.

	A	B	C	D	E
3			Model inputs		
4					
5	**Genotype**	**A1**	**A2**	**B1**	**B2**
6	*G*	1	-1	1	-1
7	Freq.	0.5	0.5	0.5	0.5
8					
9				**Population traits**	
10				**Parents**	**Offspring**
11			Average phenotype =	50.00	50.00
12			Environmental heterogeneity =	0.01	0.01

3. How does environmental variation affect heritability? Return cells D11 and E11 to 50. Hit the F9 key five times, and fill in the first column of the table below. Repeat this but this time change *both* cells D12 and E12 to 1, then to 5 and then to 20. Observe changes in cells I9 and I10. How do the graphs change? How do the V_e and V_g change? What can you conclude about the effect of the environment on estimates of heritability? Is heritability really constant for a population with a specific genetic composition? What does this exercise suggest to you about studies in which people attempt to make conclusions about heritability of traits in a natural/wild population, but they measure heritability in a greenhouse or growth chamber setting? What happens to the slope if the parental population has a different environmental variation than the offspring?

Trial	Heritability when Environmental Variation was 0.01	Heritability when Environmental Variation was 1	Heritability when Environmental Variation was 5	Heritability when Environmental Variation was 20
1				
2				
3				

4. How does heritability change if there is very little or no genetic variation in the population? Return the environmental heterogeneity to 0.01, and change the frequencies of the alleles so that they look like those below. Hit the F9 key several times to see how the heritability changes.

	A	B	C	D	E
3			Model inputs		
4					
5	**Genotype**	**A1**	**A2**	**B1**	**B2**
6	*G*	1	-1	1	-1
7	Freq.	0.001	0.999	0.001	0.999
8					
9				**Population traits**	
10				**Parents**	**Offspring**
11			Average phenotype =	50.00	50.00
12			Environmental heterogeneity =	1.00	1.00

5. What do you think might happen if we changed our model so that genotypes were no longer randomly assigned to different environments? (You do not need to try to change the model to do this, treat this instead as a thought question.)
6. Consider what might happen if there were effects of the maternal environment on the offspring (for example, if mothers in resource rich microhabitats bore larger babies). (You do not need to try to change the model to do this; treat this instead as a thought question.)

LITERATURED CITED

Falconer, D. S. 1989. *Introduction to Quantitative Genetics*, 3rd Ed. John Wiley and Sons, New York.

Hartl, D. L. 2000. *A Primer of Population Genetics*, 3rd Ed. Sinauer Associates, Sunderland, MA.

LITERATURE CITED

37 QUANTITATIVE GENETICS: EVOLUTION BY NATURAL SELECTION

In collaboration with Mary Puterbaugh

Objectives

- Set up a spreadsheet model for a population with a continuously varying trait.
- Understand the difference between selection and response to selection.
- Consider how differences in heritability and strength of selection can alter the response to selection.

Suggested Preliminary Exercises: Hardy Weinberg Equilibrium; Heritability

INTRODUCTION

To evolutionary biologists, *natural selection* and an *evolutionary response to natural selection* are different phenomena. For a population to experience **natural selection**, two conditions must be met: (1) individuals must vary from one another for a particular trait, and (2) an individual's survival and reproductive success must be affected by which of the the traits it possesses.

Some traits will be well adapted to a given environment and some will not. For example, Darwin's finches are highly variable in beak size, and individuals with larger beaks tend to survive periods of drought more successfully than those with smaller beaks (Grant and Grant 1993). This is an example of natural selection: during drought, birds with small beaks are more likely to be eliminated from the population.

Note that selection happens *within* generations. However, natural selection says nothing about what happens to beak size in subsequent generations. Since evolution can be broadly defined as a change in genetic make-up over time, we need to examine future generations to determine if natural selection is a mechanism that causes an evolutionary change in organisms. If natural selection does indeed lead to changes in future generations, then you have observed an **evolutionary response to natural selection**, or **evolution by natural selection**.

Suppose you are studying butterflies that live just one summer. You find that the caterpillars vary in weight. Some are fat and some are skinny. At the end of the summer, you are able to show that many more fat caterpillars survived pupation than did skinny caterpillars. However, when you come back the next year, there are just as many skinny caterpillars as there were the previous year. How can that be? Perhaps the caterpillar population didn't fulfilled all *three* of the criteria for evolution by natural selection:

1. Individuals in a population must vary from one another.
2. Survival and reproduction must be affected by that variation.
3. The variation must be heritable.

So in your population, caterpillars varied in weight and this variation influenced survival; but the variation in weight among individuals did not reflect *genetic* variation. Instead, it was probably due to environmental factors such as the particular plant that the caterpillar happened to eat. In other words, higher weight was not **heritable**.

Heritability is a concept best dealt with by quantitative genetics. The field of **quantitative genetics** examines quantitative (measurable) traits that vary **continuously**—over a range of values—such as beak size or caterpillar weight. All of the traits Mendel studied were *qualitative* traits in which individuals could be neatly lumped into two groups per trait, and a single gene controlled each trait. Pea color was either green or yellow; pea pods were either pinched or swollen; pea shapes were either wrinkled or smooth, and so on. What would Mendel have done if he had chosen to work with humans? Could he lump them by tall or short? Humans vary from short to tall and everything in between. Human height is a *quantitative* rather than a qualitative trait and is influenced by numerous genes and by the environment. When you consider that most traits are in fact quantitative, continuous, and affected by many genes, it is easy to understand why it took so long for scientists to understand that inheritance is caused by discrete factors called genes.

In the Hardy-Weinberg equilibrium exercise, you used a *population genetics* approach to studying evolution, where you were concerned with calculating specific changes in allele frequency over time. For example, we were interested in determining how p and q change over time. Quantitative geneticists also study evolution, but they use slightly different mathematical tools than population geneticists. In contrast to the population genetics approach, most of the mathematical equations used by a quantitative geneticist do not require knowing the genotype of individuals. A fundamental equation quantifies the **evolutionary response selection** (Falconer 1989). The formula is simple, yet wonderfully useful:

$$R = S \times h^2 \qquad \text{Equation 1}$$

where R stands for evolutionary *r*esponse to selection, a measure of how natural selection causes a population to evolve; S is the *s*trength of selection (also known as the selection differential); and h^2 is *h*eritability.

In order to understand R, let's first discuss the concepts of selection, selection differential, and heritability. Then we will return to Equation 1 and tie the concepts together.

Selection and the Strength of Selection

Natural selection occurs whenever survivorship or reproductive success is nonrandom with respect to a particular trait. Selection can be either *directional*, *stabilizing*, or *disruptive* (Figure 1). **Directional selection** occurs when the survivors are at either the high or the low end of the variation in a trait. In the caterpillar example, there was directional selection for weight of the caterpillars: the fattest caterpillars (those at the extreme high end of the population) survived.

In **stabilizing selection**, individuals with intermediate values survive best; individuals at both extremes do not survive as well. For example, suppose that small caterpillars did not survive due to insufficient resources to survive pupation, but that very large caterpillars also did not survive well because predators such as birds were better able to see and eat them. Then the best survivors would be caterpillars with an intermediate size. In this case, the caterpillar population is experiencing stabilizing selection.

The third and final type of selection is **disruptive selection**, where individuals in the population with *either* high or low extremes for a trait survive better than individuals with an intermediate-sized trait. Suppose that the caterpillars varied in their degree of melanism (pigmentation), with some caterpillars being quite dark, some being

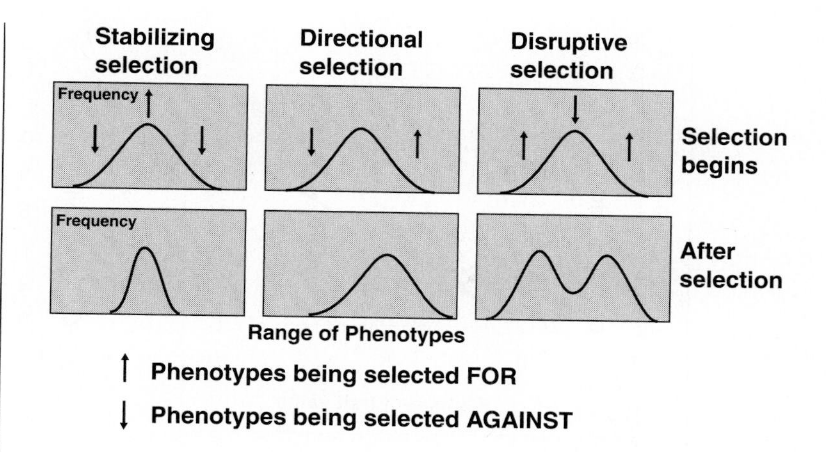

Figure 1 The effects of directional, stabilizing, and disruptive selection on a population before and after a selection event.

very light-colored, and some caterpillars having an in-between color. Now suppose that the caterpillars were found on both the white bark of birch and the darker bark of walnut trees. The light caterpillars would be protected from predators because they would be hidden on the white bark of the birch trees. The dark caterpillars would be protected on the bark of the walnut trees. However, the caterpillars with intermediate coloring would be visible to predators on both types of tree, and so they would not survive as well as either of the extreme colors.

Suppose that the only those caterpillars at the extreme high end of a trait's value in the population survived (directional selection). How would you measure the impact of natural selection on the population? You could take the mean weight of the population before any individuals died, and then compare it to the mean weight of those individuals that survived natural selection. This is what *S*, the **strength of selection** or **selection differential**, measures—the difference in a population's trait before and after natural selection. If $S = 0$, then survivors and nonsurvivors did not differ in this trait, and the offspring of survivors should not differ from the previous generation. The larger the value of *S*, the more intense the action of natural selection on the population.

Heritability

Another important component of Equation 1 is heritability, h^2. You were introduced to the concept of heritability in the previous exercise, and we will briefly review the important concepts here because knowledge of heritability is required to determine how natural selection can give rise to evolutionary change.

Heritability in a scientific sense is not the degree to which a trait is genetic, nor is it the proportion of an individual's phenotype that is controlled by genes (rather than environment). These concepts are often mistaken for h^2, which in reality has a much more specific meaning. **Heritability** is the proportion of variation for a trait that is explained by genetic variation among individuals, abbreviated V_g. The variation in a trait that is due to variation in *environmental* conditions is V_e. The total variation in population is thus $V_g + V_e$. Heritability, h^2, has the formula

$$h^2 = \frac{V_g}{(V_g + V_e)} \qquad \text{Equation 2}$$

Theoretically, h^2 can only vary between 0 and 1. When variation among individuals in a population is due entirely to differences in environmental conditions, $h^2 = 0$. When the total variation among individuals in a population is due solely to differences in the genotypes of individuals, $h^2 = 1$. Note that h^2 is a specific measure for a specific population at a specific point in time.

(A)

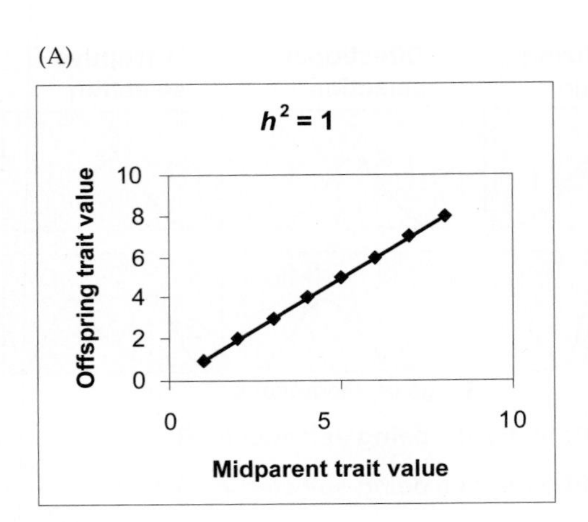

(B)

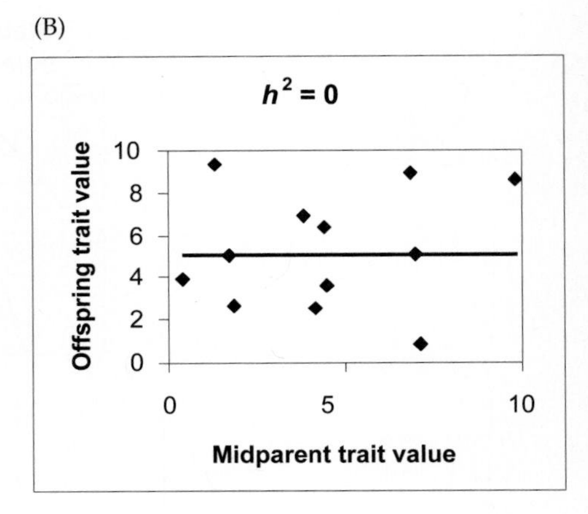

(C)

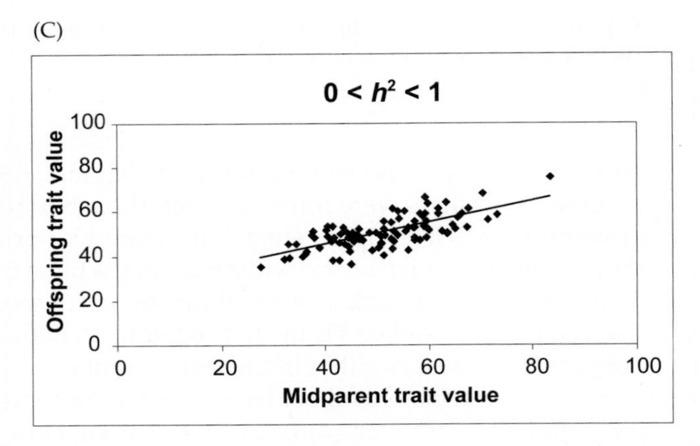

Figure 2 (A) When an offspring's trait is perfectly matched to the average of its two parents, $h^2 = 1$. Small parents will have small offspring, and large parents will have large offspring. The slope of the line is 1, and $h^2 = 1$. (B) When an offspring's trait cannot be predicted by the traits of its parents, $h^2 = 0$. Parents of any size can have offspring of any size. In this case, the slope of the regression line is 0, and $h^2 = 0$. (C) This graph shows a tendency for larger parents to have larger offspring and for smaller parents to have smaller offspring, but there is substantial scatter, suggesting that h^2 here falls between 0 and 1.

Quantitative geneticists calculate heritability in two ways. By manipulating Equation 1, you can solve for heritability as

$$h^2 = R/S$$

This is the **realized heritability**, or heritability defined by the degree to which a trait responds to selection in a population. We'll return to this equation after we learn more about *R*.

A second way to solve for heritability is to graph the trait in a set of offspring against the mean of the trait of each of their two parents (called the **midparent value**; Figure 2). The slope of the regression line for such a plot is one way to estimate heritability.

Putting *S*, h^2, and *R* Together

Now that we have a little background on *S* and h^2, let's return to Equation 1 and our discussion of how populations can evolve as a result of natural selection. Recall that

$$R = S \times h^2$$

R measures the evolutionary response to selection, or how natural selection will cause a population to evolve. Recall that there will be a response to selection only if three criteria are fulfilled:

1. Individuals in a population must vary from one another.
2. Survival and reproduction must be affected by that variation.
3. The variation must be heritable.

Equation 1 reflects all three criteria:

1. If individuals do not vary for a trait, then the denominator of Equation 2 is 0 and h^2 is undefined, so Equation 1 is undefined.
2. If S is 0, then R is 0, and natural selection did not impact the population.
3. If h^2 is 0, then R is 0.

If both S and h^2 are greater than 0, then you can expect the offspring will have a different mean trait than the previous generation's population before selection. Thus, R can be measured directly as the mean of the offspring population minus the mean of the original parental population before any individuals died.

PROCEDURES

In this exercise, you will develop a spreadsheet model of a population of 100 individuals that undergoes natural selection. You can imagine that the trait you are following is beak size in birds. (For real data on such a trait, see Grant and Grant 1993 or read Jonathan Weiner's *The Beak of the Finch*, one of our all-time favorite books.) In this exercise, you can manipulate several variables: the mean and variance of a trait in the parental population, the "quality" of breeding habitat, how the individuals are distributed across breeding habitats, the degree of environmental and genetic influence on the offspring trait (a modeling surrogate for heritability), and how natural selection favors individuals of various traits. You'll be able to manipulate these values to see how they affect S, R, and the course of evolution.

As always, save your work frequently to disk.

INSTRUCTIONS	ANNOTATION

A. Set up the model parental population.

1. Open a new spreadsheet and set up headings as shown in Figure 3.

	A	B	C	D	E
1	***Quantitative Genetics***				
2					
3	**Initial population trait**			**Offspring trait**	
4		**Mean trait value ==>**	50	**"Heritability factor" ==>**	0.5
5		**Variation in trait ==>**	10	**"Environmental factor" ==>**	0.5
6					
7	**Environmental conditions**			**Selection values**	
8		**Mean condition ==>**	50	**Select parents above ==>**	0
9		**Variation in environment ==>**	10	**Select parents below ==>**	0
10					

Figure 3

2. Enter the values shown in cells C4 and C5 for mean size and standard deviation.

Enter 50 in cell C4.
Enter 10 in cell C5.
First we'll define the parental population to have a particular mean and standard deviation for the trait. Let's suppose the birds in our study population have a mean beak size of 50 mm. Keep in mind that evolution requires variation in the parental population; this is the standard deviation of beak size. For now, enter a standard deviation of 10 in cell C5.

3. Enter the values shown in cells C8 and C9 to represent the mean and variation in evironmental conditions.

Enter 50 in cell C8.
Enter 10 in cell C9.
Cells C8 and C9 establish the environmental conditions in which the parents breed and produce offspring. Set cell C8 to 50, suggesting that on average (genetics aside), most parents nest in environments that produce offspring with 50 mm beak size. The variation in the environment is set by cell C9, which is currently set to 10. This means that the population is nesting is a very heterogeneous environment; some individuals will nest in high-quality environments that generate large offspring with big beaks, while others will nest in lower-quality environments that generate smaller offspring with smaller beaks. If the value in cell C9 were small, such as 1, it would indicate that parents are breeding in a similar (homogeneous) environment.

4. Enter the selection values shown in cells E8 and E9.

Enter 0 in cells E8 and E9.
Cells E8 and E9 establish how natural selection will "select" or pick which parents will breed. You can set these cells so that only large or small parents breed (directional selection), both small and large parents breed (disruptive selection), or only medium-sized parents breed (stabilizing selection). For now, cells E8 and E9 are set to 0, which indicates that all parents are able to breed, and natural selection will not discriminate among the parent trait size. In the Questions section, you will be asked to modify these cells to see how natural selection affects S and R.

5. Enter the values shown in cells E4 and E5 for the heritability and environmental contributions to offspring phenotype.

6. Save your work.

Enter 0.5 in cells E4 and E5.
Cells E4 and E5 define the extent to which an offspring's trait size will be controlled by its parental genotype or by the environment in which it was raised. For lack of a better term, we call these cells the "heritability factor" and the "environmental factor," respectively. Remember that *heritability* measures the amount of variation in a population that can be explained by genetic variation among individuals. In this exercise, we use the term "heritability factor" to shape each offspring's phenotype. In this sense, the word "heritability" is not correct because heritability is not a phenomenon that happens to individuals, but is a population-level measure. We trust that you have completed the heritability exercise for a true interpretation of the term.

B. Establish parental traits before and after selection.

1. Set up new column headings as shown in Figure 4.

	A	B	C	D
16	PARENTAL POPULATION		NATURAL SELECTION	
17				
18		Midparent	Survive	Midparent
19	Pair #	trait	selection?	trait

Figure 4

We will track the fates of 100 pairs of individuals (male and female breeders) and their offspring.

2. Set up a linear series from 0 to 99 in cells A20–A119.

Enter **0** in cell A20.
Enter **=1+A20** in cell A21. Copy your formula down to cell A119.

3. In cells B20–B119, use the **NORMINV** and **RAND** functions to assign average beak sizes to each pair in the parental population.

Enter the formula **=NORMINV(RAND(),$C$4,$C$5)** in cell B20. Copy this formula down to cell B119.
The formula in cell B20 tells the spreadsheet to draw a random cumulative probability (the **RAND()** portion of the formula) from a distribution whose mean is given in cell **C4** and whose standard deviation is given in cell **C5**. This probability is converted

into an actual data point by the **NORMINV** function. We'll assume that this data point represents the average of the male and female beak size. Copy the formula down to obtain midparent beak sizes for the remaining pairs in the population.

Each parent will vary in beak size, but we'll keep track of the average value from the two individuals. Remember that our population has an average beak size of 50 mm and a standard deviation of 10 mm. If our population is normally distributed (see Exercise 3, "Statistical Distributions") with respect to beak size, 68% of the population will have beak sizes between the mean and ±1 standard deviation. That is, 68% of the population will have beak sizes between 40 and 60 mm. About 95% of the population will have beak sizes between the mean and ±2 standard deviations. That is, 95% of the population will have beak sizes between 30 and 70 mm. Thus, our initial population is quite variable with respect to beak size.

4. In cells C20–C119, enter an **IF(OR)** formula to see which breeding pairs survive.

Enter the formula **=IF(OR(B20>$E$8,B20<$E$9),1,0)** in cell C20. Copy this formula down to cell C119.

We'll now subject our population to natural selection in which only certain breeding pairs survive. The **IF** formula returns one value if a condition you specify is true, and another value if the condition you specify is false. The **OR** formula returns the word "true" if any of the conditions specified are true. For example, the section **OR(B20>$E$8,B20<$E$9** tells the program to evaluate two conditions: first, Is the value in cell **B20** greater than the value in cell **E8**? and second, Is the value in cell **B20** less than the value in cell **E9**? If either of these conditions is true, the program returns the word "true"; otherwise, it returns the word "false." The **IF** formula tells the program to evaluate the **OR** function, and if it is true, return the number 1; if false, return the number 0.

Because cells E8 and E9 are both set to 0, all pairs of parents will survive natural selection, and column C should be filled with the number 1.

5. In cells D20–D119, enter a formula to return the mid-trait of those parents that survived the selection event.

Enter the formula **=IF(C20=1,B20)** in cell D20. Copy this formula down to cell D119. Cell D20 simply returns the midparent trait if the parents survived the selection event. It tells the program to evaluate cell D20, and if the value is 1, then return the traits of the parents given in cell B20.

Your spreadsheet should now look something like Figure 5, although your midparent trait values will be different due to the nature of random sampling.

6. Save your work.

	A	B	C	D
16	PARENTAL POPULATION		NATURAL SELECTION	
17				
18		Midparent	Survive	Midparent
19	Pair #	trait	selection?	trait
20	1	52.5	1	52.5
21	2	49.1	1	49.1
22	3	46.6	1	46.6
23	4	72.3	1	72.3
24	5	58.7	1	58.7

Figure 5

C. Calculate selection statistics.

1. Set up new column headings as shown in Figure 6.

	G	H	I
1	**Selection statistics**		
2		**Parents**	**Offspring**
3	**Average$_i$ =**		
4	**N_s =**		
5	**Average$_s$ =**		
6	**S =**		
7	**Average$_o$ =**		
8	**R =**		

Figure 6

2. In cell H3, use the **AVERAGE** function to obtain the mean initial parental trait (designated as Average$_i$).

Enter the formula **=AVERAGE(B20:B119)** in cell H3.

3. In cell H4, use a **SUM** formula to count the number of surviving pairs of parents, designated as N_s.

Enter the formula **=SUM(C20:C119)** in cell H4.

4. In cell H5, use the **AVERAGE** formula to obtain the mean parental trait after natural selection (designated as Average$_s$).

Enter the formula **=AVERAGE(D20:D119)** in cell H5.

5. In cell H6, calculate the strength of selection (S) as the mean trait after selection minus the mean trait before selection.

We used the formula **=H5-H3** in cell H6.
Note that since we haven't considered offspring yet, we cannot measure the response to selection, R.

6. Save your work.

D. Establish offspring traits.

Now that we've exposed the population to natural selection, we need to determine if the population *evolved* as a result. Since an evolutionary response is a change in trait over generations, we'll let the surviving pairs of parents mate and produce a single offspring, and see if the offspring beak sizes have changed as a result of natural selection. The traits in the offspring are controlled by cells E4 and E5. **The values in cells E4 and E5 must sum to 1**. If cell E4 is set to 1, an offspring will be identical to its parents. If cell E5 is set to 1, the beak size will be determined solely by the environment in which the offspring was raised.

1. Set up new column headings as shown in Figure 7.

	F	G	H
16	OFFSPRING TRAIT		
17			
18	Genetic	Environmental	Offspring
19	factor	factor	trait

Figure 7

2. In cells F20–F119, enter a formula to determine the genetic component of the offspring's beak size.

Enter the formula **=IF(C20=1,$E$4*B20)** in cell F20. Copy this formula down to cell F119.
The formula in cell F20 evaluates first whether the value in cell **C20** is **1** (the pair survived natural selection and were able to breed). If the pair survived, the spreadsheet will compute **$E$4*B20**, or the genetic "component" multiplied by the midparent beak size. Notice that the last part of the **IF** function was omitted; by default, the spreadsheet will return the word "false" if the last part of the **IF** function is not specified.

3. In cells G20–G119, enter a formula to determine how the environment affects the offspring's phenotype.

Enter the formula **=IF(C20=1,$E$5*NORMINV(RAND(),$C$8,$C$9))** in cell G20. Copy this formula down to cell G119.
How much the environment will affect the offspring's phenotype depends on three things: first, the pair must survive to breed; second, the mean of the environment in which the offspring are produced needs to be specified; and third, the standard deviation of that environment needs to be specified. The formula in cell H20 is another **IF** function that evaluates whether the pair of adults survived to reproduce. If so, the spreadsheet will use the **NORMINV** function to draw a random cumulative probability from a normal distribution whose mean is given in cell **$C$8** and whose standard deviation is given by cell **$C$9**. This number is then multiplied by the value in cell **$E$5**, which is the environmental component of the offspring's phenotype. Note again that the last part of the **IF** function was not specified, so the word "false" will be returned if the pair failed to breed.

4. In cell H20, enter a formula to compute the offspring's beak size.

Enter the formula **=IF(C20=1,F20+G20)** in cell H20. Copy this formula down to cell H119.
The offspring's final beak size is determined by the genetic component plus the environmental component. The **IF** function is used again so that only parents that survive to breed can generate offspring.

5. Save your work.

E. Construct histograms.

We'll now examine the effect of the selection event on the population visually. The most common way to depict a population's values is through a **frequency distribution**—a plot of the raw data (in this case, beak sizes) against the frequency that values appear in the population. We will calculate the frequencies of adult traits before and after natural selection, as well as offspring traits.

1. Set up new column headings as shown in Figure 8.

	J	K	L	M	N
16	FREQUENCY DISTRIBUTION				
17					
18			Frequency of parents		Frequency
19	"Bin"	Trait size	Before	After	of offspring
20	9	<10			
21	19	<20			
22	29	<30			
23	39	<40			
24	49	<50			
25	59	<60			
26	69	<70			
27	79	<80			
28	89	<90			
29		<100			

Figure 8

2. Use the **FREQUENCY** function in cells L20–L29 to count the number of adult pairs with trait sizes <10, <20, etc. before natural selection.

The **FREQUENCY** function calculates how often values occur within a range of values, and then returns a vertical array of numbers. You will use the **FREQUENCY** function to count the number of beaks in cells B20–B119 that fall below 10 mm, within 10 and 19 mm, within 20 and 29 mm, and so on. These are the "bins" in which numbers will be grouped.

The **FREQUENCY** function works best when you use the f_x key and follow the cues for entering a formula. Remember that since you will be entering this formula for an array of cells, the mechanics of entering this formula is a bit different than the typical formula entry. **Instead of selecting a single cell to enter a formula, you need to select a series of cells, then enter a formula, and then press <Control><Shift><Enter> (Windows machines) or + <Return> (Macintosh) to enter the formula for all of the cells you have selected.**

To determine the frequencies of beak lengths before selection, select cells L20–L29, then select the **FREQUENCY** function. To define the Data Array, use your mouse to highlight all 100 pairs of individuals before the selection event in cells B20–B119. To define the Bins Array, select cells J20–J28. Instead of clicking OK, press <Control><Shift><Enter>. The program will return your frequencies of beak sizes before the selection event.

After you've obtained your results, examine the formulas in cells L20–L29. Your formula should be { **=FREQUENCY(B20:B119,J20:J28)**}. The { } symbols indicate that the formula is part of an array. If for some reason you get "stuck" in an array formula, press the escape key and start over.

3. Use the **FREQUENCY** function in cells M20–M29 to count the number of adults pairs with trait sizes <10, < 20, etc. after natural selection.

The formula that calculates frequencies after the selection event is **{=FREQUENCY(D20:D119,J20:J28)}**.

4. Use the **FREQUENCY** function in cells N20:N29 to count the number of

The formula that calculates frequencies of offspring is **{=FREQUENCY(H20:H119,J20:J28)}**.

offspring with trait sizes <10, <20, etc.

Now you can visually examine strength of selection (S). (You've already calculated S in cell H6). It's the difference in the trait before and after selection—that is, the shift in the distribution as a result of natural selection. In this case, because natural selection did not kill off any adults, S is 0.

5. Graph your frequency distributions of beak sizes for parents before and after the selection event, and for offspring of surviving parents.

Use the column graph option, and label your graph fully. Your graph should resemble Figure 9, but your values may be different.

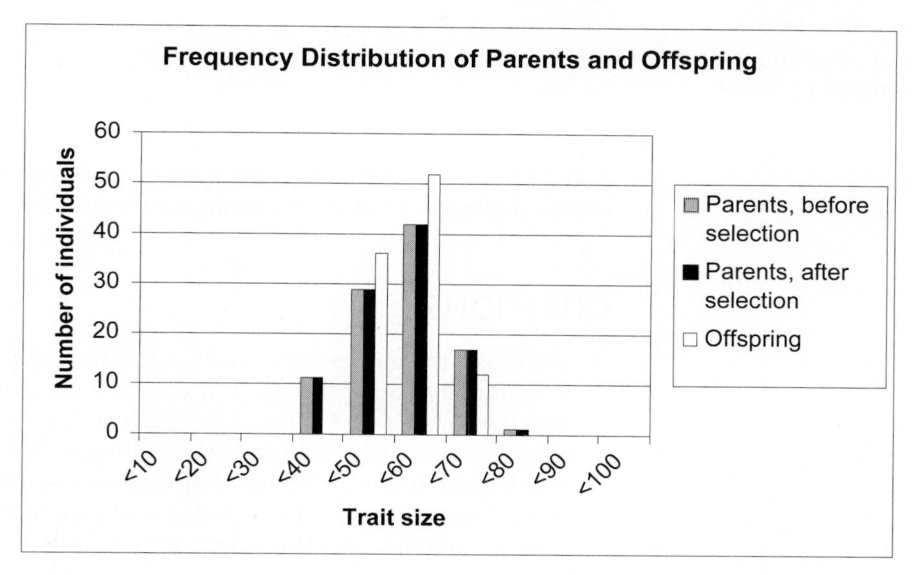

Figure 9

Interpret your graph. You should see that the frequency distribution of parents before and after natural selection is identical because all of the parents survived to breed. The offspring traits are a bit different than the parents because the environment played a role in shaping their beak sizes. Press F9, the calculate key, and you will see the offspring traits can be quite variable from calculation to calculation. This is because the environment plays an equal role in shaping beak sizes of offspring, and the environment for breeding is quite variable at the moment. If natural selection "picked" only adults with beak sizes larger than 50 mm (cell E8), our graph would look like Figure 10.

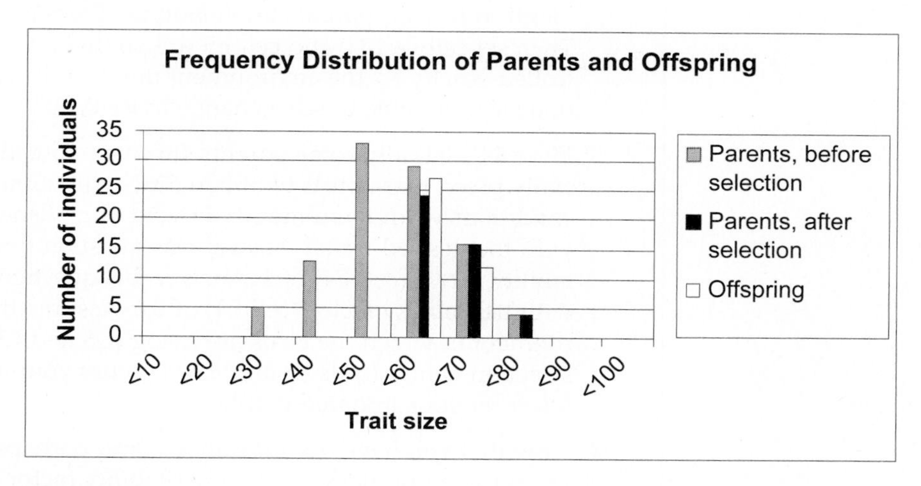

Figure 10

F. Calculate offspring statistics and R.

1. In cell I7, compute the average of the offspring's trait size.

Enter the formula **=AVERAGE(H20:H119)** in cell I7.

2. In cell I8, calculate *R*, the response to selection, as the mean offspring trait minus the mean parental trait.

Enter the formula **=I7-H3** in cell I8.

3. Save your work.

Take time now to fully interpret your graphs and calculations, keeping in mind the model entries given at the beginning of the exercise.

QUESTIONS

1. Although the model is currently set so that all parents survive to breed ($S = 0$), occasionally you will see that *R* does not equal 0. Fill in the table below by striking the F9 key 5 times. After each strike, record your results, and then describe the pattern you see. After filling in the table, continue to hit the F9 key many more times. Are your offspring ever smaller than your parents? In other words, do you ever get a negative response to selection? Are they ever larger than the parents are? Why? Interpret *S* and *R*.

Trial	Mean Offspring Trait	Response to Selection
1		
2		
3		
4		
5		

2. Let's continue to let all parents breed, but we will alter how the offspring's phenotypes are generated. Set cell E4 to 1 (and cell E5 to 0) so that offspring are identical to their parents in phenotype. Press F9 several times and interpret *R*. Then set cell E4 to 0 and cell E5 to 1 so that an offspring's phenotype is controlled strictly by the environment in which it was raised. Under what conditions is it possible to see a change in *R*? Why?

3. Now let's let only some parents survive to breed. If you were trying to commercially breed these birds to obtain birds with a mean beak size of 70 mm, what conditions would you modify in your spreadsheet to consistently generate birds with the desired traits? Answer this question first for a population with a heritability factor (cell E4) of 1. Answer this question a second time for a population with heritability factor (cell E4) of 0.6. Assume that you cannot control the environment in which the birds are living (C8 and C9), but you can change the selection values (cells E8 and E9). Discuss your answer in terms of *S* and *R*, and interpret your updated graphs.

4. Now that you have tried the above (and perhaps looked at the answer), let us try to use a population with a heritability factor of 0.6 again, but this time try to breed for birds with a mean beak size of 55 mm. Discuss your answer in terms of S and *R*.

5. In questions 3 and 4, you explored directional selection. Alter the values in cells E8–E9 to model the effects of disruptive selection. How does changing E4–E5 affect the distribution of the offspring population, *R*, and *S*? Compare your results with earlier answers from directional selection.

6. Explore your spreadsheet in new ways, and ask an interesting question and answer it. Modify parents original traits (variable or not variable, cells C4–C5), the environment of the nest (C8–C9) in which the offspring is raised, the genetic and environmental influence on offspring traits (cells E4–E5), and selection of parents (cells E8 and E9).

LITERATURED CITED

Falconer, D. S. 1989. *Introduction to Quantitative Genetics*, 3rd Edition. Longman Scientific & Technical, Essex.

Grant, B. R. and P. R. Grant. 1993. Evolution of Darwin's finches caused by a rare climatic event. *Proceedings of the Royal Society of London (B)* 251: 111–117.

Weiner, J. 1995. *The Beak of the Finch*. Vintage Books, New York.

38

SEXUAL SELECTION

In collaboration with Shelley Ball

Objectives

- Determine how female choice affects allele and genotype frequencies in a population.
- Determine how initial allele frequencies influence the evolution of allele frequencies through female choice.
- Evaluate how natural selection can counter sexual selection in the evolution of a trait.

Suggested Preliminary Exercises: Hardy-Weinberg Equilibrium and Multilocus Hardy Weinberg

INTRODUCTION

From a genetic perspective, evolution is often described as a change in allele frequency over time. What mechanisms cause changes in allele frequencies? Gene flow, mutations, and genetic drift can all spur such change. **Natural selection**—the differential survival and reproductive success of individuals in populations—is another major evolutionary force. Natural selection simply means that if some individuals have genetic characteristics that are well-suited for a particular environment, they will on average survive better and produce more offspring than other individuals in the population, thereby changing allele frequencies in subsequent generations.

In some cases, natural selection arises from differences in mating success: certain individuals possess traits that cause them to be perceived as "better" mates, and hence to mate more frequently than other individuals in the population. For example, the long, bright tails of male peacocks may have evolved because females preferentially selected males with the longest and brightest tails (the selective force was female choice). This difference in mating success due to such traits is called **sexual selection**.

Charles Darwin thought that sexual selection was different from natural selection, saying "Sexual selection … depends not on a struggle for existence, but on a struggle between the males for possession of the females; the result is not death to the unsuccessful competitor, but few or no offspring" (Darwin 1871).

The theory of sexual selection assumes the selection of traits that are purely concerned with maximizing mating success. Males can "increase the odds" of mating by having traits (such as the long, bright tail feathers in male peacocks) that attract females. Males can also maximize their mating success by the "brute force" method:

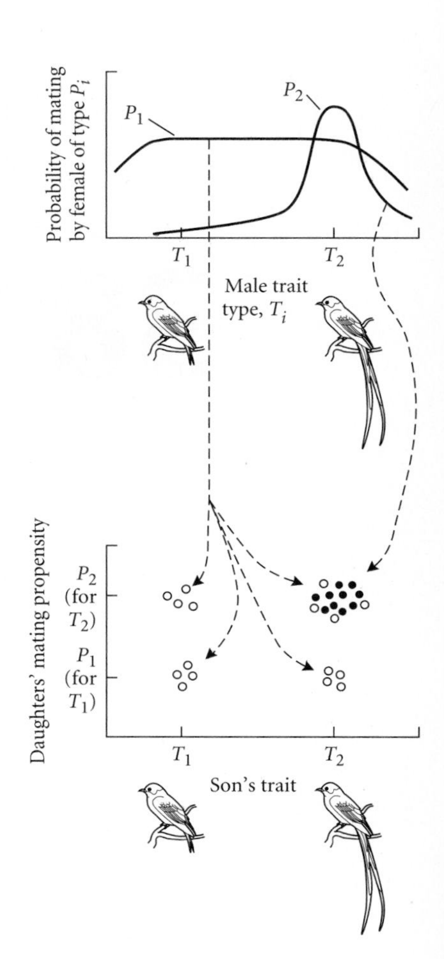

Figure 1 When some females prefer males with long tails, males with the T_2 genotypes will increase in frequency in the population in the next generation (bottom). P_1 females randomly choose to mate with both long- and short-tailed males, while P_2 females prefer males with long tails. If this preference is strong enough, and if P_2 females are sufficiently frequent in the population, long-tailed males may mate more successfully on average and thus produce more offspring than short-tailed males. These offspring will tend to inherit both the allele for long tails (from their male parents) and the allele for tail preference (from their female parents), so that as selection increases the frequency of T_2 it also increases the frequency of P_2. As P_2 becomes more frequent and an increasing proportion of females favors long tails, the advantage of having a long tail increases. Alleles T_2 and P_2 may thus both increase in frequency over time, at ever-increasing rates. The change in genotype frequency over time for males is shown in Figure 2. Note that male genotypes containing a T_2 allele increase in frequency, while male genotypes with T_1 decrease in frequency. (From Futuyma 1998.)

outcompeting other males for mating opportunities (male-male contests). Female traits may not be so visible; females maximize their fitness by selecting males that somehow enhance their own fitness or the fitness of their offspring. A female might select males that have "good genes" which enhance her offspring's fitness (an indirect benefit of mate choice), or by selecting males that are "good parents/mates," which enhance the female's own survival and reproductive success (a direct benefit of mate choice).

In these cases it's fairly easy to imagine how females that choose beneficial mates can be favored in a population, and how such choices influence the evolution of a species (Alcock 2001). But what happens when there is no direct or indirect fitness benefit associated with mate choice? Can a population still evolve due to sexual selection? The answer, in theory, is yes. Ronald Fisher introduced the theoretical argument in 1930. Fisher realized that sexual selection could cause populations to evolve when there is no fitness gain associated with mate choice, and that sometimes even traits that decrease survivorship, such as an extraordinarily long tail, can evolve in a population as a result. Fisher's model is called **runaway sexual selection**.

An important underlying assumption of Fisher's model is that both the female preference and the male trait (i.e. tail length) must be under genetic control. (Remember, traits cannot evolve unless they have a genetic basis.) So, let's imagine that males have a gene associated with tail length in which males have either a T_1 (short) or a T_2 (long) genotype (let's assume, for simplicity, that we're dealing with a haploid organism). Females also have these genes for tail length but do not express them. Let's further assume that the T_2 genotype has a fitness cost—perhaps males with long tails have higher mortality rates because predators capture them more easily. Let's also imagine that a separate, *nonlinked* gene determines mating preference, where the genotype P_1 indicates no preference for tail length but the P_2 allele indicates a strong preference for long tails. Both males and females have the P gene, but only females express the gene when they solicit matings. Thus, both sexes carry an allele for both the P and T genes. Because of this, selection for a particular allele of one gene can "drag along" a particular allele of the other gene. This association leads to a genetic correlation between the tail length gene and the mating preference gene, as shown in Figure 1.

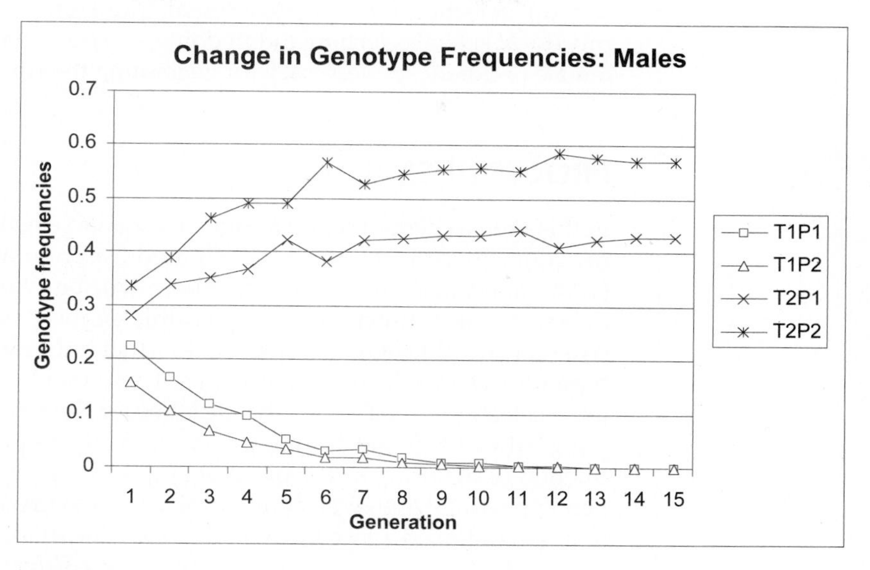

Figure 2

If runaway sexual selection actually happens in nature, why don't we commonly see birds with super-long tails? Although sexual selection for long tails increases the fitness of long-tailed males over short-tailed males, natural selection may select against long-tailed males through decreased survivorship. For example, if a tail is so long that the bird has troubles escaping from predators, there will be fewer long-tailed males in the population. Depending on the strength of selection against long-tailed males, we can expect some equilibrium level that would balance survival costs of having a long tail with the reproductive benefits of having such a tail. Figure 3 shows an example of how natural selection can drive the T_2 allele to extinction by substantially decreasing the survival probability of T_2 males.

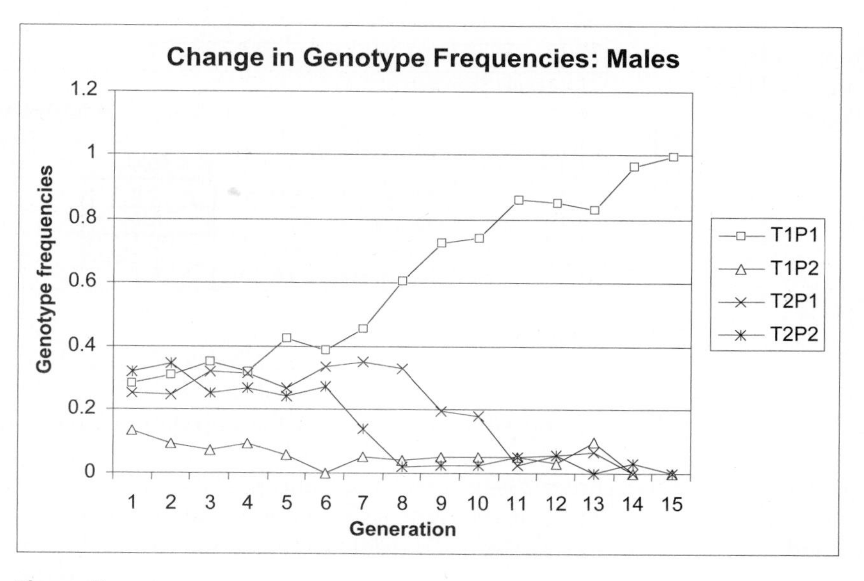

Figure 3

In Fisher's (1958) model of sexual selection, he assumed that the female's preference must confer some sort of selective advantage and that this advantage was necessary to "get the ball rolling" in the runaway process. However, later work by Kirkpatrick (1982) showed that an initial selective advantage was not a necessary prerequisite for the runaway process of sexual selection and that evolution of the male trait could occur without selection for or against female preference. We will model Fisher's runaway process of sexual selection and in doing so, show that no initial selective advantage of female preference is necessary for generating the runaway process.

PROCEDURES

In this exercise, you'll set up a runaway sexual selection model and see first-hand how the runaway process works. You'll model a population of 2000 haploid individuals (1000 males and 1000 females). There will be two alleles, P and T, as previously described, and thus there are 4 possible genotypes: P_1T_1, P_1T_2, P_2T_1, and P_2T_2. You will set up a table of mating preferences that indicate the preferences of a female genotype for the various male genotypes. These mating preferences will be converted to mate selection probabilities that account for the frequencies of male genotypes in the population. Once the mating preferences are assigned, you will simulate the reproduction of offspring as a diploid organism (by joining the male and female partner's genotypes), and then will simulate meiosis to so that organisms so that the haploid system is maintained. Once the offspring are generated, you will compute the numbers of P_1T_1, P_1T_2, P_2T_1, and P_2T_2 individuals in the offspring population, and then compute their allele frequencies. And finally, you will develop a macro to track the allele frequencies over time to see how the various genotypes evolve.

Admittedly, this is a pretty complicated spreadsheet, so take your time as you work through it and try to keep the bigger picture in mind as you develop the model. As always, save your work frequently to disk.

INSTRUCTIONS

ANNOTATION

A. Set up the model parental population.

We will start by setting up a parent population that contains 1000 males and 1000 females. The tail length locus, T, has 2 alleles, T_1 (short tail) and T_2 (long tail). The preference locus, P, has two alleles, P_1 (no preference for tail length) and P_2 (preference for long tails).

1. Open a new spreadsheet and set up headings as shown in Figure 4.

	A	B	C	D	E	F
1	***Sexual Selection***					
2			**PARENTAL GENOTYPE NUMBERS**			
3				**Tally**		**Tally**
4	**Genotype**		**Male**	**0**	**Female**	**0**
5	T1P1	*Short tail, no preference*	250		250	
6	T1P2	*Short tail, preference*	250		250	
7	T2P1	*Long tail, no preference*	250		250	
8	T2P2	*Long tail, preference*	250		250	

Figure 4

The possible genotypes for the population are given in cells A5–A8.

2. In cells C5–C8 and E5–E8, enter the number of individuals with each genotype.

Enter **250** in cells C5–C8 and E5–E8.
To begin, we will assume that the all genotypes are equally represented in the population. You will be able to change these cells later in the exercise.

3. Enter 0 in cells D4 and F4.

This is a "place holder" to tally the total number of males and females in cells D5–D8 and cells F5–F8. It is necessary so that we can assign genotypes properly to the 1000 males and females.

4. In cell D5–D8, enter formula to tally the male genotypes.

Enter **=C5** in cell D5.
Enter the formula **=D5+C6** in cell D6. Copy this formula down to cell D8.
This is a running tally that counts the total number of individuals as we consider additional genotypes. The final result in cell D8 should be 1000.

5. Set up the tally for females in cells F5–F8.

Enter **=E5** in cell F5.
Enter **=F5+E6** in cell F6. Copy the formula down to cell F8. Your total should be 1000 in cell F8.

6. Set up new spreadsheet headings as shown in Figure 5.

	A	B	C
19	**Initial population**		
20		**Male adult**	**Female adult**
21	**Individual**	**genotype**	**genotype**

Figure 5

7. Set up a linear series from 0 to 999 in cells A22–A1021.

Enter 0 in cell A22.
In cell A23, enter **=1+A22**. Copy your formula down to cell A1021. This assigns a number to each male and each female in the population.

8. In cells B22–B1021, use the **LOOKUP** function to assign genotypes to the males.

Enter the formula =**LOOKUP(A22,$D$4:$D$8,$A$5:$A$8)**. Copy this formula down to cell B1021.
The **LOOKUP** function looks up a value (**A22**) in a vector that you specify (cells **$D$4:$D$8**), and returns a genotype for the individual given in the vector **$A$5:$A$8**. (A vector is a single row or column of values). The result of this function is that genotypes are assigned to individuals in exactly the numbers that you specified in cells C5–C8.

Examine your first 10 genotypes. They should all be T_1P_1. To see how the function works, change cell C5 to 1. Now examine the genotypes of your first 10 individuals. The first male should be T_1P_1, but the rest of the males should be T_1P_2. When you feel you have a handle on how this function works, return cell C5 to 250.

9. In cells C22–C1021, enter a formula to assign genotypes to the females.

Enter the formula **=LOOKUP(A22,$F$4:$F$8,$A$5:$A$8)** in cell C22. Copy your formula down to cell C1021.
The formula for females works in the same way as that for males, using the female tallies.

10. Save your work.

B. Set up the mating preferences and mate selection probabilities.

1. Set up new column headings as shown in Figure 6.

2. Enter the female mate selection preferences shown in cells I5–L8.

	G	H	I	J	K	L
2	MATE SELECTION PREFERENCES					
3			Male genotype			
4			T1P1	T1P2	T2P1	T2P2
5		T1P1	0.25	0.25	0.25	0.25
6	Female	T1P2	0	0	0.5	0.5
7	genotype	T2P1	0.25	0.25	0.25	0.25
8		T2P2	0	0	0.5	0.5
9						
10	MATE SELECTION PROBABILITIES					
11			Male genotype			
12			T1P1	T1P2	T2P1	T2P2
13		Survival =>	1	1	1	1
14		T1P1				
15	Female	T1P2				
16	genotype	T2P1				
17		T2P2				

Figure 6

Cells H5–H8 represent the genotypes of females, and cells I4–L4 represent the genotype of a female's potential mate. The entries in cells I5–L8 establish the female's mating preferences. Thus, a female with genotype T_1P_1 has the "no preference for tail length gene," so her preferences are identical for all four male genotypes. A female with genotype T_1P_2 or T_2P_2 has the P_2 "preference for long tailed males gene," so she will prefer to mate with males that have a genotype T_2P_1 or T_2P_2, but will not prefer males with genotypes T_1P_1 or T_1P_2. **Note that the probabilities in each row in this table must sum to 1!**

3. In cell I14, enter a formula to compute the probability that a mating between the specified genotypes will take place.

Enter the formula **=I5*C5/(I5*C5+J5*C6+K5*C7+L5*C8)** in cell I14.
Although female mating *preferences* have been established, mating *probabilities* must also consider the number of males of each genotype in the population. The formula in cell I14 makes this adjustment and computes the probability that a T_1P_1 female will mate with a T_1P_1 male. The formula multiplies the preference for T_1P_1 males by the number of T_1P_1 males in the population, then adjusts this result by dividing by *preference* × *number* for all of the genotypes in the population.

4. Enter formulae to compute the remaining mate selection probabilities.

Double-check your formulae against Figure 7.

	I	J
14	=I5*C5/(I5*C5+J5*C6+K5*C7+L5*C8)	=J5*C6/(I5*C5+J5*C6+K5*C7+L5*C8)
15	=I6*C5/(I6*C5+J6*C6+K6*C7+L6*C8)	=J6*C6/(I6*C5+J6*C6+K6*C7+L6*C8)
16	=I7*C5/(I7*C5+J7*C6+K7*C7+L7*C8)	=J7*C6/(I7*C5+J7*C6+K7*C7+L7*C8)
17	=I8*C5/(I8*C5+J8*C6+K8*C7+L8*C8)	=J8*C6/(I8*C5+J8*C6+K8*C7+L8*C8)

	K	L
14	=K5*C7/(I5*C5+J5*C6+K5*C7+L5*C8)	=L5*C8/(I5*C5+J5*C6+K5*C7+L5*C8)
15	=K6*C7/(I6*C5+J6*C6+K6*C7+L6*C8)	=L6*C8/(I6*C5+J6*C6+K6*C7+L6*C8)
16	=K7*C7/(I7*C5+J7*C6+K7*C7+L7*C8)	=L7*C8/(I7*C5+J7*C6+K7*C7+L7*C8)
17	=K8*C7/(I8*C5+J8*C6+K8*C7+L8*C8)	=L8*C8/(I8*C5+J8*C6+K8*C7+L8*C8)

Figure 7

Double-check your results as well. Since there are currently equal numbers of male genotypes in the population, the mate selection probabilities should be the same as the mate selection preferences (Figure 8).

	G	H	I	J	K	L
10	MATE SELECTION PROBABILITIES					
11			Male genotype			
12			T1P1	T1P2	T2P1	T2P2
13		Survival =>	1	1	1	1
14		T1P1	0.25	0.25	0.25	0.25
15	Female	T1P2	0	0	0.5	0.5
16	genotype	T2P1	0.25	0.25	0.25	0.25
17		T2P2	0	0	0.5	0.5

Figure 8

5. Enter a survival probability for males in cells I13–L13.

Enter the number **1** in cells I13–L13.
Currently the survival probability is set to 1 so that all male genotypes have equal survival. Later in the exercise, you will be able to change these values so that males with long tails have a lower probability of survival.

6. Save your work.

C. Simulate parental matings.

Our goal is to have the spreadsheet look up the genotype of female parents (in column C) and match their genotype to genotypes listed in cells H14–H17. Ultimately, we want to determine the genotype of the female's selected mate, listed in cells I12–L12. To choose mates according to the probabilities given, we will use four different functions: **MATCH**, **INDEX**, **RAND**, and **IF**. The combination of these formulae will allow us to generate the genotype of a mate for each female in the population in column J.

1. Set up new headings as shown in Figure 9.

	D	E	F	G	H	I	J
19	Mate "selection"						
20		T1P1	T1P2	T2P1	T2P2	Random	Selected
21	Match	index	index	index	index	number	male mate

Figure 9

2. In cell D22, enter the formula **=MATCH(C22,$H$14:$H$17)**.

The **MATCH** formula returns the *relative position* of an item in a table that matches the condition you specify. The **MATCH** function has the syntax **MATCH (lookup_value,lookup_array,match_type)**, where **lookup_value** is the value you use to find the value you want in a table, **lookup_array** is a contiguous range of cells containing possible lookup values, and match_type tells the spreadsheet how to match the lookup value to the lookup array (by not specifying match-type, the default is used). In cell D22, the formula **=MATCH(C22,$H$14:$H$17)** tells the spreadsheet to find the genotype listed in cell C22, and return the *relative position* of that genotype in the $H14–$H17 table. For example, the genotype of female 1 in the spreadsheet is T_1P_1. Excel returns the value 1 to indicate that T_1P_1 individuals occupy the first position in our table. If female 1 had the genotype T_2P_2, the program would return the number 4 (fourth position).

3. In cell E22, enter the formula **=INDEX($H$14:$L$17,D22,2)**.

The **INDEX** formula returns the *value* of an element in a table, once you identify the row and column number that should be returned. The **INDEX** formula has the syntax: **INDEX(array,row_num,column_num)**, where **array** is a range of cells in a table; **row_num** selects the row in the table from which to return a value, and **column_num** selects the column in table from which to return a value. In cell E22, the formula **=INDEX($H$14:$L$17,D22,2)** tells the spreadsheet to examine the range of cells in H14–L17 and go to the row designated in cell D22 (derived from your **MATCH** formula) and column 2 (which indicates the probability of mating with an T_1P_1 male). The spreadsheet will then return the value associated with this row and column intersection. Your result should be 0.25.

4. Use the **INDEX** function to index the T_1P_2, T_2P_1, and T_2P_2 genotypes in cells F22–H22.

Enter the formula **=INDEX($H$14:$L$17,D22,3)** in cell F22.
Enter the formula **=INDEX($H$14:$L$17,D22,4)** in cell G22.
Enter the formula **=INDEX($H$14:$L$17,D22,5)** in cell H22.
These four formulae simply generate the appropriate mating probabilities for each individual in the population.

5. In cell I22, use the **RAND** function to generate a random number between 0 and 1.

Enter **=RAND()** in cell I22.
This formula randomly determines the genotype of the mate for each individual in the population. When you press F9, the calculate key, you generate a new set of random numbers.

6. In cell J22, enter a formula to establish the genotype of that female's selected mate.

Enter the formula **=IF(I22<=E22,$I$12,(IF(I22<=E22+F22,$J$12,(IF(I22<=E22+F22+G22,$K$12,$L$12)))))** in cell J22.
The formula in cell J22 looks complicated but really it's not. The formula is simply four nested **IF** statements. The formula tells the spreadsheet to examine cell **I22** (the random number). If I22 is less than or equal to the value in cell E22 (**<=E22**), then return the genotype in cell **$I$12**; otherwise walk through the next **IF** statement. The next statement examines cell **I22**, and if its value is less than or equal to the values in cells E22 + F22 (**<=E22+F22**), then return the genotype in cell **$J$12**; otherwise walk through the third **IF** statement. The third statement examines cell **I22**, and if its value is less than or equal to the sum of E22, F22, and G22 (**<=E22+F22+G22**), return the genotype in cell **$K$12**, otherwise return the value in cell **$L$12**.

7. Select cells D22–J22, and copy and the formula down to row 1021

This will establish the selected mate's genotype for each female in the population. Review your spreadsheet entries and results for the first five individuals and make sure you understand how mates were determined for the females.

8. Save your work.

D. Impose natural selection on males.

We set up the spreadsheet so that selection against a particular genotype occurs after female mating probabilities have been established. Thus, selection against a genotype does not influence the mating probabilities themselves. For now, each genotype has a survival probability of 1 (given in cells I13–L13), indicating that there is no "cost" to having a long tail. If we wished to impose selection against long-tailed males, we would alter the survival probabilities in cells I13–L13.

1. Set up new headings as shown in Figure 10.

	K
20	***Natural***
21	***selection***

Figure 10

2. In cells K22–K1021 enter a formula to compute which males survive to breed and produce offspring.

Enter the formula **=IF(RAND()<HLOOKUP(J22,$I$12:$L$13,2),J22,".")** in cell K22. Copy the formula down to cell K1021.
The formula uses an **HLOOKUP** function to find the genotype of the selected mate for female 1 (J22) in the table of cells I12–L13, and finds that male's survival probability in the second row of the table. The **RAND()** function draws a random number between 0 and 1. The **IF** function determines whether this random number is less than the appropriate survival probability. If the random number is *less than* the survival probability,

the male lives and his genotype (J22) is returned. If the random number is *greater than* the survival probability, the male dies and a period (".") is returned, indicating a death.

3. Save your work.

E. Establish offspring genotypes and allele frequencies.

1. Set up new column headings as shown in Figure 11.

	L	M	N
19	**Offspring population**		
20	**Diploid**	**Male haploid**	**Female haploid**
21	**genotype**	**genotype**	**genotype**

Figure 11

2. In cell L22, enter a formula to combine the female's haploid genotype with her mate's haploid genotype to produce a diploid offspring (only if the male survived to breed).

Enter the formula **=IF(K22=".",".",C22&J22)** in cell L22.
If the male in cell K22 is dead, the formula returns a missing value (.). If the male is not dead, the spreadsheet returns the combination of cells **C22** and **J22**; the **&** function simply concatenates the two cells.

3. In cells M22 and N22, use the **IF**, **RAND**, and **MID** functions to generate the genotypes of haploid inviduals.

Enter the formula **=IF(RAND()<0.5,MID(L22,1,2),MID(L22,5,2))&IF(RAND()<0.5,MID(L22,3,2),MID(L22,7,2))** in cells M22 and N22.
Our goal is to generate male and female offspring that have a single *T* allele and a single *P* allele. We'll let meiosis occur with random segregation of alleles.

The **MID** function returns a specific number of characters from a text string, starting at the position you specify, and based on the number of characters you specify. It has the syntax **MID(text,start_num,num_chars)** The first part of the formula, **IF(RAND()<0.5,MID(L22,1,2),MID(L22,5,2)**, tells the spreadsheet to draw a random number between 0 and 1. If that random number is <0.5, return the value associated with **MID(L22,1,2)**, otherwise returns the value associated with **MID(L22,5,2)**. The **MID(L22,1,2)** portion of the formula tells the spreadsheet to examine cell L22 and, starting at the first character, return 2 characters. The **MID(L22,5,2)** portion of the formula examines cell L22, and starting at the fifth character, returns 2 characters. This portion of the formula returns a randomly selected *T* allele. The second **IF** statement is analogous and randomly selects the *P* allele for each offspring. The two alleles are joined by the **&** symbol.

4. Select cells L22–N22 and copy their formulae down to row 1021.

5. Set up new column headings as shown in Figure 12.

	A	B	C	D	E	F
9			**OFFSPRING**			
10			**Genotype numbers**		**Genotype frequencies**	
11			**Male**	**Female**	**Male**	**Female**
12	T1P1	*Short tail, no preference*				
13	T1P2	*Short tail, preference*				
14	T2P1	*Long tail, no preference*				
15	T2P2	*Long tail, preference*				
16						

Figure 12

6. In cells C12–C15 and D12–D15, use the **COUNTIF** function to count the number of offspring male and female genotypes, respectively. Sum the totals in cells C16 and D16.

Double-check your formulae against Figure 13.

	C	D
10	Genotype numbers	
11	Male	Female
12	=COUNTIF($M$22:$M$1021,A12)	=COUNTIF($N$22:$N$1021,A12)
13	=COUNTIF($M$22:$M$1021,A13)	=COUNTIF($N$22:$N$1021,A13)
14	=COUNTIF($M$22:$M$1021,A14)	=COUNTIF($N$22:$N$1021,A14)
15	=COUNTIF($M$22:$M$1021,A15)	=COUNTIF($N$22:$N$1021,A15)
16	=SUM(C12:C15)	=SUM(D12:D15)

Figure 13

7. In cells E12–F16, compute the male and female offspring genotype frequencies.

Double-check your formulae against Figure 14.

	E	F
10	Genotype frequencies	
11	Male	Female
12	=C12/$C$16	=D12/$D$16
13	=C13/$C$16	=D13/$D$16
14	=C14/$C$16	=D14/$D$16
15	=C15/$C$16	=D15/$D$16
16	=SUM(E12:E15)	=SUM(F12:F15)

Figure 14

8. Save your work.

F. Track genotype frequencies over time.

1. Set up new headings as shown in Figure 15, but extend your generations to 15.

	N	O	P	Q	R	S	T	U	V
2		MALES				FEMALES			
3	Generation	T1P1	T1P2	T2P1	T2P2	T1P1	T1P2	T2P1	T2P2
4	1								
5	2								
6	3								
7	4								
8	5								

Figure 15

2. Open Tools | Options | Calculation and set the calculation key to manual.

3. Write a macro to track allele frequencies over time.

The macro needs to perform the following steps:

- Paste the genotype numbers of the offspring into the parental population cells.
- Press the calculate key to simulate mate selection, natural selection, and breeding.
- Record the offspring's allele frequencies in the cells O4–V18.

There are many ways to write a macro to conduct these steps. We suggest one way, but you may see other (perhaps easier) steps. Put your macro function in the "record macro" mode and assign a shortcut key (see Exercise 2). Record the following operations:

- Press F9, the calculate key, to generate new random numbers (and hence new matings and offspring for the parental population).
- Select cells E12–E15. Copy.
- Select cell O3.
- Open Edit | Find. The dialog box in Figure 16 will appear. Leave the Find What box blank and Search By Columns. Select Find Next and then Close. Your cursor should move down to the next blank cell.

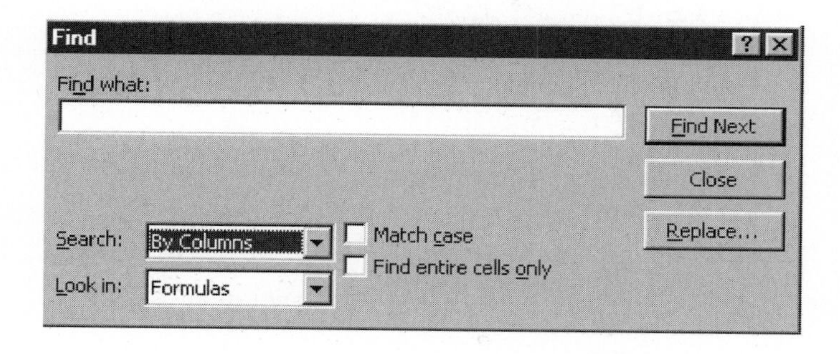

Figure 16

- Open Edit | Paste Special and select the Paste Values and Transpose options. Click OK.
- Select cells F12–F15. Copy.
- Select cell S3.
- Open Edit | Find.
- Click on Find Next and then Close.
- Open Edit | Paste Special,and select the Paste Values and Transpose options.
- Select cells C12–C15. Copy.
- Select cell C5.
- Open Edit | Paste Special and select the Paste Values option.
- Select cells D12–D15. Copy.
- Select cell E5.
- Open Edit | Paste Special and select the Paste Values option.

Stop recording. Now when you press your shortcut key, your macro will record the allele frequencies of the various genotypes for males and females.

4. Run the macro 15 times (i.e., over 15 generations).

5. Save your work.

G. Create graphs.

1. Graph the allele frequencies of males and females over time. Make a separate graph for each sex.

Use the line graph option and label your axes fully. Your graphs should resemble Figures 17 and 18.

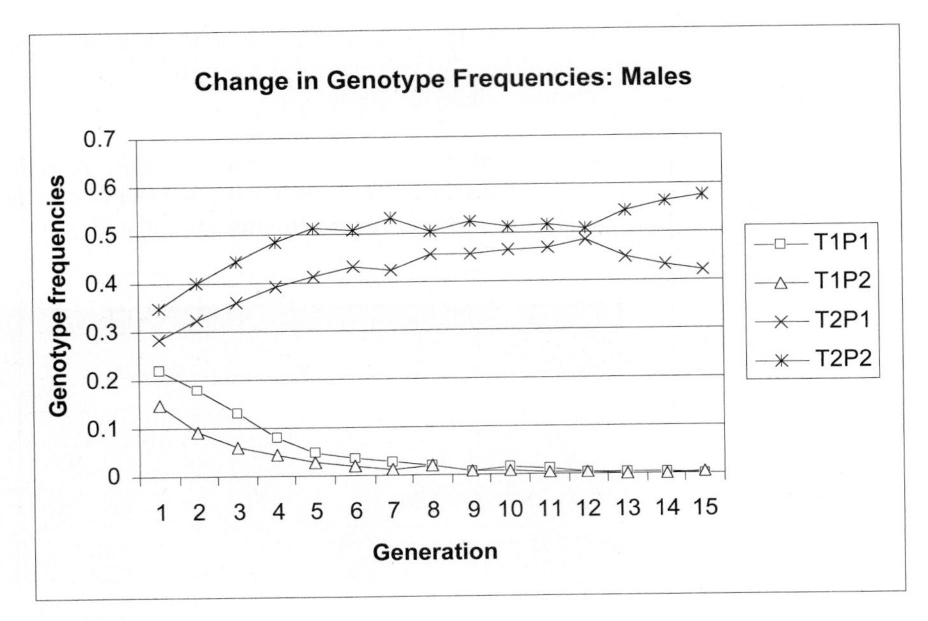

Figure 17

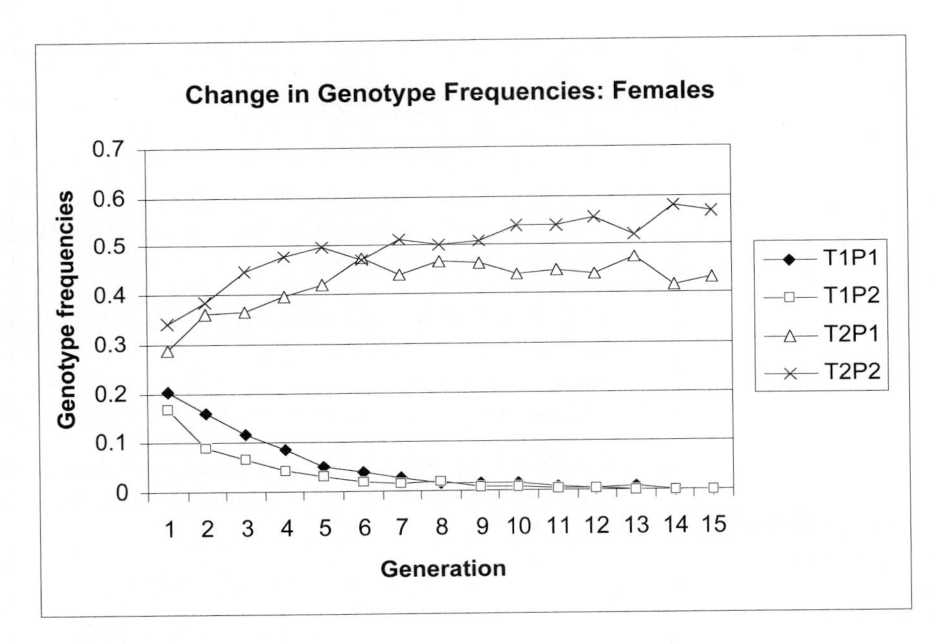

Figure 18

2. Save your work.

QUESTIONS

1. Interpret your model results. For each sex, explain how the genotypes (T_1P_1, T_1P_2, T_2P_1, T_2P_2) evolve (change in frequency) from one generation to the next. Which genotypes went extinct; which genotypes persisted? Did this differ for males and females? If so, why? What mechanism allows for the evolution of the *T* and *P* alleles?

2. In your model, females with the P_2 allele mate only with the T_2 males—no exceptions. In reality, perhaps not all females will be able to mate with T_2 males, and so some P_2 females will mate with T_1 males. Change the choice parameters

in cells I6–L6 and I8–L8 to 0.1, 0.1, 0.4, and 0.4. Reset your genotype numbers in cells C5–C8 and E5–E8 to 250. Clear your old macro results, and run your macro again. How does the "strength" of sexual selection influence the change in allele frequencies from one generation to the next?

3. How does natural selection influence the evolution of the T_2 trait even when P_2 females have full preference for T_2 male? Return the mate selection preferences to their original values as shown:

	G	H	I	J	K	L
2	**MATE SELECTION PREFERENCES**					
3			**Male genotype**			
4			**T1P1**	**T1P2**	**T2P1**	**T2P2**
5		**T1P1**	0.25	0.25	0.25	0.25
6	**Female**	**T1P2**	0	0	0.5	0.5
7	**genotype**	**T2P1**	0.25	0.25	0.25	0.25
8		**T2P2**	0	0	0.5	0.5

Decrease the survival probabilities of T_2 males in cells K13–L13 by increments of 0.1. With each incremental decrease in survival, run your macro again (clear your old results, and make sure to reset the initial genotype numbers to 250 in cells C5–C8 and E5–E8). What level of natural selection "puts the brakes" on sexual selection?

4. How do starting allele frequencies affect the outcome of a simulation? The initial genotypes we used to build the spreadsheet are admittedly very unusual; before sexual selection for tail length begins, it is much more likely that at least one (if not both) of the alleles T_2 and P_2 will be new and very rare mutations. That is, either there will be variety in tail length (long and short tails both occur with some regularity) when a mutation causes one female (or a few sisters) to prefer long tails, *or* there will already be a preference for a trait that does not yet exist, and mutation will create that trait in one male (or a few brothers). We can use this spreadsheet model to test both of these initial conditions.

Set the genotype survivals back to 1 and set the initial genotype numbers as shown below:

	A	B	C	D	E	F
3				**Tally**		**Tally**
4	**Genotype**		**Male**	**0**	**Female**	**0**
5	T1P1	*Short tail, no preference*	500	500	500	500
6	T1P2	*Short tail, preference*	0	500	0	500
7	T2P1	*Long tail, no preference*	500	1000	490	990
8	T2P2	*Long tail, preference*	0	1000	10	1000

In this population, half the males have long tails and half have short, but none of the males carry the allele for preferring long tails. Approximately half the females carry the allele for long tails, but by some unusual chance, 10 sisters in this generation all received a mutated gene that causes them to mate exclusively with long-tailed males. Clear your previous results from cells O4–V18 and run your macro to see what happens to genotype frequencies over 15 generations. Do these initial conditions result in runaway sexual selection?

5. Set the initial genotype frequencies to those shown below:

	A	B	C	D	E	F
2			PARENTAL GENOTYPE NUMBERS			
3				Tally		Tally
4	**Genotype**		**Male**	**0**	**Female**	**0**
5	T1P1	*Short tail, no preference*	900	900	900	900
6	T1P2	*Short tail, preference*	99	999	100	1000
7	T2P1	*Long tail, no preference*	0	999	0	1000
8	T2P2	*Long tail, preference*	1	1000	0	1000

In this population, 10% of the females would prefer to mate with a long-tailed male, although almost the entire population consists of short-tailed males. Approximately 10% of the males also carry the allele for preference, even though they do not express it, but almost all the males have short tails. One lone male has mutated to have a tail that is longer than the others.

To make these initial conditions a little more plausible, we need to allow that the P_2 allele does not confer *absolute* preference—otherwise the females that carried it up to this generation would not have mated, and the allele would have been lost. Resetting the mate selection preferences as shown below will give us females who would strongly prefer long-tailed males but who will settle for short-tailed males in a pinch.

	G	H	I	J	K	L
2	MATE SELECTION PREFERENCES					
3				Male genotype		
4			**T1P1**	**T1P2**	**T2P1**	**T2P2**
5		**T1P1**	0.25	0.25	0.25	0.25
6	**Female**	**T1P2**	0.01	0.01	0.49	0.49
7	**genotype**	**T2P1**	0.25	0.25	0.25	0.25
8		**T2P2**	0.01	0.01	0.49	0.49

Clear your previous results from cells O4–V18 and run your macro to see what happens to genotype frequencies over 15 generations. Do these initial conditions result in runaway sexual selection?

6. Can runaway sexual selection occur when $P_2 = 0$? Set your initial conditions so that all females and 995 males in the population have the genotype T_1P_1 and 5 males have the genotype T_2P_1. Then set the mate selection preferences as shown below:

	G	H	I	J	K	L
2	MATE SELECTION PREFERENCES					
3				Male genotype		
4			**T1P1**	**T1P2**	**T2P1**	**T2P2**
5		**T1P1**	0.3	0	0.7	0
6	**Female**	**T1P2**	0	0	0	0
7	**genotype**	**T2P1**	0.3	0	0.7	0
8		**T2P2**	0	0	0	0

Clear your previous results from cells O4–V18 and run your macro to see what happens to genotype frequencies over 15 generations. Analyze your model outputs and explain how this might occur in nature.

7. Genetic drift can also influence changes in allele frequencies over time. To examine the effects of genetic drift on this model of sexual selection, we are going to manipulate the population size by running the model using different initial genotype numbers. For example, instead of having 1000 individuals of each sex, start with 500 individuals of each. As with our initial conditions, simply start with equal numbers of each genotype (so in this case, each genotype number will be 125). Clear your results from your last simulation, set the mate selection preferences in cells I5–L8 back to their initial values, and run your model. Record your results and then run the model a few more times, each time changing the total genotype numbers (but make sure there are equal numbers of each genotype). What effect does changing the population size have on the outcome of the model?

LITERATURE CITED

Alcock, J. 2001. The evolution of reproductive success. Chaper 11, pp. 316–359 in *Animal Behavior: An Evolutionary* Approach, 7th Edition. Sinauer Associates, Sunderland, MA.

Basolo, A. L. 1990. Female preference predates the evolution of the sword in swordtail fish. *Science* 250: 808–810.

Darwin, C. 1871. *The Descent of Man, and Selection in Relation to Sex*. John Murray, London.

Fisher, R. A. 1930. *The Genetical Theory of Natural Selection*. Clarendon Press, Oxford.

Futuyma, D. 1998. *Evolutionary Biology*, 3rd Edition. Sinauer Associates, Sunderland, MA.

Kirkpatrick, M. 1982. Sexual selection and the evolution of female choice. *Evolution* 3: 1–12.

39 EVOLUTIONARILY STABLE STRATEGIES AND GROUP VERSUS INDIVIDUAL SELECTION

Objectives

- Understand the concept of game theory.
- Set up a spreadsheet model of simple game theory interactions.
- Explore the effects of different strategies on animal fitnesses.
- Understand the concept of an evolutionarily stable strategy.
- See how the concept of an evolutionarily stable strategy is a strong argument against group selection.

INTRODUCTION

Evolutionary biologists have long been interested in behavioral interactions between animals and how these interactions affect evolutionary fitness. One approach has been to model interactions using **game theory**. Game theory in its broadest sense is the mathematical analysis of conflict, and it has been applied to interactions between countries, business firms, individual humans, and animals. This exercise follows John Maynard Smith's (1976) model of behavioral interactions between animals and leads to his concept of an evolutionarily stable strategy (ESS). We will apply this model to the question of individual selection versus group selection—that is, the question of whether natural selection can act on groups as well as on individuals.

In our context, we will imagine that animals engage in contests over resource items, such as food, nest sites, or mates. We will assume that in each contest, there is only one winner, and the winner takes all of the contested resource item. Bear in mind, however, that animals engage in repeated contests, and any given animal may win on one occasion and lose on another. Our model makes several assumptions:

- We assume that winning a resource item increases an animal's fitness (in the evolutionary sense) by some amount, which we will symbolize as V (for victory).
- We assume that if an animal sustains an injury in a contest, that reduces its fitness by some amount, symbolized as W (for wound).
- Finally, we assume that if a contest continues too long, it costs both participants some amount of fitness, T (for time), representing the metabolic energy expended during the contest, and forgone opportunities to garner other resource items.

We will also assume, at least to begin with, that each animal always employs the same behavioral strategy in these contests. We will relax this assumption later.

Doves versus Hawks

By calling these behaviors "strategies," we do not necessarily imply any conscious decision-making by the animals. The word strategy in this context simply means a rigid, predictable set of behaviors that always occur in response to certain stimuli. To make this clear, we will define two strategies, called "Dove" and "Hawk" (Maynard Smith 1976). A Dove begins a contest by making a threat display but never backs up its threat with real violence. If its opponent displays, a Dove continues to display, but if its opponent attacks, a Dove retreats immediately. A Hawk wastes no time on display, but attacks immediately.

A contest between two Doves becomes a drawn-out battle of displays, with no injuries but much wasted time. In a contest between a Dove and a Hawk, the Dove retreats immediately when the Hawk attacks, and thus loses the resource item, but avoids injury. A contest between two Hawks is a violent affair, in which one party is always injured and retreats from the fray, leaving the resource item to the uninjured victor.

We can translate these descriptions into mathematical expressions using the fitness values defined above. A Dove contesting with another Dove will win half the contests and lose half, but it will always pay the time cost, T, of extended display. Thus, on average, the payoff to Doves contesting with other Doves will be $(V/2) - T$. A Dove contesting with a Hawk will always lose, but will not spend time or suffer injury. Thus, the mean payoff to Doves contesting with Hawks is zero. A Hawk will always win immediately against a Dove, and so the mean payoff to Hawks contesting with Doves is V. Finally, a Hawk fighting a Hawk will win half the time, and enjoy a fitness payoff of V, but it will also lose half the time, at a cost of W. So, the mean payoff to Hawks fighting Hawks is $(V/2) - (W/2)$, which we can simplify to $(V - W)/2$.

We can conveniently represent these outcomes in a payoff matrix in which we show all possible encounters and the fitness implications for the participants (Table 1). The payoffs are for the player on the left.

Table 1. Payoff matrix for Hawks versus Doves.

	Hawk	Dove
Hawk	$\frac{V-W}{2}$	V
Dove	0	$\frac{V}{2}-T$

We want to know which strategy confers higher fitness. To find out, we need to calculate the mean fitness of Doves and Hawks in a mixed population. Let us represent the frequency of Hawks by H, and the frequency of Doves by D. These are relative frequencies, and therefore lie between 0 and 1, and sum to 1 (i.e., $H + D = 1$).

Let us assume that encounters occur at random. If we consider all the encounters of an average Dove, the proportion of them that will involve another Dove will be D, and the proportion that will involve a Hawk will be H, or $1 - D$. The frequencies of encounters will be the same for the average Hawk.

To calculate the mean fitness of Doves, we must weight the payoffs of each kind of encounter by its frequency: the mean fitness of Doves is

$$0H + \left(\frac{V}{2} - T\right)D \qquad \text{Equation 1}$$

By the same logic, the mean fitness of Hawks is

$$\left(\frac{V-W}{2}\right)H + VD \qquad \text{Equation 2}$$

If we start with a population consisting of some mixture of Hawks and Doves, which strategy will prevail? The answer is not obvious. Hawks always win encounters with Doves, but Doves are never injured. We can approach the question by determining whether Hawk or Dove is an **evolutionarily stable strategy**, or **ESS**. An evolutionarily stable strategy is one that cannot be successfully invaded by any of the other strategies in the game.

Let us imagine a population consisting entirely of Doves. Could Hawks successfully invade? The concept of invasion in this context includes not only immigration, but also the appearance of mutations within the population. In other words, Hawks may move into the Dove population, or a genetic mutation may cause some Dove offspring to behave as Hawks.

In either case, a few invading Hawks would mean that $D \approx 1$ and $H \approx 0$. The mean fitness of Doves, Equation 1, would then be approximately

$$0(0) + \left(\frac{V}{2} - T\right)(1) \quad \text{or} \quad \frac{V}{2} - T$$

Analogously, the mean fitness of Hawks, Equation 2, would be approximately

$$\left(\frac{V-W}{2}\right)(0) + V(1) \quad \text{or} \quad V$$

Provided V and T are both greater than 0 (which is implicit in the definitions), V will be greater than $(V/2) - T$, and Hawks will increase in numbers. This is a successful invasion, and therefore Dove is not an evolutionarily stable strategy against Hawk.

PROCEDURES

But is Hawk an evolutionarily stable strategy against Dove? Could a few Doves successfully invade a population of Hawks? We will find the answer using a spreadsheet model, and it may surprise you. As always, save your work frequently to disk.

INSTRUCTIONS

ANNOTATION

A. Game Theory Model

1. Open a new spreadsheet and set up titles and column headings as shown in Figure 1. Enter only the text items for now.

These are all literals, so just select the appropriate cells and type them in.

	A	B	C	D	E	F	G	H	I
1	***Evolutionarily Stable Strategies***								
2	Based on John Maynard Smith's model of Hawks and Doves								
3	All costs and benefits are expressed in "fitness points."								
4	Model assumes that the probability of winning a fair encounter (i.e., Hawk vs. Hawk or Dove vs. Dove) is 0.50.								
5	It also assumes that a Hawk always wins against a Dove.								
6									
7	Outcome	Fitness points		Payoff matrix (payoffs to player on left)				Equilibrium mix	
8	Victory	0.50			Hawk	Dove		Proportion of Doves	
9	Wound	1.00		Hawk				Proportion of Hawks	
10	Time	0.10		Dove					
11								Fitness matrix	
12	Proportion			Fitness				Population composition	Mean fitness
13	Doves	Hawks	Doves	Hawks	Population			All Hawks	
14	0.0	1.0						All Doves	
15	0.1	0.9						Equilibrium mix	
16	0.2	0.8							

Figure 1

2. Enter the values shown in Figure 1 for V, W, and T.

In cells B8, B9, and B10 enter the values 0.50, 1.00, and 0.10, respectively. These are the values in fitness points of victory, a wound, and time lost.

3. Enter formulae to calculate values of the payoff matrix.

In cell E9, enter the formula **=0.5*(B8-B9)**. This corresponds to $(V - W)/2$, the payoff to a Hawk in an encounter with another Hawk.
In cell E10, enter the value 0. This is the payoff to a Dove in an encounter with a Hawk.
In cell F9, enter the formula **=B8**. This is the payoff to a Hawk in an encounter with a Dove. Use a formula rather than entering the value V, so that when you change V in cell B8, the change will automatically occur in cell F9 as well.
In cell F10, enter the formula **=0.5*B8-B10**. This corresponds to $(V/2) - T$, the payoff to a Dove in an encounter with another Dove.

4. Create a series in column A to represent various frequencies of Doves in the population.

In cell A14 enter the value 0.
In cell A15 enter the formula **=A14+0.1**. Copy the formula into cells A16 through A24.

5. Create a series in column B to represent various frequencies of Hawks in the population.

In cell B14 enter the formula **=1-A14**. Copy the formula into cells B15 through B24. Note that the frequency of Doves plus the frequency of Hawks must equal 1.

6. Calculate the mean fitness of Doves in a population of all Hawks.

In cell C14 enter the formula **=$E$10*B14+$F$10*A14**.
This corresponds to, Equation 1

$$0H + \left(\frac{V}{2} - T\right)D$$

and calculates the mean fitness of Doves in a population having the proportion of Doves and Hawks shown to the left in the same row.

We include **$E$10*B14** (i.e., $0H$) in the formula in case you want to change the payoff in cell E10 later in the exercise.

7. Calculate the mean fitness of Hawks in a population of all Hawks.

In cell D14 enter the formula **=$E$9*B14+$F$9*A14**.
This corresponds to Equation 2

$$\left(\frac{V - W}{2}\right)H + VD$$

and calculates the mean fitness of Hawks in a population with the same proportion of Doves and Hawks.

8. Calculate the mean fitnesses of Doves and Hawks at each of the population ratios in columns A and B. Save your work.

Copy the formulae from cells C14 and D14 into cells C15 through D24.

9. Graph the mean fitness of Doves and Hawks against the proportion of Hawks in the population.

Select cells B14 through D24 and make an XY graph. Edit your graph for readability. It should resemble the one in Figure 2.

10. Answer questions 1–5 at the end of the chapter.

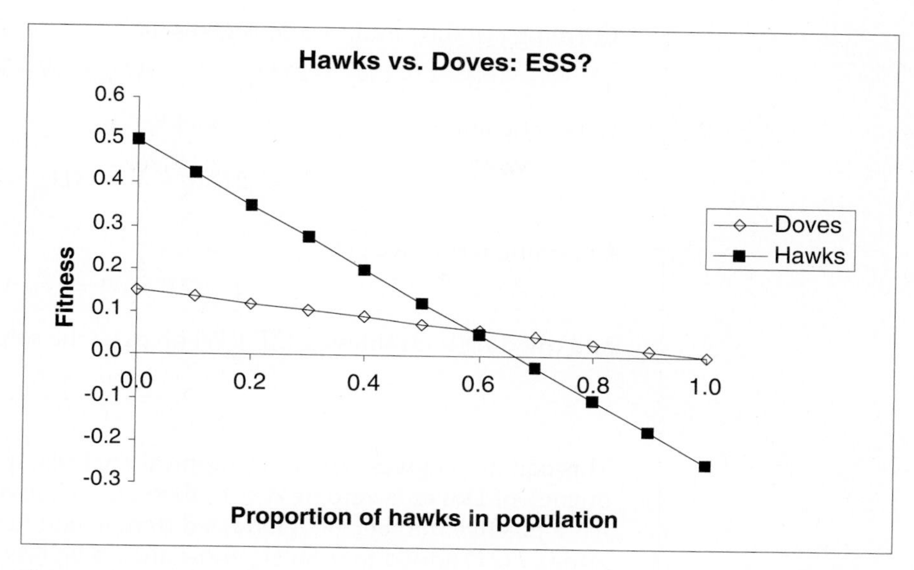

Figure 2

Equilibrium Solutions

In answering questions 1–5 at the end of the chapter, you should have discovered that if $V < W$, neither strategy is an ESS, and the equilibrium population will consist of a mixture of Hawks and Doves. In the first section of this exercise, we spoke of these strategies as being fixed patterns of behavior. However, the model may still apply even if behavior is not so rigid. We may suppose that a given animal behaves as a Hawk in some encounters and as a Dove in others. This changes our interpretation of the equilibrium result somewhat. Now we may conceive of the equilibrium as representing the optimal split in each animal's behavior. For example, if the equilibrium is 0.60 Dove and 0.40 Hawk, that would indicate that an animal achieves the greatest fitness by acting like a Dove in 60% of its encounters, and like a Hawk in 40%.

As you discovered graphically above, if wounds cost more than victory pays (i.e., if $W > V$), then neither Hawk nor Dove is an ESS. In such cases, the equilibrium population will consist of some mixture of Hawks and Doves. Can we determine what this equilibrium mixture will be?

We can, if we begin with an insight from Figure 2, our graph of fitness of Hawks and Doves at various frequencies of the two strategies. When the two strategies are at their equilibrium frequencies, their mean fitnesses are equal. This must be so, because if either strategy had a higher mean fitness, its frequency would increase, and therefore the population would not be at equilibrium.

So, if we represent the equilibrium frequency of Hawks as H_{eq} and the equilibrium frequency of Doves as D_{eq}, we can write

$$0H_{eq} + \left(\frac{V}{2} - T\right)D_{eq} = \left(\frac{V-W}{2}\right)H_{eq} + VD_{eq}$$

Because H_{eq} and D_{eq} are relative frequencies, they must add up to 1. Therefore, we can rewrite H_{eq} as $1 - D_{eq}$, and substitute:

$$0(1-D_{eq}) + \left(\frac{V}{2} - T\right)D_{eq} = \left(\frac{V-W}{2}\right)(1-D_{eq}) + VD_{eq}$$

If we eliminate the zero term on the left, and multiply both sides by 2, we get

$$(V - 2T)D_{eq} = (V - W)(1 - D_{eq}) + 2VD_{eq}$$

Carrying out the multiplications gives us

$$VD_{eq} - 2TD_{eq} = V + WD_{eq} - W - VD_{eq} + 2VD_{eq}$$

Canceling and rearranging terms yields

$$-2TD_{eq} = V + WD_{eq} - W$$

Collecting terms, we get

$$D_{eq}(2T + W) = W - V$$

and dividing both sides by $(2T + W)$ gives us the solution

$$D_{eq} = \frac{W - V}{2T + W} \qquad \text{Equation 3}$$

This equation agrees with our graphical analysis: If $W = V$, then the equilibrium frequency of Doves is zero; if $W > V$, then D_{eq} is between 0 and 1. In the numerator W has a positive number (V) subtracted from it, and in the denominator it has a positive number ($2T$) added to it, so D_{eq} must always be less than 1. Therefore, Dove is not an ESS against Hawk, regardless of the values of V, W, and T—as long as all are greater than zero.

If $W < V$, then the equation appears to predict a negative equilibrium frequency for Doves. This makes no sense, so we interpret it to mean that the frequency of Doves will decline (from any starting value) until it reaches zero. In other words, if $W < V$, then Hawk is an ESS against Dove.

For the sake of completeness, we can calculate the equilibrium frequency of Hawks as $1 - D_{eq}$, or

$$H_{eq} = 1 - \frac{W - V}{2T + W}$$

Substituting $\frac{2T + W}{2T + W}$ for 1 gives us

$$H_{eq} = \frac{2T + W}{2T + W} - \frac{W - V}{2T + W}$$

Combining the fractions, we get

$$H_{eq} = \frac{2T + W - W + V}{2T + W}$$

$$H_{eq} = \frac{2T + V}{2T + W} \qquad \text{Equation 4}$$

Although it is not as obvious, this equation makes the same predictions as Equation 3. That is, if $W = V$, then Hawk is an ESS against Dove; if $W > V$, Hawk is not an ESS against Dove (but remember, Dove is never an ESS).

Group Selection versus Individual Selection

These equilibrium solutions may not seem very interesting in themselves, but we can use them to arrive at some interesting conclusions. People often argue that some physical or behavioral trait exists because it benefits the species (or the population, or some other group). For instance, it is often said that humans (and many other animals) display cooperative behavior because cooperative groups are better at gathering food or fending off predators, or for other reasons have higher odds of survival. Such arguments are called group selection arguments, because they claim that natural selection operates on the group as a whole. **Group selection** argues that natural selection will favor a trait that confers higher fitness on the group, even if it reduces the fitness of the individuals that make it up.

The contrasting position, **individual selection**, claims that natural selection operates on individuals, not groups. Individual selection arguments predict that natural selection will favor a trait that confers higher fitness on individuals, even if it reduces the fitness of the group to which they belong.

We will use the equations for mean fitness of Doves and Hawks, and their equilibrium solutions, to investigate the contrast between group selection and individual selection. We will show that if a population consisted entirely of Doves, it would have a higher mean fitness than a population consisting entirely of Hawks or of any mixture of Hawks and Doves. A group selectionist would therefore expect the frequency of Doves in a population to increase, because that would benefit the group. However, as we will see, individual Hawks have higher fitness than individual Doves (at least when Hawks are rare). An individual selectionist would therefore expect natural selection to favor Hawks over Doves (at least when Hawks are rare), even if that reduces the fitness of the group as a whole.

PROCEDURES

Our strategy to test these ideas has five components:

- Calculate the mean fitness of the entire population, across the range of all mixtures of Hawks and Doves, from $D = 0$ and $H = 1$ to $D = 1$ and $H = 0$.
- Graphically estimate the mixture of Hawks and Doves that produces the maximum mean population fitness.
- Calculate the equilibrium mix of Doves and Hawks.
- Calculate the mean fitness of a population consisting of the equilibrium mix.
- Compare the maximum possible mean fitness of the population to its mean fitness at equilibrium.

We will repeat these steps for various values of V, W, and T, and compare the calculated values of mean fitness. We will see that this game theory model supports individual selection.

As always, save your work frequently to disk.

INSTRUCTIONS	ANNOTATION
B. Group selection versus individual selection.	
1. On the spreadsheet you prepared earlier (see Figure 1), change the values of V and W to 1, and the value of T to 0.	Enter these values into cells B8, B9, and B10, respectively.
2. Add a column heading for mean fitness of the entire population.	In cell E13 enter the label "Population."
3. Calculate the mean fitness of the entire population for each mixture of Doves and Hawks in cells A14 through B24.	In cell E14 enter the formula **=C14*A14+D14*B14**. Copy this formula into cells E15–E24. This formula multiplies the mean fitness of Doves by their frequency and the mean fitness of Hawks by their frequency, then adds the two products together. When you have finished, your spreadsheet should resemble Figure 3.

	A	B	C	D	E	F
7	Outcome	Fitness points		Payoff matrix (payoffs to player on left)		
8	Victory	1.00			Hawk	Dove
9	Wound	1.00		Hawk	0.00	1.00
10	Time	0.00		Dove	0.00	0.50
11						
12	Proportion		Fitness			
13	Doves	Hawks	Doves	Hawks	Population	
14	0.0	1.0	0.000	0.000	0.000	
15	0.1	0.9	0.050	0.100	0.095	
16	0.2	0.8	0.100	0.200	0.180	
17	0.3	0.7	0.150	0.300	0.255	
18	0.4	0.6	0.200	0.400	0.320	
19	0.5	0.5	0.250	0.500	0.375	
20	0.6	0.4	0.300	0.600	0.420	
21	0.7	0.3	0.350	0.700	0.455	
22	0.8	0.2	0.400	0.800	0.480	
23	0.9	0.1	0.450	0.900	0.495	
24	1.0	0.0	0.500	1.000	0.500	

Figure 3

4. Set up labels in column H and in cell I12, as shown in Figure 4.

These are all literals, so just select the appropriate cells and type them in.

	H	I
7	Equilibrium mix	
8	Proportion of doves	
9	Proportion of hawks	
10		
11	Fitness matrix	
12	Population composition	Mean fitness
13	All hawks	
14	All doves	
15	Equilibrium mix	

Figure 4

5. Calculate equilibrium frequencies of Doves and Hawks.

In cell I8, enter the formula =**IF((B9-B8)/(2*B10+B9)>0,(B9-B8)/(2*B10+B9),0)**. In this formula, **(B9-B8)/(2*B10+B9)** corresponds to Equation 3:

$$D_{eq} = \frac{W - V}{2T + W}$$

However, this equation can predict negative equilibrium frequencies for Doves, given some parameter values. We use the **IF()** function to restrict Dove frequencies to non-negative values. If D_{eq} is negative, we set it to zero.

In cell I9, enter the formula **=1-I8**.
This is the spreadsheet equivalent of $1 - D_{eq}$. Because $H_{eq} + D_{eq} = 1$, we do not need to use Equation 4 to calculate the equilibrium frequency of Hawks. You can, if you prefer, enter the spreadsheet equivalent of Equation 4; it should yield the same result.

6. Calculate mean fitness of a population consisting entirely of Hawks.

In cell I13 enter the formula **=E9**.
In an all-Hawk population, all encounters will be Hawk against Hawk. Therefore, all members of the population will receive the same payoff, $(V - W)/2$, which is calculated in cell E9.
You can arrive at the same result using Equation 2 to calculate the mean fitness of Hawks, bearing in mind that $H = 1$ and $D = 0$.

7. Calculate mean fitness of a population consisting entirely of Doves.

In cell I14 enter the formula **=F10**.
In an all-Dove population, all encounters will be Dove against Dove. Therefore, all members of the population will receive the same payoff, $(V/2) - T$, which is calculated in cell F10.
You can arrive at the same result using Equation 1 for mean fitness of Doves, bearing in mind that $H = 0$ and $D = 1$.

8. Calculate mean fitness of a population consisting of the equilibrium mixture of Hawks and Doves.

In cell I15 enter the formula **=$E$9*I9+$F$9*I8**.
This is the spreadsheet version of Equation 2 for the mean fitness of Hawks, this time using the equilibrium values of D and H, as calculated in cells I8 and I9. Remember that, at equilibrium, the mean fitnesses of Hawks and Doves are equal, so this is equivalent to calculating the mean fitness of all members of the population, regardless of strategy.

9. Add the data for population fitness to your existing graph.

Select the graph by clicking once anywhere in it and selecting Open Chart | Add Data. In the dialog box that appears, enter the cell addresses E13– E24. Be sure to include the label in cell E13, so that it will appear in the figure legend. Edit your graph for readability. It should resemble Figure 5.

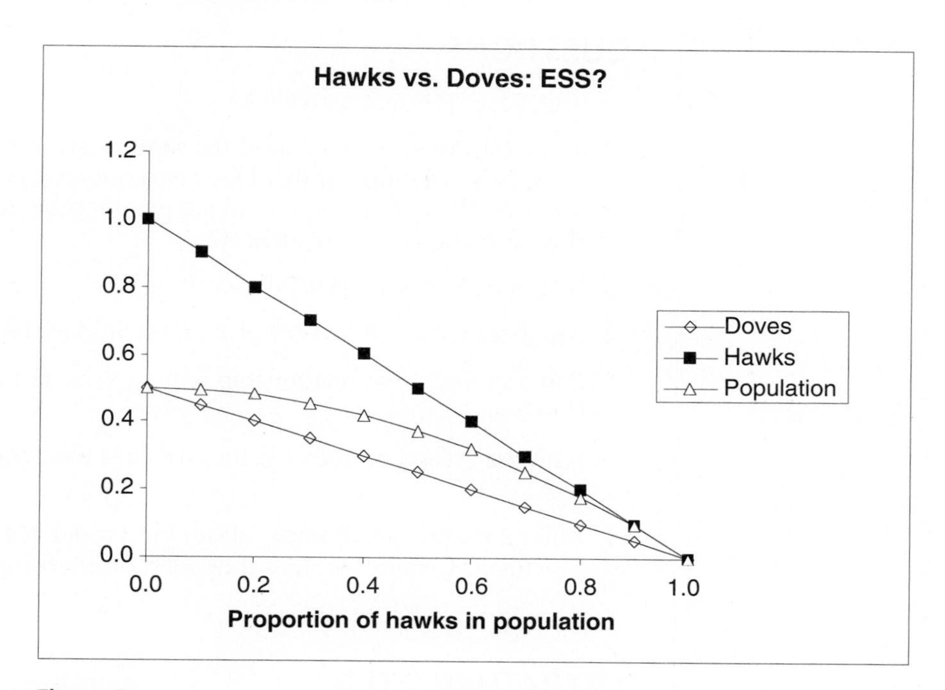

Figure 5

10. Answer questions 6 and 7 at the end of the chapter.

Conclusions

The upshot of this part of the exercise is strong support for individual selection. In every case where group and individual selection hypotheses predict different outcomes, the model produces the individual selection outcome.

One may argue, however, that this result does not prove that group selection cannot occur, only that it does not operate in this model. On the other hand, it is clearly the case that a pure population of Doves has the highest fitness in most scenarios, and yet Doves are displaced by Hawks. The matter comes down to the problem of cheaters. If everyone in the population "agrees" to behave as a Dove, the group as a whole will benefit. But if anyone "cheats" on the pact, and behaves as a Hawk, he or she will reap greater benefits than anyone behaving as a Dove. Hawkish behavior will spread through the population, either by genetic heritage, or by other Doves defecting as they see cheaters prospering. As the frequency of Hawks goes up, the fitness of each drops, because there are fewer Doves left to exploit. Even so, it still pays better to be a Hawk than a Dove. The result will be a population of Hawks, but each with lower fitness than he or she would have enjoyed if only everyone had remained a Dove. The language of "agreeing" and "cheating" should be understood metaphorically; there need be no conscious decision-making involved.

Another way to state the problem is in terms of individual interests versus group interests. If the interests of the individual are opposed to the interests of the group, individual interests are likely to dominate. Most evolutionary biologists are convinced that group selection, if it operates at all, can have noticeable effects only under very narrowly circumscribed conditions.

QUESTIONS

1. Is Dove an ESS against Hawk?
2. In the Introduction, we found the same answer without giving explicit values to V, W, or T. We implied that Dove was not an ESS against Hawk with any values of V, W, or T, as long as all are greater than zero. Can you support this conclusion using your spreadsheet?
3. Is Hawk an ESS against Dove?
4. Are there values of V, W, and T that would make Hawk an ESS against Dove?
5. Can you find what relationship among these parameters is necessary to make Hawk an ESS?
6. With the given parameter values, what is the equilibrium mixture of Hawks and Doves?
7. What does this result imply about individual versus group selection? Is this conclusion general, or does it depend on choosing parameter values carefully?

LITERATURE CITED

Maynard Smith, J. 1976. Evolution and the theory of games. *American Scientist* 64: 41–45.

40 MATING SYSTEMS AND PARENTAL CARE

Objectives

- Develop a game theory model of parental care and mating systems.
- Determine the environmental and biological conditions that lead to monogamy, polygyny, and polyandry.
- Examine which model parameters have significant impact on reproductive output for males and females.
- Verify the four evolutionarily stable strategies derived by Maynard Smith (1977).

Suggested Preliminary Exercise: Evolutionarily Stable Strategies

INTRODUCTION

You are well aware by now that there are fundamental differences between males and females of all species. From an evolutionary perspective, the goal is to produce as many offspring as possible that will, in turn, produce offspring. Males and females may have different strategies for doing this (Trivers 1972). Females, the egg producers, tend to invest a lot of energy in the production of gametes, while males invest much less in gamete production. In short, eggs are more "expensive" than sperm. For example, a human female typically produces only a few hundred viable eggs in her lifetime, whereas a human male can produce literally billions of sperm cells.

For many species, the production and propagation of gametes is the only parental investment. The fertilized egg, or zygote, is left to "sink or swim" on its own. But many other species nurture embryos through gestation and birth (almost exclusively the role of the female, though there are exceptions), and the offspring may require additional parental care in order to survive to reproductive age.

In some environments, both parents are needed to successfully rear young, while in other environments little or no care is needed. In cases where a single parent suffices to raise offspring, a male will maximize his fitness by fertilizing as many eggs as possible, leaving the parental care of his offspring to females. But if there are opportunities to mate with other, superior, males, a female should leave parental care to males to maximize her fitness! In situations where the young must be cared for, this sets up a "conflict" between the sexes because males and females differ with respect to behaviors that maximize fitness. All other things being equal, parents should maximize their fitness by fertilizing or producing as many eggs as possible, but if parental care enhances offspring survival, parents may maximize their fitness by providing care at the expense of additional matings. How can this conflict be resolved?

Mating Strategies

Mating strategies are often linked to the kind of parental care system that species employ. **Monogamy** is a mating system in which males and females form pair bonds, and often both care for the offspring. **Polygyny** is a mating system in which a male mates with several females. The female usually cares for the young while the male attempts to maximize his fitness by mating with as many females as possible. **Polyandry** is a mating system in which a female mates with several males. Males may care for the young while females attempt to maximize their fitness by mating with as many males as possible. And finally, **promiscuity** is a mating system free-for-all, in which either sex may care for the young and both males and females mate with many different individuals (Vehrencamp and Bradbury 1978; Alcock 2001).

Which mating system should be used to maximize fitness for males? Which mating system should be used to maximize fitness for females? Should parental care be given to the offspring? The answers to the questions depend, in large part, on the ecological conditions of a given environment, which affects how many parents are needed to ensure offspring survival, and how likely an individual will find another mate. But the strategy employed by a male or female also depends on the strategy adopted by the partner. For example, if the female cares for the young, and only a single parent is needed to raise offspring, the male may enhance his fitness by finding new females to mate with. But if the female does not care for the young, the male may enhance his fitness by attending the young himself. This type of conflict can be evaluated by **game theory** models, in which the different strategies played by the male and female collectively determine the evolutionary fitness gain.

A useful game theory model to resolve such conflict was developed by John Maynard Smith (1977). The model consists of two strategies: care for young (1) or desert young (0), that are chosen by both males and females. Thus, four "games" can be played: (1) both males and females care for young; (2) both males and females desert young; (3) the female cares for young and the male deserts; (4) the male cares for the young and the female deserts. Which of these games should be played depends on several parameters:

- P_0 = the probability of survival of eggs that are not cared for.
- P_1 = the probability of survival of eggs when one parent cares for young.
- P_2 = the probability of survival of eggs when two parents care for young.
- p = the probability of a deserter male finding a new mate.
- p' = the probability of a caring male finding a new mate.
- V = the number of eggs laid by a female deserter.
- v = the number of eggs laid by a female who cares for her young.

Thus, the model considers the value of parental care by one or two parents; the chance that males mate again; and how parental care affects the number of eggs the female can lay. We will assume that $P_0 \leq P_1 \leq P_2$, so that the probability of survival of eggs with parental care is never less than the probability of survival without parental care. We will also assume that $V \geq v$, so that females that care have less energy to allocate towards clutch size. Our final assumption is that p and p' do not depend on a male's parentage for a given clutch. Given these parameters, the fitness payoff for males and females can be determined as shown in Table 1.

For example, when both males and females care for the offspring, the female has a reproductive output equal to the number of eggs laid by a caring female (v) times the probability of young surviving when two parents offer care (P_2). But when a female cares but the male deserts, she has a reproductive output equal to the number of eggs laid per caring female (v) times the probability of young surviving when a single parent offers care (P_1). When both parents care for young, males have a reproductive output (fitness) equal that of the female ($v \times P_2$), but with the added benefits of remating with another female while still providing care to his first clutch ($v \times P_2 \times p'$). The equation v

Table 1. Fitness Payoff Parameters for Males and Females

	Female Fitness		Male Fitness	
	Female cares	Female deserts	Female cares	Female deserts
Male cares	$v \times P_2$	$V \times P_1$	$v \times P_2 + v \times P_2 \times p'$	$V \times P_1 + V \times P_1 \times p'$
Male deserts	$v \times P_1$	$V \times P_0$	$v \times P_1 + v \times P_1 \times p$	$V \times P_0 + V \times P_0 \times p$

$\times P_2 + v \times P_2 \times p'$ can be rewritten as $v \times P_2 \times (1 + p')$. When the female cares but the male deserts, his fitness is equal to that of a single-parent female ($v \times P_1$) plus the added benefits of remating with another female by deserting his clutch ($v \times P_1 \times p$). The equation $v \times P_1 + v \times P_1 \times p$ can be rewritten as $v \times P_1 \times (1 + p)$.

Evolutionarily Stable Mating Strategies

How do the two sexes resolve their conflicts? In this exercise, you'll set up a spreadsheet version of Maynard Smith's model and use it to explore the conditions in which different parental care systems are likely to evolve. There are four conditions that lead to a particular type of system. When these conditions are met, the strategy is called an **evolutionarily stable strategy** (**ESS** for short). In this case, the strategy played by the sexes is either "care" or "desert." A strategy is evolutionarily stable when, if all members of a population adopt it, then a mutant strategy could not invade the population and increase in frequency by natural selection (Maynard Smith 1982).

In order to arrive at ESS conditions, it's useful to first think about how the frequency of a particular strategy may change over time. We will let

- r = frequency of caring strategists (C).
- $s = 1 - r$ = frequency of deserter strategists (D).
- W(C), W(D) = fitness of caring and deserter strategists, respectively.
- E(C,C) = payoff to an individual adopting C (care) while the mate adopts C.
- E(C,D) = payoff to an individual adopting C while the mate adopts D (desert).
- E(D,D) = payoff to an individual adopting D while the mate adopts D.
- E(D,C) = payoff to an individual adopting D while the mate adopts C.

Because how well one sex fares depends on the strategies played by the opposite sex, we need to consider the fitnesses of each sex separately, while taking into account the frequency of C and D strategists in the opposite sex. Thus, calculations are needed for both sexes. For females, the fitness of players that engage in parental care is

$$W(\mathrm{C}) = [r_m \times E(\mathrm{C,C})] + [s_m \times E(\mathrm{C,D})] \qquad \text{Equation 1}$$

where r_m and s_m is the frequency of males that care and desert, respectively. The fitness of females that desert is

$$W(\mathrm{D}) = [r_m \times E(\mathrm{D,C})] + [s_m \times E(\mathrm{D,D})] \qquad \text{Equation 2}$$

Thus, you can see that the fitness of females depends on the strategies that males play as well as the frequency of each kind of strategist. The same equations work for males, except that we need to consider the frequencies of the female strategists in the population. To be clear, let's walk through an example. If we are interested in the fitness of a male that cares, we need to determine what his fitness is when he adopts a caring strategy and his mate also cares, E(C,C), and we need to determine what his fitness is when he adopts a caring strategy and his mate deserts, E(C,D). Suppose that 10% of females provide care to young while the remaining 90% desert. Thus, $r_f = 0.1$ and $s_f = 0.9$. If E(C,C) = 5 and E(C,D) = 3, then the fitness of caring males in the population is

$$W(\mathrm{C}) = [0.10 \times 5] + [0.90 \times 3] = 3.2$$

If a male adopts a deserting strategy, then we need to determine what his fitness is when he deserts and his mate also deserts, $E(\mathrm{D,D})$, and we need to determine what his fitness is when he deserts but his mate cares, $E(\mathrm{D,C})$. If $E(\mathrm{D,D}) = 0$ and $E(\mathrm{D,C}) = 3$, then the fitness of deserting males in the population is

$$W(\mathrm{D}) = [0.10 \times 3] + [0.90 \times 0] = 0.3$$

In this example, males that provide care have higher fitnesses since $W(\mathrm{C}) > W(\mathrm{D})$, but how much this strategy increases in the next generation depends on the proportion of males playing each strategy. If a lot of individuals are playing the more successful strategy, then the trait will increase more quickly. We can calculate the mean fitness for males as

$$\overline{W} = [r_{\mathrm{m}} \times W(\mathrm{C})] + [s_{\mathrm{m}} \times W(\mathrm{D})] \qquad \text{Equation 3}$$

and the mean fitness of females as

$$\overline{W} = [r_{\mathrm{f}} \times W(\mathrm{C})] + [s_{\mathrm{f}} \times W(\mathrm{D})] \qquad \text{Equation 4}$$

Once we understand Equations 1–4, we can compute the frequency of a given strategy for a given sex in the next generation as

$$r' = \frac{r \times W(\mathrm{C})}{\overline{W}} \text{ and } s' = \frac{s \times W(\mathrm{D})}{\overline{W}} \qquad \text{Equation 5}$$

and we can show the change in the frequency with which each strategy is played for both males and females over time.

PROCEDURES

As Table 2 shows, there are four possible evolutionarily stable conditions (Maynard Smith 1982). The mating strategies that evolve depend on

- the value of parental care by one or two parents
- the chance that males mate again
- how parental care affects the number of eggs the female can lay

We will explore these conditions thoroughly in the exercise and try to make sense of their logic. The goal of this exercise is to develop a spreadsheet version of Maynard Smith's model and use it to explore the conditions in which different parental care systems are likely to evolve. As always, save your work frequently to disk.

Table 2. Conditions for the Four Evolutionarily Stable Mating Strategies of Maynard Smith (1982)

Strategy	Description	Conditions[a]
ESS 1	Female cares when	$vP_2 > VP_1$
Monogamy	Male cares when	$P_2(1 + p') > P_1(1 + p)$
ESS 2	Female deserts when	$VP_1 > vP_2$
Polyandry	Male cares when	$P_1(1 + p') > P_0(1 + p)$
ESS 3	Female cares when	$vP_1 > VP_0$
Polygyny	Male deserts when	$P_1(1 + p) > P_2(1 + p')$
ESS 4	Female deserts when	$VP_0 > vP_1$
Promiscuity	Male deserts when	$P_0(1 + p) > P_1(1 + p')$

[a]Conditions for an ESS are met when the inequality for the male and the female are *both* true.

INSTRUCTIONS	ANNOTATION

A. Set up the model and payoff parameters.

1. Open a new spreadsheet and enter headings as shown in Figure 1.

	A	B	C
1	***Parental Care and Mating Systems***		
2	**Based on Maynard Smith's (1977) game theory model**		
3			
4	**# of parents**	**Probability**	
5	0	$P_0 =$	0
6	1	$P_1 =$	0.1
7	2	$P_2 =$	0.9
8			
9	**Female behavior**	**# of eggs**	
10	Deserter	V = desert =	6
11	Care	v = care =	5
12			
13	**Male behavior**	**Probability of remating**	
14	Deserter	p = desert =	0.5
15	Care	p' = care =	0.5

Figure 1

2. Enter the variable values shown in cells C5–C7, C10–C11, and C14–C15.

The variables in the model include

- the probability of survival of eggs that are not cared for (P_0)
- the probability of survival of eggs cared for by a single parent (P_1)
- the probability of survival of eggs cared for by two parents (P_2)

(Remember that probabilities range from 0 to 1.)

For males, we also include

- the probability of mating again when the male deserts a nest (p)
- the probability of mating again when the male guards a nest (p')

For females, we must include

- the number of eggs laid per female when the female deserts the nest (V)
- the number of eggs laid per female when she cares for her young (v)

3. Graph the relationship between probability of survival of eggs as a function of the number of caring adults.

Use the XY scatter graph option and label your axes fully. Your graph should resemble Figure 2.

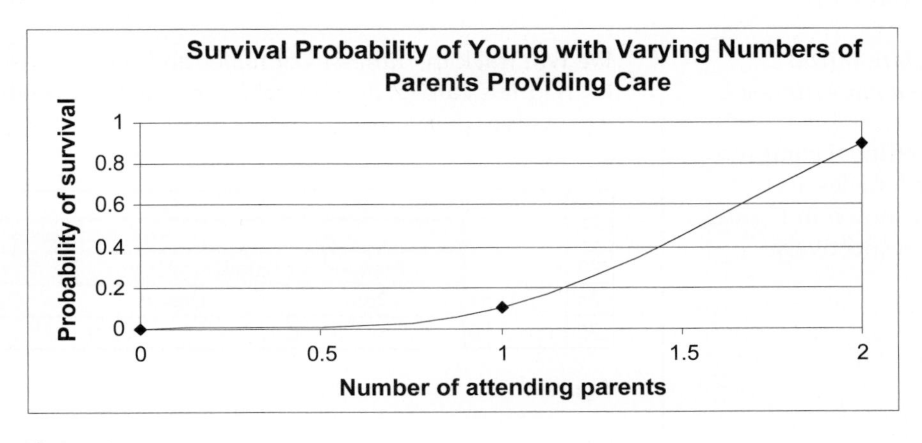

Figure 2

4. Set up new headings as shown in Figure 3.

Males and females can both employ one of two strategies: care or desert. Thus there are four fitness scenarios for each sex, depending on what strategy the mate plays. Cell I6 gives the fitness payoff for females that care when males also provide care, or *E*(C,C). Cell J6 gives the fitness payoff for females that desert while the male provides care, or *E*(D,C). Similarly, cell I11 gives the payoff for males that care when females also provide care, or *E*(C,C). Cell J12 gives the payoff to males that desert when their mates also desert, or *E*(D,D).

	H	I	J
4	**Female fitness matrix**		
5		***Female cares***	***Female deserts***
6	***Male cares***		
7	***Male deserts***		
8			
9	**Male fitness matrix**		
10		***Female cares***	***Female deserts***
11	***Male cares***		
12	***Male deserts***		

Figure 3

5. Enter formulae to compute the fitness payoffs for females in cells I6–J7 and males in cells I11–J12. Use the information in Table 1 to construct your formula.

Remember that the payoffs depend on which strategy is played by its partner. For a female that cares whose mate also cares, her payoff is the number of eggs laid per caring females × the probability of survival when both parents care for the young, or $v \times P_2$. The payoff formulae are given in Table 1, and the following formulae are based on the Table 1 equations.

Females:

- I6 **=C11*C7**
- I7 **=C11*C6**
- J6 **=C10*C6**
- J7 **=C10*C5**

Males:

- I11 **=C11*C7+C11*C7*C15** or **=I6+I6*C15**
- J11 **=C10*C6+C10*C6*C15** or **=J6+J6*C15**
- I12 **=C11*C6+C11*C6*C14** or **=I7+I7*C14**
- J12 **=C10*C5+C10*C5*C14** or **=J7+J7*C14**

6. Save your work.

B. Calculate initial female and male fitnesses.

We will track the fitnesses of males and females, as well as the frequencies in which individuals care (r) and desert (s) over a 20-year period, and determine which strategy evolves over time.

1. Set up fitness computations for females and males as shown in Figures 4 and 5, respectively.

	A	B	C	D	E
22		**Female fitness**			
23		Frequency of male strategy		Female fitness	
24	Time	Care	Desert	Care	Desert
25	0	0.1	0.9		

Figure 4

	F	G	H	I
22	**Male fitness**			
23	Frequency of female strategy		Male fitness	
24	Care	Desert	Care	Desert
25	0.9	0.1		

Figure 5

2. Set up a linear series from 0 to 20 in cells A25–A45.

Enter 0 in cell A25. Enter **=1+A25** in cell A26.
Select cell A26, and copy its formula down to cell A45.

3. Enter the starting frequencies of caring (r) males and deserting (s) males in cells B25–C25 as shown in Figure 4. Enter the starting frequencies of caring (r) and deserting (s) females in cells F25–G25 as shown in Figure 5.

Remember that r = frequency of caring (C) strategists and $s = (1 - r)$ = frequency of deserter (D) strategists. For now, enter the values shown in the figures. You will be able to change these starting frequencies later in the exercise. Cells B25–C25 give the frequency of male strategists at time 0. We need to know these frequencies in order to compute female fitness. Cells F25–G25 give the frequency of the female strategists at time 0. We need to know these frequencies in order to compute male fitness.

4. For year 0, enter formulae in cells D25 and E25 to compute the fitness, W, of females that care and desert. Refer to Equations 1 and 2 in the Introduction.

In cell D25 enter the formula **=$I$6*B25+$I$7*C25**.
In cell E25 enter the formula **=$J$6*B25+$J$7*C25**.
For the basis of these formulae, recall from Equation 1 that the fitness of females that care can be computed as

$$W(C) = [r_m \times E(C,C)] + [s_m \times E(C,D)]$$

where r_m and s_m are the frequencies of males that care and desert, respectively. The fitness of females that desert (Equation 2) is

$$W(D) = [r_m \times E(D,C)] + [s_m \times E(D,D)]$$

Your spreadsheet should now look like Figure 6.

	B	C	D	E
22	**Female fitness**			
23	Frequency of male strategy		Female fitness	
24	Care	Desert	Care	Desert
25	0.1	0.9	0.9	0.06

Figure 6

5. For year 0, enter formulae in cells H25 and I25 to compute the fitness, W, of males that care and desert.

In cell H25 enter the formula **=$I$11*F25+$J$11*G25**.
In cell I25 enter the formula **=$I$12*F25+$J$12*G25**.
Your spreadsheet should now look like Figure 7.

	F	G	H	I
22	**Male fitness**			
23	Frequency of female strategy		Male fitness	
24	Care	Desert	Care	Desert
25	0.9	0.1	6.165	0.675

Figure 7

6. Save your work.

C. Compute changes in fitnesses over time.

1. In cell B26, enter a formula to compute the frequency of a male caring strategy, r', in Year 1 for males. Refer to Equation 5 in the Introduction.

We entered the formula **=(B25*H25)/(B25*H25+C25*I25)** in cell B26.
The frequency of a caring strategy in the following generation is denoted by r'. Remember from Equation 5 that r' is calculated as

$$r' = \frac{r \times W(C)}{\overline{W}}$$

which is simply the fitness of males that care times the frequency of males that care divided by the mean fitness for males. Mean fitness of males, in turn, is calculated as

$$\overline{W} = [r_m \times W(C)] + [s_m \times W(D)]$$

2. In cell C26, enter a formula to compute the frequency of a male deserting strategy in year 1 for males.

In cell C26, enter the formula **=1-B26**.
The frequency of the deserting strategy in the following generation in denoted by s'. It can be computed as simply $1 - r'$.

3. Select cells D25–E25, and copy their formulae down 1 row.

	A	B	C	D	E
22		Female fitness			
23		Frequency of male strategy		Female fitness	
24	Time	Care	Desert	Care	Desert
25	0	0.1	0.9	0.9	0.06
26	1	0.503676471	0.496323529	2.514705882	0.302205882

Figure 8

4. In cell F26, enter a formula to compute r', the frequency of the caring strategy in year 1 for females.

Your spreadsheet should now look like Figure 8.
In cell F26 enter the formula **=(F25*D25)/(F25*D25+G25*E25)**.

5. In cell G26, enter a formula to compute s', the frequency of the deserting strategy in year 1 for females.

In cell G26 enter the formula **=1-F26**.

6. Select cells H25–I25, and copy their formulae down 1 row.

7. Select cells B26–I26, and copy their formulae down to row 45.

8. Save your work.

	F	G	H	I
22	Male fitness			
23	Frequency of female strategy		Male fitness	
24	Care	Desert	Care	Desert
25	0.9	0.1	6.165	0.675
26	0.992647059	0.007352941	6.706985294	0.744485294

Figure 9

D. Create graphs.

1. Graph the fitness of females that care and desert as a function of time (cells D24–E45).

Your spreadsheet should now look like Figure 9.

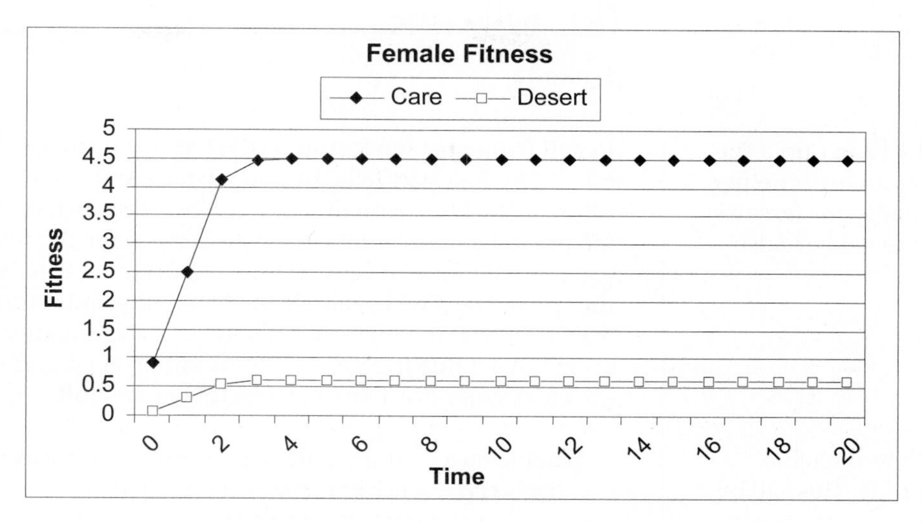

Figure 10

Use the line graph option and label your axes fully. Your graph should resemble Figure 10.

2. Graph the fitness of males that care and desert as a function of time (cells H24–I45).

Use the line graph option and label your axes fully. Your graph should resemble Figure 11.

3. Save your work. Interpret your results. Why did a two-parent caring system evolve? Play around with the model and see if you can get another kind of mating system to evolve. (Change cells C5–C7, C10–C11, and/or C14–C15.)

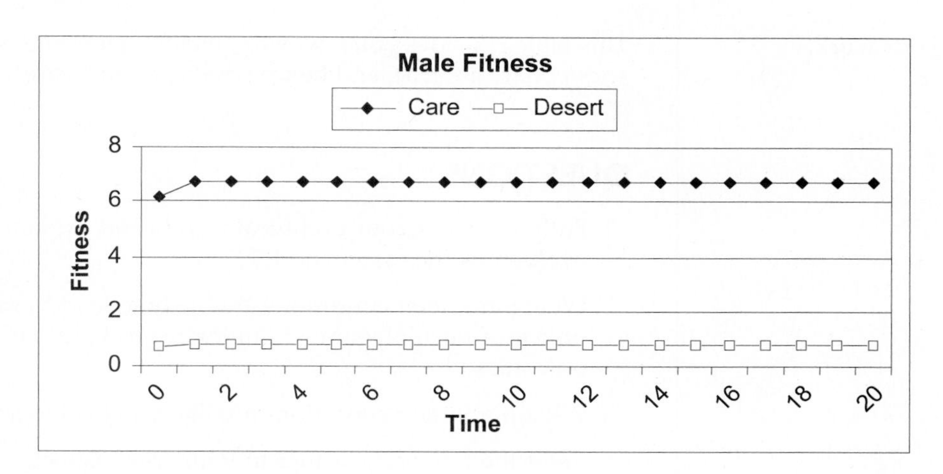

Figure 11

E. Compute the ESS inequalities.

Finally, we are able to evaluate the inequalities provided in Table 2 to determine the conditions in which an evolutionarily stable strategy evolves. Remember, the inequalitites for both the female and male must be true in order for a given mating system to evolve as an evolutionarily stable strategy. We will enter formulae in cells B19–E20 to reflect the inequalities in Table 2. If the condition is true, we will have the spreadsheet return the word TRUE; if the inequality is false, we will have the spreadsheet return the word FALSE.

1. Set up new headings as shown in Figure 12.

	A	B	C	D	E
17		**ESS 1**	**ESS 2**	**ESS 3**	**ESS 4**
18		Both care	Male cares	Female cares	Neither cares
19	**Female inequality**				
20	**Male inequality**				

Figure 12

2. In cell B19, set up a formula to evaluate whether the inequality for females for ESS 1 is true or false

In cell B19 enter the formula **=IF(C11*C7>C10*C6,TRUE)**. An **IF** formula has three parts. The first part tells the spreadsheet to evaluate a condition. In our case, the condition is the ESS inequality derived by Maynard Smith (1982) for females that care for offspring. Females will care for offspring when $vP_2 > VP_1$. The second part tells the program what value to return if the condition is true. Since the word TRUE is entered, the spreadsheet will evaluate the inequality and return TRUE if the inequality is in fact true. Note that we left the third part off of this equation, which normally tells the spreadsheet what value to return if the condition is false. If the third part is not specified, the program will return the word FALSE by default.

3. Complete the table given in cells B19–E20 by entering formulae analogous to that in Step 2. Refer to Table 2 as you enter formulae.

Double-check your results with ours. The formulae we used are:

- B20 **=IF(C7*(1+C15)>C6*(1+C14),TRUE)**
- C19 **=IF(C10*C6>C11*C7,TRUE)**
- C20 **=IF(C6*(1+C15)>C5*(1+C14),TRUE)**
- D19 **=IF(C11*C6>C10*C5,TRUE)**
- D20 **=IF(C6*(1+C14)>C7*(1+C15),TRUE)**
- E19 **=IF(C10*C5>C11*C6,TRUE)**
- E20 **=IF(C5*(1+C14)>C6*(1+C15),TRUE)**

4. Save your work.

This table provides you a way to quickly determine if the inequalities for both males and females are true, and hence which parental care system is an ESS.

QUESTIONS

1. Fully interpret your graphical results and explain how the parental care system evolved. Is the system an ESS?
2. What parameter conditions are likely to lead to single-parent care (either social polyandry or polygamy)? Enter various values in your model and explore the outcomes.
3. What parameter conditions are likely to lead to social promiscuity?
4. Enter the following values in your spreadsheet.

	A	B	C
4	**# of parents**	**Probability**	
5	0	$P_0 =$	0
6	1	$P_1 =$	0.7
7	2	$P_2 =$	0.9
8			
9	**Female behavior**	**# of eggs**	
10	Deserter	V = desert =	10
11	Care	v = care =	5
12			
13	**Male behavior**	**Probability of remating**	
14	Deserter	p = desert =	1
15	Care	p' = care =	0

Which parental care system evolves? Evaluate the conditions in cells B19–E20. You should see that two ESSs are possible. Does the initial frequencies of *r* and *s* determine which parental care system is ultimately the most successful?

5. How does the environment affect P_0, P_1, P_2? How does the environment or characteristics of the population itself affect V, v, p, and p'?
6. The model you have built assumes that $P_2 > P_1 > P_0$. Why did we assume that $V \geq v$? Are these assumptions valid? Discuss the concept of trade-offs and constraints in your answer.

LITERATURE CITED

Alcock, J. 2001. *Animal Behavior: An Evolutionary Approach*, 6th Edition. Chapters 12 and 13, pp. 360–419. Sinauer Associates, Sunderland, MA.

Maynard Smith, J. 1977. Parental investment: A prospective analysis. *Animal Behaviour* 25: 1–9.

Maynard Smith, J. 1982. *Evolution and the Theory of Games*. Cambridge University Press, Cambridge.

Trivers, R. L. 1972. Parental investment and sexual selection. *In* B. Campbell (ed.), *Sexual Selection and the Descent of Man* , pp. 136–179. Heinemann, London.

Vehrencamp, S. and J. W. Bradbury. 1978. Mating systems and ecology. *In* J. R. Krebs and N. B. Davies (eds.), *Behavioural Ecology*, pp. 251–278. Blackwell Scientific Publications, Oxford.

41 INBREEDING, OUTBREEDING, AND RANDOM MATING

Objectives

- Determine how nonrandom breeding affects allele and genotype frequencies in a population.
- Determine the effects of inbreeding on genotypic and phenotypic variation.
- Determine the effects of outbreeding on genotypic and phenotypic variation.
- Examine how assortative mating affects allele frequencies in a population.
- Explore inbreeding levels and the *F* statistic under various mating strategies.

Suggested Preliminary Exercise: Hardy-Weinberg Equlibrium

INTRODUCTION

One of the assumptions of the Hardy-Weinberg principle is that individuals in a population mate at random. In this exercise, you'll explore how violating this assumption affects the evolution of a population. **Random mating** occurs when individuals in the population pair off at random. That is, every individual has the same chance of breeding with any other individual in the population. **Inbreeding**, on the other hand, occurs when mated pairs are *more similar in genotypes* than if they were chosen at random. Because individuals of similar phenotypes will usually be somewhat similar in their genotypes, assortative mating (preferentially mating with an individual of similar phenotype) is generally thought to have the same consequences as inbreeding (Crow and Kimura 1970). **Outbreeding**, the flip side of inbreeding, occurs when mated pairs are *less similar in genotypes* than if they were chosen at random.

In this exercise, we will focus on how nonrandom mating affects the allele frequencies and genotype frequencies at a single locus. Keep in mind, however, that when organisms tend to mate nonrandomly, the entire genome is affected. Nonrandom breeding does one of two things: it either decreases the heterozygosity in the population (inbreeding) or it increases the heterozygosity of the population (outbreeding). You might think that nonrandom mating will also change the allele frequencies in the population. In fact, nonrandom mating without selection does not change the allele frequencies in a population at all. This will become apparent as you work through the exercise.

Because nonrandom mating affects heterozygosity levels, it is useful to "quantify" the level of nonrandom mating by comparing the heterozygosity observed in a population to the levels expected by Hardy-Weinberg. You might recall that if there are only two alleles, A_1 and A_2, in the population at a given locus, the frequencies of the alleles are given by ***p*** and ***q***, where p is the frequency of one kind of allele (A_1) and q is the frequency of the second kind of allele (A_2). For genes that have only two alleles,

$$p + q = 1 \qquad \text{Equation 1}$$

For example, assume that the A locus has allele frequencies of $p = A_1 = 0.6$ and $q = A_2 = 0.4$. Given the allele frequencies for a population, the Hardy-Weinberg principle allows us to predict the genotype frequencies of a population, assuming that the population is large and that mating occurs at random, and that there is no gene flow, natural selection, or mutation acting on the population. The predicted genotypes of a population in Hardy-Weinberg equilibrium are p^2:$2pq$:q^2, where p^2 is the frequency of the A_1A_1 genotype, $2pq$ is the frequency of the A_1A_2 genotype, and q^2 is the frequency of the A_2A_2 genotype. The sum of the genotype frequencies, as always, will sum to 1. In this example, a population in Hardy-Weinberg equilibrium will have roughly the following genotype frequencies:

- Freq (A_1A_1) $= p^2 = p \times p = 0.6 \times 0.6 = 0.36$
- Freq (A_1A_2) $= 2 \times p \times q = 2 \times 0.6 \times 0.4 = 0.48$
- Freq (A_2A_2) $= q^2 = 0.4 \times 0.4 = 0.16$.

Note that the genotype frequencies add to 1:

$$p^2 + 2pq + q^2 = 1 \qquad \text{Equation 2}$$

Thus, approximately 48% of the individuals are expected to be heterozygous if the population is in Hardy-Weinberg equilibrium.

A population that mates nonrandomly will deviate from the Hardy-Weinberg expectation. This deviation is often quantified through the F statistic, also called the **inbreeding coefficient**:

$$F = \frac{H_0 - H}{H_0} \qquad \text{Equation 3}$$

where H_0 is the heterozygosity level predicted by Hardy-Weinberg, and H is the observed level of heterozygosity. From an inbreeding perspective, the F statistic takes on values from 0 to 1. If the observed level of H is equal to H_0, the numerator of Equation 3 is 0, and thus F is 0, indicating a randomly breeding population. When H is less than H_0, there is a deficiency of heterozygotes in the population (due to inbreeding). Thus, positive F values indicate some level of inbreeding. The F statistic will be 1 (complete inbreeding) when the population consists of only homozygotes.

Let's walk through an example. Suppose a population has the frequencies $A_1 = 0.6$ and $A_2 = 0.4$. As we calculated earlier, the expected frequency of heterozygotes is 0.48. Assume that this population, however, consists of 0 heterozogotes. The F statistic would be

$$F = \frac{0.48 - 0}{0.48} = 1$$

This population has the highest possible F statistic, suggesting that the population is highly inbred. If the population consisted instead of 48% heterozygotes, as predicted by Hardy-Weinberg, the F statistic would be

$$F = \frac{0.48 - 0.48}{0.48} = \frac{0}{0.48} = 0$$

Although the F statistic is intended to measure inbreeding, it also measures outbreeding as well, and takes on negative values when the observed level of heterozygosity is larger than that expected by Hardy-Weinberg. The F statistic can also be calculated through pedigree analysis (Hartl 2000). Inbreeding may affect an organism's fitness, or it may not. For example, average yield in hybrid corn decreases as F increases (Neal 1935), but low levels of heterozygosity in cheetahs (*Acinonyx jubatus*) do not appear to compromise their survival (Merola 1994).

When nonrandom mating occurs in a population, the Hardy-Weinberg genotype frequencies p^2, $2pq$, and q^2 are not expected. However, if we know F, we can predict the frequencies of the A_1A_1, A_1A_2, and A_2A_2 genotypes (Hartl 2000). Let's start with the frequency of the A_1A_2 genotype. Remember that H is the observed genotype frequency of the heterozygotes in the population, so all we need to do is solve for H:

$$F = \frac{H_0 - H}{H_0}$$

Multiply both sides of the equation by H_0 to give

$$H_0 \times F = H_0 - H \qquad \text{Equation 4}$$

Then subtract H_0 from both sides:

$$-H_0 + H_0 \times F = -H \qquad \text{Equation 5}$$

Then multiply both sides of the equation by –1 and rewrite the equation so that H appears on the left side:

$$H = H_0 - H_0 \times F \qquad \text{Equation 6}$$

And finally, since $2pq$ is the same thing as H_0, or the heterozygosity expected under Hardy-Weinberg equilibrium, we can calculate H as a function of p, q, and F:

$$H = 2pq - 2pqF \qquad \text{Equation 7}$$

Thus, if you know F, Equation 7 can predict the frequency of the A_1A_2 heterozygotes in a population. The frequency of the A_1A_1 and A_2A_2 homozygotes can also be predicted if you know F. Recall that the frequency of the A_1 allele (p) in a population is simply the frequency of the homozygotes (A_1A_1) plus half the frequency of the heterozygotes (A_1A_2). For simplicity, let's call the frequency of the A_1A_1 homozygotes D:

$$p = D + (H/2) \qquad \text{Equation 8}$$

So now we need to solve for D, the frequency of the A_1A_1 homozygotes:

$$D = p - (H/2) \qquad \text{Equation 9}$$

Since we know H from Equation 7, we can substitute in Equation 9 and simplify:

$$D = p - \frac{2pq - 2pqF}{2}$$

The 2 divides out, and substracting $2pq$ and $2pqF$ from p gives us

$$D = p - pq + pqF \qquad \text{Equation 11}$$

Now we can group the first two terms and factor out a p:

$$D = p(1 - q) + pqF \qquad \text{Equation 12}$$

And finally, because $1 - q$ is the same thing as p, we arrive at

$$D = p^2 + pqF \qquad \text{Equation 13}$$

The same logic will allow you to calculate the frequency of the A_2A_2 homozygotes, which we'll call R:

$$R = q^2 + pqF \qquad \text{Equation 14}$$

Thus, Equations 7, 13, and 14 allow you to predict the genotype frequencies of a population where p, q, and F are known.

PROCEDURES

In this exercise, you will set up a spreadsheet model to explore the effects of inbreeding and outbreeding on a population. Your population will consist of 1000 individuals that select mates according to probabilities that you assign. We will consider the effects the inbreeding and outbreeding on the allele frequencies at a single locus. This locus has two alleles, A_1 and A_2. The basic model will be fairly easy to construct, but the fun will start when you begin to change mating partners and see how mate selection and breeding system affect allele and genotype frequencies.

As always, save your work frequently to disk.

INSTRUCTIONS

ANNOTATION

A. Set up the population parameters

1. Open a new spreadsheet and set up column headings as shown in Figure 1.

	A	B	C
1	***Inbreeding, Outbreeding, and Assortative Mating***		
2			
3	**Initial genotype frequencies**		**Tally**
4			0
5	A1A1	0	
6	A1A2	1	
7	A2A2	0	
8			
9			
10			
11			
12			
13			
14			
15			
16			
17			
18			
19	**Adult population**		
20			
21	**Individual**	**Random number**	**Adult genotype**

Figure 1

2. Enter values shown in cells B5–B7.

We will start with a population whose genotype frequencies are given in cells B5–B7. Our population will consist solely of A_1A_2 heterozygotes since the frequency in cell B6 is 1.

3. Create a "running tally" in cells C4–C7

Enter **=B5** in cell C5.
Enter **=SUM($B$5:B6)** in cell C6. Copy cell C6 into cell C7.
The running tally is necessary to assign genotypes to individuals in the population. It also will help you quickly verify that your genotype frequencies add to 1. Note that cell C7 must *always* equal 1. If it does not equal 1, it means that the frequencies entered in cells B5–B7 don't add to 1 (adjust accordingly).

4. Set up a linear series from 1–1000 in cells A22–A1021.

Enter **1** in cell A22.
Enter **= 1+A22** in cell A23. Copy this formula down to cell A1021.
You have now established a population of 1000 individuals. Save your work.

5. In cells B22–B1021, generate random numbers between 0 and 1.

Enter **=RAND()**. When you press F9, the calculate key, the spreadsheet generates new random numbers.

6. Use the **LOOKUP** function to assign genotypes to each of the 1000 individuals based on the frequencies you entered in cells B5–B7 and the tally of genotype frequencies in cells C4–C7. Save your work.

In cell C22, enter the formula **=LOOKUP(B22,$C$4:$C$7,$A$5:$A$7)**. Copy your formula down to cell C1021.
The **LOOKUP** function looks up a value (the random number in cell B22) in a vector that you specify (cells **$C$4:$C$7**) and returns a genotype associated with that random number in the vector **$A$5:$A$7**. (Remember that a vector is a single row or column of values.) This function is handy for assigning genotypes to individuals because if **LOOKUP** can't find the exact lookup value (the random number given in cell B22), it matches the largest value in the lookup vector (cells **$C$4:$C$7**) that is less than or equal to **lookup_value**. The result is that genotypes are assigned to individuals in approximately the proportions that you specified. Examine your first 10 genotypes. They should all be A_1A_2 if the **LOOKUP** function worked properly. To see how the function works, change cells B5 and B7 to 0.5, and set cell B6 to 0. (Remember that the final tally of genotype frequencies must equal 1 in cell C7.) Now examine the genotypes of your first 10 individuals. The genotypes should be either A_1A_1 or A_2A_2. When you feel you have a handle on how this function works, return cells B5 and B7 to 0, and return cell B6 to 1.

B. Compute allele and genotype frequencies of the population.

1. Set up new column headings as shown in Figure 2:

	D	E	F	G	H	I
1						
2						
3			Computed frequencies			
4			A1	A2	Total	F
5	Initial allele frequencies:					
6	F1 allele frequencies:					
7						
8		A1A1	A1A2	A2A2	Total	
9	Initial genotype numbers:					
10	Initial genotype frequencies:					
11	F1 genotype numbers:					
12	F1 genotype frequencies:					

Figure 2

2. In cells E9–G9, use the **COUNTIF** formula to count the number of A_1A_1, A_1A_2, and A_2A_2 genotypes in the population.

Use the Paste Function key to guide you through the formulae. The **COUNTIF** formula counts the number of cells within a range that meet the given criteria. It has the syntax **COUNTIF(range,criteria)**, where **range** is the range of cells you want to examine, and **criteria** defines what you want to count.

- E9 **=COUNTIF($C$22:$C$1021,E8)**
- F9 **=COUNTIF($C$22:$C$1021,F8)**
- G9 **=COUNTIF($C$22:$C$1021,G8)**

3. In cell H9, use the **SUM** function to sum cells E9–G9.

Enter **=SUM(E9:G9)**. Your result should be 1000.

4. In cells E10–G10, enter formulae to calculate genotype frequencies.

Remember that frequencies range from 0 to 1. To calculate the frequency of the A_1A_1 genotype in the population, write a formula that counts the number of A_1A_1 genotypes, divided by the total number of individuals in the population.
In cell E10 enter the formula **=E9/$H$9**.
In cell F10 enter the formula **=F9/$H$9**.
In cell G10 enter the formula **=G9/$H$9**.

5. In cell H10, use the **SUM** function to sum the genotype frequencies.

In cell H10 enter the formula **=SUM(E10:G10)**.
The genotype frequencies calculated in cells C9–F9 should add to 1. If they don't, double-check your formulae.

6. In cells F5 and G5, enter formulae to calculate allele frequencies.

In cell F5 enter the formula **=(E9*2+F9)/(2*H9)**.
In cell G5 enter the formula **=1-F5**.
Since our population consists of 1000 individuals, there are 2000 "gene copies" present. In order to compute frequencies we need to determine how many of those gene copies are A_1 and how many are A_2. To calculate the frequency of the A_1 allele, we multiply the number of A_1A_1 homozygotes by 2 (because each individual carries two copies of this allele) and add to this number the number of heterozygotes (each heterozygote carries one copy of this allele). This sum is then divided by the total number of gene copies in the population ($2N$) to generate the frequency of the A_1 allele. Since there are only two alleles present, and since $p + q = 1$, we can obtain the frequency of the A_2 allele by subtraction.

7. In cell H5, use the **SUM** function to sum the allele frequencies. Save your work.

Enter the formula **=SUM(F5:G5)**.

C. Select mates.

1. Set up new column headings as shown in Figure 3.

	D	E	F	G	H
14		**Mate ==>**	**A1A1**	**A1A2**	**A2A2**
15	**Parental genotype:**	**A1A1**	1	0	0
16	**Parental genotype:**	**A1A2**	0	1	0
17	**Parental genotype:**	**A2A2**	0	0	1
18					

Figure 3

2. Enter values shown in cells F15–H17.

Now we will let our population mate and produce offspring.
The parental genotypes are listed in cells E15–E17. The genotype of a potential mate is given in cells F14–H14.
Cells F15–H17 give the probabilities of mating with a particular genotype. These cells are shaded in Figure 3 to indicate that you directly enter values into these cells. For example, cell F15 gives the probability that an A_1A_1 genotype will mate with another A_1A_1 genotype. Cell G15 gives the probability that an A_1A_1 genotype will mate with a heterozygous genotype, and cell H15 gives the probability that an A_1A_1 genotype will mate with an A_2A_2 genotype. Note that *the sum of the probabilities across rows must equal 1.*

For now, enter the probabilities shown. All individuals will therefore mate with an individual of an identical genotype.

3. Set up column headings as shown in Figure 4.

	E	F	G	H	I	J
19				Mate selection		
20				Random	Preferred	Actual
21	Index 1	Index 2	Index 3	number	mate genotype	mate genotype

Figure 4

Our goal is to have the spreadsheet look up the genotype of individual 1 and match it to genotypes listed in cells E15–E17. Then we want to determine the genotype of individual 1's mate, listed in cells F14–H14. To choose mates according to the probabilities given, we will use four different functions in: **MATCH**, **INDEX**, **RAND**, and **IF**. Used in combination, these formulae will allow us to generate the genotype of a mate for each individual in the population. Note that if a preferred genotype is not present in the population, the individual does not reproduce, and a rare genotype that is present and preferred can mate more than once.

4. In cell D22, enter a **MATCH** formula and copy the formula down to cell D1021.

In cell D22, enter the formula **=MATCH(C22,$E$15:$E$17)**.
The **MATCH** formula returns the *relative position* of an item in a table that matches the condition you specify. It has the syntax **MATCH(lookup_value,lookup_array)**. The formula in cell D22 tells the spreadsheet to find the genotype listed in cell C22, and return the *relative position* of that genotype in the table **$E$15:$E$17**. For example, the genotype of individual 1 in our program is A_1A_2. The program returns the value 2, to indicate that A_1A_2 individuals occupy the second position in our array. If individual 1 had the genotype A_1A_1, it would return the number 1, and if individual 1 had the genotype A_2A_2, it would return the number 3. Copy this formula down for the remaining 999 individuals, and make sure your **MATCH** values are correct. Since your population consists solely of heterozygotes, the match values should all be equal to 2.

5. In cell E22, enter an **INDEX** formula and copy this formula down to cell E1021.

In cell E22, enter the formula **=INDEX($E$15:$H$17,D22,2)**.
Our second trick is the **INDEX** formula. This formula returns the *value* of an element in a table, once you identify the row and column number that should be returned. The **INDEX** formula has the syntax **INDEX(array,row_num,column_num)**, where **array** is a range of cells in a table; **row_num** selects the row in the table from which to return a value, and **column_num** selects the column in table from which to return a value.

The formula in cell E22 tells the spreadsheet to examine the range of cells E15–H17, and to go to the row designated in cell D22 (derived from the **MATCH** formula entered in Step 4) and column 2 (which indicates the probability of mating with an A_1A_1 individual). The program will then return the value associated with this row and column intersection. Fill this formula down for the remaining 999 individuals in the population. Make sure you understand what is going on.

6. In cells F22 and G22, enter analogous **INDEX** formulae to generate the probability of mating with heterozygote and A_2A_2 homozygote, respectively. Copy your formulae down to cells F1021 and G1021.

In cell F22 enter the formula **=INDEX($E$15:$H$17,D22,3)**.
In cell G22 enter the formula **=INDEX($E$15:$H$17,D22,4)**.

The three **INDEX** formulae generate the appropriate mating probabilities for each individual in the population. Figure 5 shows the genotypes of the first four individuals in our population, their match values, and index values.

	C	D	E	F	G	H
14			Mate ==>	A1A1	A1A2	A2A2
15		Parental genotype:	A1A1	1	0	0
16		Parental genotype:	A1A2	0	1	0
17		Parental genotype:	A2A2	0	0	1
18						
19				Mate selection		
20						Random
21	Adult genotype	Match	Index 1	Index 2	Index 3	number
22	A1A2	2	0	1	0	
23	A1A2	2	0	1	0	
24	A1A2	2	0	1	0	
25	A1A2	2	0	1	0	

Figure 5

7. In cell H22, generate a random number between 0 and 1. Copy your formula down to H1020.

In cell H22 enter the formula **=RAND()**.
The **RAND** function generates a random number between 0 and 1. We will use this formula to determine the genotype of the mate for each individual in the population. When you press F9, the calculate key, the spreadsheet generates a new set of random numbers.

8. In cell I22, enter a nested **IF** formula to select the genotype of the preferred mate. Copy your formula down to cell I1021.

In cell I22 enter the formula **=IF(H22<=E22,$F$14,IF(H22<=E22+F22,$G$14,$H$14))**. Copy the formula down cell I1021 to obtain preferred mates for the remaining individuals in the population.

Our final step in selecting the genotype of the mate is to use two nested **IF** functions. Remember that an **IF** statement returns one value if a condition you specify is true and another value if the condition you specify is false. **IF** statements have the form **IF(logical_test,value_if_true,value_if_false).**

The first portion of the **IF** formula in I22, **=IF(H22<=E22,$F$14,** tells the spreadsheet to examine cell H22 (the random number associated with individual 1). If that value is less than or equal to the value in cell E22 (the first index number), return the value in cell F14 (A_1A_1). Otherwise, go through the second **IF** statement, **IF(H22<=E22+F22,$G$14,$H$14)**. This statement tells the program to examine the random number in cell H22. If that value is less than the sum of the values in cells E22 and F22 (the first and second index numbers), return the value in cell G14 (A_1A_2). Otherwise, return the value in cell H14 (A_2A_2).

9. In cell J22, enter an **IF** formula to generate the actual mate genotype, and copy this formula down to cell J1021.

Enter the formula **=IF(VLOOKUP(I22,$A$5:$B$7,2)>0,I22,".")**. Although we have established the mating preferences, we now need to ensure that an individual with the preferred genotype actually exists in the population for mating. The formula in cell J22 is a VLOOKUP function nested within an IF function. It tells the spreadsheet to look up the preferred mate's genotype given in cell I22 in the table of cells A5–B7 and return the associated value in the second column, which is the frequency of the preferred genotype. If the frequency of the preferred genotype is greater than 0, preferred individuals exist in the population for mating, and the spreadsheet returns the genotype listed in cell I22. If the preferred genotype does not exist in the population, its genotype frequency is 0, so the formula returns a period to indicate that the individual will not mate.

Take some time to make sure you can see how the formulae in cells I22 and J22 are working. In the example shown in Figure 6, individual 1 (A_1A_2) prefers to mate with an A_1A_2 genotype because its random number is greater than Index 1 (0) and less than the sum of Index 1 and Index 2 (which is 1). Since the preferred genotype is present in the population (its frequency is greater than 0), the actual mate genotype (A_1A_2) is given in cell J22.

10. Save your work.

	D	E	F	G	H	I	J
19	Mate Selection						
20					Random	Preferred	Actual
21	Match	Index 1	Index 2	Index 3	number	mate genotype	mate genotype
22	2	0	1	0	0.905419	A1A2	A1A2
23	2	0	1	0	0.079056	A1A2	A1A2
24	2	0	1	0	0.793008	A1A2	A1A2
25	2	0	1	0	0.85476	A1A2	A1A2

Figure 6

D. Obtain genotypes of offspring.

1. Set up column headings as shown in Figure 7.

	K	L
19	Offspring	
20		
21	Genotype	Genotype

Figure 7

2. In cell K22, enter a formula that will produce an "offspring" by randomly combining a genotype from each of the two parents. Copy the formula down to cell K1021.

In cell K22 enter the formula **=IF(RAND()<0.5,LEFT(C22,2),RIGHT(C22,2))&IF(RAND()<0.5,LEFT(J22,2),RIGHT(J22,2))**.
You are already familiar with the **RAND** function. The **LEFT** and **RIGHT** functions return either the *leftmost* or *rightmost* characters in a string of characters. For example, **LEFT(C22,2)** returns the leftmost two characters listed in cell C22.

The formula in cell K22 draws a random number for individual 1; if the random number is less than or equal to 0.5, individual 1 contributes the "left" allele in its genotype as a gamete; otherwise, it contributes the "right" allele in its genotype as a gamete. A second **IF** statement is used to determine the gamete contributed by individual 1's mate in column J. The gamete from individual 1 and its mate are joined with an **&** symbol, which produces the genotype of the offspring.

3. Enter a formula so all heterozygotes will be listed as A_1A_2. Copy the formula down to L1021.

In cell L22, enter the formula **=IF(K22="*A2A1*","*A1A2*",K22)**.
This formula is necessary because some of the heterozygous offspring will be listed as A_1A_2 and some will be listed as A_2A_1. For simplicity, we will make all the heterozygotes be listed as A_1A_2.

4. Save your work.

E. Calculate the new genotype and allele frequencies and the F statistic.

1. In cells E11–G11, use the **COUNTIF** formula to count the number of genotypes in the offspring population.

In cell E11 enter the formula **=COUNTIF($L$22:$L$1021,E8)**.
In cell F11 enter the formula **=COUNTIF($L$22:$L$1021,F8)**.
In cell G11 enter the formula **=COUNTIF($L$22:$L$1021,G8)**.

2. In cell H11, sum the offspring genotypes.

Enter the formula **=SUM(E11:G11)**. Double-check your formulae. Your results should total to 1000.

3. In cells E12–G12, calculate genotype frequencies of the offspring population.

In cell E12 enter the formula **=E11/$H$11**.
In cell F12 enter the formula **=F11/$H$11**.
In cell G12 enter the formula **=G11/$H$11**.

4. In cell H12, sum the offspring genotype frequencies.

Enter the formula **=SUM(E12:G12)**. Your results should total to 1.

5. In cell F6, enter a formula to calculate the frequency of the A_1 allele in the offspring population.

In cell F6 we entered the formula **=(E11*2+F11)/(2*H11)**
Remember that in a population of 1000 individuals, there are 2000 "gene copies" present because each individual carries two alleles. We just need to know how many of those gene copies are A_1 and how many are A_2.

6. In cell G6, calculate the frequency of the A_2 allele.

In cell G6 enter the formula **=1-F6.**
Since there are only two alleles at the A locus, $p + q = 1$. Since you already calculated p, q can be obtained by subtraction.

7. In cell I5, calculate F, the inbreeding coefficient of the offspring. Save your work.

Now we are ready to calculate the inbreeding coefficient of our offspring population. Remember that

$$F = \frac{H_0 - H}{H_0}$$

where H_0 is the heterozygosity level predicted by Hardy-Weinberg, and H is the observed level of heterozygosity. We used the formula **=((2*F6*G6)-F12)/(2*F6*G6)**. Your result should be close to 0, since the offspring population will consist of approximately 25% A_1A_1, 50% A_1A_2, and 25% A_2A_2 genotypes, as predicted by Hardy-Weinberg. Take a moment to consider your results.

F. Create graphs.

1. Graph the genotype frequencies of the parental population and the offpsring population.

Use a column graph. Select cells E10–G10 to graph the parental genotypes and cells E12–G12 to graph the offspring genotypes. Your graph should resemble Figure 8.

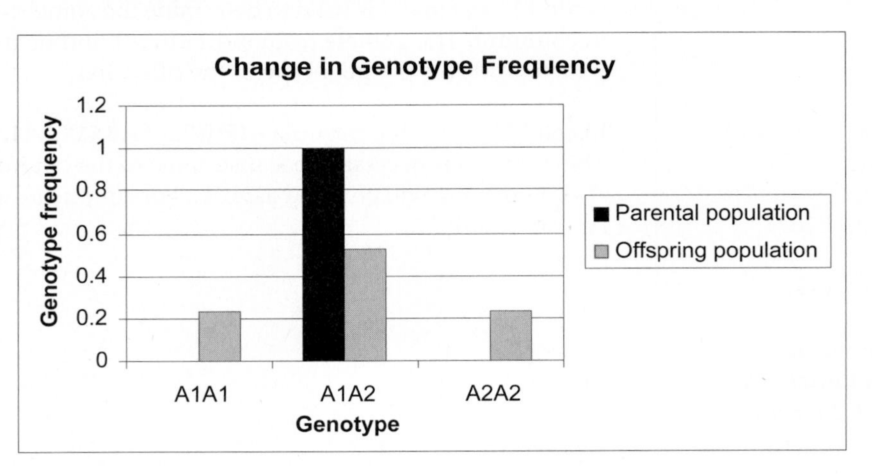

Figure 8

2. Graph the allele frequencies of the parental population and the offspring population for the A1 allele.

Use a line graph and select cells F5 and F6. Label your graph fully. Your graph should resemble Figure 9. Set the scale of the y-axis to range between 0 and 1.

3. Save your work.

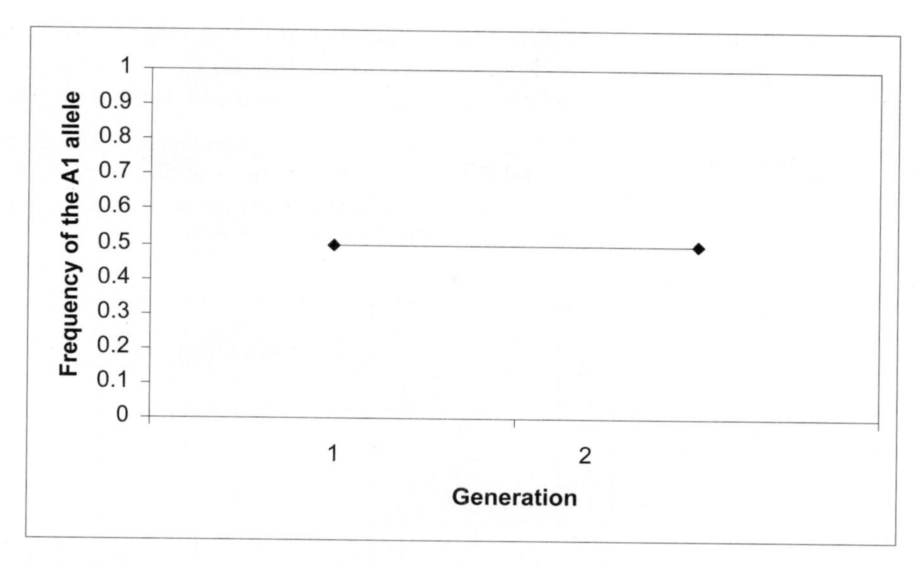

Figure 9

QUESTIONS

1. How does the allele frequency change from the parental population to the offspring population? How does the genotype frequency change from the parental population to the offspring population? Change the parental genotype frequencies in cells B5–B7 to 0.33, 0.34, and 0.33. How did the allele frequency change from the parental population to the offspring population? How did the genotype frequency change from the parental population to the offspring population?

2. Press F9, the calculate key, to generate new results. Why do your results vary from trial to trial?

3. Assume your offspring population will now breed and produce the next generation. How do F, p , and the genotype frequencies change over time with complete inbreeding? Set up new column headings as shown.

	J	K	L	M	N	O
3		Generation				
4		**1**	**2**	**3**	**4**	**5**
5	**A1A1**	0	0.247			
6	**A1A2**	1	0.498			
7	**A2A2**	0	0.255			
8	***p***	0.5	0.496			
9	***F***		0.003936			

Enter the genotype frequencies of your parental population in cells K5–K7. Enter the frequency of the A_1 allele, p, in cell K8. Enter your genotype frequencies of the offspring population in cells L5–L7. Enter p and F for the offspring population in cells L8–L9. (Your values will likely be a bit different than shown. If you copy and paste your results into the cells, make sure you choose paste special | paste values). Now let the offspring genotypes be the parental genotypes. Enter the offspring genotype values in cells B5–B7. Record the genotype frequencies of the new offspring population in cells M5–M7. Repeat until 5 generations have elapsed. How do *F*, *p* , and the genotype frequencies change over time with complete inbreeding? Graph your results.

4. Set the initial genotype frequencies to 0.25, 0.5, and 0.25 (cells B5–B7). What is the relationship between the probability of mating with the same genotype and *F*? Set up new column headings as shown, where Probability is the probability of mating with the same genotype:

	K	L	M
10		**Offspring**	
11	Probability	*p*	*F*
12	**0**		
13	**0.2**		
14	**0.4**		
15	**0.6**		
16	**0.8**		
17	**1**	0.5	0.5

You have already examined the case where $p = 1$. Enter the offspring p and F values in cells L17–M17. Now change the mating probabilities in cells F15–H17. Start with strict outbreeding, where the probability is 0. Enter 0 in cells F15, G16, and H17. Set the other mating probabilities so that the probability of mating with a dissimilar genotype is the same for the two remaining alternative genotypes (e.g., for Probability = 0, set the probability of mating with the other two kinds of genotypes to 0.5 so that they have equal changes of being selected for mating). For example, when the probability of mating with a similar genotype is 0.4, your spreadsheet should look like this:

	D	E	F	G	H
14		**Mate ==>**	**A1A1**	**A1A2**	**A2A2**
15	**Parental genotype:**	**A1A1**	0.4	0.3	0.3
16	**Parental genotype:**	**A1A2**	0.3	0.4	0.3
17	**Parental genotype:**	**A2A2**	0.3	0.3	0.4

Record p (the frequency of the A_1 allele in the offspring population) and F in cells L12 and M12. Repeat the process for the remaining probabilities. Graph the relationship F and the probability of mating with the same genotype. Graph the relationship between the frequency of the A_1 allele in the offspring population and the probability of mating with the same genotype. Interpret your results.

5. Assume that A_1 is dominant to A_2, and that individuals breed with the same phenotype. Set the mating probabilities in cells F15–H17 accordingly (e.g., A_1A_1 individuals are equally likely to mate with A_1A_1 or A_1A_2 individuals, but are not likely to mate with A_2A_2 individuals). How does assortative mating differ from inbreeding effects on genotype and allele frequencies of the offspring population? How does it differ from random mating? To simulate random mating, enter the parental genotype frequencies in the cells; since mates are drawn at random, an individual should encounter a random mate proportionally to the parental frequencies.

LITERATURE CITED

Crow, J. F. and M. Kimura. 1970. *An Introduction to Population Genetics Theory*. Harper & Row, New York.

Hartl, D. 2000. *A Primer of Population Genetics*, 3rd Edition. Sinauer Associates, Sunderland, MA.

Merola, M. 1994. A reassessment of homozygosity and the case for inbreeding depression in the cheetah, *Acinonnyx jubatus*: Implications for conservation. *Conservation Biology* 8: 961–971.

Neal, N. P. 1935. The decrease in yielding capacity in advanced generations of hybrid corn. *Journal of the American Society of Agronomy* 27: 666–670.

42 GENETIC DRIFT

Objectives

- Set up a spreadsheet model of genetic drift.
- Determine the likelihood of allele fixation in a population of 10 individuals.
- Evaluate how initial allele frequencies in a population of 10 individuals affect probability of fixation.
- Compare the effects of genetic drift on small versus large populations.

Suggested Preliminary Exercise: Hardy-Weinberg Equilibrium

INTRODUCTION

Random events play a strong role in evolution, especially in small populations. Genetic drift is a random process; it is the chance fluctuations in allele frequencies within a populations as a result of random sampling among gametes (Hartl 2000). To understand what genetic drift is, we start with a very brief refresher in population genetics.

For diploid organisms such as vertebrates, each *individual* carries two alleles in their genetic makeup at each locus (one allele was inherited from the mother and one allele was inherited from the father). Let's assume that there are two types of allele, A_1 and A_2, for a given gene in a population. If the two alleles in an individual are of the same type, the individual is said to be *homozygous* (A_1A_1 or A_2A_2). If the alleles are of different types, the individual is said to be *heterozygous* (A_1A_2). Although *individuals* are either homozygous or heterozygous at a particular gene, *populations* are described by their **genotype frequencies** and **allele frequencies**. The word "frequency" in this case means the proportion of occurrence in a population. To obtain the genotype frequencies of a population, simply count up the number of each kind of genotype and divide by the total number of individuals in the population. For example, if we study a population of 55 individuals, and 8 individuals are A_1A_1, 35 are A_1A_2, and 12 are A_2A_2, the genotype frequencies (f) are

$$f(A_1A_1) = 8/55 = 0.146$$

$$f(A_1A_2) = 35/55 = 0.636$$

$$f(A_2A_2) = 12/55 = 0.218$$

$$\text{Total} = 1.00$$

The sum of the genotype frequencies of a population always equals 1.

Allele frequencies, in contrast, describe the proportion of all alleles in the population that are of a specific type (Hartl 2000). For our population of 55 individuals, there are a total of 110 gene copies present in the population (each of 55 individuals has 2 copies, so $55 \times 2 = 110$). To calculate the allele frequencies of the population, we need to calculate how many of those allele copies are of type A_1 and how many are of type A_2. To calculate how many copies are A_1, count the number of A_1A_1 homozygotes and multiply that number by 2 (each homozygote has two A_1 copies), then add to it the number of A_1A_2 heterozygotes (each heterozygote has one A_1 copy). The number of A_1 alleles in the population is then divided by the total number of gene copies in the population to generate an allele frequency. Thus, the total number of A_1 alleles in the population is $(2 \times 8) + (1 \times 35) = 51$. The frequency of A_1 is calculated as $51/(2 \times 55) = 51/110 = 0.464$. Similarly, the total number of A_2 alleles in the population is $(2 \times 12) + (1 \times 35) = 59$. The frequency of A_2 is calculated as $59/(2 \times 55) = 59/110 = 0.536$. As with genotype frequencies, the total of the allele frequencies of a population always equals 1. By convention, frequencies are designated by letters. If there are only two alleles in the population, these letters are conventionally $\boldsymbol{p}$ and $\boldsymbol{q}$, where p is the frequency of one kind of allele and q is the frequency of the other. For genes that have only two alleles,

$$p + q = 1 \qquad \text{Equation 1}$$

If there were more than two kinds of alleles for a particular gene, we would calculate allele frequencies for the other kinds of alleles in the same way. For example, if three alleles were present, A_1, A_2, and A_3, the frequencies would be p (the frequency of the A_1 allele), q (the frequency of the A_2 allele) and r (the frequency of the A_3 allele). No matter how many alleles are present in the population, the frequencies should always add to 1. Note that when we describe a population in terms of its allele frequencies, we don't necessarily know the genetic makeup of *individuals* in the population. For instance, all individuals can be homozygous (A_1A_1, A_1A_1, A_2A_2, A_2A_2, A_2A_2) or individuals can be a mix of homozygous and heterozygous genotypes (A_1A_2, A_1A_2, A_1A_1, A_2A_2, A_2A_2); the allele frequencies are the same in both situations.

In summary, for a population of N individuals, suppose the number of A_1A_1, A_1A_2, and A_2A_2 genotypes are n_{A1A1}, n_{A1A2}, and n_{A2A2}, respectively. If p represents the frequency of the A_1 allele, and q represents the frequency of the A_2 allele, the estimates of the allele frequencies in the population are

$$f(A_1) = p = (2n_{A1A1} + n_{A1A2}) \; / \; 2n \qquad \text{Equation 2}$$

$$f(A_2) = q = (2n_{A2A2} + n_{A1A2}) \; / \; 2n \qquad \text{Equation 3}$$

Genetic Drift and Evolution

Evolution is often described as a change in allele frequencies in a population over time (Hartl 2000). For example, we may notice that the frequency of the A_1 allele in our population changed from a value of 0.4 at time t to a value of 0.5 at time $t + 1$. There are several evolutionary forces that could have produced this change, such as natural selection, mutation, and gene flow. **Genetic drift**, the change in allele frequencies in populations that occurs by chance, without direction, is another kind of evolutionary force that can alter allele frequencies over time. Its impact is often greatest in small populations, and results in a loss of genetic diversity for a given (single) population.

Suppose, for example, that a population of 5 individuals has two alleles, A_1 and A_2, at a given locus, with frequencies p and q, respectively. Suppose further that in a certain generation, $p = q = 0.5$ (in other words, the frequency of allele A_1 is equal to the frequency of allele A_2). We will let this population mix and breed randomly to produce 5 new offspring that make up the next generation. Thus, the birth rates will remain low in this population. We can simulate random breeding by using a random number generator, where the random numbers 0, 1, 2, 3, and 4 represent the passing down of the A_1 allele to the next generation, and random numbers 5, 6, 7, 8, and 9 represent the passing down of the A_2 allele to the next generation.

Note that the two alleles are each represented by five numbers because the allele frequencies are initially equal. By drawing 10 random numbers to represent the 10 alleles making up the "new" generation, we can assign genotypes to the 5 new offspring and then calculate the new gene frequencies. For example, if the random numbers 0, 1, 5, 3, 9, 8, 3, 4, 8, and 2 are drawn, the 10 alleles in the next generation are A_1, A_1, A_2, A_1, A_2, A_2, A_1, A_1, A_2, and A_1, with genotypes taken in the order A_1A_1, A_2A_1, A_2A_2, A_1A_1, and A_2A_1. If you count how many alleles in this new population are A_1 and how many are A_2 (out of 10 total alleles), you find that this new generation has allele frequencies of $p = 0.6$ and $q = 0.4$. The population has evolved due to genetic drift.

We can continue this process for several generations to examine how the allele frequencies will continue to fluctuate over time. We used this method to track the frequency of the A_1 allele in 5 different populations, each consisting of 10 individuals, as shown below (Figure 1). In all populations, the frequency of A_1 was 0.5 to begin with. Inspection of Figure 1 shows that the frequency of the A_1 allele is 1 after 20 generations in two populations (populations 2 and 5). This means, by definition, that the frequency of the A_2 allele is 0. In contrast, the frequency of the A_1 allele is 0 after 20 generations in two other populations (populations 1 and 3).

In the first situation, we say that the A_1 allele has become **fixed** in the population, so that its frequency is 1. In the second situation, the A_2 allele has become fixed and the A_1 allele has been **lost** from the population. In both cases, allelic diversity has been lost from the population because there is now only one allele where previously there were two. Population 4 was also subjected to drift, but both the A_1 and A_2 alleles remained present in the population for at least 20 generations.

The important point is that when populations are very small, and are kept small over time, genetic drift tends to eliminate alleles from within a population, ultimately fixing the population at a frequency of either $p = 1$ or $q = 1$. You'll see how this happens as you work through the exercise. We can also think of the effects of drift *across* all five populations. Taking this larger view, genetic drift results in different populations becoming genetically different from each other because by chance, different alleles will become

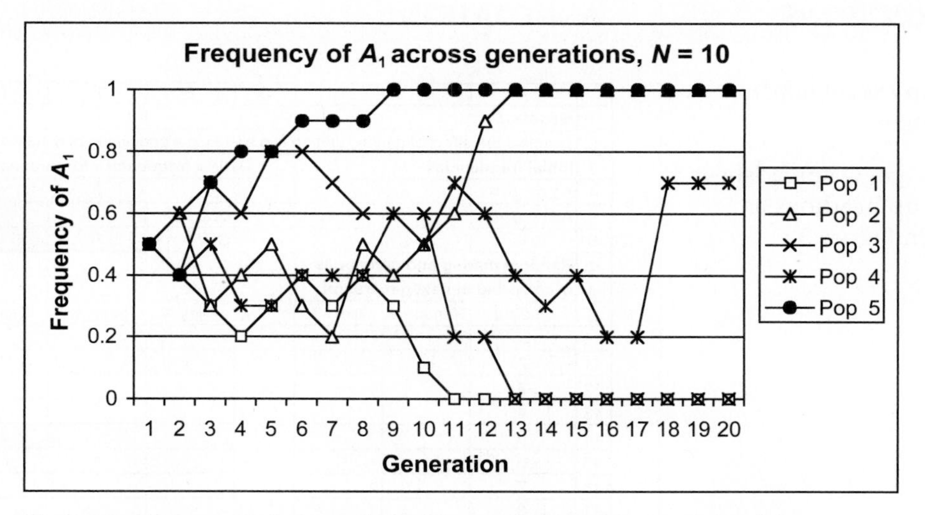

Figure 1 In five populations of size $N = 10$, the initial frequencies of A_1 and A_2 were each 0.5. After 20 generations, allele A_1 has become fixed in populations 2 and 5 and lost in populations 1 and 3. Only in population 4 do both alleles still exist.

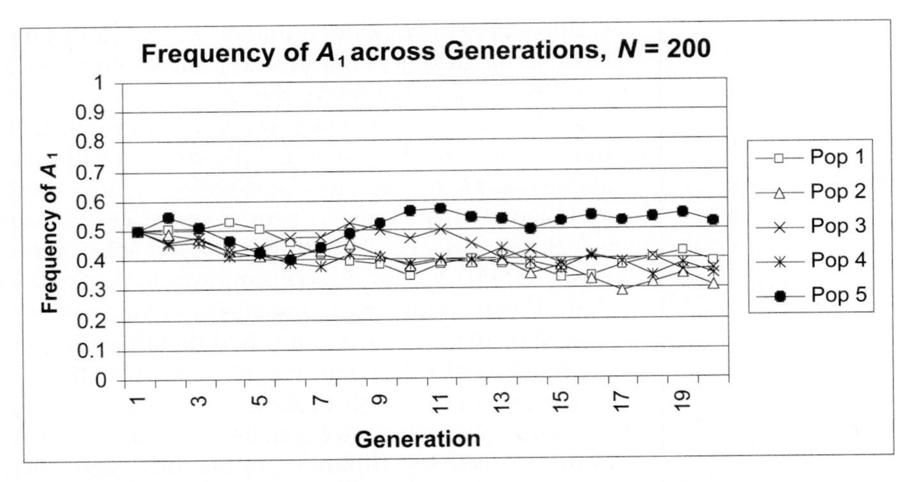

Figure 2 The effects of genetic drift are less dramatic when N is large (population size is large).

fixed in different populations. Some populations will be fixed at one allele, while another population will be fixed at a different allele.

The effects of drift become less important as population size increases. Figure 2 shows five populations, each consisting of 200 individuals and with an initial frequency of 0.5 for the A_1 allele. For the larger population sizes, drift is still apparent, but in no case did the A_1 or A_2 allele become fixed.

PROCEDURES

In this exercise, you'll set up a spreadsheet model to explore the effects of genetic drift. In doing so, you should learn why drift occurs and how it affects genetic diversity.

As always, save your work frequently to disk.

INSTRUCTIONS

ANNOTATION

A. Set up the population parameters.

1. Open a new spreadsheet and set up headings as shown in Figure 3.

	A	B	C	D	E	F	G	H	I
1	***Genetic Drift***								
2	Examine the effect of genetic drift on two alleles in a population of 5 individuals.								
3	**Initial frequencies**				**New frequencies for subsequent generations**				
4	A1 => *p* =					*p*	*q*		
5	A2=> *q* =				G(2) =>				
6					G(*N*) =>				
7	**Random mating and genotypes**								
8	**of offspring in next generation**								
9	Allele #	Random #	Allele ID		Allele #	Random #	Allele ID		
10	1								
11	2								
12	3								
13	4								
14	5								
15	6								
16	7								
17	8								
18	9								
19	10				10				
20									
21						**New A1 =**	**New A2 =**		
22									

Figure 3

2. Enter 0.5 in cells B4 and B5.

Set the initial population's allele frequencies in cell B4 (frequency of the A_1 allele) and B5 (frequency of the A_2 allele). These frequencies are the gene frequencies of your initial generation, or generation 1.

3. Set up a linear series from 1 to 10 in cells A10–A19.

Enter 1 in cell A10.
Enter **=1+A10** in cell A11. Copy your formula down to cell A19.

4. In cells B10–B19, use the **RAND** function to generate a random number between 0 and 1.

In cell B10 enter the formula **=RAND()**. Copy the formula down to cell B19.
Press F9, the calculate key, to generate new random numbers. These random numbers will be used to assign an allele that is inherited by the next generation in the next step.

5. In cells C10–C19, use the **IF** function to simulate which alleles are passed down from the parental generation as a result of random mating.

In cell C10, enter the formula **=IF(B10<$B$4,"A1","A2")**. Copy the formula down to cell C19.
The initial population of 5 individuals mates randomly and produce 5 new offspring that will make up generation 2. Each offspring in the population will inherit 2 alleles at the locus. The first offspring in generation 2 will inherit alleles given in cells A10–A11. The second offspring in generation 2 will inherit alleles given in cells A12–A13, and so on. The formula in C10 uses an **IF** function to determine whether the random number is associated with the A_1 allele or the A_2 allele. The formula tells the spreadsheet to evaluate cell B10; if the random number in cell B10 is less than the frequency of the A_1 allele designated in cell B4, then allele number 1 in the next generation will be an A_1 allele. Otherwise, allele number 1 in the next generation will be an A_2 allele.

6. Note the genotypes of the 5 offspring in generation 2.

The genotypes of our 5 offspring (Figure 4) were A_2A_2, A_1A_1, A_2A_1, A_2A_2, and A_1A_2. Your genotypes will likely be different than ours. Press F9, the calculate key, to generate new random numbers, and hence new offspring genotypes.

	A	B	C	D
3	**Initial frequencies**			
4	A1 => p =	0.5		
5	A2=> q =	0.5		
6				
7	**Random mating and genotypes**			
8	**of offspring in next generation**			
9	Allele #	Random #	Allele ID	
10	1	0.97142	A2	
11	2	0.572015	A2	
12	3	0.300821	A1	
13	4	0.438254	A1	
14	5	0.658612	A2	
15	6	0.447047	A1	
16	7	0.6244	A2	
17	8	0.699563	A2	
18	9	0.180683	A1	
19	10	0.505648	A2	
20				

Figure 4

7. In cell F5, use the **COUNTIF** function to count the number of A_1 alleles in the second generation (labeled G2), and calculate the new frequency of the A_1 allele (p).

Enter the formula **=COUNTIF(C10:C19,"A1")/10** in cell F5.
The **COUNTIF** function counts the number of cells within a range that meet the given criteria. It has the syntax **COUNTIF(range,criteria)**, where range is the range of cells you want to examine, and criteria is the item that will be counted. Since you entered **=COUNTIF(C10:C19,"A1")**, the program will examine cells C10–C19 and count the number of times A_1 appears. This number, when divided by the total alleles in the population, **/10**, gives the new A_1 allele frequency p for generation 2.

8. In cell G5, enter a **COUNTIF** formula to calculate the new frequency of the A_2 allele (q).

In cell G5 enter the formula **=COUNTIF(C10:C19,"A2")/10.** This equation is analogous to the one in Step 7.
After these formulas have been entered, each time you press F9 the spreadsheet will generate a new set of random numbers and will automatically compute the new allele frequencies in cells F5 and G5. We obtained allele frequencies of p = 0.4 and q = 0.6 for generation 2 (see Figure 4; you probably obtained different results; that's fine).

9. Manually type whatever frequencies you obtained in cells F5 and G5 into cells F6 and G6. Save your work.

Your frequencies will change each time the spreadsheet is calculated. By entering the frequencies in cells F6 and G6 by hand, you are "fixing" the frequencies for future generations.

B. Project allele frequencies to generation 3.

Now we'll repeat the entire process over time by letting generation 2 grow and reproduce 5 new individuals that will make up generation 3. To simulate the third generation, set up a new set of alleles, random numbers, and allele identifications in columns E, F, and G, as you did for generation 2.

1. Set up a linear series from 1 to 10 in cells E10–E19.

Enter the number 1 in cell E10.
Enter the formula **=1+E10** in cell E11. Copy the formula down to cell E19.

2. In cells F10–F19, generate a random number between 0 and 1.

Enter **=RAND()** in cells F10–F19 to assign a random number to each allele in generation 3.

3. In cells G10–G19, use an **IF** formula to determine whether the first allele in generation 3 is A_1 or A_2.

In cell G10 enter the formula **=IF(F10<$F$6,"A1","A2")**. Copy your formula down to cell G19.
This **IF** formula tells the spreadsheet to examine the random number in cell F10 and assign it a value of A_1 if it is less than the allele frequency designated in F6. If the random number is greater than the allele frequency designated in F6, the program assigns it an A_2 allele.

We determine the results of random mating in generation 2 by assigning an allele (A_1 or A_2) to each new random number in generation 3. Remember that the assignment of random numbers now depends on the allele frequencies in the *second generation* (listed in F6 and G6), and no longer depend on the initial population.

4. In cells F22 and G22, calculate the new allele frequencies inherited by the third generation.

In cell F22 enter the formula **=COUNTIF(G10:G19,"A1")/10** to compute the frequency of the A_1 allele.
In cell G22 enter the formul **=COUNTIF(G10:G19,"A2")/10** to compute the frequency of the A_2 allele.
As before, we use the **COUNTIF** formula to count the total number of A_1 and A_2 alleles..

5. Examine the change in allele frequencies over generations 1–3.

In our version of the exercise, generation 1 had initial allele frequencies of p = 0.5 and q = 0.5; generation 2 had allele frequencies of p = 0.4 and q = 0.6; and generation 3 had 0.4 and 0.6 (given in cells B4 and B5). You will almost certainly obtain different results from your own spreadsheet, and that's fine!

6. Obtain allele frequencies for generation 4.

You can quickly obtain the allele frequencies for generation 4 by copying the frequencies of generation 3 in cells F22 and G22 and pasting these values into cells F6 and G6, replacing the frequencies you used for generation 2. (This is why in Figure 3, this cell is labeled generation N, or G(N) for short.) So you can:

- Copy cells F22–G22.
- Select cells F6–G6.
- Open Edit | Paste Special. Select Paste Values and OK.
- Press F9 to automatically calculate new allele frequencies for generation 4 in cells F22 and G22.

What happened? Because of the way you typed in formulas for designating allele types in cells G10-G19, your assignment of alleles to the next generation depends on the parental generation that preceded it. Now the frequencies from generation 4 have been automatically counted in cells F22 and G22.

7. Save your work.

C. Track allele frequencies over time.

1. Set up some new headings as shown in Figure 5, but extend your generations to 20.

	J	K	L	M	N
1	**Change in frequencies over generations**				
2	**Generation**	**A1**	**A2**		
3	1	0.5	0.5		
4	2				
5	3				
6	4				
7	5				
8	6				
9	7				
10	8				
11	9				
12	10				

Figure 5

2. Enter 0.5 in cells K3 and K4.

Ultimately, you will track the fate of the frequencies of the A_1 and A_2 alleles over 20 generations. We will start again with generation 1, which has allele frequencies of p = 0.5 and q = 0.5.

3. Enter 0.5 in cells F6 and G6.

The values in cells F6 and G6 now represent the allele frequencies for G(1), or generation 1. We can now track how these frequencies change over 20 generations.

4. Write a macro to track allele frequencies for 20 generations.

From the menu, open Tools | Options | Calculations and select Manual Calculation. Then open the Macro function (see Exercise 2) to Record and assign a shortcut key. Perform the following steps:

- Press F9 to generate a new set of random numbers.
- Highlight cells F22 and G22, the new gene frequencies for the second generation.
- Go to Edit | Copy.
- Select cell K2, then go to Edit | Find | Find What. Leave the Find What cell *completely* blank, but make sure the Search by Columns option is selected.

- Select Find Next. The first blank cell in column K should be highlighted. Close the Find box.
- Open Edit | Paste Special, and paste in Values.
- Select cells F6 and G6
- Open Edit | Repeat Paste Special. This action will paste the new frequencies into cells F6 and G6, and will ensure that the spreadsheet uses these new frequencies to assign allele types to the offspring that make up the next, new generation.

Stop recording. Press your shortcut key until you have obtained allele frequencies for 20 generations.

5. Save your work.

D. Create graphs.

1. Graph the frequencies of the A_1 and A_2 allele over time.

Use the Line Graph option and make sure your axes are clearly labeled. Your graph should resemble Figure 6.

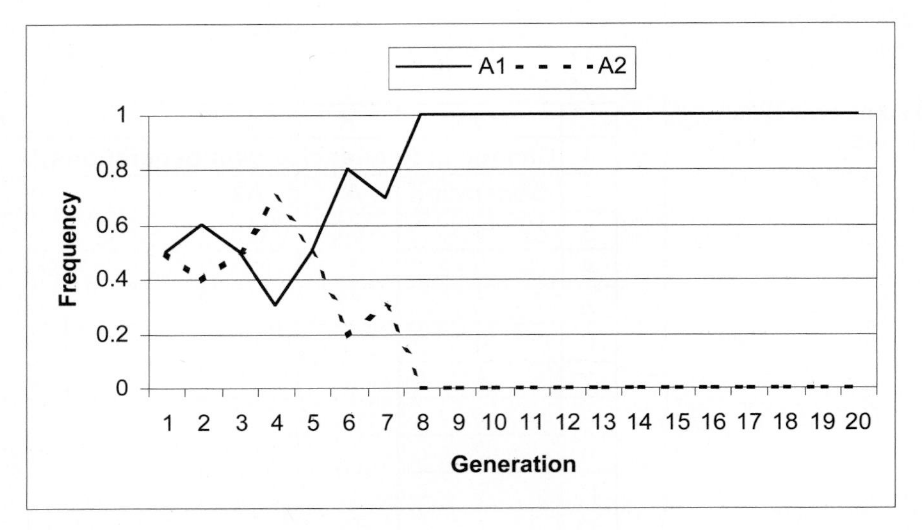

Figure 6

2. Save your work. Answer Question 1 at the end of the exercise before proceeding.

E. Run 100 trials.

How likely is it that a given allele would become fixed in your population? To know the probability of fixation in your population of 5 with initial gene frequencies of p = 0.5 and q = 0.5, you will need to repeat your entire simulation a minimum of 100 times (more would be better) and examine the outcomes of a variety of different simulations.

1. Make sure you are in the automatic calculation mode.

Open Tools | Options | Calculation and select Automatic.

2. Set up your spreadsheet as shown in Figure 7, but allow for 20 generations and 100 trials (extend your generations to cell U28 and your trials to cell A128).

	A	B	C	D	E	F	G
27	*Trial*	*Generation*					
28		*1*	*2*	*3*	*4*	*5*	*6*
29	*1*						
30	*2*						
31	*3*						
32	*4*						
33	*5*						

Figure 7

3. Copy results from Trial 1 (cells K3–K22, the frequency of A_1 over 20 generations) into cells B29–U29.

You've run 1 "trial" so far, with the results listed in cells K3–K22. We need to put those values into cells B29–U29.

Highlight cells K3–K22.
Open Edit | Copy. Select cell B29.
Open Edit | Paste Special | Paste Values, and select Paste Transpose. The transpose option will paste in the allele frequencies in row 29, automatically filling in the frequency of the A_1 allele across generations.

4. Switch to manual calculation.

Open Tools | Options | Calculation, and select Manual.

5. Develop a new macro to run 100 trials.

Try writing this macro on your own. If you get stuck, here are the steps we recorded to perform the task:

- Enter 0.5 in cells F6–G6. This will re-set the initial allele frequencies to 0.5 in generation 1.
- Press F9 to generate a new set of random numbers. The spreadsheet automatically calculates the new frequencies listed in cells F22 and G22.
- Use your mouse to highlight cells K4–L22.
- Press the delete key. The results of generation 2–20 from your first trial will be wiped out.
- Press the shortcut key (usually <Control> + some key) that runs your *first* macro (Step 4 in Section C), until you have generated allele frequencies for 20 generations.
- Select cells K3 to K22, and open Edit | Copy.
- Select cell B28, and go to Edit | Find. At the Find What prompt, leave the cell completely blank.
- Select the Search by Columns option. Select Find Next and then Close. This action will move your cursor to the next open cell in Column B.
- Open Edit | Paste Special. Select the Paste Values and Transpose options.

Stop recording. Press your shortcut key until you have run *a minimum* of 100 trials.

F. Calculate probability of fixation.

Now that you have run a number of trials, you can determine how likely it is that an allele would become fixed in the population after 20 generations. First, we'll count the number of times the A_1 allele went "extinct" (the frequency of the A_1 allele = 0, and the A_2 allele was fixed at 1). Then we'll count the number of times the A_1 allele was fixed at 1 (the A_2 allele went extinct).

1. Return to automatic calculation.

Open Tools | Options | Calculation and select Automatic.

2. Set up column labels as shown in Figure 8.

	V	W
27	**A1 extinct?**	**A1 fixed?**
28	**1 = yes**	**1 = yes**

Figure 8

3. In cells V29–V128, use the **IF** function to calculate how many times the A_1 allele went extinct.

Enter the formula **=IF(U29=0,1,0)** in cell V29. Copy this formula down to cell V128. The **IF** statement in V29 tells the spreadsheet to examine the contents of cell U29 (the allele frequency of the twentieth generation in trial 1). If cell U29 = 0, then assign it a value of 1; otherwise assign it a value of 0. Thus, if the A_1 allele went extinct in the course of 20 generations for a particular trial, the value in column V is scored as 1.

4. In cells W29–W128, use the **IF** function to calculate how many times the A_2 allele went extinct.

Enter the formula **=IF(U29=1,1,0)** in cell W29. Copy this formula down to cell W128. The spreadsheet will return a "1" if the A_1 allele became fixed at 1 (and thus the A_2 allele went extinct).

5. Sum the number of times the A_1 allele went extinct in cell V129. Sum the number of times the A_1 allele was fixed in cell W129.

Enter the formula **=SUM(V29:V128)** in cell V129.
Enter the formula **=SUM(W29:W128)** in cell W129.
In this step you simply add the number of times the A_1 allele went extinct ($p = 0$) and the number of times the A_1 allele became fixed at $p = 1$ for your trials.

6. In cell V130, enter a formula to calculate the probability of fixation as the probability that either the A_1 or A_2 allele will be fixed in the population. Label this value in U130.

We entered the formula **=(V129+W129)/100** in cell V130.
Now you can estimate the probability of fixation of an allele for a population of size 5 with initial gene frequencies of $p = 0.5$ and $q = 0.5$. These probabilities are simply the total number of times the A_1 allele went extinct *or* became fixed at 1, divided by the total number of trials you ran.

7. Save your work.

QUESTIONS

1. Trace the fate of the frequency of the A_1 allele over time. Did it vary dramatically? What was its frequency in the 20th generation? Was the frequency of the A_1 allele ever 1 or 0 at any time during your simulation? If so, did it bounce back to a new frequency, or did it remain fixed at a given level over time? Why?

2. How do the initial frequencies in the population affect the probability of extinction or of fixation? Change your initial allele frequencies to $p = 0.8$, $q = 0.2$. Set cell K3 to 0.8, and cell L3 to 0.2. Open Tools | Macro | Macros, then edit your Trials macro. You should see the Visual Basic for Applications Code that Excel "wrote" as you recorded your macro. Modify the values from 0.5 to 0.8 and 0.2. Close out of the edit box and return to your spreadsheet. Clear the results of your 100 trials, then run your 100 trials again. Graph and explain your results.

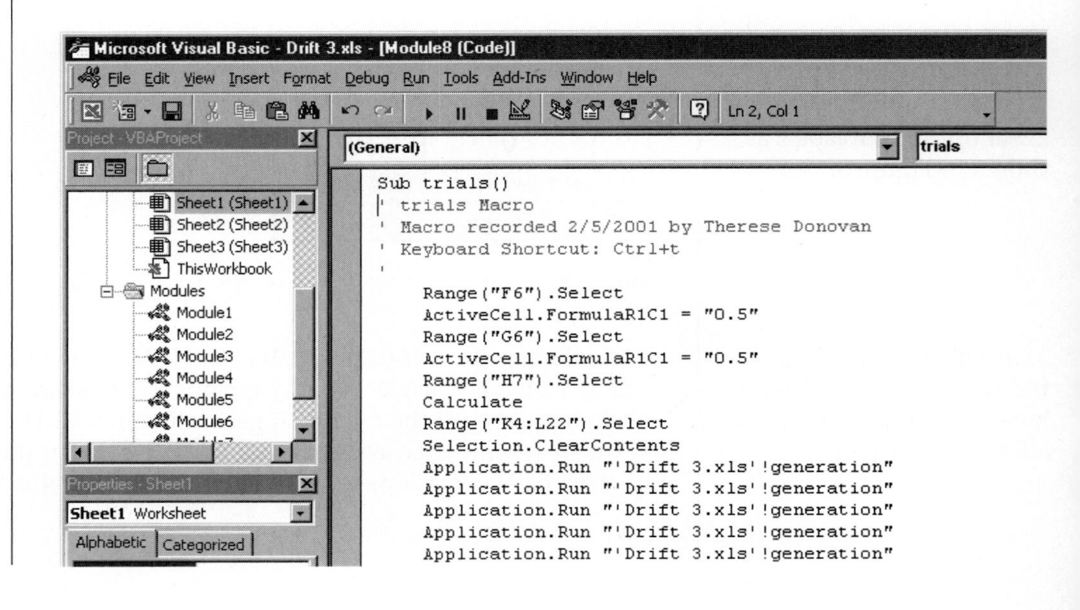

*3. (Advanced) What are the effects of genetic drift in a much larger population (say $N = 50$ or $N = 100$), where the initial allele frequencies are $p = 0.5$ and $q = 0.5$? Expand your model to compare the results of the effects of drift on small versus large populations. Copy the entire spreadsheet to a new page, and make your modifications on the new sheet.

4. What are some possible consequences of drift in populations, particularly if drift leads to fixation of alleles? Should this be of concern to wildlife managers? Could you use your model to estimate the minimum population size required to minimize the effects of drift?

LITERATURE CITED

Hartl, D. 2000. *A Primer of Population Genetics*, 3rd Edition. Sinauer Associates, Sunderland, MA.

43

EFFECTIVE POPULATION SIZE

In collaboration with Allan Strong

Objectives

- Explore how allele frequencies drift over time with stable populations of different sizes.
- Explore how allele frequencies drift over time when population sizes fluctuate.
- Calculate and interpret the effective population size of the population.

Suggested Preliminary Exercises: Hardy-Weinberg Equilibrium; Genetic Drift

INTRODUCTION

The Hardy-Weinberg principle states that when populations are infinitely large, mate randomly, and experience no selection, mutation, or gene flow, both the allele and genotype frequencies can be predicted for the next generation. From a genetic perspective, infinitely large Hardy-Weinberg populations are considered "ideal" populations. That is, the number of males and females are equal, mating occurs randomly, all individuals contribute more or less equally to the next generation, and population size is large and does not vary over time. Thus, in a population with N number of breeding individuals, each parent has a $1/N$ probability of producing a gamete that will be incorporated into future offspring.

But most, if not all, populations violate at least some of these assumptions. Population numbers fluctuate over time, have unequal sex ratios, or have mating systems where only a few dominant individuals breed, or disperse in such a way that not all individuals contribute equally to the next generation's genetic makeup. In other words, all of these "violations" can influence the way gametes are passed down to future generations.

How can we characterize populations that are *not* ideal? It is useful to directly compare the actual censused population size, N_t, to its **effective population size**, $\boldsymbol{N_e}$. The effective population size tells you how large the observed population is based on its genetic behavior. Because all populations have a finite size, they will experience some degree of genetic drift and inbreeding, even if the population is ideal in every other sense. The degree of drift and inbreeding in an ideal population with a finite size can be used as a baseline to which other, nonideal populations can be compared. You might recall from the preceding exercise that genetic drift is the change in allele frequency over generations that occurs because, by

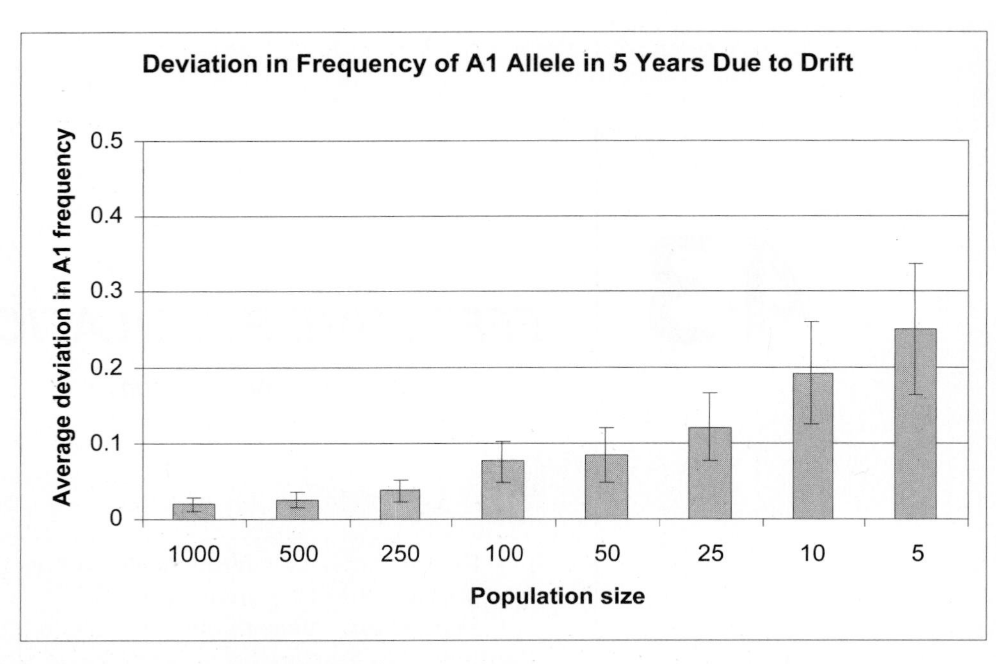

Figure 1 In all cases, the starting frequency of the A_1 allele = 0.5. After 5 generations, the deviation in the allele frequency from 0.5 was recorded. You can see that small populations experience a significant amount of drift (change in allele frequency due to sampling error) compared to larger populations.

chance, alleles are not passed down to subsequent generations as predicted by Hardy-Weinberg. The smaller the population, the more drift occurs and the more likely alleles will become fixed. Figure 1 shows how much drift occurs over 5 generations in populations ranging in size from 1000 down to 5 individuals.

The concept of effective population relates directly to the concepts of genetic drift and inbreeding (Wright 1931). The effective size of a population, N_e, is the number of individuals that will contribute genes equally to the next generation. For example, suppose we count 270 turtles in a population (the censused population), and would like to know how those 270 turtles "behave" from a genetic standpoint. The effective population size tells us that number. If N_e for this population equals 50, that means that our turtle population (N_t = 270) behaves or experiences changes in its genetic makeup like an "ideal" population of 50 individuals (that is, a population where mating is random, sex ratios are even, individuals contribute gametes equally to the next generation, and population size does not vary over time, but that nonetheless experiences drift and inbreeding because the population is not infinite).

Often N_e is less than N_t, suggesting that many natural populations behave genetically like a smaller population. A fluctuation in population size from year to year is one way that effective population size is reduced in nature. For example, suppose a population consists of 1000 individuals in generation 1, 10 individuals in generation 2, and 1000 individuals in generation 3. Generation 2 is considered a "bottleneck" generation for the population because only a handful of individuals actually survived through that period. Although we can count 1000 individuals in generation 3, the effective population size will be less than 1000 because the bottleneck has made the 1000 individuals in generation 3 more genetically related than the 1000 individuals in generation 1. In fact, this population will behave genetically more like an "ideal" population of 29 individuals (N_e = 29). Therefore, the number of individuals contributing genetically to the next generation is less than the actual population size.

You may ask, "How did we arrive at the number 29 in the above example?" The number 29 is the **harmonic mean** of the numbers 1000, 10, and 1000, or the reciprocal of the

average of the reciprocals of these three numbers. In other words, one way of calculating N_e is to compute the harmonic mean (see Crow and Kimura 1970 for greater detail). By using reciprocals to compute the harmonic mean, small numbers have a much greater effect than larger numbers. If $a = 10$ and $b = 2000$, then a has much more influence on the harmonic mean than b because 1/10 is much greater than 1/2000. Conceptually, this is exactly why computations of N_e are based on harmonic means: The importance of inbreeding and genetic drift is much greater when the population is small than when it is large, so the smaller population numbers should be emphasized in any computation of N_e.

The **harmonic mean,** N_e, for populations that fluctuate in number can be calculated as

$$\frac{1}{N_e} = \frac{1}{t} \times \left(\frac{1}{N_1} + \frac{1}{N_2} + \ldots + \frac{1}{N_t}\right)$$

where t is the number of years under consideration, and $N_1, N_2, \ldots, N_t$ are the censused population sizes over time.

To be clear, let's walk through an example. Suppose we censused a population for 6 consecutive years, and counted 1000, 5, 5, 1000, 5, and 1000 individuals over time. The effective population size, N_e, is equal to the harmonic mean of 1000, 5, 5, 1000, 5, and 1000, and is calculated as

$$\frac{1}{N_e} = \frac{1}{t} \times \left(\frac{1}{N_1} + \frac{1}{N_2} + \ldots + \frac{1}{N_t}\right)$$

$$\frac{1}{N_e} = \frac{1}{6} \times \left(\frac{1}{1000} + \frac{1}{5} + \frac{1}{5} + \frac{1}{1000} + \frac{1}{5} + \frac{1}{1000}\right) = 10$$

$$\frac{1}{N_e} = 0.167 \times 0.603$$

$$N_e = \frac{1}{0.167 \times 0.603} = 10$$

This means that although we can count 1000 individuals in year 6, *genetically* the population is behaving like an ideal population of size 10.

In addition to fluctuating population size, effective population sizes are affected by **sex ratio, dispersal distances,** and **variation in offspring produced per female.** It's fairly straightforward to understand how mating systems and sex ratio can affect N_e. If a censused population of 100 individuals consists of only 2 female breeders and 10 male breeders, the gametes that are passed down to the future generation are strongly influenced by the genetic makeup of those breeders. Disperal distance affects N_e because it determines how close or far siblings establish breeding sites from each other, which in turn affects the probability of mating with relatives. And variation in the number of offspring produced affects N_e by altering which genes are incorporated into the next generation. For example, all females may breed in a given year, but if one or two females have "boom" years (reproduce a lot) while others have "bust" years, the variance in reproductive output is high. Obviously, these females do not contribute gametes equally to the next generation. It is beyond the scope of this exercise to discuss all of these factors (see Crow and Kimura 1970), but you should be aware that the effective size of natural populations is influenced in a variety of ways.

PROCEDURES

The derivations for the various effective population size formulae are complicated, and therefore this exercise is devoted less to the math and more to explaining the genetic behavior of populations conceptually. In this exercise, we will simulate the effects of changes in gene frequencies for a population over the course of 6 generations. The first part of the exercise focuses on how much genetic drift occurs in populations with a constant size. In each generation, the genotypes of individuals will be drawn according the Hardy-Weinberg theory, based on the genetic makeup of the parents in the preceding generation. We will assume that generations do not overlap and that individuals can self-fertilize—that is, the same parent can contribute both egg and sperm to produce an offspring. We will then allow populations to fluctuate so that you can observe the how much drift occurs when population sizes change over time. Additionally, we will construct a simple model to examine graphically the relationship between N_t and N_e over 6 generations. This part of the exercise will enable us to evaluate the effect of bottlenecks in N_t on the effective populations size.

As always, save your work frequently to disk.

INSTRUCTIONS

ANNOTATION

A. Set up the model population.

1. Open a new spreadsheet and set up column headings as shown in Figure 2.

	A	B	C	D	E	F	G	H	I	J	K	L	M
1	*Effective Population Size Simulation*												
2							POPULATION SIZE						
3			Gen. 1	Parents	Gen. 2	Parents	Gen. 3	Parents	Gen. 4	Parents	Gen. 5	Parents	Final
4	Allele freq.	Initial	1000		1000		1000		1000		1000		1000
5	A1	0.5											
6	A2	0.5											

Figure 2

2. Enter 0.5 in cells B5 and B6.

We'll consider a population whose initial allele frequencies are p = frequency of the A_1 allele = 0.5 and q = frequency of the A_2 allele = 0.5. Remember that $p + q$ must equal 1 for loci that have only two alleles.

3. Enter the number 1000 in cells C4, E4, G4, I4, K4, and M4.

The cells C4, E4, G4, I4, K4, and M4 give the population size over generations. The final generation is given in cell M4. To begin, our population will have a constant size of N_t = 1000. Later in the exercise we will vary these numbers. Shade these cells to remind you that they can be directly manipulated in the exercise.

4. In cells D4, enter the formula **=2*E4**. Enter analogous formulae into cells F4, H4, J4, and L4.

Cell D4 "controls" the *maximum* number of individuals from generation 1 that will survive and potentially produce offspring in generation 2. For example, generation 2 will consist of 1000 individuals, so up to 2000 randomly selected parents from generation 1 will produce them (i.e., 2000 gametes will be passed down from generation 1 to generation 2, and all 1000 individuals in generation 1 potentially contribute to the next generation's gene pool). If generation 2 consisted of only 10 individuals, we would let only 20 randomly selected parents potentially produce them (the first 20 individuals listed in the spreadsheet). If generation 2 consisted of 4000 individuals (for example), then all of the individuals in generation 1 would potentially produce offspring. Cell F4 "controls" the number of individuals from generation 2 that will contribute offspring to generation 3, etc.

By copying the D4 formula over to cells F4, H4, J4, and L4, the maximum number of parents will be determined by the population size in the next generation. Your formulae in those cells should be:

- F4 **=2*G4**
- H4 **=2*I4**
- J4 **=2*K4**
- L4 **=2*M4**

5. Save your work.

6. Set up new headings as shown in Figure 3.

	A	B	C	D	E	F	G	H	I	J	K	L
8	Genotype #'s											
9	A1A1											
10	A1A2											
11	A2A2											
12	SUM											
13		Individual	Gen. 1	Parents	Gen. 2	Parents	Gen. 3	Parents	Gen. 4	Parents	Gen. 5	Parents

Figure 3

7. Set up a linear series from 1 to 1000 in cells B14–B1013.

Enter **1** in cell B14.
Enter **=1+B14** in cell B15. Copy this formula down to cell B1013.
We will simulate the population dynamics over 6 generations. For any generation, the maximum population size can be 1000 (assuming the environment's carrying capacity will support 1000 individuals).

8. In cells C14–C1013, enter a formula to assign a genotype to individual 1 in generation 1 based on the frequencies given in cells B5–B6.

In cell C14 enter the formula **=IF(B14<=$C$4,IF(RAND()<$B$5,$A$5,$A$6)&IF(RAND()<$B$5,$A$5,$A$6),"")**. Copy the formula down to cell C1013

Use the **IF** function as you did in the Hardy-Weinberg exercise, with one IF function nested within another to control the population size according to the value in cell B14. Remember that the IF formula returns one value if a condition you specify is true, and another value if the condition you specify is false.

The first part of the formula in cell C14 tells the spreadsheet to determine if cell B14 is less than or equal to (<=) the value in cell C4. If so, carry out the function **IF(RAND()<$B$5,$A$5,$A$6)&IF(RAND()<$B$5,$A$5,$A$6)** to assign a genotype to the individual. If cell B14 is greater than the value in cell C14, return a double quote mark, "" (which will return as a blank cell). This portion of the formula controls the population size. The genotype assignment is the same as you did in the Hardy-Weinberg exercise: The function tells the program to choose a random number between 0 and 1 (the **RAND()** part of the formula). If that random number is less than the value designated in cell B5 (the frequency of the A_1 allele), then assign it an allele of A_1; otherwise, assign it a value of A_2. Since all individuals have two alleles for a given locus, the formula is repeated again and genotype is generated by joining the two alleles with an **&** symbol. Once you've obtained genotypes for individual 1, copy this formula down to cell C1013 to obtain genotypes for all 1000 individuals in the population in generation 1.

9. Enter a formula in cells C9–C11 to count the number of individuals of each genotype in generation 1.

In cell C9 enter the formula **=COUNTIF(C14:C1013,"A1A1")**.
In cell D9 enter the formula **=COUNTIF(C14:C1013,"A1A2")+COUNTIF(C14:C1013,"A2A1")**.
In cell E9 enter the formula **=COUNTIF(C14:C1013,"A2A2")**.
You are using the **COUNTIF** function to count the various genotypes in generation 1. Don't forget that heterozygotes can be either A_1A_2 or A_2A_1. Double-check your results in the next step.

10. Sum the genotypes in Generation 1 in cell C12.

In cell C12 enter the formula **=SUM(C9:C11)**. Your result should be 1000.

11. Enter formulae in cells C5 and C6 to compute the actual allele frequencies in generation 1.

In cell C5 enter the formula **=(2*C9+C10)/(2*C12)**.
In cell C6 enter the formula **=1-C5** or **=(2*C11+C10)/(2*C12)**.
Remember from the Hardy-Weinberg exercise that you can compute the allele frequencies easily if you know the genotype frequencies. The equations are freq(A_1) = p = $(2N_{A1A1} + N_{A1A2})$ / $2N$, where N is the total number of individuals in the population. The frequency of the A_2 allele can be computed either by subtraction (= $1 - p$), or by freq(A_2) = q = $(2N_{A2A2} + N_{A1A2})$ / $2N$.

12. In cells D14–D1013, enter a formula to select the parents that can potentially produce offspring in the next generation.

In cell D14 enter the formula **=IF(B14<=$D$4,C14,"")**. Copy this formula down to cell D1013.
The formula in cell D14 identifies the parents. The allele frequencies of this parental population will be used to assign genotypes to individuals in generation 2. If cell B14 (individual 1) is less than or equal to the maximum number of parents in generation 1, the program will return individual 1's genotype. Otherwise, it will return a blank cell (the double-quote marks).

13. Copy cells C5–C6 and C9–C12 across to cells L5–L6 and L9–L12.

14. In cells E14–L14, enter formulae for the remaining generations, and copy your formulae down to row 1013 of each column as you go. Save your work.

This action will allow you to obtain genotype numbers and allele frequencies of the parents in generation 1, as well as future generations and parents. The entries for future generations will not make sense until you have completed the next step.

Follow the examples from generation 1, but make sure you update the formulae appropriately. Pay attention to absolute and relative references, and make sure that the new generation is based on the allele frequencies of the parental generation preceding it. Double-check your formulae.

We used the following formulae:

- Cell E14 **=IF(B14<=$E$4,IF(RAND()<$D$5,$A$5,$A$6)&IF(RAND()<$D$5,$A$5,$A$6),"")**
- Cell F14 **=IF(B14<=$F$4,E14,"")**
- Cell G14 **=IF(B14<=$G$4,IF(RAND()<$F$5,$A$5,$A$6)&IF(RAND()<$F$5,$A$5,$A$6),"")**
- Cell H14 **=IF(B14<=$H$4,G14,"")**
- Cell I14 **=IF(B14<=$I$4,IF(RAND()<$H$5,$A$5,$A$6)&IF(RAND()<$H$5,$A$5,$A$6),"")**
- Cell J14 **=IF(B14<=$J$4,I14,"")**
- Cell K14 **=IF(B14<=$K$4,IF(RAND()<$J$5,$A$5,$A$6)&IF(RAND()<$J$5,$A$5,$A$6),"")**
- Cell L14 **=IF(B14<=$L$4,K14,"")**

Review your formulae and double-check your work. Make sure you understand the formulae (and model) before proceeding.

B. Compute changes in A_1 due to genetic drift.

1. In cell M5, compute the deviation in the A_1 allele as the difference between the initial frequency in cell B5 and the final frequency in cell L5.

In cell M5 enter the formula **=ABS(L5-B5)**. Enter a label for this value in cell N5 as shown in Figure 4.
This is simply the **absolute value** of the difference between the initial and final frequency of the A_1 allele. It merely quantifies how far the A_1 allele drifted—we don't care about which direction the allele drifted.

	M	N
3	**Final**	
4	1000	
5		<= deviation

Figure 4

Remember that so far our population is ideal, except that it is finite—it consists of 1000 individuals over the generations. Any change in allele frequencies is due solely to genetic drift because the model does not include gene flow, natural selection, mutation, or nonrandom mating.

2. Press F9 to run a new simulation. What level of drift did the population experience?

You should see that the level of drift varies each time you press F9, the calculate key. This is because of the random way in which genotypes are assigned to individuals in each generation based on the Hardy-Weinberg principle. In order to "quantify" the level of drift, we will run 100 simulations, each time recording the deviation in frequency of the A_1 allele from the initial conditions. The average and standard deviation of these simulations will give a better indication (quantification) of the level of drift the population experienced after five generations and a constant population size of $N_t = 1000$.

3. Set up new headings as shown in Figure 5, except extend your trials to 100 (cell O103).

	O	P	Q
2		**Drift of A1**	
3	Trial	*N* = 1000	*N* = 10
4	1		
5	2		
6	3		
7	4		
8	5		

Figure 5

4. Develop a macro to track drift over 100 simulations – track your results in cells P4–P103.

Open the macro program and assign a shortcut key (refer to Exercise 2 for details on building macros). In Record mode, perform the following steps:

- Press F9 to obtain a new set of random numbers, and hence a new set of genotypes for the populations.
- Select cell M5, the change in frequency of the A_1 allele due to drift, then open Edit | Copy.
- Select cell P3, the column labeled "N = 1000".
- Open Edit | Find. In the dialog box, leave the Find What box empty, searching by columns and formulas, and then select Find Next and Close.
- Open Edit | Paste Special | Paste Values. Click OK.
- Open Tools | Macro | Stop Recording.

Now press your shortcut key until 100 simulations have been recorded.

5. In cell P104, enter a formula to compute the average deviation in the A_1 allele due to drift.

In cell P104, enter the formula **=AVERAGE(P4:P103)**.

6. In cell P105, compute the standard deviation of the 100 simulations.

In cell P105, enter the formula **=STDEV(P4:P103)**.

7. In cell P106, enter **=P105/2**.

For graphing purposes, we will divide the standard deviation by 2 so that when the standard error bars are added to our graph (next section), half of the line will be above the mean and half will be below it.

8. Change your population numbers so that each generation consists of 10 individuals, as in Figure 6.

	C	D	E	F	G	H	I	J	K	L	M
3	**Gen. 1**	Parents	**Gen. 2**	Parents	**Gen. 3**	Parents	**Gen. 4**	Parents	**Gen. 5**	Parents	**Final**
4	10		10		10		10		10		10

Figure 6

Now we will compare drift for a fixed population size of $N_t = 10$.

9. In column Q, develop a new macro to record deviations in the A_1 allele for this population.

See Step 4.

10. Copy cells P104-P106 to cells Q104–Q106.

This will generate means and standard deviations for this population, whose size is fixed at 10 individuals across generations.

C. Create graphs.

1. Graph the average deviation of the A_1 allele due to drift for the population when N = 1000 versus N = 10.

Use the column graph option. Under the Series tab, select cells P3 and Q3 as x-axis labels. Your graph should resemble Figure 7.

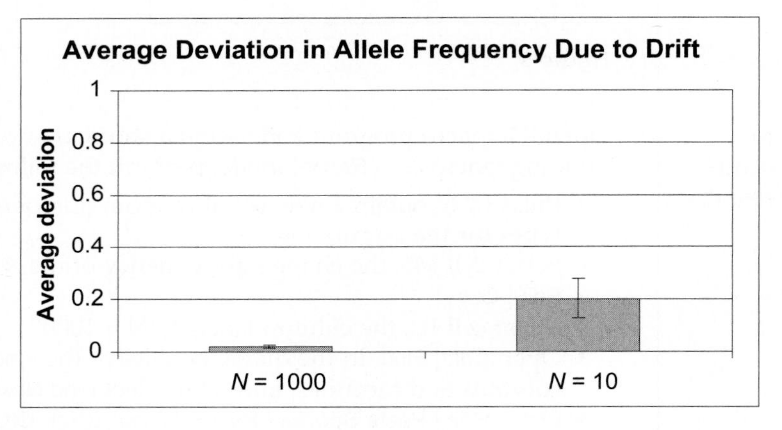

Figure 7

2. Add error bars to your graph.

3. Save your work. We will interpret your model results and explore how fluctuating population size affects the level of drift in a population in the Questions section.

To add error bars to your graph, click once somewhere in one of the columns in your graph. Go to Format | Selected Data Series. In the dialog box (Figure 8), select Y-Error Bars, then select the Display Both option for displaying error bars. Under Error Amount, select the Custom option. Select cells P106–Q106 in the + box, and repeat for the – box. Click OK and error bars will be added to your graph.

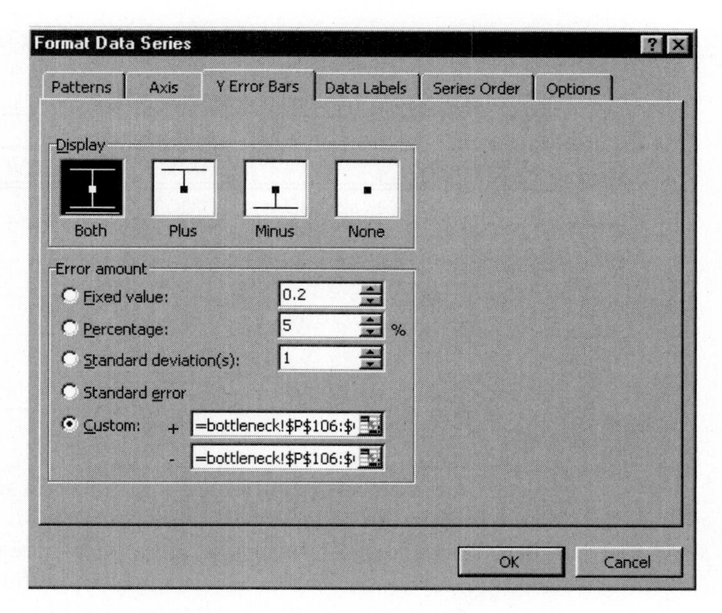

Figure 8

QUESTIONS

1. Compare the drift in the A_1 allele for the population of $N = 1000$ (constant over time) and the population of $N = 10$ (constant over time). Which population shows a greater level of drift? Why?

2. When populations fluctuate, they "behave" like smaller populations that have a constant population in that they experience genetic drift in similar ways. Alter your spreadsheet so that the population size for generations is

- Generation 1 = 1000
- Generation 2 = 5
- Generation 3 = 5
- Generation 4 = 1000
- Generation 5 = 5
- Final generation = 1000.

The final generation consists of 1000 individuals, yet the effective population size, as computed with the formula is 10:

$$\frac{1}{N_e} = \frac{1}{6} \times \left(\frac{1}{1000} + \frac{1}{5} + \frac{1}{5} + \frac{1}{1000} + \frac{1}{5} + \frac{1}{1000} \right) = 10$$

This means that the fluctuating population will change in allele frequencies through drift in a way a constant population of size 10 will. Prove this to yourself by running a new macro (record the results in column R) and comparing your results to the constant, small population size. Graph your results.

3. Directly compute N_e for your 6 generations. Set up the following new headings:

	S	T	U	V	W	X
2	Generation	*Nt*	1/*Nt*	Sum 1/*Nt*	*Ne*	HARMEAN
3	1	1000	0.001	0.001	1000	1000
4	2	5				
5	3	5				
6	4	1000				
7	5	5				
8	6	1000				

Enter formulae in cells T3–T8 to link population sizes given in cells C4, E4,…, M4. Enter a formula in cells U3–U8 to compute $1/N$. In cells V3–V8, enter formulae to track the sum of $1/N$ as more generations are considered. Finally, enter a formula in cell W3 to compute N_e. Refer back to the introduction for your computations. Graph how N_e and N_t change over time, and fully interpret your graph.

4. Explore the spreadsheet function **HARMEAN**, which computes the harmonic mean of a series of numbers directly in column X. For any given series of numbers, when is the harmonic mean the highest possible value? When is it the lowest possible value? For any given series of numbers, under what conditions is $N_e > N_t$? Explore you model by changing values of N_t, increasing and decreasing the variation in numbers over time. Pay attention to how N_e is affected by bottlenecks both in the current generation and in subsequent generations.

LITERATURE CITED

Crow, J. F., and M. Kimura. 1970. *An Introduction to Population Genetics Theory.* Harper & Row, New York.

Lande, R., and G. F. Barrowclough. 1987. Effective population size, genetic variation, and their use in population management. *In* M. E. Soulé (ed.), *Viable Populations for Conservation*, pp. 87–123. Cambridge University Press, Cambridge.

Wright, S. 1931. Evolution in Mendelian populations. *Genetics* 16: 97–159.